Electrical

Level One

Trainee Guide
Tenth Edition

 Pearson

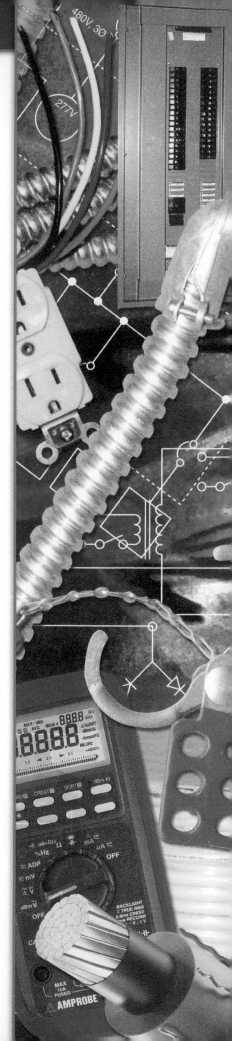

NCCER

Chief Executive Officer: Don Whyte
President: Boyd Worsham
Chief Operations Officer: Katrina Kersch
Electrical Curriculum Project Manager: Veronica Westfall
Director, Product Development: Tim Davis
Senior Production Manager: Erin O'Nora
Senior Manager of Projects: Chris Wilson
Testing/Assessment Project Manager: Elizabeth Schlaupitz
Project Assistant: Lauren Corley
Lead Technical Writer: Veronica Westfall
Technical Writers: Gary Ferguson, Troy Staton, Karyn Payne, John Mueller, Chris Wilson

Managing Editor: Graham Hack
Desktop Publishing Manager: James McKay
Art Manager: Kelly Sadler
Multimedia Project Manager: Alan Youngblood
Digital Content Coordinator: Rachael Downs
Production Specialists: Gene Page, Eric Caraballoso
Production Assistance: Adrienne Payne, David Gregoire, Edward Fortman, Joanne Hart, Olga Trofymenko, Karyn Payne
Editors: Jordan Hutchinson, Karina Kuchta, Hannah Murray
Product Development Program Specialist: Tim Douglas

Composition: NCCER
Printer/Binder: LSC Communications
Cover Printer: LSC Communications
Text Fonts: Palatino and Univers
Comtent Technologies: Gnostyx

Credits and acknowledgments for content borrowed from other sources and reproduced, with permission, in this textbook appear at the end of each module.

Pearson

Perfect Bound: 978-0-13-690853-1
Case Bound: 978-0-13-690865-4

Preface

To the Trainee

Electricity powers the applications that make our daily lives more productive and efficient. The demand for electricity has led to vast job opportunities in the electrical field. Electricians constitute one of the largest construction occupations in the United States, and they are among the highest-paid workers in the construction industry. According to the U.S. Bureau of Labor Statistics, the demand for trained electricians is projected to increase, creating more job opportunities for skilled craftspeople in the electrical industry.

Electricians install electrical systems in structures such as homes, office buildings, and factories. These systems include wiring and other electrical components, such as circuit breaker panels, switches, and lighting. Electricians follow blueprints, the *National Electrical Code®*, and state and local codes. They use specialized tools and testing equipment, such as ammeters, ohmmeters, and voltmeters. Electricians learn their trade through craft and apprenticeship programs. These programs provide classroom instruction and on-the-job learning with experienced electricians.

We wish you success as you embark on your first year of training in the electrical craft, and hope that you will continue your training beyond this textbook. There are more than 700,000 people employed in electrical work in the United States, and there are many opportunities awaiting those with the skills and desire to move forward in the construction industry.

New with *Electrical Level One*

NCCER and Pearson are pleased to present *Electrical Level One*, which has been updated to meet the 2020 *National Electrical Code®* and includes revisions to the Module Examinations.

In addition to the 2020 *NEC®* changes, this edition of *Electrical Level One* features several updated modules. *Occupational Overview: The Electrical Industry* (Module ID 26101-20), *Safety for Electricians* (Module ID 26102-20), *Introduction to the National Electrical Code®* (Module ID 26105-20), *Basic Electrical Construction Documents* (Module ID 26110-20), and *Electrical Test Equipment* (Module ID 26112-20) have all been enhanced to reflect the latest best practices in the industry, including the addition of dozens of new photographs and wiring diagrams.

Our website, **www.nccer.org**, has information on the latest product releases and training.

Your feedback is welcome. You may email your comments to **curriculum@nccer.org** or send general comments and inquiries to **info@nccer.org**.

NCCER Standardized Curricula

NCCER is a not-for-profit 501(c)(3) education foundation established in 1996 by the world's largest and most progressive construction companies and national construction associations. It was founded to address the severe workforce shortage facing the industry and to develop a standardized training process and curricula. Today, NCCER is supported by hundreds of leading construction and maintenance companies, manufacturers, and national associations. The NCCER Standardized Curricula was developed by NCCER in partnership with Pearson, the world's largest educational publisher.

Some features of the NCCER Standardized Curricula are as follows:

- An industry-proven record of success
- Curricula developed by the industry, for the industry
- National standardization providing portability of learned job skills and educational credits
- Compliance with the Office of Apprenticeship requirements for related classroom training (*CFR 29:29*)
- Well-illustrated, up-to-date, and practical information

NCCER also maintains the NCCER Registry, which provides transcripts, certificates, and wallet cards to individuals who have successfully completed a level of training within a craft in NCCER's Curricula. *Training programs must be delivered by an NCCER Accredited Training Sponsor in order to receive these credentials.*

Special Features

In an effort to provide a comprehensive and user-friendly training resource, this curriculum showcases several informative features. Whether you are a visual or hands-on learner, these features are intended to enhance your knowledge of the construction industry as you progress in your training. Some of the features you may find in the curriculum are explained below.

Introduction

This introductory page, found at the beginning of each module, lists the module Objectives, Performance Tasks, and Trade Terms. The Objectives list the knowledge you will acquire after successfully completing the module. The Performance Tasks give you an opportunity to apply your knowledge to real-world tasks. The Trade Terms are industry-specific vocabulary that you will learn as you study this module.

Figures and Tables

Photographs, drawings, diagrams, and tables are used throughout each module to illustrate important concepts and provide clarity for complex instructions. Text references to figures and tables are emphasized with *italic* type.

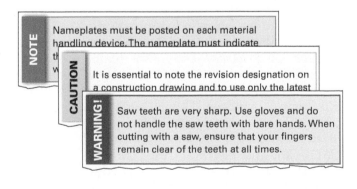

Notes, Cautions, and Warnings

Safety features are set off from the main text in highlighted boxes and categorized according to the potential danger involved. Notes simply provide additional information. Cautions flag a hazardous issue that could cause damage to materials or equipment. Warnings stress a potentially dangerous situation that could result in injury or death to workers.

Trade Features

Trade features present technical tips and professional practices based on real-life scenarios similar to those you might encounter on the job site.

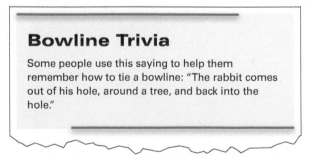

Bowline Trivia

Some people use this saying to help them remember how to tie a bowline: "The rabbit comes out of his hole, around a tree, and back into the hole."

> **NOTE**
> Nameplates must be posted on each material handling device. The nameplate must indicate the...

> **CAUTION**
> It is essential to note the revision designation on a construction drawing and to use only the latest...

> **WARNING!**
> Saw teeth are very sharp. Use gloves and do not handle the saw teeth with bare hands. When cutting with a saw, ensure that your fingers remain clear of the teeth at all times.

Case History

Case History features emphasize the importance of safety by citing examples of the costly (and often devastating) consequences of ignoring best practices or OSHA regulations.

Going Green

Going Green features present steps being taken within the construction industry to protect the environment and save energy, emphasizing choices that can be made on the job to preserve the health of the planet.

GOING GREEN

Reducing Your Carbon Footprint

Many companies are taking part in the paperless movement. They reduce their environmental impact by reducing the amount of paper they use. Using email helps to reduce the amount of paper used,

Did You Know

Did You Know features introduce historical tidbits or interesting and sometimes surprising facts about the trade.

Did You Know?

Safety First

Safety training is required for all activities. Never operate tools, machinery, or equipment without prior training. Always refer to the manufacturer's instructions.

Step-by-Step Instructions

Step-by-step instructions are used throughout to guide you through technical procedures and tasks from start to finish. These steps show you how to perform a task safely and efficiently.

Perform the following steps to erect this system area scaffold:

Step 1 Gather and inspect all scaffold equipment for the scaffold arrangement.

Step 2 Place appropriate mudsills in their approximate locations.

Step 3 Attach the screw jacks to the mudsills.

Trade Terms

Each module presents a list of Trade Terms that are discussed within the text and defined in the Glossary at the end of the module. These terms are presented in the text with **bold, blue** type upon their first occurrence. To make searches for key information easier, a comprehensive Glossary of Trade Terms from all modules is located at the back of this book.

During a rigging operation, the **load** being lifted or moved must be connected to the apparatus, such as a crane, that will provide the power for movement. The connector—the link between the load and the apparatus—is often a sling made of synthetic, chain, or **wire rope** materials. This section focuses on three types of slings:

Section Review

Each section of the module wraps up with a list of Additional Resources for further study and Section Review questions designed to test your knowledge of the Objectives for that section.

1.0.0 Section Review

1. For material handling tasks, it is just as important to be mentally fit as it is to be _____.
   a. physically fit
   b. physically aggressive
   c. closely supervised
   d. over 200 pounds

2. Which of the following is a type of knot that is often used to join the ends of two ropes in non-critical, low-strain applications?
   a. Bowline
   b. Clove hitch
   c. Half hitch
   d. Square knot

Review Questions

The end-of-module Review Questions can be used to measure and reinforce your knowledge of the module's content.

Review Questions

1. Identification tags for slings must include the _____.
   a. type of protective pads to use
   b. type of damage sustained during use
   c. color of the tattle-tail
   d. manufacturer's name or trademark

2. The type of wire rope core that is susceptible to heat damage at relatively low temperatures is the _____.
   a. fiber core
   b. strand core
   c. independent wire rope core
   d. metallic link supporting core

3. Synthetic slings must be inspected _____.
   a. once every month
   b. visually at the start of each work week
   c. before every use
   d. once wear or damage becomes apparent

4. An alloy steel chain sling must be removed from service if there is evidence that _____.
   a. the sling has been used in different hitch configurations
   b. replacement links have been used to repair the chain
   c. the sling has been used for more than one year
   d. strands in the supporting core have weakened

5. A piece of rigging hardware used to couple the end of a wire rope to eye fittings, hooks, or other connections is a(n) _____.
   a. eyebolt
   b. hitch
   c. shackle
   d. U-bolt

6. A lifting clamp is most likely to be used to move loads such as _____.
   a. steel plates
   b. piping bundles
   c. concrete blocks
   d. plastic tubing

7. Chain hoists are able to lift heavy loads by utilizing a _____.
   a. rope and pulley system
   b. rigger's strength
   c. stationary counterweight
   d. gear system

8. Before attempting to lift a load with a chain hoist, make sure that the _____.
   a. hoist is secured to a come-along
   b. load is properly balanced
   c. tag lines are properly anchored
   d. tackle is connected to its power source

9. A hitch configuration that allows slings to be connected to the same load without using a spreader beam is a _____.
   a. double-wrap hitch
   b. choker hitch
   c. bridle hitch
   d. basket hitch

10. To make the emergency stop signal that is used by riggers, extend both arms _____.
    a. horizontally with palms down and quickly move both arms back and forth
    b. directly in front and then move both arms up and down repeatedly
    c. vertically above the head and wave both arms back and forth
    d. horizontally with clenched fists and move both arms up and down

NCCER Standardized Curricula

NCCER's training programs comprise more than 80 construction, maintenance, pipeline, and utility areas and include skills assessments, safety training, and management education.

Boilermaking
Cabinetmaking
Carpentry
Concrete Finishing
Construction Craft Laborer
Construction Technology
Core Curriculum: Introductory
  Craft Skills
Drywall
Electrical
Electronic Systems Technician
Heating, Ventilating, and Air
  Conditioning
Heavy Equipment Operations
Heavy Highway Construction
Hydroblasting
Industrial Coating and Lining
  Application Specialist
Industrial Maintenance Electrical
  and Instrumentation Technician
Industrial Maintenance Mechanic
Instrumentation
Ironworking
Manufactured Construction
  Technology
Masonry
Mechanical Insulating
Millwright
Mobile Crane Operations
Painting
Painting, Industrial
Pipefitting
Pipelayer
Plumbing
Reinforcing Ironwork
Rigging
Scaffolding
Sheet Metal
Signal Person
Site Layout
Sprinkler Fitting
Tower Crane Operator
Welding

Maritime

Maritime Industry Fundamentals
Maritime Electrical
Maritime Pipefitting
Maritime Structural Fitter
Maritime Welding
Maritime Aluminum Welding

Green/Sustainable Construction

Building Auditor
Fundamentals of Weatherization
Introduction to Weatherization
Sustainable Construction
  Supervisor
Weatherization Crew Chief
Weatherization Technician
Your Role in the Green
  Environment

Energy

Alternative Energy
Introduction to the Power Industry
Introduction to Solar Photovoltaics
Power Generation Maintenance
  Electrician
Power Generation I&C
  Maintenance Technician
Power Generation Maintenance
  Mechanic
Power Line Worker
Power Line Worker: Distribution
Power Line Worker: Substation
Power Line Worker: Transmission
Solar Photovoltaic Systems Installer
Wind Energy
Wind Turbine Maintenance
  Technician

Pipeline

Abnormal Operating Conditions,
  Control Center
Abnormal Operating Conditions,
  Field and Gas
Corrosion Control
Electrical and Instrumentation
Field and Control Center
  Operations
Introduction to the Pipeline
  Industry
Maintenance
Mechanical

Safety

Field Safety
Safety Orientation
Safety Technology

Supplemental Titles

Applied Construction Math
Tools for Success

Management

Construction Workforce
  Development Professional
Fundamentals of Crew Leadership
Mentoring for Craft Professionals
Project Management
Project Supervision

Spanish Titles

Acabado de concreto: nivel uno
  (*Concrete Finishing Level One*)
Aislamiento: nivel uno
  (*Insulating Level One*)
Albañilería: nivel uno
  (*Masonry Level One*)
Andamios (*Scaffolding*)
Carpintería: Formas para
  carpintería, nivel tres
  (*Carpentry: Carpentry Forms, Level
  Three*)
Currículo básico: habilidades
  introductorias del oficio
  (*Core Curriculum: Introductory Craft
  Skills*)
Electricidad: nivel uno
  (*Electrical Level One*)
Herrería: nivel uno
  (*Ironworking Level One*)
Herrería de refuerzo: nivel uno
  (*Reinforcing Ironwork Level One*)
Instalación de rociadores: nivel uno
  (*Sprinkler Fitting Level One*)
Instalación de tuberías: nivel uno
  (*Pipefitting Level One*)
Instrumentación: nivel uno, nivel
  dos, nivel tres, nivel cuatro
  (*Instrumentation Levels One through
  Four*)
Orientación de seguridad
  (*Safety Orientation*)
Paneles de yeso: nivel uno
  (*Drywall Level One*)
Seguridad de campo
  (*Field Safety*)

Acknowledgments

This curriculum was revised as a result of the farsightedness and leadership of the following sponsors:

ABC of Iowa
ABC of Western Pennsylvania
Beacon Electrical Contractors
Cianbro Corporation
Elm Electrical, Inc.
Gaylor Electric, Inc.
Gould Construction Institute

Industrial Management and Training Institute
National Field Services
Madison Comprehensive High School
Putnam Career and Technical Center
Tri-City Electrical Contractors

This curriculum would not exist were it not for the dedication and unselfish energy of those volunteers who served on the Authoring Team. A sincere thanks is extended to the following:

Paul Asselin
David Coelho
Tim Dean
Tim Ely
Mark Kozloski
Dan Lamphear

David Lewis
John Lupacchino
Gerard McDonald
Scott Mitchell
John Mueller
Steve Newton

Ron Otts
Mike Powers
Josiha Schuh
Wayne Stratton
James Westfall

NCCER Partners

American Fire Sprinkler Association
Associated Builders and Contractors, Inc.
Associated General Contractors of America
Association for Career and Technical Education
Construction Industry Institute
Construction Users Roundtable
Gulf States Shipbuilders Consortium
ISN Software Corporation
Manufacturing Institute
Mason Contractors Association of America
Merit Contractors Association of Canada
NACE International
National Association of Women in Construction
National Insulation Association
National Technical Honor Society
NAWIC Education Foundation
North American Crane Bureau
North American Technician Excellence
Pearson
Prov

SkillsUSA®
Steel Erectors Association of America
University of Florida, M. E. Rinker Sr., School of Construction Management

Contents

Module One
Occupational Overview: The Electrical Industry

Skilled people in the electrical field are essential to maintain electrical systems and equipment in residential, commercial, and industrial settings. This module describes the various career paths in the electrical industry. It also covers the apprenticeship requirements for electricians and discusses employer/employee responsibilities. (Module ID 26101-20; 2.5 Hours)

Module Two
Safety for Electricians

In order to work safely, electricians must be aware of potential hazards and stay constantly alert to them. This includes taking the proper precautions and practicing basic rules of safety. This module discusses hazards and describes the various types of personal protective equipment (PPE) used to reduce injuries. It also covers the standards related to electrical safety and the Occupational Safety and Health Administration (OSHA) lockout/tagout rule. (Module ID 26102-20; 10 Hours)

Module Three
Introduction to Electrical Circuits

All kinds of instruments use electrical circuitry to function. This module discusses basic atomic theory and electrical theory, which are the fundamental concepts behind electricity in every setting. It also covers electrical units of measurement and explains how Ohm's law and the power equation can be used to determine unknown values. This module also introduces electrical schematic diagrams. (Module ID 26103-20; 7.5 Hours)

Module Four
Electrical Theory

Knowledge of electrical circuits is essential in the electrical field. A sound understanding of basic circuits, as well as the methods for calculating the electrical energy within them, forms the foundation for utilizing these principles in practical applications. This module explains how to apply Ohm's law to series, parallel, and series-parallel circuits. It also covers Kirchhoff's voltage and current laws. (Module ID 26104-20; 7.5 Hours)

Module Five
Introduction to the National Electrical Code®

The *NEC*® is one of the most important tools for electricians. When used together with the applicable electrical code for your local area, the *NEC*® provides the minimum requirements for the installation of electrical systems. This module describes the purpose of the *NEC*® and explains how to use it to find the installation requirements for various electrical devices and wiring methods. It also provides an overview of the National Electrical Manufacturers Association and Nationally Recognized Testing Laboratories. (Module ID 26105-20; 7.5 Hours)

Module Six
Device Boxes

Electricians work with device boxes almost every day on every project, making a thorough understanding of the types of boxes available and their applications essential. This module describes the various types of boxes and explains how to calculate the *NEC*® fill requirements for outlet and junction boxes under 100 cubic inches (1,650 cubic centimeters). (Module ID 26106-20; 10 Hours)

Module Seven

Hand Bending

The art of conduit bending is dependent upon the skills of the electrician and requires a working knowledge of basic terms and proven procedures. Practice, knowledge, and training will help you gain the skills necessary for proper conduit bending and installation. This module describes methods for hand bending conduit. It covers 90-degree bends, back-to-back bends, offsets, and saddle bends. It also describes how to cut, ream, and thread conduit. (Module ID 26107-20; 10 Hours)

Module Eight

Wireways, Raceways, and Fittings

Electrical raceways present challenges and requirements involving proper installation techniques, general understanding of raceway systems, and applications of the *NEC®* to raceway systems. Acquiring quality installation skills for raceway systems requires practice, knowledge, and training. This module describes various types of raceway systems, along with their installation and *NEC®* requirements. It also describes the use of various conduit bodies. (Module ID 26108-20; 20 Hours)

Module Nine

Conductors and Cables

As an electrician, you will be required to select the proper wire and/or cable for a job. You will also be required to pull this wire or cable through conduit runs in order to terminate it. This module discusses conductor types, cable markings, color codes, and ampacity derating. It also describes how to install conductors using fish tape and power conduit fishing systems. (Module ID 26109-20; 10 Hours)

(continued)

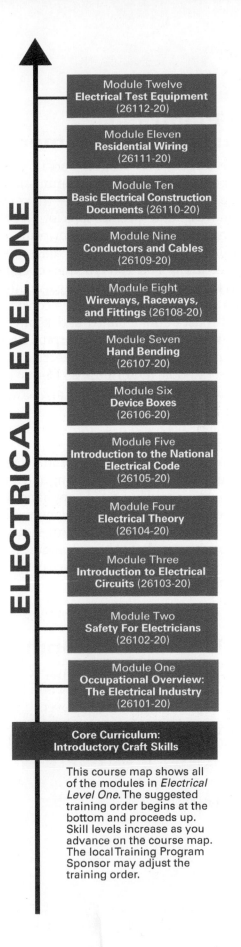

ELECTRICAL LEVEL ONE

Module Twelve
Electrical Test Equipment
(26112-20)

Module Eleven
Residential Wiring
(26111-20)

Module Ten
Basic Electrical Construction Documents (26110-20)

Module Nine
Conductors and Cables
(26109-20)

Module Eight
Wireways, Raceways, and Fittings (26108-20)

Module Seven
Hand Bending
(26107-20)

Module Six
Device Boxes
(26106-20)

Module Five
Introduction to the National Electrical Code
(26105-20)

Module Four
Electrical Theory
(26104-20)

Module Three
Introduction to Electrical Circuits (26103-20)

Module Two
Safety For Electricians
(26102-20)

Module One
Occupational Overview: The Electrical Industry
(26101-20)

Core Curriculum: Introductory Craft Skills

This course map shows all of the modules in *Electrical Level One.* The suggested training order begins at the bottom and proceeds up. Skill levels increase as you advance on the course map. The local Training Program Sponsor may adjust the training order.

Occupational Overview:
The Electrical Industry

OVERVIEW

Skilled people in the electrical field are essential to maintain electrical systems and equipment in residential, commercial, and industrial settings. This module describes the various career paths in the electrical industry. It also covers the apprenticeship requirements for electricians and discusses employer/employee responsibilities.

Module 26101-20

Trainees with successful module completions may be eligible for credentialing through the NCCER Registry. To learn more, go to **www.nccer.org** or contact us at 1.888.622.3720. Our website, **www.nccer.org**, has information on the latest product releases and training.

Your feedback is welcome. You may email your comments to **curriculum@nccer.org**, send general comments and inquiries to **info@nccer.org**, or fill in the User Update form at the back of this module.

This information is general in nature and intended for training purposes only. Actual performance of activities described in this manual requires compliance with all applicable operating, service, maintenance, and safety procedures under the direction of qualified personnel. References in this manual to patented or proprietary devices do not constitute a recommendation of their use.

26101-20 V10.0

26101-20
OCCUPATIONAL OVERVIEW: THE ELECTRICAL INDUSTRY

Objectives

When you have completed this module, you will be able to do the following:

1. Identify the various sectors and trade options in the electrical industry.
   a. Describe the typical components in a residential wiring system.
   b. Describe the typical components in a commercial wiring system.
   c. Describe the typical components in an industrial wiring system.
   d. List various career paths and opportunities in the electrical trade.
2. Understand the apprenticeship/training process for electricians.
   a. List Department of Labor (DOL) requirements for apprenticeship.
   b. Describe various types of training in the electrical field.
3. Understand the responsibilities of the employee and employer.
   a. Identify employee responsibilities.
   b. Identify employer responsibilities.

Performance Tasks

This is a knowledge-based module; there are no performance tasks.

Trade Terms

Electrical service
Occupational Safety and Health Administration (OSHA)
On-the-job learning (OJL)
Raceway system
Rough-in
Substation
Trim-out

Industry Recognized Credentials

If you are training through an NCCER-accredited sponsor, you may be eligible for credentials from NCCER's Registry. The ID number for this module is 26101-20. Note that this module may have been used in other NCCER curricula and may apply to other level completions. Contact NCCER's Registry at 888.622.3720 or go to **www.nccer.org** for more information.

> **NOTE**
>
> NFPA 70®, *National Electrical Code*® and *NEC*® are registered trademarks of the National Fire Protection Association, Quincy, MA.

Contents

Figures

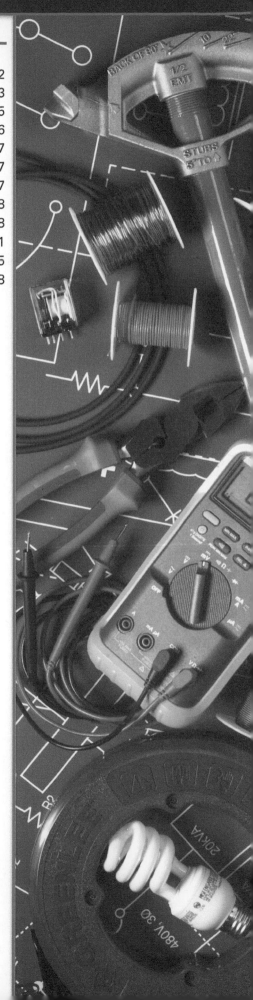

This page is intentionally left blank.

1.0.0 IDENTIFYING SECTORS AND CAREER OPTIONS IN THE ELECTRICAL INDUSTRY

Objective

Identify the various sectors and trade options in the electrical industry.

a. Describe the typical components in a residential wiring system.
b. Describe the typical components in a commercial wiring system.
c. Describe the typical components in an industrial wiring system.
d. List various career paths and opportunities in the electrical trade.

Trade Terms

Electrical service: The electrical components that are used to connect the serving utility to the premises wiring system.

Raceway system: Enclosures that house the conductors in an electrical system (such as fittings, boxes, and conduit).

Rough-in: The beginning stage of wiring that involves the installation of the panelboard, raceway system, wiring, and boxes.

Substation: An enclosed assembly of high-voltage equipment, including switches, circuit breakers, buses, and transformers, that connects the power generation facility to the grid and through which electrical energy is passed in order to change its characteristics, such as stepping voltage up or down, changing control frequency, or other characteristics.

Trim-out: The final stage of wiring that involves the installation and termination of devices and fixtures.

The modern world is one of electrical dependency, where most people take the availability of electricity for granted until experiencing an unexpected power failure or power outage. It takes an army of electrically skilled individuals to generate, transmit, distribute, and maintain electrical systems and equipment in order to provide the convenience of continuous electrical energy.

The electrical field can be divided into three broad categories: residential, commercial, and industrial. When studying to become an electrician,

all trainees reach a point at which he or she must decide which area of electrical work to pursue. Many skilled electricians become comfortable in residential or commercial wiring, whereas others feel at home in large industrial facilities, such as petrochemical plants, installing or maintaining huge electrical systems including motors and control devices. An electrician may decide to start a contracting company or to teach the craft to others.

1.1.0 Residential Wiring Systems

Components of a residential electrical system include an electrical supply, electrical service, nonmetallic-sheathed cable, nail-on device boxes, panelboards, and fixtures. Phases of residential electrical wiring include rough-in, trim-out, testing, and troubleshooting. Rough-in is the beginning stage of wiring that involves the installation of the panelboard, raceway system, wiring, and boxes, while trim-out is the final stage of wiring that involves the installation and termination of devices and fixtures. The following primary components of residential wiring systems are shown in *Figure 1*:

- Pad-mounted transformer
- Residential electrical service
- Nail-on device box
- Nonmetallic-sheathed cable
- Interior panel (subpanel)
- Luminaire (lighting fixture)

Interior panel enclosures, such as the one shown in *Figure 1 (E)*, are typically installed and partially terminated during the rough-in stage.

1.2.0 Commercial Wiring Systems

Electrical installations in commercial structures, such as office buildings and stores, contain many of the same elements as residential installations. One exception is that, in commercial and industrial electrical installations, conductors are typically installed in metal raceways, requiring the installing electricians to be skilled in conduit bending. A well-trained electrician can install a metal raceway system with little or no waste in conduit, while an inexperienced beginner will typically go through several pieces of conduit before acquiring the necessary bend.

Some of the elements that make up a commercial electrical system are shown in *Figure 2* and include the following:

- Pad-mounted commercial transformer
- Commercial electrical service
- Conduit system
- Fire alarm system
- Office and outdoor lighting

(A) PAD-MOUNTED TRANSFORMER

(B) RESIDENTIAL ELECTRICAL SERVICE

(C) NAIL-ON DEVICE BOX

(D) NONMETALLIC-SHEATHED CABLE

(E) INTERIOR PANEL (SUBPANEL)

(F) LUMINAIRE (LIGHTING FIXTURE)

Figure 1 Primary components of residential wiring.

(A) PAD-MOUNTED COMMERCIAL TRANSFORMER

(B) COMMERCIAL ELECTRICAL SERVICE

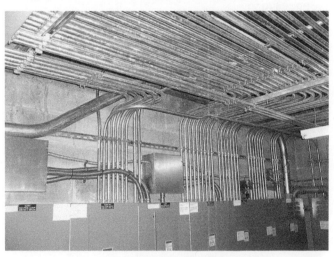

(C) CONDUIT SYSTEM

(D) FIRE ALARM SYSTEM

(E) OFFICE LIGHTING

(F) OUTDOOR LIGHTING

Figure 2 Commercial electrical system components.

1.3.0 Industrial Wiring Systems

Because of the hazardous materials that exist in many industrial facilities, the installation and maintenance of electrical systems in these volatile environments must follow rigid requirements governed by the *National Electrical Code®* (*NEC®*). For similar reasons, commercial and residential installations also have strict code requirements that electricians must obey.

Conduit systems in potentially explosive (classified) locations must be sealed to outside vapors and gases, and any potential sparking or arcing device must be contained within a special enclosure or casing to prevent the ignition of hazardous vapors that might be present.

Industrial electricians are generally split into two groups: installers and maintenance personnel. In large industrial facilities, electrical systems are typically installed by contract electricians who do not work directly for the facility but rather for a contractor hired by the facility. It is the responsibility of these electricians to install conduit systems, conductors, motors, and equipment. The completed system is then turned over to maintenance electricians who work directly for the facility (on-site personnel). These electricians maintain the system once it is energized and operating. In smaller facilities, industrial plant electricians may both install and maintain the electrical equipment.

Figure 3 illustrates some of the electrical equipment that may be found in industrial facilities, including distribution switchgear, a rigid metallic conduit (RMC) system, and a motor control center.

1.4.0 Career Paths and Opportunities in the Electrical Industry

Every electrical system in the United States and all over the world must be installed and maintained by someone qualified to do the work. This creates a great opportunity for work that is rewarding both personally and financially. Growth in the industry, new technology, upgrading and retrofitting of existing equipment, and retirement of current workers create openings for trained and skilled electricians. Examples of electrical occupations include residential electricians, commercial electricians, industrial electricians, service electricians, remodel electricians, electrical maintenance technicians, utility/substation electricians, and maritime electricians.

Upgrading the Lights Over Chicago

Things may have gotten a little brighter and more colorful in the city of Chicago after a retrofit to the lighting at the John Hancock Center, which is the tallest building in this photo. In 2017, contractors replaced the building's famed Crown of Lights, which surrounds the building on its 99th floor and can be seen for miles.

The original Crown of Lights consisted of 552 fluorescent tubes that produced a white crown around the 99th floor. When building owners wanted to change the lighting color, workers had to manually remove each tube to slip a colored covering over each one, which was a week-long endeavor.

The electricians replaced the fluorescent tubes with light-emitting diode (LED) luminaires mounted inside white reflective cabinets around the 99th floor windows. The touchscreen controller in the engineer's office can store more than 500 preset scenes to light the building in a variety of colors.

The LED luminaires offer much more efficiency. The fluorescent tubes consumed 57,408 watts (W) of power, while the new power usage is only 9,400W. In addition to the lower operating costs, the maintenance cost of $80,000 per year to change the fluorescent fixture coverings has been eliminated.

Figure Credit: iStock@pawel.gaul

(A) DISTRIBUTION SWITCHGEAR

(B) RIGID METALLIC CONDUIT (RMC) SYSTEM

(C) MOTOR CONTROL CENTER

Figure 3 Industrial electrical equipment.

1.4.1 Residential Electrician

The primary goal of a residential electrician is to provide a complete electrical system in a residential structure. Elements of a residential wiring installation include installing the electrical service entrance equipment, branch circuit conductors, device boxes, panel enclosures, overcurrent protective devices (circuit breakers), and lighting, smoke detectors, and other fixtures.

Residential electrical contractors are the employers of residential electricians. Various methods of employment are available, depending on the policies of the contractor. Residential electricians often work directly for the contractor. Under other circumstances, electricians work as individual contractors who are responsible for filing their own taxes and providing their own insurance, as well as supplying their own equipment. The latter type of electrician is generally paid by piece work, which is a fixed price for each house completed. Tract homes (subdivisions), similar to those shown in *Figure 4*, are frequently wired by residential electricians.

Through licensure, a residential electrician can also be a business owner of a one-person or multi-person contracting business. Specific licenses are required to become a residential electrical contractor.

1.4.2 Commercial Electrician

Commercial electricians install power, light, and control wiring in various locations including apartment buildings, stores, offices, service stations, and hospitals.

Commercial electricians are often employed by electrical contractors who work as subcontractors for a general contractor. Many electrical contractors install both residential and commercial wiring, such as in metal-frame commercial buildings (shown in *Figure 5*). However, they often employ electricians who are specialists in one or the other type of wiring.

1.4.3 Service Electrician

Service electricians can work in the residential or light commercial markets. Service electricians generally respond to calls for repairs or upgrades to existing electrical systems. These electricians are required to have a thorough knowledge of electrical theory so that they can troubleshoot problems in electrical systems. They also install

Figure 4 Tract homes.

wiring, new receptacles, and new lighting fixtures in existing buildings or outdoors. At times, service electricians may change out electrical services to a building. Unlike construction electricians, service electricians typically do not work on installations in new construction. Service electricians often operate their business as a sole proprietorship.

1.4.4 Remodeling Contractor

Some electrical contractors focus all or a portion of their business on remodeling existing electrical systems in homes or businesses. Remodeling contracts can be quite extensive, or they can be simple installations. During renovations of older buildings, remodeling contractors might rewire the entire structure to replace aging wiring and system components. They might also handle the wiring of any new additions to a home or building, such as the installation of new lighting or ceiling fans, switches or dimmers, smart home appliances or controls, pool or hot tub wiring, and panelboard changes.

1.4.5 Industrial Electrician

Electricians who specialize in installing electrical systems in industrial facilities require additional training due to the amount of specialty equipment that must be installed and tested. Electricians working in hazardous locations must understand the special code requirements associated with these locations. These craft workers must differentiate between the hazardous classes and divisions and know the requirements for each type of location. In addition, industrial electricians must be familiar with three-phase power, motors, and motor control systems. They may also be

Electricians Connect the Atlanta Falcons to the 21st Century

Atlanta's Mercedes-Benz Stadium, home to the NFL's Atlanta Falcons, is the first professional sports venue to achieve a Leadership in Energy and Environmental Design (LEED) Platinum rating from the US Green Building Council. Electricians helped the builders achieve that status with state-of-the-art lighting and electrical systems. The stadium uses an LED lighting system that can change the colors of the lights inside and outside of the arena. Additionally, the lights allow for low maintenance and energy costs.

The lighting is not the only energy saving feature. The building also boasts more than 4,000 solar photovoltaic panels to supply power to the building and its surrounding parking lots. Electric vehicle connections also are available for up to 50 cars.

Figure 5 Metal-frame commercial building.

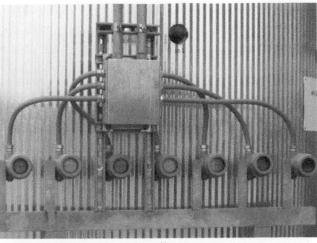

Figure 6 Instrumentation installation.

responsible for installing the conduit and wiring for process control instrumentation, such as the installation shown in *Figure 6*. Finally, industrial electricians must be able to troubleshoot any of these systems should they fail during initial testing.

Large corporations and contractors that specialize in building industrial facilities typically employ industrial electricians.

1.4.6 Electrical Maintenance Technician

Electrical maintenance technicians work in commercial and industrial facilities. These technicians typically work directly for the owner or management of the facility. In large facilities, they are usually members of a maintenance group supervised by a maintenance manager or supervisor.

In industrial facilities, maintenance electricians are frequently responsible for both the electrical and instrumentation systems and equipment and are referred to as *E&I technicians*. Instrumentation is a craft of its own, and it requires additional training over and above the electrical skills training found in this course.

Electrical maintenance electricians are usually employees of the facility; however, there are contract maintenance groups that provide maintenance personnel who work side-by-side with full-time plant personnel in maintaining electrical and instrumentation systems.

A common component in industrial facilities with which all electrical maintenance personnel must be familiar is the magnetic motor starter, illustrated in *Figure 7*. In industrial environments, magnetic motor starters frequently fail, so maintenance electricians must be able to disassemble and reassemble, troubleshoot, and repair these components.

1.4.7 Utility Substation Electrician

Utility substation electricians are trained and qualified in the construction, maintenance, and operation of substation facilities to ensure reliable electrical service to the public. A substation

Figure 7 Magnetic motor starter.

Figure 8 Utility power substation.

electrician specializes in work on transmission and distribution substations (*Figure 8*), including the installation and commissioning of equipment used for power delivery. The work of a substation electrician includes monitoring, troubleshooting, repairing, and testing circuit breakers, transformers, voltage regulators, reclosers, and other equipment. The substation electrician will also install grounding systems, cable tray, conduit, and other raceways. In addition, the substation electrician must understand how to size, select, and install the various types of cable used in substations, from control cables up to medium-voltage cable. This type of work is conducted for utility companies, municipal power providers, and some independent power providers.

1.4.8 *Maritime Electrician*

While the maritime industry is different from the construction industry, shipbuilders and ship repair facilities are also big employers of electricians. Electricians in the maritime industry do not follow the *National Electrical Code®*. Instead, they must adhere to standards developed by the American Bureau of Shipping (ABS), the Institute of Electrical and Electronics Engineers (IEEE), and US Navy standards.

Figure 9 A maritime electrician works on shipboard electrical systems.

Maritime electricians work on a variety of electrical systems on a vessel (*Figure 9*), including ship steering systems, alarms and monitoring systems, and power distribution systems. Maritime electricians install batteries and generators and work on motors, pumps, and navigation systems.

1.0.0 Section Review

1. The termination of devices and fixtures is completed during a construction phase known as _____.

   a. rough-in
   b. trim-out
   c. final finish
   d. cleanup

2. Which of the following is true with regard to commercial electrical work?

   a. Wiring is rarely installed in raceways.
   b. Most wiring is installed in classified (hazardous) locations.
   c. Fire alarm systems are commonly required.
   d. It is exactly the same as residential work.

3. In industrial facilities, the electrical systems are most likely maintained by _____.

   a. on-site maintenance personnel
   b. external electrical contractors
   c. senior management
   d. the local utility

4. Employers of commercial electricians are often electrical contractors who work as subcontractors for _____.

   a. larger electrical contractors
   b. standards organizations
   c. general contractors
   d. architects

2.0.0 TRAINING AND APPRENTICESHIP PROCESS FOR ELECTRICIANS

Objective

Understand the apprenticeship/training process for electricians.

a. List Department of Labor (DOL) requirements for apprenticeship.
b. Describe various types of training in the electrical field.

Trade Term

On-the-job learning (OJL): Job-related learning an apprentice acquires while working under the supervision of journey-level workers. Also called *on-the-job training (OJT)*.

The demand for skilled electricians is high. New homes, schools, office buildings, malls, airports, industrial plants, and many other types of structures are being constructed every day. According to the US Bureau of Labor Statistics, this new construction creates a great demand for skilled workers every year.

Wide World of Sports

Every fall, 600 men and women from 47 countries participate in the US Open tennis tournament held in New York. The United States Tennis Association's National Tennis Center includes 33 outdoor courts, 9 indoor courts, and 3 stadium-style courts, making it the largest public tennis facility in the world. When the tennis center was expanded, electricians laid more than 415 miles of high-performance cable with more than 80,000 terminations and 5 miles of fiber optic cable. The scope of work included multimedia systems covering network broadcast, internal cable television distribution, scoring systems, and audio capabilities. The broadcast system includes 6 television studios, 3 interview rooms, and 30 broadcast booths. During the Open, the scores from all the courts are posted live throughout the complex and to the US Open website.

Pay in the construction industry is very good, and the pay for electricians is close to the top of the scale for all construction occupations. There are many ways to increase your skills and grow professionally in construction. There are also many opportunities to try different types of jobs and earn more money within the electrical trade. You will find that electrical work is demanding but fulfilling. There is a large variety of work to be done. You may be indoors installing boxes or hooking up motor controllers, or you may be outside climbing a ladder or running conduit on a pipe rack high in the air.

Electricity and electrical equipment are needed everywhere. This means electricians are also needed everywhere.

2.1.0 DOL Apprenticeship Standards

Apprenticeship training goes back thousands of years, and its basic principles have not changed. First, it is a means for a person entering the craft to learn from those who have mastered the craft. Second, it focuses on learning by doing—real skills versus theory. Some theory is presented in the classroom. However, it is always presented in a way that helps the trainee understand the purpose behind the skill that is to be learned.

The US Department of Labor (DOL) Office of Apprenticeship sets the minimum standards for training programs across the country. These programs rely on mandatory classroom instruction and **on-the-job learning (OJL)**. When you are in an OJL program, you are being paid to learn the trade. This is a huge advantage compared to a typical college student who may be paying a large tuition bill to attend class. Another advantage of learning on the job is the hands-on experience you get, which is invaluable in any job. The DOL apprenticeship standards require at least 144 hours of classroom instruction per year and 2,000 hours of OJL per year. In a typical electrical apprenticeship program, trainees spend at least 576 hours in classroom instruction and 8,000 hours in supervised OJL before receiving journeyman certificates issued by registered apprenticeship programs. The OJL must be supervised by a journeyman or master electrician.

NCCER uses the minimum DOL standards as a foundation for developing comprehensive curricula that provide trainees with in-depth classroom and OJL experience. This four-year NCCER electrical training program provides trainees with industry-driven training and education using a purely competency-based teaching approach. This means that trainees must show the instructor that they possess the knowledge and skills needed to safely perform the hands-on tasks that are covered in each module.

When a certified instructor is satisfied that a trainee has the required knowledge and skills for a given module, that information is sent to NCCER and kept in the Registry system. NCCER's Registry system can then confirm training and skills for workers as they move from state to state, from company to company, or even within a company. See the *Appendix* for examples of the credentials issued by NCCER.

Whether you enroll in an NCCER program or another apprenticeship program, make sure you work for an employer or sponsor who supports a nationally standardized training program that includes credentials to confirm your skill development.

2.2.0 Types of Training

All apprenticeship standards prescribe certain work-related or on-the-job learning (OJL). This training may begin after graduation from high school or before graduation as a part of a youth apprenticeship program. After the training is completed, electricians may sit for one or more licensing exams.

2.2.1 On-the-Job Learning

OJL is broken down into specific tasks in which the apprentice receives hands-on training. In addition, a specified number of hours is required in each task. The total number of OJL hours for an apprenticeship program is traditionally 8,000, which amounts to four years of training.

In a competency-based program, it may be possible to shorten this time by testing out of specific tasks through a series of performance exams. In a traditional program, the required OJL may be acquired in increments of 2,000 hours per year.

The apprentice must log all work time (*Figure 10*) and turn it in to the apprenticeship committee so that accurate time control can be maintained. After each 1,000 hours of related work, the apprentice will typically receive a pay increase as prescribed by the apprenticeship standards.

For those entering an apprenticeship program, a high school or technical school education is desirable. Courses in shop, mechanical drawing, and general mathematics are helpful. Manual dexterity, good physical conditioning, and quick reflexes are important. The ability to solve problems quickly and accurately and to work closely with others is essential. You must also have high awareness of safety concerns.

The prospective apprentice must submit certain information to the apprenticeship committee. This may include the following:

- Aptitude test (General Aptitude Test Battery or GATB Form Test) results (usually administered by the local Employment Security Commission)
- Proof of educational background (candidate should have school transcripts sent to the committee)
- Letters of reference from past employers and friends
- Proof of age
- If the candidate is a veteran, a copy of Form DD214
- A record of technical training received that relates to the construction industry and/or a record of any pre-apprenticeship training

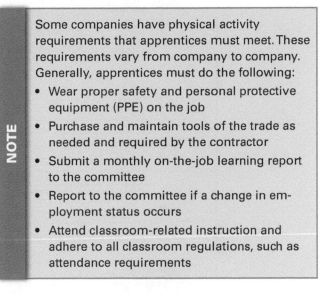

NOTE

Some companies have physical activity requirements that apprentices must meet. These requirements vary from company to company. Generally, apprentices must do the following:
- Wear proper safety and personal protective equipment (PPE) on the job
- Purchase and maintain tools of the trade as needed and required by the contractor
- Submit a monthly on-the-job learning report to the committee
- Report to the committee if a change in employment status occurs
- Attend classroom-related instruction and adhere to all classroom regulations, such as attendance requirements

Note that informal OJL provided by employers is usually less thorough than OJL provided through a formal apprenticeship program. The degree of training and supervision in this type of program often depends on the size of the employer. A small contractor may provide training in only one area, whereas a large company may be able to provide training in several areas.

2.2.2 Youth Apprenticeship Program

Also available is the Youth Apprenticeship Program, which allows students to begin their apprenticeship training while still in high school. Students entering the program in eleventh grade can complete as much as two years of the four-year NCCER standardized craft training program by high school graduation. In addition, the program, in cooperation with local craft employers, allows students to work in the trade and earn money while still in school. Upon graduation, students can enter the industry at a higher level and with more pay than someone just starting the apprenticeship program.

This training program is similar to the one used by NCCER learning centers, contractors, and colleges across the country. Students are recognized through official transcripts and can enter the next year of the program wherever it is offered. Students may also have the option of applying the credits at a two-year or four-year college that offers degree or certification programs in the construction trades.

2.2.3 Licensing

After completing your training, you will probably want to take your state or local licensing exam. The purpose of licensing is to provide assurance that you are qualified to install and/or maintain electrical systems. Licensing will allow you to work independently and earn a higher income. As a licensed electrician, you are not only responsible for your work, but you are liable for that work as well. If someone is working for you, then you are also responsible and liable for that person's work.

ABC Associated Builders and Contractors, Inc.

Southern California Chapter

Electrical Apprentice Monthly Record

New OJT Fax# (714) 463-1710 as of April 2015

Print Name: ______________________

Apprentice Signature: ______________________

SSN: ______________________

Month/ Year:

	Dates	A: Residential	B: Commercial	C: Industrial	D: Special	Total Hours
Week 1						
Week 2						
Week 3						
Week 4						
Week 5						
	Total Hours					

Hours will be checked & verified with the copies of your pay-stubs that are submitted with this form.

Company: ______________________ ☐ **Unemployed** ☐ **Leave of Absence** ☐ **Missing Pay-stub(s)**

For additional copies please visit our website at: **www.abcsocal.org/Apprenticeship/OJT_Record_Sheets.aspx**

ABC Associated Builders and Contractors, Inc.

Southern California Chapter

Electrical Apprentice Monthly Record

New OJT Fax# (714) 463-1710 as of April 2015

Print Name: ______________________

Apprentice Signature: ______________________

SSN: ______________________

Month/ Year:

	Dates	A: Residential	B: Commercial	C: Industrial	D: Special	Total Hours
Week 1						
Week 2						
Week 3						
Week 4						
Week 5						
	Total Hours					

Hours will be checked & verified with the copies of your pay-stubs that are submitted with this form.

Company: ______________________ ☐ **Unemployed** ☐ **Leave of Absence** ☐ **Missing Pay-stub(s)**

For additional copies please visit our website at: **www.abcsocal.org/Apprenticeship/OJT_Record_Sheets.aspx**

Figure 10 Apprentice monthly record page.

Licensing requirements vary from state to state and may vary by municipality. Contact your local building department for the requirements in your area. After you receive your license, your state or locality may require continuing education in order to renew your license.

Think About It

Licensing

What are the licensing requirements in your area?

2.0.0 Section Review

1. The NCCER electrical training program applies DOL standards and is a _____.

    a. one-year program
    b. two-year program
    c. three-year program
    d. four-year program

2. Entry into an apprenticeship program is likely to require a(n) _____.

    a. GATB Form Test
    b. license
    c. Scholastic Aptitude Test
    d. OSHA 40-hour course

3.0.0 EMPLOYEE AND EMPLOYER RESPONSIBILITIES

Objective

Understand the responsibilities of the employee and employer.

  a. Identify employee responsibilities.
  b. Identify employer responsibilities.

Trade Term

Occupational Safety and Health Administration (OSHA): The federal government agency established to ensure a safe and healthy environment in the workplace.

The safe and cost-effective installation and service of electrical systems requires close collaboration between the employer and the employees. All workers must understand the responsibilities of providing a safe and productive workplace.

3.1.0 Employee Responsibilities

To be successful, you must be able to use current trade materials, tools, and equipment to finish the task quickly and efficiently. You must keep up-to-date on technical advancements and continually gain the skills to use them. A professional never takes chances with regard to personal safety or the safety of others.

3.1.1 Professionalism

The term *professionalism* broadly describes the desired overall behavior and attitude expected in the workplace. Professionalism is too often absent from the construction site. Most people would argue that professionalism must start at the top in order to be successful. It is true that management support of professionalism is important to its success in the workplace, but it is just as important that individuals recognize personal responsibilities for professionalism.

Professionalism includes honesty, productivity, safety, civility, cooperation, teamwork, clear and concise communication, being on time, and coming prepared to work. It can be demonstrated in a variety of ways every minute you are on the job.

Professionalism goes beyond face-to-face interactions. It also extends into written communication. A great deal of work communication is conducted over email. *Figure 11* shows an example of a carefully written email. The following are some general rules to keep your email professional:

- Always start with a clear subject line that indicates the purpose of the message.
- Begin the email by addressing the recipient.
- Try to keep the email brief and to the point.
- Be sure to clarify what response or action is required of the recipient.
- Write in a positive tone and avoid blaming language.
- Do not type in all capital letters.
- Avoid sarcasm, as it can be easily misunderstood.

Text messaging is also common in the workplace. When texting on the job, it is important to keep messages professional. Texting is best when used for brief messages that should be viewed immediately and do not require much interaction. It should not be used for complex messages or messages that are meant to convey an emotion, such as an apology or criticism. Here are some tips for maintaining professional text communications:

- Always proofread messages and make sure words have not been autocorrected to something you did not mean.
- Use clear and concise language. Avoid using shorthand abbreviations unless the recipient is a friend, close colleague, or somebody who you know understands the shorthand.
- Never text while driving or operating machinery.

Professionalism is a benefit to both the employer and the employee. It is a personal responsibility. The construction industry is what each individual chooses to make of it—choose professionalism and the industry image will follow.

3.1.2 Honesty

Honesty and personal integrity are important traits of successful professionals. Professionals pride themselves on performing a job well and being punctual and dependable. Each job is completed in a professional way, never by cutting corners or reducing materials. A valued professional maintains work attitudes and ethics that protect tools, materials, and other property belonging to employers, customers, and other trades from damage or theft at the shop or job site.

Honesty and success go hand-in-hand for both the employer and the professional electrician. It is not simply a choice between good and bad but

Dear Mr. Jones,

The paint colors and faucets available for your bathroom are listed below (photos of faucets and paint colors are attached to this email). Please let me know what you decide by 5:00 pm on Friday, March 21st. If you have any questions, please do not hesitate to contact me at 703-555-1212.

Paint Colors (Available in semi-gloss or eggshell finish)
- #1415 – Soft Jade
- #1416 – Garden Moss
- #1417 – Forest Glen

Faucet Sets (Available in polished brass or polished chrome)

Model	Price	Handle Style
Meridian	$109.88	Single
Mermaid	$83.50	Dual
Monitor	$95.75	Dual

Regards,
John Q. Smith
Smith Contracting

Figure 11 Sample email that shows professionalism.

a choice between success and failure. Dishonesty will always catch up with you. Whether you steal materials, tools, or equipment from the job site or simply lie about your work, it will not take long for your employer to find out.

If you plan to be successful and enjoy continuous employment, consistent earnings, and being sought after as opposed to seeking employment, then start out with the basic understanding of honesty in the workplace. You will reap the benefits.

Honesty means more, however, than simply not taking things that do not belong to you. It also means giving a fair day's work for a fair day's pay. Employers place a high value on employees who display honesty. Being honest about a mistake you may have made on the job will build trust with your employer and provide a learning experience for you. Errors in judgement can be forgiven, but dishonesty is much harder to forgive.

3.1.3 Loyalty and Respect

Employees expect employers to look out for their interests, to provide them with steady employment, and to promote them to better jobs as openings occur. Employers feel that they, too, have a right to expect loyalty from their employees— to keep their interests in mind, to speak well of

them to others, to keep any minor troubles strictly within the plant or office, and to keep absolutely confidential all matters that pertain to the business. Both employers and employees should keep in mind that loyalty is not something to be demanded; rather, it is something to be earned.

While loyalty is something to be earned, respect should always exist. Respect is showing regard for the feelings and rights of others. Respect establishes a sense of fairness among coworkers. Showing respect for other individuals involves treating them how you'd like to be treated. It means not insulting or demeaning others. Respect can be shown through your tone and your body language.

3.1.4 Willingness to Learn

Every company and job site has its own way of doing things. Employers expect their workers to be willing to learn these ways. You must be willing to adapt to change and learn new methods and procedures as quickly as possible. Sometimes, a change in safety regulations or the purchase of new equipment makes it necessary for even experienced employees to learn new methods and operations. Successful people take every opportunity to learn more about their trade.

Electricians Are Key Players for the NBA and NHL

Electricians are critical players in building modern sports complexes for professional sports franchises. The Pepsi Center was built in Denver, Colorado for the NBA's Denver Nuggets, the NHL's Colorado Avalanche, and several other teams. During the two-year construction cycle, electricians installed more than 120 miles of conduit, 569 miles of wire, 13,000 fixtures, and 280 panels. The lighting systems are computer controlled and can be preset for basketball games and concerts. In addition to high-quality sound for concerts, the state-of-the-art sound and security system includes closed-circuit television (CCTV) monitoring and card access security controls.

Figure Credit: iStock@AndreyKav

3.1.5 Taking Responsibility

Most employers expect their employees to see what needs to be done and do it. After an assignment is received and the procedure and safety guidelines are fully understood, you should assume the responsibility for that task without further reminders.

3.1.6 Cooperation

To cooperate means to work together. In our modern business world, cooperation is the key to getting things done. Learn to work as a member of a team with your employer, supervisor, and fellow workers in a common effort to get the work done efficiently, safely, and on time. Look for the positive characteristics of your co-workers and supervisors. Helpful suggestions and compliments are much more effective than negative ones.

3.1.7 Rules and Regulations

Employees can work well together only if there is some understanding about the nature of the work to be done, when and how it will be done, and who will do it. Rules and regulations are a necessity in any work situation and must be followed by all employees.

3.1.8 Tardiness and Absenteeism

Tardiness means being late for work, and absenteeism means being off the job for one reason or another. While occasional absences are unavoidable, consistent tardiness and frequent absences are an indication of poor work habits, unprofessional conduct, and a lack of commitment.

Although workers may not be paid when they are absent or tardy, there is still a cost to the employer. For example, the worker's health care insurance must still be paid, even though the worker is not on site. In addition, jobs are bid and scheduled based on a certain workforce size. If you are not there, work is not being done and schedules are not being met. It is important for you to be at work, on time, every day. If you must be absent, call in as soon as possible so that your employer can find a replacement.

3.1.9 Safety

In exchange for the benefits of your employment and your own well-being, you are obligated to work safely. You are also obligated to make sure anyone you supervise or work with is working safely. Your employer is obligated to maintain a safe workplace for all employees.

> **NOTE**
>
> Safety is everyone's responsibility. Everybody on a job site has a responsibility toward their own safety as well as the safety of others.

Ethical Principles for Members of the Construction Trades

Honesty—Be honest and truthful in all dealings. Conduct business according to the highest professional standards. Faithfully fulfill all contracts and commitments. Do not deliberately mislead or deceive others.

Integrity—Demonstrate personal integrity and the courage of your convictions by doing what is right even if there is pressure to do otherwise. Do not sacrifice your principles because it seems easier.

Loyalty—Be worthy of trust. Demonstrate fidelity and loyalty to companies, employers and sponsors, co-workers, trade institutions, and other organizations.

Fairness—Be fair and just in all dealings. Do not take advantage of another's mistakes or difficulties. Fair people are open-minded and committed to justice, equal treatment of individuals, and tolerance for and acceptance of diversity.

Respect for others—Be courteous and treat all people with equal respect and dignity.

Obedience—Abide by laws, rules, and regulations relating to all personal and business activities.

Commitment to excellence—Pursue excellence in performing your duties, be well-informed and prepared, and constantly try to increase your proficiency by gaining new skills and knowledge.

Leadership—By your own conduct, seek to be a positive role model for others.

You have a responsibility to maintain a safe working environment. This means two things:

- Follow your company's rules for proper working procedures and practices.
- Report any unsafe equipment and conditions directly to your supervisor.

If you see something unsafe while on the job, report it! Do not ignore it. It will not correct itself. In the end, even if you do not think an unsafe condition affects you, it does. Always report unsafe conditions. Do not think your employer will be angry because your productivity suffers while the condition is being reported. On the contrary, your employer will be more likely to criticize you for not reporting a problem.

WARNING!

For the safety of yourself and others, always report unsafe conditions. Ignoring them could cause serious injury or even death.

Your employer knows that the short time lost in making conditions safe again is nothing compared with shutting down the whole job because of a major disaster. If that happens, you are out of work anyway. In fact, Occupational Safety and Health Administration (OSHA) regulations require you to report hazardous conditions. This applies to every part of the construction industry. Whether you work for a large contractor or a small contractor, you are obligated to report unsafe conditions.

In addition to the OSHA standards, there are specific standards related to electrical systems and devices. The *National Electrical Code*® (*NEC*®) sets the minimum standards for the safe installation of electrical systems. For example, *NEC*® temporary power requirements are more stringent than OSHA standards and can be enforced by the inspector and OSHA. You will become very familiar with the *NEC*® as you progress through your training.

Another standard with which you should become familiar is NFPA 70E®, *Standard for Electrical Safety in the Workplace*. This standard covers safe work practices that must be applied when working on or near exposed energized parts. It describes in detail the steps required for putting a circuit or electrical system into an electrically safe working condition. It also identifies safe approach distances to exposed energized parts and describes the PPE required to protect against electric shock and arc flash hazards. *Figure 12* shows an arc flash hood that is used to protect against a potential arc flash hazard.

3.2.0 Employer Responsibilities

Just as the employee has responsibilities on the job, the employer also has responsibilities. These are set out in the *Occupational Safety and Health Act of 1970*. The job of OSHA is to set occupational safety and health standards for all places of employment, enforce these standards, ensure that employers provide and maintain a safe workplace for all employees, and provide research and educational programs to support safe working practices.

OSHA was adopted with the stated purpose to assure as best as possible every worker in the nation with safe and healthful working conditions and to preserve our human resources.

OSHA requires each employer to provide a safe and hazard-free working environment. OSHA also requires that employees comply with OSHA rules and regulations that relate to worker conduct on the job. To gain compliance, OSHA can perform spot inspections of job sites, impose fines for violations, and even stop work from proceeding until the job site is safe.

According to OSHA standards, you are entitled to on-the-job safety training. Your employer must do the following:

- Show you how to do each job safely
- Provide you with the required personal protective equipment
- Warn you about specific hazards
- Supervise you for safety while performing the work

The enforcement for this Act of Congress is provided by the federal and state safety inspectors, who have the legal authority to impose fines for safety violations. The law allows states to have their own safety regulations and agencies to enforce them, but the US Secretary of Labor must first approve each state's programs. In states that do not have their own regulations and agencies, federal OSHA standards are mandatory.

OSHA standards are listed in *29 CFR 1926, OSHA Safety and Health Standards for the Construction Industry* (sometimes called *OSHA Standards 1926*). Other safety standards that apply to construction are published in *29 CFR 1910, OSHA Safety and Health Standards for General Industry*, in NFPA 70E®, and in the *NEC*®.

The most important general requirements that OSHA places on employers in the construction industry are as follows:

- The employer must post, in an easily seen area, signs informing employees of their rights and responsibilities.
- The employer must ensure there are no serious hazards on the job site and make sure the workplace complies with OSHA rules and regulations.
- Warning signs, posters, and labels must be posted in all required areas.

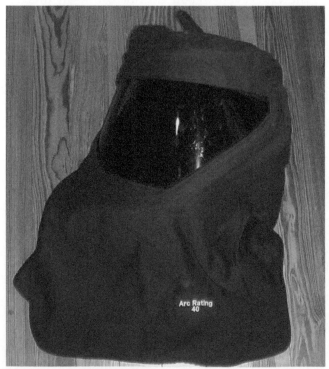

Figure 12 Arc flash hood.

- The employer must perform frequent and regular job-site inspections of equipment.
- The employer must instruct all employees to recognize and avoid unsafe conditions and to know the regulations that pertain to the job so employees may control or eliminate any hazards.
- No one may use any tools, equipment, machines, or materials that do not comply with *29 CFR, Part 1926*.
- The employer must ensure that only qualified individuals operate tools, equipment, and machines.
- The employer must provide medical training and examinations when required by OSHA.

- Employers with more than 10 employees must keep records of work-related injuries and illnesses. These records must be available to employees.
- Employers must not discriminate against employees who are exercising their rights under OSHA regulations.

Additional employer responsibilities are described in the Americans with Disabilities Act of 1990 (ADA). If a worker has a disability but is qualified to do the job, that worker has equal rights to employment. The US Equal Employment Opportunity Commission, along with state and local civil rights agencies, enforce ADA regulations.

3.0.0 Section Review

1. The minimum standards for the safe installation of electrical systems can be found in _____.

    a. the *National Electrical Code® (NEC®)*
    b. *29 CFR, Part 1910*
    c. *29 CFR, Part 1926*
    d. NFPA 70E®

2. OSHA standards for general industry are covered in _____.

    a. *29 CFR, Part 1910*
    b. *29 CFR, Part 1926*
    c. *29 CFR, Part 1965*
    d. *29 CFR, Part 1970*

---

1. Panelboards in residential wiring are typically installed during _____.
   a. rough-in
   b. trim-out
   c. service installation
   d. planning

2. A residential wiring system typically uses _____.
   a. medium-voltage cable
   b. tray cable
   c. flat conductor cable
   d. nonmetallic-sheathed cable

3. In which of the following environments would you find a commercial electrician at work?
   a. Shipyard
   b. Tract home construction
   c. Office building
   d. Petrochemical plant

4. Which of the following would most likely require special knowledge of hazardous locations?
   a. Residential wiring
   b. Industrial wiring
   c. Commercial wiring
   d. Service work

5. Tract homes are typically wired by _____.
   a. residential electricians
   b. commercial electricians
   c. homeowners
   d. utility workers

6. Minimum standards for apprenticeship training programs are established by _____.
   a. OSHA
   b. the DOL
   c. NCCER
   d. the employer

7. Which of the following is true with regard to your apprenticeship?
   a. Once you have finished your apprenticeship, your training is over.
   b. Licensing requirements vary from state to state.
   c. OSHA sets the minimum standards for training programs across the country.
   d. After completing your apprenticeship, you will automatically receive a license.

8. Which of the following is true with regard to tardiness and absenteeism?
   a. It is never acceptable to be absent, even when you are contagious.
   b. It is okay to be a little late for work as long as you make up the time.
   c. If you must be absent, call in early so that your employer can find a replacement.
   d. If you are not being paid, your absence does not cost the company anything.

9. If you see a safety violation at your job site, you should _____.
   a. ignore it unless it affects you directly
   b. make a mental note to avoid the area in future
   c. report it to your supervisor
   d. assume it is okay as long as no one has been injured

10. The primary mission of OSHA is to _____.
    a. inspect job sites for safety violations
    b. fine companies that violate safety regulations
    c. distribute safety equipment to workers
    d. ensure that employers maintain a safe workplace

Trade Terms Quiz

Fill in the blank with the correct term that you learned from your study of this module.

1. _____ is the federal government agency established to ensure a safe and healthy environment in the workplace.

2. Job-related learning acquired while working is known as _____.

3. Raceways would probably be installed in the _____ stage.

4. The _____ connects the serving utility to the premises wiring system.

5. Devices and fixtures would be installed during _____.

6. An enclosure that houses conduit, boxes, and fittings in an electrical system is called a _____.

7. An enclosure that includes transformers, circuit breakers, buses, and other components involved in the transmission and distribution of electricity is known as a _____.

Trade Terms

Electrical service
Occupational Safety and Health
   Administration (OSHA)

On-the-job learning (OJL)
Raceway system
Rough-in

Substation
Trim-out

1. Phases of residential electrical wiring include

   _____

2. True or False? A major difference between residential and commercial wiring is that commercial wiring is usually installed in metal conduit.

3. Industrial electricians are usually split into two groups: _____ and _____.

4. Your apprenticeship program requires _____ hours of OJL per year.
   a. 500
   b. 1,000
   c. 2,000
   d. 4,000

5. True or False? Competency-based training means that you must demonstrate the skills necessary to perform hands-on tasks before advancing to the next stage of the curriculum.

6. A _____ allows you to complete as much as two years of your apprenticeship before you have finished high school.

7. True or False? It's okay to take scrap pieces home as long as you don't think they'll be needed on the job.

8. Many employees feel that sick days can be treated as floating holidays to be taken whenever they would like a day off. How is this unfair to an employer or co-workers?

   _____

9. True or False? OSHA requires that employers provide a safe and hazard-free job site.

10. In addition to the OSHA safety standards, name two other standards that relate to electrical systems and devices.

    _____

Tim Dean

Electrician/Electrical Trades Instructor
Central Ohio ABC/Madison Comprehensive
High School

Provide a summary of how you got started in the construction industry.

Upon exiting The University of Toledo, I took a job with my brother-in-law, who worked as an electrician in Akron, Ohio. I knew little or nothing about electricity but needed a job to support myself and my wife.

Who inspired you to enter the industry? Why?

I suppose my brother-in-law, Tom Argenio, was my inspiration. He possessed a work ethic and craftsmanship that is very rare in our society today. He taught me not only how to be a good tradesman but also to understand the pride of quality workmanship.

What do you enjoy most about your job?

In the early days of my career, I simply appreciated having a job. But as time passed and my experience grew, I acquired a thirst for knowledge and understanding. Knowing the whys, whens, wheres, and hows brought new meaning to the skills I was attaining.

Do you think training and education are important in construction? If so, why?

The electrical trade is one of the most diverse and challenging of all the construction trades. Training and education are not only important but mandatory to stay safe, efficient, qualified, and prepared for the challenges of new technology.

How important are NCCER credentials to your career?

Being a part of NCCER has been one of the most rewarding experiences of my life. Acquiring credentials for performing as a Subject Matter Expert has benefited not only me personally by recognition and professional development, but the organization I work for as well. I enjoy the challenges required to stay abreast of the continuous evolution of our industry, and NCCER provides the vehicle and resources for me to keep up with cutting-edge technological changes.

How has training/construction impacted your life and your career?

When I entered the trade, I was clueless. I had no idea what I was getting myself into, but as I continued, I found a challenge in seeking an understanding of how and why things worked the way they did. I was engaged in the process and discovered that the more I knew and learned, the more I was worth to my employer.

Would you suggest construction as a career to others? If so, why?

I would recommend a career in the electrical trade to anyone who has a desire to learn and a thirst for understanding the new technologies that continue to evolve in our society. The electrical construction trade is both rewarding and challenging. Being a part of this great industry opens doors to new and exciting opportunities in many different areas. One only has to look around at the world we live in to know that being an electrician is much more than a job. Career opportunities abound, and there is truly no limit to what you can attain.

How do you define craftsmanship?

To me, craftsmanship is the result of acquiring knowledge, developing skills, infusing moral and ethical values, and blending personal pride to construct a product or process of recognizable and enduring quality.

Trade Terms Introduced in This Module

Electrical service: The electrical components that are used to connect the serving utility to the premises wiring system.

Occupational Safety and Health Administration (OSHA): The federal government agency established to ensure a safe and healthy environment in the workplace.

On-the-job learning (OJL): Job-related learning an apprentice acquires while working under the supervision of journey-level workers. Also called *on-the-job training (OJT)*.

Raceway system: An enclosure that houses the conductors in an electrical system (such as fittings, boxes, and conduit).

Rough-in: The beginning stage of wiring that involves the installation of the panelboard, raceway system, wiring, and boxes.

Substation: An enclosed assembly of high-voltage equipment, including switches, circuit breakers, buses, and transformers, that connects the power generation facility to the grid and through which electrical energy is passed in order to change its characteristics, such as stepping voltage up or down, changing control frequency, or other characteristics.

Trim-out: The final stage of wiring that involves the installation and termination of devices and fixtures.

SAMPLES OF NCCER TRAINING CREDENTIALS

NCCER

Board of Trustees confers upon

Sample Student

this certificate of completion for

Electrical Level One

in the Standardized Craft Training program
on this Twenty-first day of July, 2014.

Donald E. Whyte
Donald E. Whyte
President, NCCER

Additional Resources

This module presents thorough resources for task training. The following reference material is recommended for further study.

OSHA 29 CFR 1910, Standards for General Industry, Occupational Safety and Health Administration US Department of Labor. *www.ecfr.gov.*

OSHA 29 CFR 1926, Standards for the Construction Industry, Occupational Safety and Health Administration US Department of Labor. *www.ecrf.gov.*

National Electrical Code® Handbook, Latest Edition. Quincy, MA: National Fire Protection Association.

Wright, Maury. "Hancock Skyscraper gets crown of dynamic LED lighting to celebrate Chicago events." LEDs Magazine. *https://www.ledsmagazine.com/architectural-lighting/article/16700599/hancock-skyscraper-gets-crown-of-dynamic-led-lighting-to-celebrate-chicago-events.*

Lang, Justin. "NBC Sports' New Energy-Efficient Headquarters Facility." *Projection Lights and Staging News.*

"Mercedes-Benz Stadium." *www.wsp.com.*

Figure Credits

Section Review Answer Key

SECTION 1.0.0

Answer	Section Reference	Objective
1. b	1.1.0	1a
2. c	1.2.0	1b
3. a	1.3.0	1c
4. c	1.4.2	1d

SECTION 2.0.0

Answer	Section Reference	Objective
1. d	2.1.0	2a
2. a	2.2.1	2b

SECTION 3.0.0

Answer	Section Reference	Objective
1. a	3.1.9	3a
2. a	3.2.0	3b

This page is intentionally left blank.

NCCER CURRICULA — USER UPDATE

NCCER makes every effort to keep its textbooks up-to-date and free of technical errors. We appreciate your help in this process. If you find an error, a typographical mistake, or an inaccuracy in NCCER's curricula, please fill out this form (or a photocopy), or complete the online form at **www.nccer.org/olf**. Be sure to include the exact module ID number, page number, a detailed description, and your recommended correction. Your input will be brought to the attention of the Authoring Team. Thank you for your assistance.

Instructors – If you have an idea for improving this textbook, or have found that additional materials were necessary to teach this module effectively, please let us know so that we may present your suggestions to the Authoring Team.

NCCER Product Development and Revision
13614 Progress Blvd., Alachua, FL 32615

Email: curriculum@nccer.org
Online: www.nccer.org/olf

❏ Trainee Guide ❏ Lesson Plans ❏ Exam ❏ PowerPoints Other _____

Craft / Level: _____ Copyright Date: _____

Module ID Number / Title: _____

Section Number(s): _____

Description: _____

_____

_____

_____

_____

Recommended Correction: _____

_____

_____

_____

_____

Your Name: _____

Address: _____

_____

Email: _____ Phone: _____

This page is intentionally left blank.

Safety for Electricians

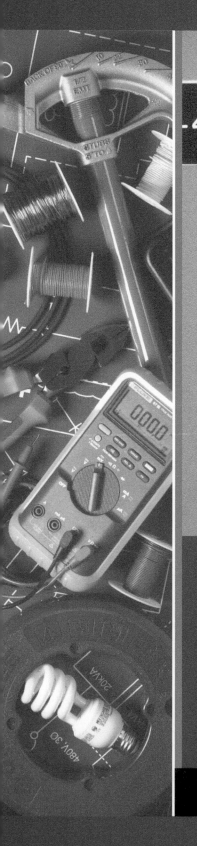

OVERVIEW

In order to work safely, electricians must be aware of potential hazards and stay constantly alert to them. This includes taking the proper precautions and practicing basic rules of safety. This module discusses hazards and describes the various types of personal protective equipment (PPE) used to reduce injuries. It also covers the standards related to electrical safety and the Occupational Safety and Health Administration (OSHA) lockout/tagout rule.

Module 26102-20

Trainees with successful module completions may be eligible for credentialing through the NCCER Registry. To learn more, go to **www.nccer.org** or contact us at 1.888.622.3720. Our website, **www.nccer.org**, has information on the latest product releases and training.

Your feedback is welcome. You may email your comments to **curriculum@nccer.org**, send general comments and inquiries to **info@nccer.org**, or fill in the User Update form at the back of this module.

This information is general in nature and intended for training purposes only. Actual performance of activities described in this manual requires compliance with all applicable operating, service, maintenance, and safety procedures under the direction of qualified personnel. References in this manual to patented or proprietary devices do not constitute a recommendation of their use.

Objectives

When you have completed this module, you will be able to do the following:

1. Identify electrical hazards and their effects.
   a. Understand the effects of electrical shock on the human body.
   b. Verify that circuits are de-energized.
   c. Identify causes of electrical incidents.
   d. Explain the hierarchy of risk controls.
2. Use PPE to reduce the risk of injury.
   a. Identify OSHA requirements for protective equipment.
   b. Select and use protective equipment.
3. Identify the standards that relate to electrical safety.
   a. Apply OSHA requirements in the workplace.
   b. Understand the purpose of NFPA 70E®.
4. Recognize the safety requirements for various hazards.
   a. Identify the safety hazards associated with ladders, scaffolds, and lift equipment.
   b. Avoid back injuries by practicing proper lifting techniques.
   c. Demonstrate basic tool safety.
   d. Identify confined space entry procedures.
   e. Work safely with dangerous materials.
   f. Select and use appropriate fall protection.

Performance Tasks

Under the supervision of the instructor, you should be able to do the following:

1. Properly select and use PPE.

2. Describe the safety requirements for an instructor-supplied task, such as replacing the lights in your classroom.

   - Discuss the work to be performed and the hazards involved.
   - If a ladder is required, perform a visual inspection of the ladder and set it up properly.
   - Ensure that local emergency telephone numbers are either posted or known by you and your partner(s).
   - Plan an escape route from the location in the event of an accident.

Trade Terms

Arc flash boundary (AFB)
Arc flash risk assessment
Arc rating
Double-insulated/ungrounded tools
Error precursors

Fibrillation
Grounded tool
Ground fault circuit interrupter (GFCI)
Hot stick
Incident energy

Limited approach boundary
Polychlorinated biphenyls (PCBs)
Qualified person
Restricted approach boundary
Unqualified person

Industry Recognized Credentials

If you are training through an NCCER-accredited sponsor, you may be eligible for credentials from NCCER's Registry. The ID number for this module is 26102-20. Note that this module may have been used in other NCCER curricula and may apply to other level completions. Contact NCCER's Registry at 888.622.3720 or go to **www.nccer.org** for more information.

> **NOTE**
>
> NFPA 70®, *National Electrical Code*® and *NEC*® are registered trademarks of the National Fire Protection Association, Quincy, MA.

Contents

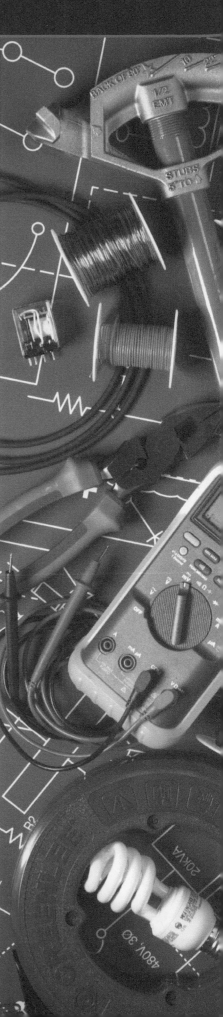

Contents (continued)

Figures and Tables

This page is intentionally left blank.

1.0.0 ELECTRICAL HAZARDS

Objective

Identify electrical hazards and their effects.
   a. Understand the effects of electrical shock on the human body.
   b. Verify that circuits are de-energized.
   c. Identify causes of electrical incidents.
   d. Explain the hierarchy of risk controls.

Trade Terms

Double-insulated/ungrounded tools: Electrical tools that are constructed so that the case is insulated from electrical energy. The case is made of a nonconductive material.

Fibrillation: Very rapid irregular contractions of the muscle fibers of the heart that result in the muscle being unable to contract and pump blood properly.

Ground fault circuit interrupter (GFCI): A protective device that functions to de-energize a circuit or portion thereof within an established period of time when a current to ground exceeds some predetermined value. This value is less than that required to operate the overcurrent protective device of the supply circuit.

In order to work safely, you must understand potential hazards and stay alert to them. You must take the proper precautions and practice the basic rules of safety. You must be safety-conscious at all times and report any unsafe conditions to your supervisor and co-workers. Safety should become a habit. Keeping a safe attitude on the job will go a long way in reducing the number and severity of accidents. Remember that your safety is up to you.

As an apprentice electrician, you need to be especially careful. You should only work under the direction of experienced personnel who are familiar with the various job-site hazards and the means of avoiding them.

The most life-threatening hazards on a construction site are often referred to as the Occupational Safety and Health Administration (OSHA) Focus Four and include the following:

- Falls when you are working in high places
- Electric shock and arc-related burns caused by coming into contact with live electrical circuits
- The possibility of being crushed by falling materials or equipment
- The possibility of being struck by flying objects or moving equipment/vehicles such as trucks, forklifts, and other construction equipment

These exposures account for approximately 90 percent of fatalities on construction sites. Other hazards include cuts, burns, back sprains, and getting chemicals or objects in your eyes. Most injuries, from minor to deadly, are preventable if the proper precautions are taken.

1.1.0 Electrical Shock

Electricity can be described as a potential that results in the movement of electrons in a conductor. This movement of electrons is called *electrical current*. Some substances, such as silver, copper, steel, and aluminum, are excellent conductors. The human body is also a conductor. The conductivity of the human body greatly increases when the skin is wet or moistened with perspiration.

Electrical current flows along any path in which the voltage can overcome the resistance. If the human body contacts an electrically energized point and is also in contact with the ground or another point in the circuit, the human body becomes a path for the current. *Table 1* shows the effects of current passing through the human body.

> **NOTE**
> The unit of measure represented by mA is known as a *milliampere* (or *milliamp*) and is equal to one one-thousandth of an ampere.

As shown in *Table 1*, a minor shock of 5mA results in an involuntary movement away from the source. This can result in injuries as the shocked worker jumps back from the electrical shock source, only to rip open or break an arm or hand on the way out of a cabinet or work area. When the current is between 6mA and 30mA, the shock causes loss of muscular control. This may result in the worker falling from an elevated position or cause the worker to fall into a more dangerous electrical source. As the current levels increase beyond about 20mA, muscular contractions can prevent the victim from pulling away. At 50mA, respiratory paralysis may result in suffocation. Current levels above 1A may cause the heart to go into a state of **fibrillation**, which causes very rapid irregular contractions of the muscle fibers

---

Table 1 Current Level Effects on the Human Body

Current Value	Typical Effects
1mA	Perception level. Slight tingling sensation.
5mA	Slight shock. Involuntary reactions can result in serious injuries such as falls from elevations.
6 to 30mA	Painful shock, loss of muscular control.
50 to 150mA	Extreme pain, respiratory arrest, severe muscular contractions. Death possible.
1,000mA to 4,300mA	Ventricular fibrillation, severe 4,300mA muscular contractions, nerve damage. Typically results in death.

Source: U.S. Department of Labor

of the heart that result in the muscle being unable to contract and pump blood properly. This condition is fatal unless the heart rhythm is corrected using a defibrillator. Current levels of 4A or more may stop the heart, resulting in death unless immediate medical attention is provided.

Other effects of electrical shock include entry and exit wounds from high-voltage contact, and thermal burns from current flow of a few amps and up. Thermal burns are often not apparent at first; however, the tissue in the current path may be destroyed and necrotize (rot away) from the inside over time. This is why it is critical to have a medical exam if you receive even a minor shock.

The amount of current measured in amperes that passes through a body determines the outcome of an electrical shock. The higher the voltage, the greater the chance for a fatal shock. In a one-year study in California, the following results were observed by the State Division of Industry Safety:

- Thirty percent of all electrical accidents were caused by contact with conductors. Of these accidents, 66 percent involved low-voltage conductors (those carrying 1,000 volts [V] or less).
- Portable, electrically operated hand tools made up the second largest number of injuries (15 percent). Almost 70 percent of these

injuries happened when the frame or case of the tool became energized. These injuries could have been prevented by following proper safety practices, using properly maintained grounded or double-insulated/ungrounded tools, and using ground fault circuit interrupter (GFCI) protection (described later in this module).

> **NOTE**
> Electric shocks or burns are a major cause of accidents in the construction industry. According to the National Institute for Occupational Safety and Health, workers in the construction industry are four times more likely to be electrocuted at work than all other industries combined.

In one ten-year study, investigators found 9,765 electrical injuries in the United States. A little more than 13 percent of the high-voltage injuries (over 1,000V) resulted in death. These high-voltage totals included limited-amperage contacts, which are often found on electronic equipment. When tools or equipment touch high-voltage overhead lines, the chance that a resulting injury will be fatal climbs to 28 percent. Of the low-voltage injuries, 1.4 percent were fatal.

Electrical Safety in the Workplace

According to OSHA, there are approximately 350 electrical-related fatalities per year in the United States, nearly one per day. Common electrical hazards include contact with overhead lines, lack of ground fault protection, path to ground missing or discontinuous, improper use of extension or flexible cords, and equipment not being used in the manner prescribed. If you see an unsafe condition on a job site, whether it's a tool cord with the ground pin removed or a fabricated extension cord, stop working immediately and report it. Failure to do so may result in death.

Severity of Shock

In *Table 1*, how many milliamps separate a mild shock from a potentially fatal one? What is the fractional equivalent of this in amps? How many amps are drawn by a 60W light bulb?

WARNING!

High voltage, identified in the *NEC®* as more than 1,000V, is almost ten times more likely to kill than low voltage. However, on the job you spend most of your time working on or near lower voltages. Due to the frequency of contact, most electrocution deaths actually occur at low voltages. Lack of caution when working with lower voltages is undoubtedly a factor in these deaths. These facts have been included to help you gain respect for the environment where you work and to stress how important safe working habits really are.

1.1.1 Body Resistance

Electricity travels in closed circuits, and its normal route is through a conductor. Shock occurs when the body becomes part of the electric circuit (*Figure 1*). As shown, the body offers a resistance of about 1,000 ohms (Ω), and a current of only 50mA can cause death. This means that protection from contact should begin at 50V.

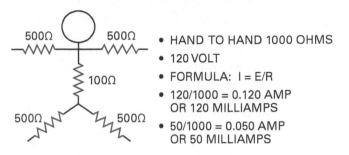

- HAND TO HAND 1000 OHMS
- 120 VOLT
- FORMULA: $I = E/R$
- 120/1000 = 0.120 AMP OR 120 MILLIAMPS
- 50/1000 = 0.050 AMP OR 50 MILLIAMPS

Figure 1 Body resistance.

The current must enter the body at one point and leave at another. Shock normally occurs in one of three ways: the person must come in contact with both wires of the electric circuit; one wire of the electric circuit and the ground; or a metallic part that has become live by being in contact with an energized wire while the person is also in contact with the ground.

To understand the physical harm electrical shock does to the body, you need to understand something about the physiology of certain body parts: the skin, the heart, and muscles.

Skin covers the body and is made up of three layers. The most important layer, as far as electric shock is concerned, is the outer layer of dead cells referred to as the *stratum corneum* or *horny layer*. This layer is composed mostly of a protein called *keratin*, which provides the largest percentage of the body's electrical resistance. When it is dry, the outer layer of skin may have a resistance of several thousand ohms. When it is moist, there is a radical drop in resistance, as is also the case if there is a cut or abrasion that pierces this layer. The amount of resistance provided by the skin will vary widely from person to person. A worker with a thick horny layer will have a much higher resistance than a child. The resistance will also vary widely at different parts of the body. For instance, the worker with high-resistance hands may have low-resistance skin on the back of his calf.

The heart is the pump that sends life-sustaining blood to all parts of the body. The blood flow is caused by the contractions of the heart muscle, which is controlled by electrical impulses. The electrical impulses are delivered by an intricate system of nerve tissue with built-in timing mechanisms, which make the chambers of the heart contract at exactly the right time. An outside electric impulse can upset the rhythmic, coordinated beating of the heart muscle. When this happens, the heart is said to be in *fibrillation*, and the pumping action stops. Death will occur quickly if the normal beat is not restored. Remarkable as it may seem, what is needed to defibrillate the heart is a DC shock of a high intensity.

The other muscles of the body are also controlled by electrical impulses delivered by nerves. Nerves can be defined as biological conductors of electricity. Electric shock can cause loss of muscular control, resulting in the inability to let go of an electrical conductor. Electric shock can also cause injuries of an indirect nature in which involuntary muscle reaction from the electric shock can cause bruises, fractures, and even death resulting from collisions or falls.

The severity of shock received when a person becomes a part of an electric circuit is affected by three primary factors: the amount of current flowing through the body (measured in amperes), the path of the current through the body, and the length of time the body is in the circuit. Other factors that may affect the severity of the shock are the frequency of the current, the phase of the heart cycle when shock occurs, and the general health of the person prior to the shock. Effects can range from a barely perceptible tingle to immediate cardiac arrest. Although there are no absolute limits, or even known values that show the exact injury at any given amperage range, *Table 1* lists the general effects of electric current on the body for different current levels. As this table illustrates, a difference of only 100 mA exists between a current that is barely perceptible and one that is likely to kill you.

A severe shock can cause considerably more damage to the body than is visible. For example, a person may suffer internal hemorrhages and destruction of tissues, nerves, and muscle. In addition, shock is often only the beginning in a chain of events. The final injury may well be from a fall, cuts, burns, or broken bones.

1.1.2 Burns

The most common injury associated with electrical shock is a burn. Burns suffered in electrical accidents may be of three types: electrical burns, arc burns, and thermal contact burns.

Electrical burns are the result of electric current flowing through the tissues or bones. Tissue damage is caused by the heat generated by the current flow through the body. An electrical burn is one of the most serious injuries you can receive and should be given immediate attention. Since the most severe burning is likely to be internal, a small surface wound could actually be an indication of severe internal burns.

Arc burns make up a substantial portion of the injuries from electrical malfunctions. The electric arc between metals can be up to 35,000°F, which is about four times hotter than the surface of the sun. Workers several feet from the source of the arc can receive severe or fatal burns. Since most electrical safety guidelines recommend safe working distances based on shock considerations, workers can be following these guidelines and still be at risk from arc. Electric arcs can occur due to poor electrical contact or failed insulation. Electrical arcing is caused by the passage of substantial amounts of current through the vaporized terminal material (usually metal or carbon).

> **WARNING!**
>
> Since the heat of the arc is dependent on the short circuit current available at the arcing point and the time it takes to clear (trip), arcs generated by low-voltage systems can be just as dangerous (or even more dangerous) than those at higher voltages.

The third type of burn is a thermal contact burn. It is caused by contact with heated objects thrown during the blast associated with an electric arc (*Figure 2*). This blast comes from the pressure developed by the near-instantaneous heating of the air surrounding the arc, and from the expansion of the metal as it is vaporized. (Copper expands by a factor in excess of 67,000 times when it is vaporized.) The pressure wave can be great enough to hurl people, switchgear, and cabinets considerable distances. It can stop your heart, impale you with shrapnel, blow off limbs, cause deafness, and cause you to inhale vaporized metal. Another hazard associated with the blast is molten metal projectiles, which can also cause thermal contact burns and associated damage.

Many things can be done to reduce the chance of receiving an electrical shock. Always comply with your company's safety policy and all applicable rules and regulations, including job-site rules. In addition, OSHA publishes the *Code of Federal Regulations* (*CFR*). *CFR Part 1910* covers the OSHA standards for general industry and *CFR Part 1926* covers the OSHA standards for the construction industry.

Do not approach any electrical conductors closer than indicated in *Table 2* unless they are de-energized and your company has designated you as a qualified individual for that task. The values given in the table are minimum safe clearance distances; your company may have more restrictive requirements. *Table 2* is based on information provided in NFPA 70E®, *Standard for Electrical Safety in the Workplace, Table 130.4(D)(a)* for AC.

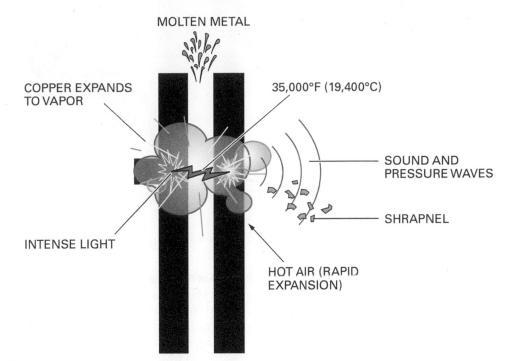

MOLTEN METAL

COPPER EXPANDS
TO VAPOR

35,000°F (19,400°C)

SOUND AND
PRESSURE WAVES

SHRAPNEL

INTENSE LIGHT

HOT AIR (RAPID
EXPANSION)

Figure 2 Arc flash.

Table 2 Limited Approach Boundaries to Live Parts (AC)

Nominal System Voltage Range (Phase-to-Phase)	Limited Approach Boundary from Fixed Circuit Component
50 to 750	3 ft 6 in
751 to 15kV	5 ft 0 in
15.1kV to 36kV	6 ft 0 in
36.1kV to 121kV	8 ft 0 in
138kV to 145kV	10 ft 0 in
161kV to 169kV	11 ft 8 in
230kV to 242kV	13 ft 0 in
345kV to 362kV	15 ft 4 in
500kV to 550kV	19 ft 0 in
765kV to 800kV	23 ft 9 in
Source: U.S. Department of Labor	

1.1.3 Electrical Safety Precautions

There are several precautions you can take to help make your job safer. They include the following:

- Always remove all jewelry (e.g., rings, watches, exposed body piercings, bracelets, and necklaces) before working on electrical equipment. Most jewelry is made of conductive material and wearing it can result in a shock and other injuries if the jewelry is caught in moving components. It can also increase the severity of a burn during an arc flash event.

- When working on energized equipment, it is safer to work in pairs. In doing so, if one of the workers experiences a harmful electrical shock, the other worker can release the victim and call for help.
- Plan each job before you begin it. Make sure you understand exactly what it is you are going to do. If you are not sure, ask your supervisor.
- You will need to look over the appropriate prints and drawings to locate isolation devices and potential hazards. Never defeat safety interlocks. Remember to plan your escape route before starting work. Know where the nearest phone is and the emergency number to dial for assistance.
- If you realize that the work will go beyond the scope of what was planned, stop and get instructions from your supervisor before continuing. Do not attempt to plan as you go. A large percentage of electrical incidents are caused by changes in scope.

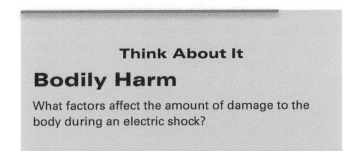

Think About It

Bodily Harm

What factors affect the amount of damage to the body during an electric shock?

- It is critical that you stay alert. Workplaces are dynamic, and situations relative to safety are always changing. If you leave the work area to pick up material, take a break, or have lunch, reevaluate your surroundings when you return. Always remember to plan ahead. Complete a checklist or pre-task safety plan and/or permit before beginning work.

1.2.0 Verifying That Circuits Are De-Energized

You should always assume that all the circuits are energized until you have verified that the circuit is de-energized. This is called a *live-dead-live test*. Follow these steps to verify that a circuit is de-energized:

Step 1 Ensure that the circuit is properly tagged and locked out (*CFR 1910.333/1926.417*).

Step 2 Verify the test instrument operation on a known voltage source using the appropriately rated tester, such as the proving unit shown in *Figure 3*. To operate a proving unit, the leads of the meter to be tested are connected to the positive and negative terminals on the proving unit, and then the meter is checked to make sure that it displays the expected value.

Step 3 Using the test instrument, check the circuit to be de-energized while wearing the appropriate PPE. The voltage should be zero.

Step 4 Verify the test instrument operation, once again on a known power source.

> **WARNING!**
> Under all circumstances, always test before touching any electrical conductor, even if it has been locked out and proven de-energized.

1.3.0 Causes of Electrical Incidents

There are three basic causes of incidents: unsafe conditions, unsafe equipment, and unsafe acts. Safety regulations are designed to prevent unsafe equipment and conditions. In addition, employer/employee awareness, new technologies, GFCIs, touch-safe (encased) terminals, and insulated tools have greatly reduced incidents due to unsafe conditions and unsafe equipment.

Figure 3 Proving unit with meter under test.

So why do electrical incidents continue to happen? Most electrical injuries result from failure to follow safe work practices and not being aware of or ignoring a hazard. OSHA lists the following activities as some of the most frequent causes of electrical injuries:

- Failure to place the circuit or equipment in an electrically safe work condition
- Contact with power lines by ladders, powered construction equipment, earth-moving equipment, and construction tools such as long-handled cement finishing floats
- Lack of ground-fault protection on outlet receptacles, powered hand tools, extension cords, and installed electrical equipment
- Damaged equipment such as cut, nicked, or pinched power cords and cables, worn insulation on power cords or cables, missing ground prongs, and damaged tool casings
- Path to ground missing or discontinuous due to loose or broken ground wires, improperly grounded equipment, improper grounds, or extremely dry conditions around existing grounds
- Equipment not being used in the prescribed manner, such as fabricating extension cords from multi-receptacle boxes, fabricating extension cords from nonmetallic (NM) cable, using power tools with damaged or modified cords, and using oversized fuses or circuit breakers

1.4.0 Hierarchy of Risk Controls

Following a hierarchy of risk controls leads to the implementation of safer systems and can be used to reduce or eliminate hazards (*Figure 4*). They include the following:

- Elimination

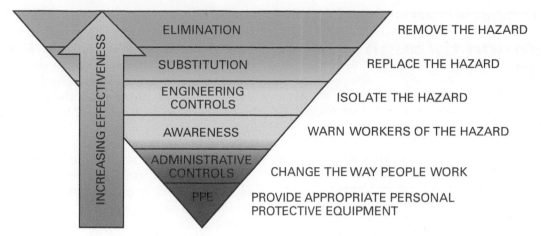

Figure 4 Hierarchy of risk control methods.

- Substitution
- Engineering controls
- Awareness
- Administrative controls
- Personal protective equipment

Elimination, substitution, and engineering controls are the most effective methods of reducing risk. These controls are usually applied at the source of the hazard and are therefore less likely to be affected by human error. Awareness, administrative controls, and PPE are the least effective methods of reducing risk as they are more likely to be affected by human error.

1.4.1 Elimination

The best method of reducing a risk is to eliminate the hazard. For example, this can be done by establishing an electrically safe work condition using appropriate lockout/tagout procedures. Another example would be to schedule off-shift maintenance to eliminate the hazards associated with working near operating equipment.

1.4.2 Substitution

Substitution is used to replace a hazard with a less hazardous or nonhazardous material, system, or process. Examples include choosing a nontoxic coating over a volatile coating, reducing energy by replacing a 120V control circuit with 24V control circuitry, replacing one large storage tank with several smaller cylinders, and using meters with remote reading capability. Another example is the use of robotic systems that allow workers to perform remote racking from a safe distance. *Figure 5* shows a remotely controlled robotic device used to control a circuit breaker carriage.

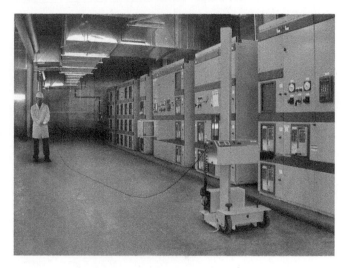

Figure 5 Robotic device used to control a circuit breaker carriage.

1.4.3 Engineering Controls

Engineering controls include methods of isolating the hazard from the worker. Common types of engineering controls include barricades, safety nets, machine guards, interlocks, and hand switches located outside the danger zone that must be touched and/or held down in order for the machine to operate. Engineering controls require compliance on the part of workers. When engineering controls are used, the company must develop and enforce policies and procedures mandating their use.

1.4.4 Awareness

Awareness can help to reduce risk by warning workers of various jobsite hazards. Examples include signs, such as electrical hazard warning labels. Signs must be highly visible and easy to understand. Symbols may be used to help any workers with language barriers.

Prevention through Design

The National Institute for Occupational Safety and Health (NIOSH) leads a national initiative known as *Prevention through Design (PtD)*. The goal of the initiative is to prevent or reduce occupational injuries, illnesses, and fatalities by eliminating the hazard at the source. For example, many power tools are very loud and present a hearing hazard that must be managed through the use of hearing protection. The NIOSH Buy Quiet initiative encourages the use of power tools that are designed to operate at a much lower noise level.

Warning devices are another form of awareness control. These devices include horns, bells, whistles, and lights. In some cases, workers hear alarms or signals so often, they unconsciously ignore them. In order for these warning devices to be effective, they must be distinctive and audible over ambient noise levels.

1.4.5 Administrative Controls

Administrative controls use practices and policies to limit an employee's exposure to a hazard.

Safety policies and procedures, operating procedures, and maintenance procedures are examples of administrative controls. They are based on procedures that have been documented and formalized by management. Lockout/tagout, confined-space entry procedures, and work permits are also examples of administrative controls.

Worker rotation is also an administrative control. Workers should be rotated if the work involves hot or cold environments, constant or repetitive motion, or is either very stressful or tedious. This can be accomplished by cross-training workers for different jobs on the site, adjusting the work schedule, or providing frequent breaks.

Administrative controls require training and enforcement in order to be effective. When administrative controls are specified, they should be audited to verify their effectiveness.

1.4.6 Personal Protective Equipment

Personal protective equipment is designed to prevent workers from coming into contact with a hazard. Examples include shock and arc flash PPE. PPE protects individuals, but does not create a safe work environment, and is therefore the least effective method of risk prevention. PPE is an important component of risk control but should be considered the last line of defense in reducing hazards.

1.0.0 Section Review

1. Which of the following is true regarding electrical shock?
   a. Most electrical incidents involve contact with high-voltage conductors (over 1,000V).
   b. The majority of electrical accidents involving electrically operated tools happen when the frame or case of the tool becomes energized.
   c. Construction workers are no more likely to be electrocuted than workers in other industries.
   d. CPR can be used to correct heart fibrillation following a shock.

2. A live-dead-live test is used to verify _____.
   a. that a circuit is operating properly
   b. the operation of a motor
   c. that a circuit is de-energized
   d. the voltage of a known power source

3. Most electrical injuries result from failure to _____.
   a. wear appropriate PPE
   b. document procedures
   c. coordinate work activities
   d. follow safe work practices

4. A barricade is an example of _____.
   a. an administrative control
   b. an engineering control
   c. worker awareness
   d. hazard elimination

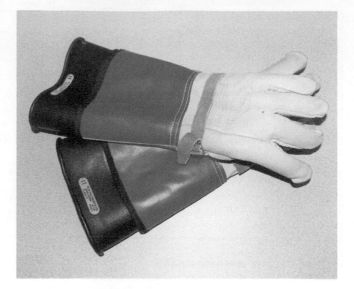

2.0.0 USING PPE TO REDUCE THE RISK OF INJURY

Objective

Use PPE to reduce the risk of injury.
a. Identify OSHA requirements for protective equipment.
b. Select and use protective equipment.

Trade Term

Hot stick: An insulated tool designed for the manual operation of disconnecting switches, fuse removal and insertion, and the application and removal of temporary grounds.

Figure 6 Rubber gloves, leather protectors, and glove dust.

Personal protective equipment is the last line of defense against injury. If a hazard cannot be eliminated through engineering (modifications to equipment, processes, or systems), administratively eliminated (e.g., a maintenance shutdown), or physically guarded against, PPE is the only protection remaining between workers and hazards.

PPE is used to protect against exposure to physical, chemical, radiological, electrical, mechanical, or other workplace hazards. For example, rubber gloves are used to prevent the skin from coming into contact with energized circuits. A separate leather cover protects the rubber glove from punctures and other damage (*Figure 6*). The leather protectors also provide the wearer with a certain level of protection against arc flash. Glove dust can be used to absorb moisture and prevent the gloves from sticking.

2.1.0 OSHA Requirements for Protective Equipment

OSHA addresses the safeguards for personal protection in *CFR 1910.335*. The design requirements for specific types of electrical protective equipment are covered in *CFR 1910.137*.

CFR 1910.137(a) presents the design requirements for electrical protective equipment, while *CFR 1910.335* presents safeguards for the use of this equipment. Some of these requirements include the following:

- Employees working in areas where there are potential electrical hazards shall be provided with, and shall use, electrical protective equipment that is appropriate for the specific parts of the body to be protected and for the work to be performed.
- Protective equipment shall be maintained in a safe, reliable condition and shall be periodically inspected or tested, as required by *CFR 1910.137*.

Arc Blast

An electrician in Louisville, KY was cleaning a high-voltage switch cabinet. He removed the padlock securing the switch enclosure and opened the door. He used a voltage meter to verify absence of voltage on the three load phases at the rear of the cabinet. However, he did not test all potentially energized parts within the cabinet, and some components were still energized. He was wearing standard work boots and safety glasses but was not wearing protective rubber gloves or arc-rated apparel. As he used a paintbrush to clean the switch, an arc blast occurred, which lasted approximately one-sixth of a second. The electrician was knocked down by the blast. He suffered third-degree burns and required skin grafts on his arms and hands. The investigation determined that the blast was caused by debris, such as a cobweb, falling across the open switch.

The Bottom Line: Always wear the appropriate personal protective equipment and test all components for the absence of voltage before working in or around electrical devices.

- If the insulating capability of protective equipment may be subject to damage during use, the insulating material shall be protected.
- Employees shall wear nonconductive head protection wherever there is a danger of head injury from electric shock or burns due to contact with exposed energized parts.
- Employees shall wear protective equipment for the eyes and face wherever there is danger of injury to the eyes or face from electric arcs or flashes or from flying objects resulting from an electrical explosion.
- When working near exposed energized conductors or circuit parts, each employee shall use insulated tools or handling equipment if the tools or handling equipment might make contact with such conductors or parts. If the insulating capability of insulated tools or handling equipment is subject to damage, the insulating material shall be protected.
- Fuses must only be removed using appropriate fuse-handling equipment that is insulated for the circuit voltage.

WARNING!
Fuses should never be installed or removed when energized. This could cause an electrical arc, which can result in death or a serious injury.

- Ropes and handlines used near exposed energized parts shall be nonconductive.
- Protective shields, protective barriers, or insulating materials shall be used to protect each employee from shock, burns, or other electrically related injuries while that employee is working near exposed energized parts that might be accidentally contacted or where dangerous electric heating or arcing might occur.

- When normally enclosed live parts are exposed for maintenance or repair, they shall be guarded to protect unqualified persons from contact with the live parts.

The types of electrical safety equipment, protective apparel, and protective tools available for use are quite varied. This module will discuss the most common types of safety equipment. These include the following:

- Currently tested rubber protective equipment, including gloves and blankets
- Arc-rated apparel
- Natural fiber clothing
- Electrical hot sticks
- Fuse pullers
- Footwear rated to protect against electrical hazards
- Safety glasses
- Face shields

2.2.0 Selecting and Using Electrical Safety Equipment

In addition to the traditional PPE required on construction sites (hearing protection, safety footwear, eye protection (including prescription eyewear), electrical workers require specialized PPE to protect against exposure to energized circuits or equipment.

OSHA requires that all employees who may be exposed to high-voltage hazards use electrical rated (Class E) hard hats. Class E hard hats have the highest rating and are tested to withstand 20,000VAC for a period of three minutes. The other electrical rated hard hat is a Class G hard hat, which is tested to withstand 2,200VAC for one minute. Class E and G climbing-style hard hats are also available. These hard hats can be equipped with eye protection, rated face shields (including arc-rated face shields), along with chin straps, lights, and ear protection. A rule of thumb is to replace a hard hat every five years. However, hard hats exposed to severe conditions (e.g., chemicals, sunlight, or high temperatures) should be replaced every two years.

In addition to hard hats, two other important articles of protection for electrical workers are insulated rubber gloves and rubber blankets, both of which must be matched to the voltage rating of the circuit or equipment.

> **WARNING!**
> Rubber protective equipment must be used, stored, inspected, and tested properly. If it fails during use, a serious injury or death could occur.

Rubber protective equipment is available in two types. Type 1 designates rubber protective equipment that is manufactured of natural or synthetic rubber that is properly vulcanized, and Type 2 designates equipment that is ozone resistant, made from any elastomer or combination of elastomeric compounds. Ozone is a form of oxygen that is produced from electricity and is present in the air surrounding a conductor under high voltages. Normally, ozone is found at voltages of 10kV and higher, such as those found in electric utility transmission and distribution systems. Type 1 protective equipment can be damaged by corona cutting, which is the cutting action of ozone on natural rubber when it is under mechanical stress. Type 1 rubber protective equipment can also be damaged by ultraviolet rays. However, it is very important that the rubber protective equipment in use today be natural rubber (Type 1). Type 2 rubber protective equipment is very stiff and is not as easily worn as Type 1 equipment.

2.2.1 Classes of Protective Equipment

The American National Standards Institute (ANSI) and the American Society for Testing and Materials International (ASTM) have designated a specific classification system for rubber protective equipment. The maximum AC use voltage and equipment tag colors are shown in *Table 3*.

Before rubber protective equipment can be worn by personnel in the field, all equipment must have a current test date stenciled on the equipment. Insulating gloves must be inspected each day by the user before they can be used.

Table 3 Rubber Protective Equipment Classification

Classification (Tag Color)	Max Voltage
Class 00 (beige tag)	500V
Class 0 (red tag)	1,000V
Class 1 (white tag)	7,500V
Class 2 (yellow tag)	17,000V
Class 3 (green tag)	26,500V
Class 4 (orange tag)	36,000V

They must also be electrically tested every six months and any time the insulating value is in question. Because rubber protective equipment is used for personal protection and serious injury could result from its misuse or failure, it is important that an adequate safety factor be provided between the voltage on which it is to be used and the voltage at which it was tested.

All rubber protective equipment must be marked with the appropriate voltage rating and last inspection date. The markings that are required to be on rubber protective equipment must be applied in a manner that will not interfere with the protection that is provided by the equipment.

> **WARNING!**
> Never work on anything energized without direct instruction from your employer and appropriate safety equipment.

2.2.2 Rubber Gloves

Voltage-rated rubber gloves are of the gauntlet type and are available in various sizes. For the best possible protection and service life, observe the following general guidelines/considerations when using rubber gloves in electrical work:

- Always wear leather protectors over your gloves. Leather protectors shield the gloves against contact with sharp or pointed objects that may cut, snag, or puncture the rubber. Leather protectors also provide burn protection not provided by the gloves themselves.
- Always wear rubber gloves right side out (serial number and size to the outside).
- Always keep the gauntlets up. Rolling them down sacrifices a valuable area of protection. Tuck the sleeves of your shirt or protective suit under the glove cuffs to prevent an arc blast from coming inside your clothing.

- Always inspect and field check gloves before using them. Always check the inside for any debris.
- Use light amounts of manufacturer-approved glove dust or cotton liners with the rubber gloves. This helps to absorb some of the perspiration that can damage the gloves over years of use.
- Wash the rubber gloves in lukewarm, clean, fresh water after each use. Dry the gloves inside and out prior to returning to storage. Never use any type of cleaning solution on the gloves.
- Once the gloves have been properly cleaned, inspected, and tested, they must be properly stored. Store them in a cool, dry, dark place that is free from ozone, chemicals, oils, solvents, or other materials that could damage the gloves. Do not store gloves near hot pipes or in direct sunlight. Store both gloves and sleeves in their natural shape in a bag or box inside their leather protectors. They should be undistorted, right side out, and unfolded. Storing gloves inside out will reduce the life of the gloves.
- Gloves can be damaged by many different chemicals, especially petroleum-based products such as oils, gasoline, hydraulic fluid inhibitors, hand creams, pastes, and salves. If contact is made with these or other petroleum-based products, the contaminant should be wiped off immediately. If any signs of physical damage or chemical deterioration are found (e.g., swelling, softness, hardening, stickiness, ozone deterioration, or sun checking), the protective equipment must not be used.
- Never wear watches or rings while wearing rubber gloves; this can cause damage from the inside out and defeats the purpose of using rubber gloves. Never wear anything conductive.
- Rubber gloves must be electrically tested every six months by a certified testing laboratory. Always check the test date before using gloves.
- Use rubber gloves only for their intended purpose, not for handling chemicals or other work. This also applies to the leather protectors.

Before rubber gloves are used, a visual inspection and an air test should be made. This should be done prior to use and as many times during the day as you feel necessary. To perform a visual inspection, stretch a small area of the glove, checking to see that no defects exist, such as the following:

- Embedded foreign material
- Deep scratches
- Pinholes or punctures
- Snags or cuts

Gloves and sleeves can be inspected by rolling the outside and inside of the protective equipment between the hands. This can be done by squeezing together the inside of the gloves or sleeves to bend the outside area and create enough stress to the inside surface to expose any cracks, cuts, or other defects. When the entire surface has been checked in this manner, the equipment is then turned inside out, and the procedure is repeated. It is very important not to leave the rubber protective equipment inside out as that places stress on the preformed rubber.

Remember that even the smallest amount of damage to a rubber glove compromises its insulating ability. Look for signs of deterioration from age, such as hardening and slight cracking. Also, if the glove has been exposed to petroleum products, it should be considered suspect because deterioration can be caused by such exposure. If the gloves are suspect, turn them in for evaluation. If the gloves are defective, turn them in for disposal. Never leave a damaged glove lying around; someone may think it is a good glove and not perform an inspection prior to using it.

After visually inspecting the glove, other defects may be observed by applying the air test (*Figure 7*). The following general steps can be used to perform an air test:

Step 1 Stretch the glove and look for any defects.

Step 2 Twirl the glove around quickly or roll it down from the glove gauntlet to trap air inside.

Step 3 Trap the air by squeezing the gauntlet with one hand. Use the other hand to squeeze the palm, fingers, and thumb to check for weaknesses and defects.

Step 4 Hold the glove up to your ear to try to detect any escaping air.

Step 5 If the glove does not pass this inspection, it must be turned in for disposal.

Gloves may also be inspected using a glove inflator (*Figure 8*). Glove inflators offer the advantage of more precise field testing. To use a glove inflator, the glove is secured to the housing using a nylon strap and inflated by pumping the bellows of the inflator against a firm surface. The inflated glove is then checked for leaks.

> **CAUTION**
> Never use compressed gas for the air test or inflate gloves beyond the manufacturer's recommendations, as this can damage the gloves.

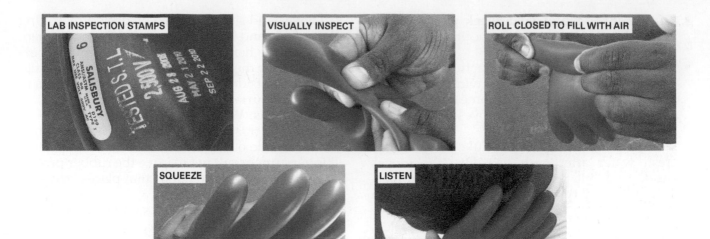

LAB INSPECTION STAMPS

VISUALLY INSPECT

ROLL CLOSED TO FILL WITH AIR

SQUEEZE

LISTEN

Figure 7 Glove inspection.

2.2.3 Insulating Blankets

An insulating blanket is a versatile cover-up device best suited for the protection of maintenance technicians against accidental contact with energized electrical equipment.

These blankets are designed and manufactured to provide insulating quality and flexibility for use in covering. Insulating blankets are designed only for covering equipment and should not be used on the floor. Special rubber floor mats, called *switchboard matting*, are available for floor use. Use caution when installing these over sharp edges or when covering pointed objects.

Blankets must be tested yearly and inspected before each use. To check rubber blankets, place the blanket on a flat surface and roll the blanket from one corner to the opposite corner. If there are any irregularities in the rubber, this method will expose them. After the blanket has been rolled from each corner, it should then be turned over and the procedure repeated.

Insulating blankets are cleaned in the same manner as rubber gloves. Once the protective equipment has been properly cleaned, inspected, and tested, it must be properly stored. It should be stored in a cool, dry, dark place that is free from ozone, chemicals, oils, solvents, or other materials that could damage the equipment. Avoid storage near hot pipes or in direct sunlight. Blankets may be stored rolled in containers that are designed for this use; the inside diameter of the roll should be at least 2" (50 mm).

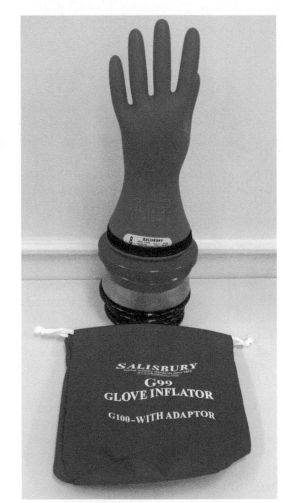

Figure 8 Glove inflator.

2.2.4 Other Protective Apparel

Besides rubber gloves, there are other types of special-application protective apparel, such as arc-rated suits, face shields, and rubber sleeves.

Manufacturing plants should have other types of special-application protective equipment available for use, such as high-voltage sleeves, high-voltage boots, nonconductive protective helmets, nonconductive eyewear and face protection, and switchboard blankets.

All equipment must be inspected before use and during use, as necessary. The equipment used and the extent of the precautions taken depend on each individual situation; however, it is better to be overprotected than underprotected when you are trying to prevent electric shock, arc blast, and burns.

When working with energized equipment, arc-rated suits with rated face and head protection may be required in some applications.

The clothes worn beneath arc-rated protective apparel must also be carefully selected. Never wear synthetic-fiber (acrylic, polyester, or nylon) clothing during energized work; these types of materials will melt when exposed to high temperatures and because they burn hotter, will actually increase the severity of a burn. Wear non-flammable clothing, leather boots or shoes, safety glasses, and hard hats. Use hearing protection where needed.

2.2.5 Hot Sticks

A **hot stick** is an insulated tools designed for the manual operation of disconnecting switches, fuse removal and insertion, and the application and removal of temporary grounds.

A hot stick is made up of two parts, the head or hood and the insulating rod. The head can be made of metal or hardened plastic, while the insulating section may be wood, plastic, laminated wood, or other effective insulating materials. Telescoping sticks are also available.

Most plants have hot sticks available for different purposes. Select a stick of the correct type and size for the application and inspect it before use. Look for signs of obvious damage, deep scratches, dust, or surface contaminants. Never use a damaged hot stick.

Storage of hot sticks is important. They should be hung up vertically on a wall to prevent any damage. They should also be stored away from direct sunlight and prevented from being exposed to petroleum products.

Per *OSHA 1910.269(j)(2)(i)*, each live line tool (such as a hot stick) must be wiped clean and inspected for defects before each use.

Live line tools must be removed from service every two years and tested in accordance with *OSHA 1910.269(j)(2)(iii)*.

2.2.6 Fuse Pullers

Use the plastic or fiberglass style of fuse puller for removing and installing low-voltage cartridge fuses. All fuse pulling and replacement operations must be done using fuse pullers.

WARNING!

Fuses should only be installed or removed when they are safely de-energized. Failure to do so can result in a serious injury or death.

Arc-Flash Hazard Warnings

In other than dwelling units, *NEC Section 110.16* requires that all switchboards, switchgear, panelboards, industrial control panels, meter socket enclosures, and motor control centers be clearly marked to warn qualified persons of potential electric arc flash hazards. *NEC Section 110.16(B)* requires service equipment rated 1,200A or more to be permanently marked with the nominal system voltage, available fault current at the service overcurrent protective devices, clearing time of the service overcurrent protective devices based on the available fault current at the service equipment, and the date the label was applied.

Using a Phasing Tester

Using a phasing tester or other voltage detection equipment can expose you to a considerable arc flash hazard. Always take the time to use the appropriate personal protective equipment. Remember, you have only one chance to protect yourself.

Figure Credit: Tim Ely

The best type of fuse puller is one that has a spread guard installed. This prevents the puller from opening if resistance is met when installing fuses.

2.2.7 Shorting Probes

Before working on de-energized circuits that have capacitors installed, you must discharge the capacitors using a safety shorting probe. This procedure requires special training and may only be performed by qualified individuals.

2.2.8 Eye and Face Protection

NFPA 70E® and OSHA require that you wear safety glasses at all times whenever you are working on or near energized circuits. Face protection is worn over your safety glasses, and is required whenever there is danger of electrical arcs or flashes, or from flying or falling objects resulting from an electrical explosion. NFPA 70E® is discussed in detail later in this module.

2.0.0 Section Review

1. Periodic inspection and testing of protective equipment is required by _____.
   a. ANSI
   b. OSHA
   c. NFPA
   d. NEMA
2. Rubber protective equipment with a beige tag is rated for _____.
   a. 500V
   b. 1,000V
   c. 7,500V
   d. 17,000V

3.0.0 ELECTRICAL SAFETY STANDARDS

Objective

Identify the standards that relate to electrical safety.

a. Apply OSHA requirements in the workplace.
b. Understand the purpose of NFPA 70E®.

Trade Terms

Arc flash boundary (AFB): An approach limit at a distance from exposed energized electrical conductors or circuit parts within which a person could receive a second-degree burn if an electrical arc flash were to occur.

Arc flash risk assessment: A study investigating a worker's potential exposure to arc flash energy, conducted for the purpose of injury prevention and the determination of safe work practices and appropriate levels of PPE.

Arc rating: The maximum incident energy resistance demonstrated by a material (or a layered system of materials) prior to material breakdown, or at the onset of a second-degree skin burn. Expressed in joules/cm^2 or calories/cm^2.

Error precursors: Situations that put a worker at risk due to the demands of the task, conditions, worker attitude, and/or environment.

Grounded tool: An electrical tool with a three-prong plug at the end of its power cord or some other means to ensure that stray current travels to ground without passing through the body of the user. The ground plug is bonded to the conductive frame of the tool.

Incident energy: The amount of thermal energy impressed on a surface at a certain distance from the source of an electrical arc. Incident energy is typically expressed in calories per square centimeter (cal/cm^2).

Limited approach boundary: An approach limit at a distance from an exposed energized electrical conductor or circuit part within which a shock hazard exists.

Qualified person: One who has demonstrated the skills and knowledge related to the construction and operation of the electrical equipment and installations, and has received safety training to identify and avoid the hazards involved.

Restricted approach boundary: An approach limit at a distance from an exposed energized electrical conductor or circuit part within which there is an increased likelihood of electric shock.

Unqualified person: A person who is not a qualified person.

All electricians must become familiar with both national safety standards and local requirements. The main national standards applicable to electrical work are the *National Electrical Code* (NFPA 70®), *Standard for Electrical Safety in the Workplace* (NFPA 70E®), *Recommended Practice for Electrical Equipment Maintenance* (NFPA 70B), and *OSHA 29 CFR, Parts 1910 and 1926*. This module provides a brief overview of NFPA 70E® and OSHA requirements.

3.1.0 Applying OSHA Requirements in the Workplace

The purpose of OSHA is "to ensure safe and healthful working conditions for working men and women." OSHA is authorized to enforce standards and assist and encourage the states in their efforts to ensure safe and healthful working conditions. OSHA assists states by providing for research, information, education, and training in the field of occupational safety and health.

It is everybody's responsibility to ensure that workplaces are free of hazards. Employers must ensure that their workers are properly trained and have the required PPE to perform the job safely. Employees must follow established procedures, stay within the job plan, inspect and use the appropriate PPE, and avoid shortcuts.

The law that established OSHA specifies the duties of both the employer and employee with respect to safety. Some of the key requirements are outlined below; note that this list does not include everything, nor does it override the procedures called for by your employer.

3.1.1 Employer Responsibilities

OSHA Standard 29, Part 1910 and *OSHA Standard 29, Part 1926* cover occupational safety and health standards. Within each standard are subparts that deal with specific areas of safety. For example, *OSHA Standard 29, Part 1910, Subpart S, Sections 331 through 335* address electrical safety-related work practices. *OSHA Standard 29, Section 1910.332* requires training for all employees exposed to electrical shock.

If individual OSHA standards do not address a specific safety issue, they are covered under *Section 5(a)(1)* of the *Occupational Safety and Health Act of 1970*, which is often called the *General Duty Clause*. *Section 5(a)(1)* requires each employer to provide a workplace that is free from recognized hazards that are likely to cause death or serious physical harm. It also states that the employer must comply with all applicable occupational safety and health standards as determined by OSHA.

When OSHA inspectors check a facility for compliance, they normally examine the safety training records and standard operating procedures (SOPs) or other company policies that ensure that employees are properly trained. If there has been a reportable incident, especially an electrical incident, the inspectors will want to see all training records, SOPs, work permits, and any other documents associated with the incident.

Employers are responsible for ensuring that their workers are properly trained, have a safe work environment, are informed of hazards present in the work, have clearly written instructions, and are given all the necessary tools to do their jobs safely. Employers are also responsible for ensuring that workers follow established safe work procedures and use all required PPE.

3.1.2 Employee Responsibilities

Employees are responsible for reading, understanding, and following all company safety policies. They must also inspect, use, and properly store all required PPE, look out for the safety of themselves and of others, and work only within the job plan. Note that PPE must be inspected before each use.

> **CAUTION**
>
> OSHA states that employees have a duty to follow the safety rules laid down by the employer. Additionally, some states can reduce the amount of benefits paid to an injured employee if that employee was not following known, established safety rules. Your company may also terminate you if you violate an established safety rule.

In the past, workers were expected to simply address situations as they arose. That is no longer acceptable. NFPA 70E®, *Standard for Electrical Safety in the Workplace*, requires job safety planning and a job briefing before the start of each task. In other words, you must have a specific goal in mind and perform only the planned job.

Many electrical incidents can be traced to task creep. For example, if you are taking a current reading at a motor starter and notice a loose connection, do not grab your screwdriver and tighten it. You will make contact with an exposed energized part and greatly increase the likelihood of initiating an arc fault.

> **CAUTION**
>
> Always keep a clean work area when installing or removing equipment. Any scrap wire or even insulation could cause a fault or prevent protective devices from tripping. One cause of electrical faults is loose or uninstalled fasteners (bolts and screws) and tools left inside an electrical enclosure. Be sure to properly install doors and covers and account for all parts and tools. Do not leave a piece of electrical equipment unless all bolts and fasteners have been installed. Never use an electrical enclosure as a storage locker for spare parts and tools. Also, do not use the top surfaces of enclosures as storage or work areas and make sure to maintain required working space in front of the equipment per *NEC Section 110.26*.

The OSHA standards are split into several sections. As discussed earlier, the two that affect you the most are *CFR 1926*, the standard for the construction industry, and *CFR 1910*, the standard for general industry. Either or both may apply depending on where you are working and what you are doing. Your company may also have its own policies and procedures. In addition, you will be required to follow the safety procedures of any plant or facility where you are working.

3.1.3 OSHA Requirements for Electrical Safety

The most important piece of safety equipment required when performing work in an electrical environment is common sense. All areas of electrical safety precautions and practices draw upon common sense and attention to detail. One of the most dangerous conditions in an electrical work area is a poor attitude toward safety.

As stated in *CFR 1910.333(a)/1926.416*, safety-related work practices shall be employed to prevent electric shock or other injuries resulting from either direct or indirect electrical contact when work is performed near or on equipment or circuits that are or may be energized. The following are considered some of the basic and necessary attitudes and electrical safety precautions that lay the groundwork for a proper safety program:

Safety on the Job Site

Uncovered openings present several hazards, including the following:

- Workers may trip over them.
- If they are large enough, workers may fall through them.
- If there is a work area below, tools or other objects may fall through them, causing serious injury to workers below.

Any hole more than 2" (50 mm) in any direction must be protected with a cover or with guardrails, as shown here. A cover must be able to handle twice the anticipated load, be secured in place, and have the word "HOLE" or "COVER" on it.

Figure Credit: Mike Powers

- OSHA requires the posting of hard hat areas. Be alert to those areas and always wear your hard hat properly, with the bill in front. Hard hats should be worn whenever overhead hazards exist, or there is the risk of exposure to electric shock or burns.
- Wear safety shoes on all job sites. Keep them in good condition.
- Do not wear clothing with exposed metal zippers, buttons, or other metal fasteners. Avoid wearing loose-fitting or torn clothing.
- Always wear safety glasses with full side shields. In addition, the job may also require protective equipment such as face shields or goggles.
- All electrical work shall be in compliance with the latest *NEC®* and OSHA standards.
- The noncurrent-carrying metal parts of fixed, portable, and plug-connected equipment shall be grounded. Choose either a **grounded tool** or a double-insulated tool. An example of a double-insulated tool is shown in *Figure 9*.

Working Around Openings

Did you know that OSHA could cite your company for working around the open pipe in the photo shown here, even if you weren't responsible for creating it? If you are working near uncovered openings, you have only two options:

- Cover the opening properly (either you may do it or you may have the responsible contractor cover it).
- Stop work and leave the job site until the opening has been covered.

Figure Credit: Mike Powers

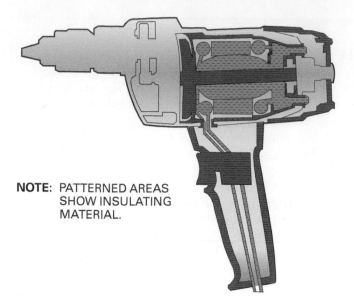

NOTE: PATTERNED AREAS SHOW INSULATING MATERIAL.

Figure 9 Double-insulated electric drill.

- Extension cords shall be the three-wire type, shall be protected from damage, and if hung overhead, shall not be fastened with staples or bare wire or hung in a manner that could cause damage to the outer jacket or insulation. Never run an extension cord through a doorway or window that can pinch the cord. Also, never allow vehicles or equipment to drive over cords.
- Exposed lamps in temporary lights shall be guarded to prevent accidental contact, except where lamps are deeply recessed in the reflector. Temporary lights must be supported in accordance with their listing or labeling instructions.
- All receptacles for attachment plugs shall be of an approved type and properly installed. Installation of the receptacle will be in accordance with the listing and labeling for each receptacle.
- *NEC Section 590.6* requires that all 125V, single-phase, 15A, 20A, and 30A receptacle outlets used to provide temporary or permanent power for construction and maintenance activities must be ground fault protected.
- Each disconnecting means for motors and appliances and each service feeder or branch circuit at the point where it originates shall be legibly marked with its purpose and voltage.
- Flexible cords shall be used in continuous lengths without splices (unless permitted by *NEC Section 400.13*) and shall be of a type listed in *NEC Table 400.4*.

- GFCI protection is required per *NEC Section 590.6(B)(1)*. If GFCI protection is not used, a written assured equipment grounding conductor program must be continuously enforced at the site for all tools and cord sets per *NEC Section 590.6(B)(2)*. A typical ground fault circuit interrupter (GFCI) is shown in *Figure 10*. A GFCI monitors the current imbalance between the circuit's ungrounded (hot) conductor and the grounded (neutral) conductor. If the circuit is operating properly, the current returning to the power supply should equal the current leaving the power supply (with a very small tolerance to account for leakage current). If the difference exceeds a predetermined level (e.g., 5mA), then the current is going somewhere unintended and the GFCI opens the contacts to de-energize the circuit.

Figure 10 Typical GFCI receptacle.

Potential Hazards

A self-employed builder was using a metal cutting tool on a metal carport roof and was not using GFCI protection. The male and female plugs of his extension cord partially separated, and the active pin touched the metal roofing. When the builder grounded himself on the gutter of an adjacent roof, he received a fatal shock.

The Bottom Line: Always use GFCI protection and be on the lookout for potential hazards.

All work on electrical equipment should be done with circuits de-energized and cleared or grounded. Work on energized electrical equipment should be avoided if at all possible. *CFR 1910.333(a)(1)* states that live parts to which an employee may be exposed shall be de-energized before the employee works on or near them, unless the employer can demonstrate that de-energizing introduces additional or increased hazards or is not possible because of equipment design or operational limitations. Live parts that operate at less than 50 volts to ground need not be de-energized if there will be no increased exposure to electrical burns or to explosion due to electric arcs.

All conductors, buses, and connections should be considered energized until proven otherwise. As stated in *1910.333(b)(1)*, conductors and parts of electrical equipment that have not been locked out or tagged out in accordance with this section should be treated as energized. Routine operation of the circuit breakers and disconnect switches contained in a power distribution system can be hazardous if not approached in the right manner. Several basic precautions that can be observed in switchgear operations are:

- Wear proper clothing made of natural fiber or fire-resistant fabric in accordance with NFPA 70E®.
- Wear eye, face, and head protection.
- Whenever operating circuit breakers in low-voltage or medium-voltage systems, always stand off to the side of the unit with the minimum amount of your body facing the circuit breaker.

Think About It

GFCIs

Explain how GFCIs protect people. Where should a GFCI be installed in the circuit to be most effective?

- Always try to operate disconnect switches and circuit breakers under a no-load condition.
- Never intentionally force an interlock on a system or circuit breaker.

Often, a circuit breaker or disconnect switch is used for providing lockout on an electrical system. Perform the following procedures when using the device as a lockout point:

- Breakers must always be locked out and tagged when you are planning to work on a circuit that is tied to that breaker. Always follow the standard rack-out and removal procedures that were supplied with the switchgear. Once removed, a sign must be hung on the breaker identifying its use as a lockout point, and approved safety locks must be installed when the breaker is used for isolation. Breakers equipped with closing springs should be discharged to release all stored energy in the breaker mechanism.
- Some of the circuit breakers used are equipped with keyed interlocks (Kirk locks) for protection during operation. These locks are used to ensure proper sequence of operation only. They are not used to lock out a circuit or system. When opening or closing a disconnect manually, it should be done quickly with a positive force. Lockout/tagout should be used when the disconnects are open.
- Whenever performing switching or fuse replacements, always use the protective equipment necessary to ensure personnel safety. Always prepare for the worst-case accident when performing switching.
- Whenever re-energizing circuits following maintenance or removal of a faulted component, use extreme care. Always verify that the equipment is in a condition to be re-energized safely. All connections should be insulated and all covers and screws installed. Have all personnel stand clear of the area for the initial re-energization. Never assume everything is in perfect condition. Verify the conditions.

The following procedure is provided as a guideline for ensuring that equipment and systems will not be damaged by reclosing low-voltage circuit breakers into faults. If a low-voltage circuit breaker has opened for no apparent reason, perform the following:

Step 1 Verify that the equipment being supplied is not physically damaged and shows no obvious signs of overheating or fire.

Step 2 Make all appropriate tests to locate any faults.

Step 3 Reclose the feeder breaker. Stand off to the side when closing the breaker. Whenever possible, avoid reaching across the front of the device.

Step 4 If the circuit breaker trips again, do not attempt to reclose the breaker. In a plant environment, Electrical Engineering should be notified, and the cause of the trip must be isolated and repaired.

The same general procedure should be followed for fuse replacement, with the exception of transformer fuses. If a transformer fuse blows, the transformer and feeder cabling should be inspected and tested before re-energizing. A blown fuse to a transformer is very significant because it normally indicates an internal fault. Transformer failures are catastrophic in nature and can be extremely dangerous. If applicable, contact the in-plant Electrical Engineering Department prior to commencing any effort to re-energize a transformer.

Power must always be removed from a circuit when removing and installing fuses. The air break disconnects (or quick disconnects) provided on the upstream side of a large transformer must be opened prior to removing the transformer's fuses. Otherwise, severe arcing will occur as the fuse is removed or installed. This arcing can result in personnel injury and equipment damage.

WARNING!	When replacing fuses, always wear appropriate PPE, which may include arc-rated suits and rated hand, face, and head protection.

To replace fuses servicing circuits *below* 1,000 volts, observe the following steps:

Step 1 Turn off the power to the disconnect.

Step 2 Verify that the fuses are de-energized.

Step 3 Remove the blown fuse.

Step 4 Install the new fuse. Push it in firmly and verify that it is seated properly.

Step 5 Turn the power back on.

To replace fuses servicing systems *above* 1,000 volts, observe the following steps:

Step 1 Open and lock out the disconnect switches.

Step 2 Unlock the fuse compartment.

Step 3 Verify that the fuses are de-energized.

Step 4 Attach the fuse removal hot stick to the fuse and remove it.

3.1.4 OSHA Lockout/Tagout Rule

The OSHA lockout/tagout rule covers the specific procedure to be followed for the "servicing and maintenance of machines and equipment in which the unexpected energization or startup of the machines or equipment, or releases of stored energy, could cause injury to employees." This standard establishes minimum performance requirements for the control of such hazardous energy.

The first step to be completed before working on a circuit is to ensure that equipment is isolated from all potentially hazardous energy (for example, electrical, mechanical, hydraulic, chemical, or thermal), and tagged and locked out before employees perform any servicing or maintenance activities in which the unexpected energization, startup, or release of stored energy could cause injury. All employees shall be instructed in the lockout/tagout procedure.

CAUTION	Although 99 percent of your work may be electrical, be aware that you may also need to lock out mechanical and other types of energy equipment. The following is an example of a lockout/tagout procedure. Make sure to use the procedure that is specific to your employer or job site.

I. Introduction

A. This lockout/tagout procedure has been established for the protection of personnel from potential exposure to hazardous energy sources during construction, installation, service, and maintenance of electrical energy systems.

B. This procedure applies to and must be followed by all personnel who may be potentially exposed to the unexpected startup or release of hazardous energy (e.g., electrical, mechanical, pneumatic, hydraulic, chemical, thermal, kinetic, or other).

Exceptions:

1. This procedure does not apply to process and/or utility equipment or systems with cord and plug power supplies when the cord and plug are the only source of hazardous energy, are removed from the source, and remain under the exclusive control of the authorized employee.

2. This procedure does not apply to troubleshooting (diagnostic) procedures and installation of electrical equipment and systems when the energy source cannot be de-energized because continuity of service is essential or shutdown of the system is impractical. Additional personal protective equipment for such work is required and the safe work practices identified for this work must be followed.

II. Definitions

- *Affected employee* – Any person working on or near equipment or machinery when maintenance or installation tasks are being performed by others during lockout/tagout conditions.

- *Appointed authorized employee* – Any person appointed by the job-site supervisor to coordinate and maintain the security of a group lockout/tagout condition.

- *Authorized employee* – Any person authorized by the job-site supervisor to use lockout/tagout procedures while working on electrical equipment.

- *Authorized supervisor* – The assigned job-site supervisor who is in charge of coordination of procedures and maintenance of security of all lockout/tagout operations at the job site.

- *Energy isolation device* – An approved electrical disconnect switch capable of accepting approved lockout/tagout hardware for the purpose of isolating and securing a hazardous electrical source in an open or safe position.

- *Lockout/tagout hardware* – A combination of padlocks, danger tags, and other devices designed to attach to and secure electrical isolation devices.

III. Training

A. Each authorized supervisor, authorized employee, and appointed authorized employee shall receive initial and as-needed user-level training in lockout/tagout procedures.

B. Training includes recognition of hazardous energy sources, the type and magnitude of energy sources present, and procedures for energy isolation and control.

C. Retraining must be conducted whenever lockout/tagout procedures are changed or there is evidence that procedures are not being followed properly.

IV. Protective Equipment and Hardware

A. Lockout/tagout devices shall be used exclusively for controlling hazardous energy sources.

Think About It

Working on Energized Systems

Some electricians commonly work on energized systems because they think it's too much trouble to turn off the power. What practices have you seen around your workplace that could be deadly?

B. All padlocks must be numbered and assigned to one employee only.

C. No duplicate or master keys will be made available to anyone except the site supervisor.

D. A current list with the lock number and authorized employee's name must be maintained by the site supervisor.

E. Danger tags must be of the standard white, red, and black DANGER—DO NOT OPERATE design and shall include the authorized employee's name, the date, and the appropriate network company (use permanent markers).

F. Danger tags must be used in conjunction with padlocks, as shown in *Figure 11*.

V. Procedures

A. Preparation for lockout/tagout:

1. Check the procedures to ensure that no changes have been made since you last used a lockout/tagout.

2. Identify all authorized and affected employees involved with the pending lockout/tagout.

B. Sequence for lockout/tagout:

1. Notify all authorized and affected personnel that a lockout/tagout is to be used and explain the reason why.

2. Shut down the equipment or system using the normal Off or Stop procedures.

3. Lock out energy sources and test disconnects to be sure they cannot be moved to the On position and open the control cutout switch. If there is no cutout switch, block the magnet in the Switch Open position before working on electrically operated equipment/apparatus such as motors, relays, etc. Remove the control wire.

4. Lock and tag the required switches in the Open position. Each authorized employee must affix a separate lock and tag. An example is shown in *Figure 12*.

5. Dissipate any stored energy by attaching the equipment or system to ground.

6. Verify that the test equipment is functional via a known power source.

7. Confirm that all switches are in the Open position and use test equipment to verify that all parts are de-energized.

(A) ELECTRICAL LOCKOUT

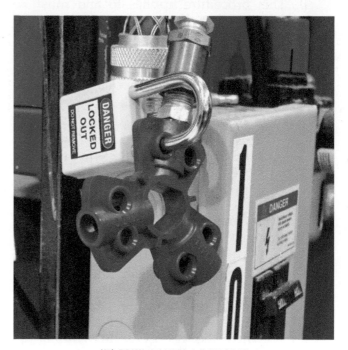

(B) PNEUMATIC LOCKOUT

Figure 11 Lockout/tagout devices.

8. If it is necessary to temporarily leave the area, upon returning, retest to ensure that the equipment or system is still de-energized.

C. Restoration of energy:

1. Confirm that all personnel and tools, including shorting probes, are accounted for and removed from the equipment or system.

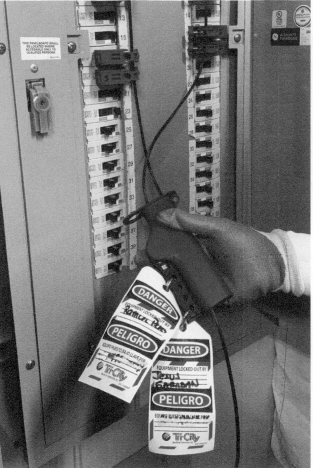

(A) LOCKOUT/TAGOUT WITH SEPARATE LOCKS

(B) LOCKOUT/TAGOUT CORD DEVICE

Figure 12 Multiple lockout/tagout device.

2. Completely reassemble and secure the equipment or system.
3. Replace and/or reactivate all safety controls.
4. Remove locks and tags from isolation switches. Authorized employees must remove their own locks and tags.
5. Notify all affected personnel that the lockout/tagout has ended and the equipment or system is energized.
6. Operate or close isolation switches to restore energy.

VI. Emergency Removal Authorization
   A. In the event a lockout/tagout device is left secured, and the authorized employee is absent, or the key is lost, the authorized supervisor can remove the lockout/tagout device.

   B. The authorized employee must be informed that the lockout/tagout device has been removed.
   C. Written verification of the action taken, including informing the authorized employee of the removal, must be recorded in the job journal.

Think About It

Lockout/Tagout – Who Does It and When?

What situations are likely to require lockout/tagout? Who is responsible for performing the lockout/tagout? When would more than one person be responsible?

What's wrong with this picture?

Figure Credit: Mike Powers

3.2.0 NFPA 70E®

The Occupational Safety and Health Administration (OSHA) issues regulations addressing safety in the workplace. While OSHA is a federal agency and its regulations are law, OSHA also relies on national consensus standards for certain requirements. For electrical safety, OSHA recognizes several standards from the National Fire Protection Association (NFPA). The *National Electrical Code* (NFPA 70®) provides the minimum requirements for the installation of electrical systems, the *Recommended Practice for Electrical Equipment Maintenance* (NFPA 70B) details preventive maintenance for electrical equipment, and the *Standard for Electrical Safety*

in the Workplace (NFPA 70E®) provides practical safe working requirements relative to the hazards arising from the use of electricity.

NFPA 70E® is divided into three chapters:

- Chapter 1, Safety-Related Work Practices
- Chapter 2, Safety-Related Maintenance Requirements
- Chapter 3, Safety Requirements for Special Equipment

Like the *NEC®*, each chapter is further subdivided into articles and sections. There are also numerous annexes that provide calculation methods, forms, and other references. This module focuses on the safety-related work practices covered in *70E Chapter 1*.

70E Article 90 describes the scope and structure of the standard. *70E Article 100* defines terms used throughout the standard. Take the time to review these definitions; many of them are specific to the hazards discussed in this standard.

70E Article 105 covers the application of safety-related work practices and procedures. *70E Section 105.3* discusses the responsibilities of the employer and employee.

General requirements for electrical safety-related work practices are covered in *70E Article 110*. The purpose of this section is to provide for employee safety relative to electrical hazards in the workplace. The responsibilities of owners or prime contractors and outside service personnel or subcontractors are covered in *70E Section 110.3*.

70E Section 110.2 identifies safety training requirements for employees who could encounter electrical hazards not reduced to a safe level through applicable electrical installation requirements. This applies to any employee who may approach nearer than a safe distance or who is expected to test, troubleshoot, or repair electrical equipment. It also applies to operators who may perform switching operations. The degree of training must be appropriate to the hazards and risk encountered by the employee. *70E Section 110.2(A)(3)* also identifies the requirements for retraining. The intent of retraining is to address changes to NFPA 70E® as well as changes in the employer's policies and procedures concerning electrical safety. The interval for retraining must not exceed three years.

NFPA eLibrary App

The NFPA eLibrary app is available at *catalog.nfpa. org* and provides instant, searchable access to purchased eBooks, including NFPA 70E®.

70E Section 110.2(A) addresses training requirements and characteristics of both qualified and unqualified personnel. It is important to note that there is no value judgment attached to these terms and they have no relationship to license status or time in the trade. A qualified person has demonstrated skills and knowledge related to the construction and operation of electrical equipment and installations and has received safety training to identify the hazards and reduce the associated risk. An unqualified person is simply one who is not a qualified person.

70E Articles 120 and 130 address work involving electrical hazards. These sections identify requirements such as creating an electrically safe work condition, performing an electrical risk assessment, and use of test equipment. The procedures to meet the requirements of *70E Article 130* are discussed in the remainder of this module and supplemented by *70E Article 120* and the Informative Annexes found in the back of the standard.

3.2.1 Recognizing Hazard Boundaries

Ideally, work on or near electrical equipment would always be performed with no electrical power applied (also known as an *electrically safe work condition*), but that is not always possible. NFPA 70E® requires special safety procedures when working on or near circuits having voltage levels of more than 50V line-to-line.

Distance is the best protection against electrical hazards. NFPA 70E® establishes specific limits of approach to exposed energized parts. These limits are for shock protection and are called *approach boundaries*. The approach boundaries and required PPE and other tools or equipment are determined by performing a shock risk assessment.

When working with electrical equipment, assume that the equipment is energized until personally verifying otherwise. As noted earlier, electrical equipment presents a potential shock hazard, arc flash hazard, and blast hazard. *Figure 13* shows a diagram indicating the approach limits defined by NFPA 70®. Note that the arc flash boundary (AFB) is determined independently and may be at a greater or lesser distance than shock protection boundaries.

The exposed energized component can be a wire or a mechanical component inside the electrical equipment. All boundary distances are measured from that point. When establishing boundaries, exposed movable conductors are treated differently than exposed fixed circuit parts or conductors.

Every possible electrical hazard within a work area must be analyzed and documented. Specific PPE for a given situation are based on the information gathered from the analysis of a given hazard. That documented data includes all the electrical hazards (arc flash, blast, and shock). After

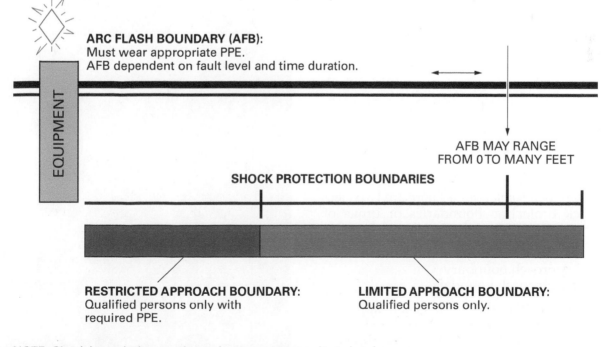

Figure 13 Approach limits.

---

all hazards have been documented, all personnel (qualified and unqualified) working in the area must be trained to recognize and avoid the identified hazards. *70E Section 130.4(E)* establishes rules for unqualified persons who must enter the **limited approach boundary**. Only qualified persons using all required PPE are allowed to enter and work inside the **restricted approach boundary** per *70E Section 130.4(F)*.

Arc flash boundary – When an arc flash hazard is present, an arc flash boundary must be established. This boundary is determined by how far away a person would need to be located to avoid receiving serious burns in the event of an arc flash. Anyone within the arc flash boundary is exposed to the possibility of second-degree (blistering) burns or worse and must therefore use arc-rated protective apparel as determined through **arc flash risk assessment** and other PPE identified in *70E Article 130*.

Depending on the **incident energy**, an arc flash boundary might be within a shock protection boundary or outside of (exceed) a shock protection boundary. Many electrical safety programs establish both the arc flash boundary and the outer shock protection boundary at whatever distance is greater, as determined by the risk assessment.

When an electrical fault causes an arc flash, the explosion produces both a fireball and a shock wave extending away from the arc flash location (the conductor or circuit). Anyone within the arc flash boundary will be exposed to searing heat as well as an extremely bright light that may cause temporary loss of vision and pain. The heat from arc flashes is often hot enough to melt metal fixtures inside the enclosure. *Figure 14* shows an arc flash.

The arc flash boundary is determined for thermal energy, but blast accompanies the arc flash. The blast creates a shock wave that can blow equipment apart and people away from the blast. Shrapnel, toxic gases, and copper vapor explode in all directions. The blast also creates sound waves that can damage hearing.

Shock protection boundaries – There are two electrical shock protection boundaries or limits of approach. NFPA 70E® identifies these electrical shock boundaries as follows:

- Limited approach boundary
- Restricted approach boundary

Table 4 shows the shock protection approach boundaries to exposed energized parts listed in *70E Table 130.4(D)(a)*. Column 1 shows the different voltage levels, measured phase-to-phase. Columns 2 and 3 cover the limited approach boundary. Column 2 shows the required distance from an exposed movable conductor (such as an overhead line), while Column 3 shows the distance from an exposed fixed circuit part or conductor. Column 4 covers the restricted approach boundary. See *70E Informative Annex C* for additional information on limits of approach. *70E Table 130.4(D)(b)* identifies the shock protection approach boundaries to energized parts of direct-current voltage systems.

The limited approach boundary is a shock protection boundary at a specified distance from an exposed energized part that can be crossed only by qualified persons. All unqualified persons in the area must be made aware of the hazards and warned not to cross the boundary. Where there is a need for an unqualified person to cross the limited approach boundary, a qualified person must advise the unqualified person of the possible hazards and continuously escort the unqualified person while inside the limited approach boundary.

WARNING!

Any worker exposed to electrical hazards must be qualified to manage the hazard and use appropriate PPE. If any worker is not exposed to an electrical hazard but has the potential to be, the worker must be informed of the hazard and instructed in how to avoid it.

Figure 14 Arc flash.

Arc Flash Relays

The energy discharged in an arc is calculated by multiplying the square of the short-circuit current by the time the arc takes to develop (energy = I^2t). Therefore, the energy and consequently the power and size of an arc flash are directly related to both short-circuit current and time. The higher the arcing current or the longer the duration of the arc, the more damage is likely to occur. An arc flash relay uses light sensors or a combination of light and current sensors to detect the light/current produced by an arc flash and trip upstream devices. This reduces the incident energy and limits the potential damage. See *70E Informative Annex O.2.3(5)*.

Table 4 Shock Protection Approach Boundaries to Energized Electrical Conductors or Circuit Parts for Alternating-Current Systems [Data from *70E Table 130.4(D)(a)*]

	Limited Approach Boundary[b]		
Nominal System Voltage Range, Phase to Phase[a]	**Exposed Movable Conductor**[c]	**Exposed Fixed Circuit Part**	**Restricted Approach Boundary**[b] **Includes Inadvertent Movement Adder**
Less than 50V	Not specified	Not specified	Not Specified
50V to 150V[d]	3.0m (10 ft 0 in)	1.0m (3 ft 6 in)	Avoid contact
151V to 750V	3.0m (10 ft 0 in)	1.0m (3 ft 6 in)	0.3m (1 ft 0 in)
751V to 15kV	3.0m (10 ft 0 in)	1.5m (5 ft 0 in)	0.7m (2 ft 2 in)
15.1kV to 36kV	3.0m (10 ft 0 in)	1.8m (6 ft 0 in)	0.8m (2 ft 7 in)
36.1kV to 46kV	3.0m (10 ft 0 in)	2.5m (8 ft 0 in)	0.8m (2 ft 9 in)
46.1kV to 72.5kV	3.0m (10 ft 0 in)	2.5m (8 ft 0 in)	1.0m (3 ft 3 in)
72.6kV to 121kV	3.3m (10 ft 8 in)	2.5m (8ft 0 in)	1.0m (3 ft 4 in)
138kV to 145kV	3.4m (11 ft 0 in)	3.0m (10 ft 0 in)	1.2m (3 ft 10 in)
161kV to 169kV	3.6m (11 ft 8 in)	3.6m (11 ft 8 in)	1.3m (4 ft 3 in)
230kV to 242kV	4.0m (13 ft 0 in)	4.0m (13 ft 0 in)	1.7m (5 ft 8 in)
345kV to 362kV	4.7m (15 ft 4 in)	4.7m (15 ft 4 in)	2.8m (9 ft 2 in)
500kV to 550kV	5.8m (19 ft 0 in)	5.8m (19 ft 0 in)	3.6m (11 ft 10 in)
765kV to 800kV	7.2m (23 ft 9 in)	7.2m (23 ft 9 in)	4.9m (15 ft 11 in)

Note (1): For arc flash boundary, see *70E Section 130.5(A)*.

Note (2): All dimensions are distance from exposed energized electrical conductors or circuit part to employee.

a. For single-phrase systems above 250 volts, select the range that is equal to the system's maximum phase-to-ground voltage multiplied by 1.732.

b. See definition in *70E Article 100* and text in *70E Section 130.4(D)(2)* and *Informative Annex C* for elaboration.

c. *Exposed movable conductors* describes a condition in which the distance between the conductor and a person is not under the control of the person. The term is normally applied to overhead line conductors supported by poles.

d. This includes circuits where the exposure does not exceed 120 volts nominal.

Reprinted with permission from NFPA 70E®-2018, *Standard for Electrical Safety in the Workplace®*, Copyright © 2017, National Fire Protection Association, Quincy, MA. This reprinted material is not complete and official position of the NFPA on the referenced subject, which is represented only by the standard in its entirety which may be obtained through the NFPA website at *www.nfpa.org*.

The restricted approach boundary is a shock protection boundary that, due to its proximity to exposed energized parts, requires the use of shock protection techniques and equipment when crossed. The restricted approach boundary may be crossed only by qualified persons using the required PPE (*Figure 15*) and authorized by an energized electrical work permit.

Work within the restricted approach boundary requires that rubber-insulating equipment be used inside that boundary. It also requires the use of insulated tools rated for the voltages on which they are used.

The restricted approach boundaries in *70E Tables 130.4(D)(a) and (b)* include an added safety margin to compensate for inadvertent movement of the worker. The interaction of exposed energized parts and test equipment or hand tools is the initiator of many shock and arc flash incidents when the tool or test lead becomes part of the circuit path. Some estimates state that 75% or more of arc incidents begin in this manner.

Electrical risk assessment – Part of an employer's responsibility involves having an electrical risk assessment completed on all electrical hazards in the workplace. Those analyses include a shock risk assessment and an arc flash risk assessment. After each hazard is analyzed and documented, PPE is identified for tasks performed on or near the hazard, and warnings are posted on the exterior of the equipment per *70E Section 130.5(H)*. Labels should be readable from outside any boundary, and must list the following data:

- Nominal system voltage
- Arc flash boundary
- At least one of the following:
  - Available incident energy and the corresponding working distance, *or* the arc flash PPE category in *70 E Table 130.7(C)(15)(a)* or *70E Table 130.7(C)(15)(b)* for the equipment
  - Minimum arc rating of clothing
  - Site-specific level of PPE

Figure 16 shows a typical warning label. The label lists the distances and exposure levels for both arc flash and shock, and the recommended type of gloves to use. The equipment name and number identify the bus.

3.2.2 Human Performance and Workplace Electrical Safety

Human performance must be considered when performing an electrical risk assessment. *70E Informative Annex Q* provides information on human performance and workplace electrical safety. The principles of human performance are based on the following concepts:

- All people make mistakes.
- Error-prone situations and conditions are predictable, manageable, and preventable.
- Individual performance is influenced by organizational processes and values.
- High levels of worker performance are the result of positive reinforcement from leaders, peers, and subordinates.
- Incidents can be avoided by identifying why mistakes occur and applying the lessons learned from past incidents.

Figure 15 Worker using appropriate PPE.

Did You Know?

Insulated tools are intended to protect against providing an unintended path that becomes an arc fault. Anyone working close to energized parts must be extremely careful in their movements and use only insulated or insulating tools within the restricted approach boundary.

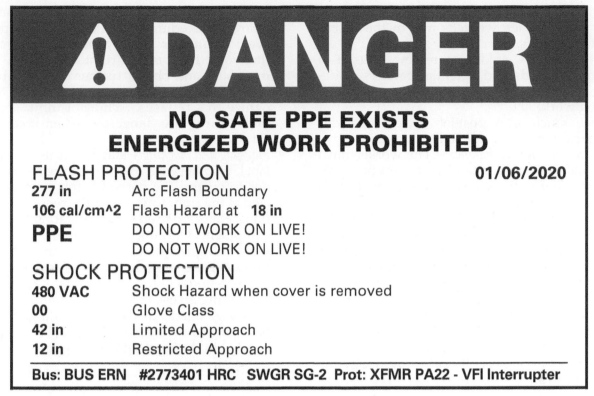

DANGER

NO SAFE PPE EXISTS
ENERGIZED WORK PROHIBITED

FLASH PROTECTION   01/06/2020

277 in   Arc Flash Boundary

106 cal/cm^2   Flash Hazard at   **18 in**

PPE   DO NOT WORK ON LIVE!
                     DO NOT WORK ON LIVE!

SHOCK PROTECTION

480 VAC   Shock Hazard when cover is removed

00   Glove Class

42 in   Limited Approach

12 in   Restricted Approach

Bus: BUS ERN   #2773401 HRC   SWGR SG-2 Prot: XFMR PA22 - VFI Interrupter

Figure 16 Electrical hazard warning label.

Analyzing human performance helps to pinpoint potential error points. Workers operate in one or more human performance modes: rule-based, skill-based, and knowledge-based. These modes and the errors associated with them are described in *70E Informative Annex Q.4*:

- *Rule-based mode* – Workers operate in rule-based mode when the task has been done before or is covered by a procedure. It is called the *rule-based mode* because it applies the use of written or practiced rules or procedures. Common types of errors in this mode include deviating from an approved procedure or applying the correct procedure to the wrong situation. Due to the predictable and tested rules and outcomes, the rule-based mode is the most desirable.
- *Knowledge-based mode* – Workers operate in knowledge-based mode when a task has not been encountered before and there is no specific procedure to follow. In this case, the worker must rely on knowledge and past experience in order to make decisions. The most common errors in knowledge-based mode occur when decisions are based on an inaccurate assessment of the situation or focus on a single aspect of the problem rather than the big picture.

- *Skill-based mode* – Workers operate in skill-based mode when performing common and familiar tasks (e.g., operating a low-voltage circuit breaker). Common errors in skill-based mode are due to lack of attention and a perceived reduction in risk.

Error precursors are situations that put a worker at risk due to the demands of the task, conditions, worker attitude, and/or environment. *70E Informative Annex Table Q.5* groups error precursors into the following categories:

- *Task demands* – High workload, monotony, time pressure, multiple tasks, critical/irreversible tasks
- *Work environment* – Distractions, interruptions, workarounds, unexpected conditions, new routine
- *Individual capabilities* – New task or technique, unsafe attitude, poor communication skills
- *Human nature* – Stress, habits, assumptions, complacency, overconfidence, shortcuts

Each of these precursors can be countered with one or more human performance tools. Human performance tools reduce the likelihood of error when applied to error precursors. *70E Informative Annex Q.6* lists the following human performance tools:

- *Job planning and pre-job briefing* – These meetings identify each worker's role in the execution of the tasks.
- *Job site review* – A job site review is used to identify hazards and potential barriers or delays. It can be performed any time prior to or during work.
- *Post-job review* – A post-job review provides feedback that can be applied to future jobs.
- *Procedure use and adherence* – The worker must read and fully understand each step in a procedure before attempting to complete it. It is helpful to track progress by checking off each step as it is completed. This verifies that each step has been done in the specified sequence and ensures that the procedure can be resumed at the correct point if it is interrupted. If a procedure cannot be used as written or the expected result cannot be predicted, stop and resolve any issues before proceeding.
- *Self-check with verbalization* – A verbal self-check is a valuable tool when performing a critical or irreversible procedure. It is also known by the acronym STAR: Stop, Think, Act, and Review. Before, during, and after performing a task that cannot be reversed, the worker should stop, think, and verbalize the intended action.
- *Three-way communication* – When a statement is made by a sender, it is then repeated by the receiver to confirm the accuracy of the message, and again validated by the sender. Whenever possible, any letters should be stated using the phonetic alphabet.
- *Stop when unsure* – When a worker is unable to follow a procedure as written, if an unexpected event occurs, or if the worker has a feeling that something is not right, then the worker must stop and obtain further direction before proceeding. All workers must be trained to recognize that phrases such as "I think" or "I'm pretty sure," are dangerous when it comes to safety.
- *Flagging and blocking* – Flagging is used to mark, label, or otherwise identify components in order to ensure the correct component is operated at the right time under the required conditions. Flags are used when operating look-alike equipment, working on multiple components, performing frequent operations in a short period of time, or interrupting process-critical equipment. Blocking is a method of physically preventing access to an area or equipment controls (e.g., barricades, fences, or hinged covers on switches or control buttons). Blocking can be used in conjunction with flagging.

The reduction or elimination of electrical incidents requires that all members of an organization work together to promote a culture that values error prevention and the use of human performance tools to identify and prevent error-prone situations and conditions.

3.0.0 Section Review

1. The OSHA lockout/tagout rule _____.
   a. applies only to those employees performing electrical work
   b. applies to any stored energy that could cause injury (electrical, mechanical, hydraulic, chemical, thermal, kinetic, or other)
   c. represents the most stringent requirements and supersedes site procedures
   d. only applies to government work sites
2. The NFPA 70E® standard covers the _____.
   a. minimum installation requirements for electrical equipment
   b. manufacturing specifications for electrical equipment
   c. practical safe working requirements relative to electrical hazards
   d. operating instructions for electrical equipment

4.0.0 HAZARDS AND SAFETY REQUIREMENTS

Objective

Recognize the safety requirements for various hazards.

a. Identify the safety hazards associated with ladders, scaffolds, and lift equipment.
b. Avoid back injuries by practicing proper lifting techniques.
c. Demonstrate basic tool safety.
d. Identify confined space entry procedures.
e. Work safely with dangerous materials.
f. Select and use appropriate fall protection.

Performance Tasks

1. Properly select and use PPE.
2. Describe the safety requirements for an instructor-supplied task, such as replacing the lights in your classroom.
   - Discuss the work to be performed and the hazards involved.
   - If a ladder is required, perform a visual inspection on the ladder and set it up properly.
   - Ensure that local emergency telephone numbers are either posted or known by you and your partner(s).
   - Plan an escape route from the location in the event of an accident.

Trade Term

Polychlorinated biphenyls (PCBs): Toxic chemicals that may be contained in liquids used to cool certain types of large transformers and capacitors.

There are many hazards on a job site, including fall hazards, equipment hazards, chemical hazards, and tool hazards. The best prevention is safety awareness. If an accident occurs, know where first aid is available and follow company procedures for reporting it. Emergency first aid telephone numbers must be readily available to everyone on the job site. Refer to *CFR 1910.151/1926.23* and *1926.50* for specific requirements.

4.1.0 Ladders, Scaffolds, and Lift Equipment

About half of the injuries that result from contact with energized components involve ladders and scaffolds. In addition to electrical injuries, the involuntary recoil that can occur when a person is shocked can result in a fall from a ladder or elevated surface.

4.1.1 Ladders

Many job-site accidents involve the misuse of ladders. In fact, ladders present enough of a hazard that many contractors have a "Ladders Last" policy, meaning that they would prefer to use personnel lifts, construct scaffolding, or move the work to ground level rather than have employees exposed to the hazards of working from a ladder. Ladders are covered in *CFR 1910.23* and *CFR 1926, Subpart X.* Adhere to the following general rules every time you use any ladder—they can prevent serious injury or even death:

- Select the correct ladder for the application. This involves selecting the right type of ladder for the job (e.g., straight, extension, stepladder, platform, etc.), along with the correct material (fiberglass vs. aluminum), height, and duty rating (maximum safe load capacity).
- Before using any ladder, inspect it. Look for loose or missing rungs, cleats, bolts, or screws. Also check for cracked, bent, broken, or badly worn rungs, cleats, or side rails. Various types of ladder damage are shown in *Figure 17*.
- Before you climb a ladder, make sure you clear any debris from around the base of the ladder so you do not trip over it when you descend.
- If you find a ladder in poor condition, do not use it. Report it and tag it for repair or disposal.
- Never modify a ladder by cutting it or weakening its parts.
- Do not set up ladders where they may be run into by others, such as in doorways or walkways. If it is absolutely necessary to set up a ladder in such a location, protect the ladder with barriers.
- Do not increase a ladder's reach by putting it in a mechanical lift or standing it on boxes, barrels, or anything other than a flat, solid surface.
- Check your shoes for grease, oil, or mud before climbing a ladder. These materials could make you slip.
- Always face the ladder and maintain three-point contact with the ladder (either have both feet and one hand on the ladder or both hands and one foot as you climb).

- Never lean out from the ladder. Keep your belt buckle centered between the rails. If something is out of reach, get down and move the ladder.
- Never pull from a ladder. The action of pulling with one hand while attempting to hold on with the other will cause you to shift your balance and may result in a fall. Only pull with both feet on a solid platform.

Straight and extension ladders – There are also some specific hazards related to straight ladders and extension ladders. For this reason, the following rules must be observed when working with them:

- Always place a straight ladder at the proper angle. The horizontal distance from the ladder feet to the base of the wall or support should be about one-fourth the working height of the ladder (*Figure 18*).
- Secure straight ladders to prevent slipping. Use ladder shoes or hooks at the top and bottom. Another method is to secure a board to the floor against the ladder feet. For brief jobs, someone can hold the straight ladder.
- Side rails should extend above the top support point by at least 36" (0.9 m). This provides stability as well as a handhold when stepping around the ladder.
- It takes two people to safely extend and raise an extension ladder. Extend the ladder only after it has been raised to an upright position.
- Never carry an extended ladder.

(A) CRUMBLING RAIL

(B) CRACKED STEP

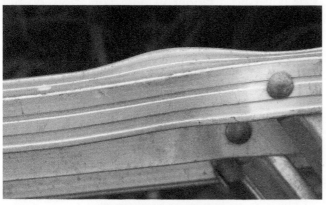

(C) BENT BACK BRACE

Figure 17 Types of ladder damage.

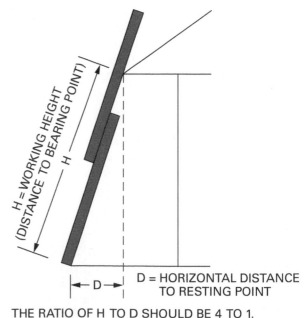

H = WORKING HEIGHT (DISTANCE TO BEARING POINT)

H

D

D = HORIZONTAL DISTANCE TO RESTING POINT

THE RATIO OF H TO D SHOULD BE 4 TO 1.

Figure 18 Straight ladder positioning.

- Never use two ladders spliced together.
- Ladders should not be painted because paint can hide defects.

Stepladders – The following specific rules must be followed when working with stepladders:

- Always open the stepladder all the way and lock the spreaders to avoid collapsing the ladder accidentally.
- Use a stepladder that is high enough for the job so that you do not have to reach. It is a good idea to have someone hold the ladder if it is more than 10' (3 m) high.
- Never use a stepladder as a straight ladder.
- Never stand on or straddle the top two rungs of a stepladder.
- Do not use ladders as shelves.

Fireman's Rule

To set up a straight ladder, place your feet against the side rails, stand straight up, and put your hands at right angles to your body directly in front of you. If you can grab the side rails, the ladder is at the correct angle. This is called the *Fireman's Rule*.

Figure Credit: Mike Powers

What's wrong with this picture?

Figure Credit: Mike Powers

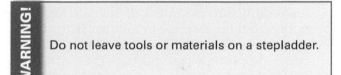

WARNING!

Do not leave tools or materials on a stepladder.

Sometimes you will need to move or remove protective equipment, guards, or guardrails to complete a task using a ladder. Always wear appropriate fall protection and replace what you moved or removed before leaving the area.

4.1.2 Scaffolds

Working on scaffolds (*Figure 19*) also involves being safe and alert to hazards. Scaffolds are covered in *CFR 1926, Subpart L*. In general, keep scaffold platforms clear of unnecessary material or scrap. These can become deadly tripping hazards or falling objects. Carefully inspect each part of the scaffold as it is erected. Your life may depend on it! Makeshift scaffolds have caused many injuries and deaths on job sites. Use only scaffolding and planking materials designed and marked for their specific use. When working on a scaffold, follow the established specific requirements set by OSHA for the use of fall protection. When appropriate, wear an approved harness with a lanyard properly anchored to the structure.

Figure 19 Scaffolding.

NOTE

The following requirements represent a compilation of the more stringent requirements of both *CFR 1910* and *CFR 1926*.

The following are some of the basic OSHA rules for working safely on scaffolds:

- Scaffolds must be erected on sound, rigid footing (referred to as a mud sill) that can carry the maximum intended load. If the scaffolding is not erected on concrete or another firm surface, you can use 2×10 or 2×12 scaffold-grade lumber to support the base plates of the scaffolding.
- Scaffolds must be erected straight and plumb with no bent or deformed pieces. Correctly erected scaffolding will be symmetrical with the same parts on both sides (unless it has an outrigger attachment).
- Guardrails and toe boards must be installed on the open sides and ends of platforms higher than 10' (3 m) above the ground or floor. Note that many companies lower this to 6' (1.8 m) or even 4' (1.2 m).

- There must be a screen with ½" (13 mm) maximum openings between the toe board and the mid rail where persons are required to work or pass under the scaffold.
- Each end of a platform, unless cleated or otherwise restrained by hooks or equivalent means, must extend over the centerline of its support by at least 6" (150 mm), but no more than 12" (300 mm) for a platform 10' feet (3 m) or less in length or 18" (460 mmm) for a platform longer than 10' feet (3 m).
- If the scaffold does not have built-in ladders that meet the standard, then it must have an attached ladder access.
- All employees who erect or dismantle scaffold must be trained to do so. All employees who use a scaffold must be trained in its safe use.
- Unless it is impossible, fall protection must be worn while building or dismantling all scaffolding.

What's wrong with this picture?

Figure Credit: Mike Powers

Case History

Scaffolds and Electrical Hazards

Remember that scaffolds are excellent conductors of electricity. Recently, a maintenance crew needed to move a scaffold and although time was allocated in the work order to dismantle and rebuild the scaffold, the crew decided to push it instead. They did not follow OSHA recommendations for scaffold clearance and did not perform a job site survey. During the move, the five-tier scaffold contacted a 12,000V overhead power line. All four members of the crew were killed and the crew chief received serious injuries.

The Bottom Line: Never take shortcuts when it comes to your safety and the safety of others. Trained safety personnel should survey each job site prior to the start of work to assess potential hazards. Safe working distances should be maintained between scaffolding and power lines.

- Work platforms must be completely decked for use by employees.
- Use trash containers or other similar means to keep debris from falling and never throw or sweep material from above.

4.1.3 Lift Equipment

On the job, you may be working in the operating area of lifts, hoists, or cranes. The following safety rules are for those who are working in the area with overhead equipment but are not directly involved in its operation:

- Stay alert and pay attention to the warning signals from operators.
- Never stand or walk under a load, regardless of whether it is moving or stationary.
- Always warn others of moving or approaching overhead loads.
- Never attempt to distract signal persons or operators of overhead equipment.
- Obey warning signs.

Vertical Towers

This electrician is installing a light fixture using a vertical tower or Genie® lift. This lift is designed to fold up small enough so that it can be maneuvered through house doorways. Some models can be driven.

Figure Credit: Mike Powers

- Do not use equipment that you are not qualified to operate.
- Cranes that are operated in areas with places in which a person can become trapped or pinched must have barricades placed around them to warn away workers.
- Hoists rigged in shafts or the outside of buildings must be secured to prevent them from being pulled down.
- Never overload a hoist, lift, or crane. Always follow lift ratings—there is no such thing as a safety factor that can be used to cheat a load.
- Cranes, lifts, and hoists must be inspected daily.
- Never lift a person using a crane, hoist, or material lift.
- Never use equipment improperly. For example, never rig a lift to hoist a load.
- Only people who have been trained in rigging should ever do rigging for a lift. It is extremely easy for a load to fall out of inappropriately rigged lifts.
- Personnel hoists require gates on every landing that can only be opened from the hoist side. This ensures that they are not opened onto a fall hazard.

4.2.0 Proper Lifting Techniques

Back injuries cause many lost working hours every year. That is in addition to the misery felt by the person with the hurt back! Learn how to lift properly and size up the load. To lift, first stand close to the load. Then, squat down and keep your back straight and your chin up. Get a firm grip on the load and keep the load close to your body. Lift by straightening your legs. Make sure that you lift with your legs and not your back. Do not be afraid to ask for help if you feel the load is too heavy. An example of proper lifting is shown in *Figure 20*.

Keep the following precautions in mind when lifting:

- Make the lift smoothly and under control.
- Look straight ahead while lifting.
- Move your feet to pivot; do not twist or you may injure yourself.
- Constantly scan the path ahead for obstructions. If you cannot see your path over or around the object being carried, then you must have help to transport the object.
- Avoid lifting objects over your head.
- Never lift over the side or tailgate of a pickup truck.
- Don't twist your body when lifting an object or setting it down.

- Never reach over an obstacle to lift a load.
- Don't step over objects in your way.

Lifting puts an extraordinary amount of pressure on your back. For example, if you bend from the waist to pick up an object weighing X, you are applying 10 times the amount of pressure (10X) to your lower back. Lower back injuries are one of the most common workplace injuries because it's so easy to be careless about lifting, especially when you are in a hurry. Remember, it is much easier to ask for help than it is to nurse an injured back.

Lift Safely to Preserve Your B.A.C.K.

B – Balance: keep your stance wide and get a good grip on the object.

A – Alignment: keep your back relaxed and upright.

C – Contract and close: contract your stomach muscles and hold loads close.

K – Knees: make sure you bend them, not your waist.

1

2

3

4

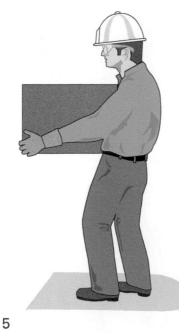
5

6

Figure 20 Proper lifting.

Figure Credit: Mike Powers

4.3.0 Basic Tool Safety

When using any tools for the first time, read the operator's manual to learn the recommended safety precautions. If you are not certain about the operation of any tool, ask the advice of a more experienced worker. Before using a tool, you should know its function and how it works.

4.3.1 Hand Tool Safety

Hand tools are non-powered tools and may include anything from screwdrivers to cable strippers (*Figure 21*). Hand tools are dangerous if they are misused or improperly maintained.

Keep the following precautions in mind when using hand tools:

- Use tools only for their designated purpose.
- Always maintain your tools properly. If the wooden handle on a tool such as an axe or hammer is loose, splintered, or cracked, the head of the tool may fly off and strike the user or another person. Tag the tool out of service and do not use it.
- Repair or replace damaged or worn tools. A wrench with its jaws sprung might easily slip, causing hand injuries. If the wrench flies, it may strike the user or another person.
- Impact tools such as chisels, wedges, and drift pins are unsafe if they have mushroomed

heads. The heads might shatter when struck, sending sharp fragments flying.

- Use bladed tools with the blades and points aimed away from yourself and other people to avoid injury if the tool slips during use.
- Store bladed tools properly; use the sheath or protective covering if there is one.
- Keep blades sharp and inspect them regularly. Dull blades are difficult to use and control and can be far more dangerous than well-maintained blades.
- Never hold material with one hand while using the other hand to cut it. The hand on the material is subject to severe injury when the cutting tool slips.
- Never leave tools on top of ladders or scaffolding.
- When working at elevation, use tool tethers to keep tools from dropping onto lower walking or working surfaces.

In general, your risks are greatly reduced by inspecting and maintaining tools regularly and always wearing appropriate personal protective equipment such as safety goggles, hard hats, and filtering masks. Also, keep floors dry and clean to prevent accidental slips that can result in injuries caused by the tools you may be using.

4.3.2 Power Tool Safety

Power tools can be hazardous when they are improperly used or not well maintained. Most of the risks associated with hand tools are also risks when using power tools. When you add a power source to a tool, however, the risk factors increase.

Power tools are powered by different sources. Some examples of power sources for power tools include the following:

- Batteries
- Electricity
- Pneumatics (air pressure)
- Liquid fuel (gasoline or propane)
- Hydraulics (fluid pressure)

You must know the safety rules and proper operating procedures for each tool you use. Specific operating procedures and safety rules for using a tool are provided in the operator's/user's manual supplied by the manufacturer. Before operating any power tool for the first time, always read the manual to familiarize yourself with the tool. If the manual is missing, contact the manufacturer for a replacement.

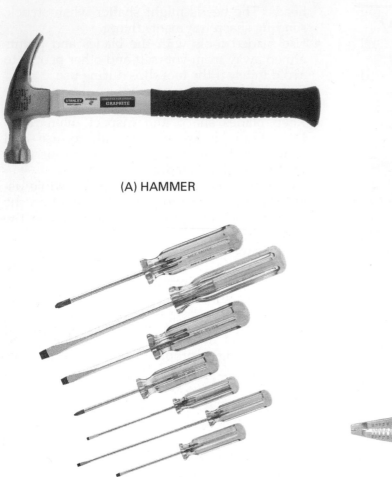

(A) HAMMER

(C) SCREWDRIVERS

(B) RATCHET CABLE CUTTER

(D) MULTI-PURPOSE TOOL

Figure 21 Hand tools.

Follow these general guidelines to prevent accidents and injury:

- Inspect all tools for damage before use. Remove damaged tools from use and tag DO NOT USE.
- Never carry or lower a tool by the cord or hose.
- Keep cords and hoses away from heat, oil, and sharp edges.
- Do not attempt to operate any power tool before being checked out by your instructor or supervisor on that particular tool.
- Always wear eye protection, a hard hat, and any other required personal protective equipment when operating power tools.
- Wear face and hearing protection when required.

- Wear proper respiratory equipment when necessary.
- Wear the appropriate clothing for the job being done. Wear close-fitting clothing that cannot become caught in moving tools. Roll up or button long sleeves, tuck in shirttails, and tie back long hair. Do not wear any jewelry, including watches or rings.
- Do not distract others or let anyone distract you while operating a power tool.
- Do not engage in horseplay.
- Do not run or throw objects.
- Consider the safety of others, as well as yourself. Observers should be kept at a safe distance away from the work area.
- Never leave a power tool running while it is unattended.
- Assume a safe and comfortable position before using a power tool. Be sure to maintain good footing and balance in order to respond to kickbacks, jumps, or sudden shifts.

- Secure work with clamps or a vise, freeing both hands to safely operate the tool.
- To avoid accidental starting, never carry a tool with your finger on the switch.
- Be sure that a power tool is properly grounded and connected to a GFCI before using it.
- Ensure that power tools are disconnected before performing maintenance or changing accessories.
- Use a power tool only for its intended use.
- Keep your feet, fingers, and hair away from the blade and/or other moving parts of a power tool.
- Never use a power tool with guards or safety devices removed or disabled.
- Never operate a power tool if your hands or feet are wet.
- Keep the work area clean at all times.
- Become familiar with the correct operation and adjustments of a power tool before attempting to use it.
- Keep tools sharp and clean for the best performance.
- Follow the instructions in the user's manual for lubricating and changing accessories.

Figure 22 Pneumatic nail gun.

Battery-Operated Tools

Battery-operated tools always have the power to cause injury unless the battery has been removed. It is easy to inadvertently start a battery-operated tool if you hold it with your finger on the trigger.

- Keep a firm grip on the power tool at all times.
- Use electric extension cords of sufficient size to power the particular tool you are using.
- Many jobs require extension cords to be suspended. Use only nonconductive supports or listed products.
- Do not run extension cords across walkways where they will pose a tripping hazard.
- Do not run extension cords in areas of vehicle traffic or where wheeled carts are in use.
- Report unsafe conditions to your instructor or supervisor.
- Tools that shoot nails (*Figure 22*), rivets, or staples, and operate at pressures greater than 100 pounds per square inch (psi) or 689.5 kilopascals (kPa), must be equipped with a safety device that won't allow fasteners to be shot unless the muzzle is pressed against a work surface.
- Compressed-air guns should never be pointed toward anyone and the muzzle should never be pressed against a person.
- The use of stud-type guns (including both powder-actuated tools and tools powered by gasoline, other fuel, and battery/spring power) requires special training and certification.
- Never use explosive or flammable materials around stud-type guns.
- Never point stud-type guns at anybody.
- Never pick up an unattended stud-type gun. Instead, tell your supervisor that the tool has been left unattended.
- Never play with stud-type guns. These tools are as dangerous as a loaded firearm.

Don't Remove the Ground Pin

An employee was climbing a metal ladder to hand an electric drill to the journeyman installer on a scaffold above him. When the victim reached the third rung from the bottom of the ladder, he received an electric shock that killed him. The investigation revealed that the extension cord had a missing ground pin and that a conductor on the green grounding wire was making intermittent contact with the energized black wire, thereby energizing the entire length of the grounding wire and the drill's frame. The drill was not double insulated.

    The Bottom Line: Do not disable any safety device on a power tool. A ground fault can be deadly.

    Source: The Occupational Safety and Health Administration (OSHA)

4.4.0 Confined Space Entry Procedures

Occasionally, you may be required to do your work in a confined space. Confined spaces in construction are covered in *CFR 1926, Subpart AA*. If this is the case, you need to be aware of some special safety considerations. For details on the subject of working in manholes and vaults, refer to *CFR 1926.1200* through *1926.1213*. The general precautions are listed in the following paragraphs.

4.4.1 General Guidelines

A confined space includes (but is not limited to) any of the following: a manhole (*Figure 23*), boiler, tank, tunnel, hopper, bin, sewer, vat, pipeline, vault, pit, air duct, or vessel. A confined space is identified as follows:

- It has restricted entry and exit.
- It is not intended for continued human occupancy.
- It has the potential for other hazards, such as electrical, explosive, and animal.
- It has the potential for entrapment/engulfment.

Figure 23 Confined space entry with proper ventilation and PPE.

- It has the potential for accumulating a dangerous atmosphere.

Entry into a confined space occurs when any part of the body crosses the plane of entry. No employee shall enter a confined space unless the employee has been trained in confined space entry procedures. Other requirements for confined spaces include the following:

- All hazards must be eliminated or controlled before a confined space entry is made.
- The air quality in the confined space must be continually monitored.
- All appropriate personal protective equipment shall be worn at all times during confined space entry and work.
- Ladders used for entry must be secured.
- A rescue retrieval system must be in use when entering confined spaces and while working in permit-required confined spaces (discussed later). Each employee must be capable of being rescued by the retrieval system, or a trained, practiced rescue squad with all appropriate PPE must be available and able to respond immediately.
- The area outside the confined space must be properly barricaded, and appropriate warning signs must be posted. *Figure 24* shows a confined space cover with a warning sign.
- Entry permits can only be issued and signed by a qualified person such as the job-site supervisor. Permits must be kept at the confined space while work is being conducted. At the end of the shift, the entry permits must be made part of the job journal and retained for one year.

4.4.2 Confined Space Hazard Review

Before determining the proper procedure for confined space entry, a hazard review shall be performed. The hazard review shall include, but not be limited to, the following conditions:

Figure 24 Ventilated confined space cover with warning sign.

- The past and current uses of the confined space
- The physical characteristics of the space including size, shape, air circulation, etc.
- Proximity of the space to other hazards
- Existing or potential hazards in the confined space, such as atmospheric conditions (oxygen levels, flammable/explosive levels, and/or toxic levels), presence/potential for liquids, presence/potential for particulates, or potential for engulfment
- Potential for mechanical/electrical hazards in the confined space (including work to be done)

Once the hazard review is completed, the supervisor, in consultation with the project managers and/or safety manager, shall classify the confined space as one of the following:

- A non-permit confined space
- A permit-required confined space

Permit-required confined spaces must be posted. Once the confined space has been properly classified, the appropriate entry and work procedures must be followed.

The contractor must list all confined spaces on site and designate them as permit-required or non-permit confined spaces.

4.5.0 Dangerous Materials

The hazard communication requirements for dangerous materials are covered in *CFR 1910.1200*. You must be prepared in case an accident occurs on the job site. First aid training that includes certification classes in CPR and artificial respiration could be the best insurance you and your fellow workers ever receive. Make sure that you know where first aid is available at your job site. Also, make sure you know the accident reporting procedure. Each job site should also have a first aid manual or booklet giving easy-to-find emergency treatment procedures for various types of injuries. Emergency first aid telephone numbers should be readily available to everyone on the job site. Refer to *CFR 1910.151/1926.23* and *1926.50* for specific first aid requirements.

4.5.1 Solvents

The solvents that are used by electricians may give off vapors that are toxic enough to make people temporarily ill or even cause permanent injury. Many solvents are skin and eye irritants. Solvents can also be systemic poisons when they are swallowed or absorbed through the skin.

Solvents in spray or aerosol form are dangerous in another way. Small aerosol particles or solvent vapors mix with air to form a combustible mixture with oxygen. The slightest spark could cause an explosion in a confined area because the mix is perfect for fast ignition. There are procedures and methods for using, storing, and disposing of most solvents and chemicals. These procedures are normally found in the safety data sheets (SDSs) available at your facility.

An SDS is required for all materials that could be hazardous to personnel or equipment. These sheets contain information on the material, such as the manufacturer and chemical makeup. As much information as possible is kept on the hazardous material to prevent a dangerous situation; or, in the event of a dangerous situation, the information is used to rectify the problem in as safe a manner as possible. See *Figure 25* for an example of SDS information you may find on the job.

It is always best to use a nonflammable, nontoxic solvent whenever possible. However, any time solvents are used, it is essential that your work area be adequately ventilated and that you wear the appropriate personal protective equipment:

- Wear a chemical face shield with chemical goggles to protect the eyes and skin from sprays and splashes.
- Wear a chemical apron to protect your body from sprays and splashes. Remember that some solvents are acid-based. If they come into contact with your clothes, solvents can eat through your clothes to your skin.
- A paper filter mask does not stop vapors; it is used only for nuisance dust. In situations where a paper mask does not supply adequate protection, chemical cartridge respirators might be needed. These respirators can stop many vapors if the correct cartridge is selected. In areas where ventilation is a serious problem, a self-contained breathing apparatus (SCBA) must be used.
- Make sure that you have been given a full medical evaluation and that you are properly trained in using respirators at your site.

The best respiratory protection when using solvents (or other materials that present inhalation hazards) is to avoid the hazard entirely. Off-shift work, ventilation, and/or rescheduled work schedules should always be used to eliminate the need for working in areas with poor air quality. For example, in an area where hazardous solvents are used, the electrical work can be done off schedule when the solvents are not being used. When this cannot be done, protection against high concentrations of dust, mist, fumes, vapors, and gases is provided by appropriate respirators.

Appropriate respiratory protective devices should be used for the hazardous material involved and the extent and nature of the work performed.

An air-purifying respirator is, as its name implies, a respirator that removes contaminants from air inhaled by the wearer. The respirators may be divided into the following types: particulate-removing (mechanical filter), gas- and vapor-removing (chemical filter), and a combination of particulate-removing and gas- and vapor-removing.

Particulate-removing respirators are designed to protect the wearer against the inhalation of particulate matter in the ambient atmosphere. They may be designed to protect against a single type of particulate, such as pneumoconiosis-producing dust, nuisance dust, toxic dust, metal fumes or mist, or against various combinations of these types.

Gas- and vapor-removing respirators are designed to protect the wearer against the inhalation of gases or vapors in the ambient atmosphere. They are designated as gas masks, chemical-cartridge respirators (nonemergency gas respirators), and self-rescue respirators. They may be designed to protect against a single gas such as chlorine; a single type of gas, such as acid gases; or a combination of types of gases, such as acid gases and organic vapors.

_____ **Section 1 – Product & Company Identification** _____

Product Name.........................: RIDGID Dark Thread Cutting Oil
Product Catalog No...............: 41590, 70830, 41610, 41600

Recommended Use...............: Thread Cutting

Company Name.....................: Ridge Tool Company
Address................................: 400 Clark Street
                                                   Elyria, Ohio 44035-6001
Telephone.............................: 1-800-519-3456 (USA) (8:00 am – 5:00 pm EST, M-F)
Emergency Telephone..........: call 9-1-1 or local emergency number
Website.................................: www.RIDGID.com

Issue Date.............................: May 29, 2019

_____ **Section 2 – Hazards Identification** _____

This product is classified as not hazardous per US OSHA *29CFR 1910.1200* (HazCom 2012) and Canada's Hazardous Products Regulations (WHMIS 2015).

GHS Label Elements: Not applicable

_____ **Section 3 – Composition / Information On Ingredients** _____

Component:	CAS #	% By Weight
Mineral Oil	Confidential	40-100%

This product does not contain silicone or chlorinated additives.

Specific chemical identities and/or exact percentages have been withheld as trade secrets

_____ **Section 4 – First Aid Measures** _____

INGESTION:
    Rinse mouth thoroughly. Call a Poison Center or doctor if you feel unwell. Do NOT induce vomiting.

INHALATION:
    Move to fresh air. Call a Poison Center or doctor if you feel unwell.

Figure 25 Portion of an SDS.

Chemical Safety

The first line of defense with chemicals is to read and follow the directions found on the container. If you follow these instructions, you should be safe from chemical exposure. Be aware that everyone reacts differently to chemicals and you may be hypersensitive to a particular chemical that does not bother your co-workers. Leave the area at the first sign of an allergic reaction and seek medical attention.

If you are required to use a respiratory protective device, you must be evaluated by a physician to ensure that you are physically fit to use a respirator. You must then be fitted and thoroughly instructed in the respirator's use.

Any employee whose job entails having to wear a respirator must keep his face free of facial hair in the seal area.

> **WARNING!**
>
> Do not use any respirator unless you have been fitted for it and thoroughly understand its use. As with all safety rules, follow your employer's respiratory program and policies. Respiratory protective equipment must be inspected regularly and maintained in good condition. Respiratory equipment must be properly cleaned on a regular basis and stored in a sanitary, dustproof container.

4.5.2 Asbestos

Asbestos is a mineral-based material that is resistant to heat and corrosive chemicals. Depending on the chemical composition, asbestos fibers may range in texture from coarse to silky. The properties that make asbestos fibers so valuable to industry are its high tensile strength, flexibility, heat and chemical resistance, and good frictional properties.

Asbestos fibers enter the body by inhalation of airborne particles or by ingestion and can become embedded in the tissues of the respiratory or digestive systems. Exposure to asbestos can cause numerous disabling or fatal diseases. Among these diseases are asbestosis, an emphysema-like condition; lung cancer; mesothelioma, a cancerous tumor that spreads rapidly in the cells of membranes covering the lungs and body organs; and gastrointestinal cancer. The use of asbestos was banned in 1978.

Because asbestos was still in the manufacturing pipeline for a while after it was banned, you need to assume that any facility constructed before 1980 has asbestos in it. The owner must have a survey with any work rules needed to work safely around the asbestos. Common products that contain asbestos include thermal pipe insulation, mastic for ducts and insulation, spray-on fireproofing, floor tiles, ceiling tiles, roof insulation, exterior building sheathing, old wire insulation, and even pipe. As an electrician, you must not drill through or otherwise work with asbestos—you can be trained to work around it only when it can be done safely. Asbestos work is a trade that requires special training and protective equipment.

The signs shown in *Figure 26* must be placed in areas containing asbestos.

Case History

Altered Respiratory Equipment

A self-employed man applied a solvent-based coating to the inside of a tank. Instead of wearing the proper respirator, he used nonstandard air supply hoses and altered the face mask. All joints and the exhalation slots were sealed with tape. He collapsed and was not discovered for several hours.

The Bottom Line: Never alter or improvise safety equipment.

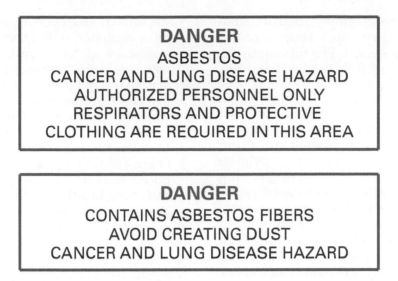

DANGER
ASBESTOS
CANCER AND LUNG DISEASE HAZARD
AUTHORIZED PERSONNEL ONLY
RESPIRATORS AND PROTECTIVE
CLOTHING ARE REQUIRED IN THIS AREA

DANGER
CONTAINS ASBESTOS FIBERS
AVOID CREATING DUST
CANCER AND LUNG DISEASE HAZARD

Figure 26 Danger signs for areas containing asbestos.

4.5.3 Batteries

Working around wet cell batteries can be dangerous if the proper precautions are not taken. Batteries often give off hydrogen gas as a byproduct. When hydrogen mixes with air, the mixture can be explosive in the proper concentration. For this reason, smoking is strictly prohibited in battery rooms, and only insulated tools should be used. Proper ventilation also reduces the chance of explosion in battery areas. Follow your company's procedures for working near batteries. Also, ensure that your company's procedures are followed for lifting heavy batteries.

WARNING!

Battery-powered scissor lifts may have unsealed batteries that require water levels to be checked and topped off. Never use a flame to look inside a battery—it can cause an explosion. Charging batteries with low water levels can damage them.

4.5.4 Acids

Batteries also contain acid, which will eat away human skin and many other materials. Personal protective equipment for battery work typically includes chemical aprons, sleeves, gloves, face shields, and goggles to prevent acid from contacting skin and eyes. Follow your site procedures for dealing with spills of these materials. Also, know the location of first aid when working with these chemicals.

Because of the chance that battery acid may contact someone's eyes or skin, wash stations are located near battery rooms. Do not connect or disconnect batteries without proper supervision. Everyone who works in the area should know where the nearest wash station is and how to use it. Battery acid should be flushed from the skin and eyes with large amounts of water or with a neutralizing solution.

CAUTION

If you come in contact with battery acid, flush the affected area with water and report it immediately to your supervisor.

4.5.5 PCBs and Vapor Lamps

Polychlorinated biphenyls (PCBs) are chemicals that were marketed under various trade names as a liquid insulator/cooler in older transformers. In addition to being used in older transformers, PCBs are also found in some large capacitors and in the small ballast transformers used in street lighting and ordinary fluorescent light fixtures. Disposal of these materials is regulated by the Environmental Protection Agency (EPA) and must be done through a regulated disposal company; use extreme caution and follow your facility procedures.

WARNING!

Do not come into contact with PCBs. They present a variety of serious health risks, including lung damage and cancer.

In addition, any vapor lamps, such as fluorescent, halide, or mercury vapor lamps, contain mercury and must be recycled. The tubes must be packaged and handled carefully to avoid breakage.

4.5.6 Lead Safety

In 2010, the Environmental Protection Agency (EPA) enacted the Renovation, Repair, and Painting (RRP) rule. This standard is designed primarily to protect young children living in or spending time in buildings containing lead-based paint. It requires lead-safe certification and work practices to be followed during renovation work in all homes or other child-occupied facilities built prior to 1978. The rule is triggered by disturbing more than 6 sq ft (0.6 sq m) per room of an interior painted surface or more than 20 sq ft (1.8 sq m) of an exterior painted surface.

Housing and Urban Development (HUD) projects have a lower trigger of only 2 sq ft (0.2 sq m) of interior lead-based paint, more than 20 sq ft (1.8 sq m) of exterior lead-based paint, or 10 percent of the total surface area on an interior or exterior painted component that contains lead-based paint. This can include relatively small areas such as window sills, baseboard, and trim.

The RRP rule contains the following basic requirements:

- At least one RRP certified renovator is required at each job site. Certification involves lead-safe work training by an EPA-accredited training provider.
- In addition to individual certification, the contracting firm or agency must also be certified.
- Contractors must give the client a copy of the "Renovate Right" pamphlet (available for download at *www.epa.gov*).

4.5.7 Silica Safety

Exposure to silica can cause lung cancer, silicosis, chronic obstructive pulmonary disease, and kidney disease. Silica hazards are covered in *CFR 1926.1153* and *CFR 1910.1053*. According to OSHA, approximately two million construction workers are exposed to respirable crystalline silica on the job each year. This includes those who drill, cut, crush, or grind materials such as concrete and stone. To combat the adverse health effects of occupational exposure to silica, OSHA has introduced a silica rule. This rule reduces the permissible exposure limit (PEL) for respirable crystalline silica to 50 micrograms per cubic meter of air, averaged over an 8-hour shift. It also requires employers to use the following controls to limit worker exposure:

- Implement engineering controls (for example, tools equipped with shrouds or dust collection systems and tools equipped with integrated water delivery systems) to limit worker exposure to the PEL.
- Provide respirators when engineering controls cannot adequately limit exposure.
- Limit worker access to high exposure areas.
- Develop a written exposure control plan.
- Offer medical exams to monitor highly exposed workers and supply information about their lung health.
- Train workers on silica risks and how to limit exposures.

OSHA estimates that this rule will save over 600 lives and prevent more than 900 new cases of silicosis each year. *Table 5A* and *Table 5B* provide OSHA exposure controls for various types of equipment/tasks. Following the guidelines in this table will ensure compliance with the standard.

4.6.0 Fall Protection

The OSHA requirements for fall protection can be found in *CFR 1926, Subpart M*. All employees must receive documented training before working in any area where there is the possibility of exposure to a fall of 6' (1.8 m) or more. This rule does not apply to ladders, scaffolds, stairways, and mechanical lifts, which have their own standards.

Fall protection must be used when employees are on a walking or working surface that is 6' (1.8 m) or more above a lower level and has an unprotected edge or side. The areas covered include, but are not limited to the following:

> **WARNING!**
>
> An edge or side where there is no guardrail system at least 39" (1 m) high is considered unprotected.

- Equipment
- Finished and unfinished floors or mezzanines
- Temporary or permanent walkways/ramps
- Finished or unfinished roof areas
- Elevator shafts and hoistways
- Floor, roof, or walkway holes
- Work areas that are 6' (1.8 m) or more above dangerous equipment (*Exception*: If the dangerous equipment is unguarded, fall protection must be used at all heights regardless of the fall distance.)

Table 5A OSHA Exposure Controls for Silica

Equipment/Task	Engineering and Work Practice Control Methods	Required Respiratory Protection and Minimum Assigned Protection Factor (APF)	
		Less Than or Equal to 4 hours/shift	More Than 4 hours/shift
Stationary masonry saws	Use saw equipped with integrated water delivery system that continuously feeds water to the blade. Operate and maintain tool in accordance with manufacturer's instructions to minimize dust emissions.	None	None
Handheld power saws (any blade diameter)	Use saw equipped with integrated water delivery system that continuously feeds water to the blade. Operate and maintain tool in accordance with manufacturer's instructions to minimize dust emissions.	None	None
Handheld power saws for cutting fiber-cement board (with blade diameter of 8 inches or less)	For tasks performed outdoors only: Use saw equipped with commercially available dust collection system. Operate and maintain tool in accordance with manufacturer's instructions to minimize dust emissions. Dust collector must provide the air flow recommended by the tool manufacturer, or greater, and have a filter with 99% or greater efficiency.	None when used outdoors. APF 10 when used indoors or in an enclosed area.	None when used outdoors. APF 10 when used indoors or in an enclosed area
Walk-behind saw	Use saw equipped with integrated water delivery system that continuously feeds water to the blade. Operate and maintain tool in accordance with manufacturer's instructions to minimize dust emissions.	None	None
Rig-mounted core saws or drills	Use tool equipped with integrated water delivery system that supplies water to cutting surface. Operate and maintain tool in accordance with manufacturer's instructions to minimize dust emissions.	None	None
Handheld and stand-mounted drills (including impact and rotary hammer drills)	Use drill equipped with commercially available shroud or cowling with dust collection system. Operate and maintain tool in accordance with manufacturer's instructions to minimize dust emissions. Dust collector must provide the air flow recommended by the tool manufacturer, or greater, and have a filter with 99% or greater efficiency and a filter-cleaning mechanism. Use a HEPA-filtered vacuum when cleaning holes.		

Equipment/Task	Engineering and Work Practice Control Methods	Required Respiratory Protection and Minimum Assigned Protection Factor (APF)	
		Less Than or Equal to 4 hours/shift	More Than 4 hours/shift
Jackhammers and handheld powered chipping tools	Use tool with water delivery system that supplies a continuous stream or spray of water at the point of impact. OR Use tool equipped with commercially available shroud and dust collection system. Operate and maintain tool in accordance with manufacturer's instructions to minimize dust emissions. Dust collector must provide the air flow recommended by the tool manufacturer, or greater, and have a filter with 99% or greater efficiency and a filter-cleaning mechanism.	None when used outdoors. APF 10 when used indoors or in an enclosed area	APF 10 when used outdoors or indoors/in an enclosed area
Handheld grinders for uses other than mortar removal	For tasks performed outdoors only: Use grinder equipped with integrated water delivery system that continuously feeds water to the grinding surface. Operate and maintain tool in accordance with manufacturer's instructions to minimize dust emissions. OR Use grinder equipped with commercially available shroud and dust collection system. Operate and maintain tool in accordance with manufacturer's instructions to minimize dust emissions. Dust collector must provide 25 cubic feet per minute (cfm) or greater of airflow per inch of wheel diameter and have a filter with 99% or greater efficiency and a cyclonic pre-separator or filter-cleaning mechanism.	None	None when used outdoors. APF 10 when used indoors or in an enclosed area.
Heavy equipment and utility vehicles for tasks such as grading and excavating but not including: demolishing, abrading, or fracturing silica-containing materials.	Apply water and/or dust suppressants as necessary to minimize dust emissions. OR When the equipment operator is the only employee engaged in the task, operate equipment from within an enclosed cab.	None	None

Figure 27 shows a simple temporary guardrail installed on the stairway of a residence under construction. *Figure 28* shows a complex guardrail system used during the construction of a department store. The large central opening will eventually house the building escalators.

> **NOTE**
> Walking/working surfaces do not include ladders, scaffolds, vehicles, or trailers.

According to OSHA, an employee can never be exposed to a fall of more than 6' (1.8 m). This is called *100 percent fall protection*. The first order of protection is always to eliminate the hazard if possible. If it is not possible to eliminate the hazard, the employee must be protected by one of the following in this order of preference:

1. Guardrail system
2. Controlled access zone or other administrative system
3. Fall restraint (e.g., limiting the length of a lanyard so that it does not reach the edge of a building or roof)
4. Personal fall arrest system (PFAS)

4.6.1 Guardrail System

Guardrail systems must be constructed as follows:

- The top rails must be 42" (+/–3") or 1.1 m (+/–75 mm) and must be capable of withstanding 200 pounds (90.7 kg).

Figure 27 Temporary guardrail on stairs.

Figure 28 Complex guardrail system.

- The mid-rails must be at 21" (+/–3") or 530 mm (+/–75 mm) and must be capable of withstanding 150 pounds (68.0 kg).
- The toe board must be 4" tall (100 mm) and no more than ¼" (6 mm) above the floor for drainage.
- Guardrails can be made of 2 × 4s, pipes, chains, or cables.
- Chains and cables require flags every 6' (1.8 m). If cables are used, they must be secured to avoid deflection greater than 3" (75 mm) in, out, or down from the 42" (1.1 m) requirement.
- Banding material is not allowed for guardrail construction. Cable guardrails require the use of cable clamps on the cable. Clamps must be forged and not malleable. They must be torqued and installed properly. Ensure that the clamp does not damage the loadbearing line.

> **NOTE**
> These ratings apply to all portions of the rail system such as anchors, anchorage material, and clamps. Guardrails cannot be used to secure a PFAS unless they are designed to support 5,000 pounds (2,268 kg) per person.

Warning lines, signs, or barricades must be installed back from the edge at least 6' (1.8 m). That way, even if someone falls over the barrier, they will not fall to the level below.

4.6.2 Controlled Access Zone

There are times when a guardrail cannot be attached to the building. In these cases, a controlled access zone may be the solution. A controlled access zone may consist of guards, barricades, badge systems, or other administrative measures.

Figure Credit: Mike Powers

Roofers are allowed to be within 6' (1.8 m) of the edge of a roof. All other workers must remain 15' (4.6 m) from the edge. *Figure 29* shows a controlled access zone on a roof. With a controlled access zone, even a worker who trips over the barricade won't fall off the roof.

4.6.3 Personal Fall Arrest System

A personal fall arrest system (PFAS) provides fall arrest after a worker falls. This equipment must be selected, inspected, donned, anchored, and maintained to be effective. The complete system usually consists of a full-body harness, lanyard, and anchorage device.

Full-body harnesses – Full-body harnesses are the only acceptable equipment to wear for PFAS. Select the appropriate harness based on size and gender. Inspect the equipment before use. Harnesses must be worn snug (but not tight) with all required straps attached. When properly applied, you should be able to slide two fingers under the straps with little difficulty. The D-ring in the back of the harness must be centered between the shoulder blades. After donning the harness, have a co-worker pull sharply up on the D-ring.

SIX FEET FROM EDGE
(FOR ROOFERS ONLY; ALL OTHER WORKERS MUST REMAIN 15 FEET FROM THE EDGE)

Figure 29 Controlled access zone.

You should feel the grab around the thighs, chest, and buttocks. Jobs that require positioning must be accomplished using a full-body harness with side D-rings. Safety belts are not allowed.

Lanyards – Lanyards are used to connect the harness to the attachment point. As no employee can be exposed to a fall of more than 6' (1.8 m), standard lanyards must be no longer than 6' (1.8 m). You can be exposed to 1,800 pounds of force (816.5 kg) in a properly worn harness. The use of shock absorbing lanyards or retractable lanyards can reduce that force to as low as 400 to 600 pounds (181.4 to 272.2 kg). All lanyards must have locking snap hooks.

> **WARNING!**
> Never attach two locking snap hooks to the same D-ring as they can foul each other, causing the relatively weak gates to break.

Figure 30 shows an electrician making an adjustment to an outdoor lighting fixture while hanging out of a 15th-story window. For extra protection, he is attached to two lanyards on two separate D-rings. The lanyard attached to the back of his harness is used to provide primary fall protection, while the positioning lanyard attached to his waist is used to hold him in place as he completes the task.

A twin-tailed lanyard is required when climbing. While climbing, you cannot unhook your lanyard to move it to another anchorage and still have 100 percent fall protection. Thus, with two lanyards, you can "walk" to where you are working.

Consideration should be given to where the lanyard may be exposed (e.g., rebar, sharp metal edges, etc.) so that it will not be cut if a fall occurs and the lanyard is put under stress. Lanyard choices include materials that are resistant to

TIE OFF SEPARATE LANYARDS TO DIFFERENT D-RINGS

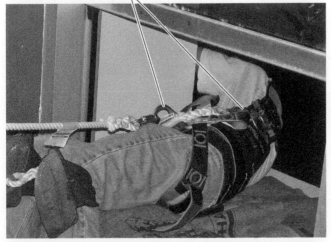

Figure 30 Electrician tied off to two lanyards.

What's wrong with this picture?

Figure Credit: Mike Powers

cutting and, when used on a horizontal surface, have increased shock absorbance if a fall occurs. These are called *leading edge lanyards*. They are thicker and have a shock absorber to prevent the lanyard from breaking. Shock absorbers work by slowing the employee to a stop by ripping stitches while elongating up to 42" (1.1 m).

Retractable lanyards come in a variety of sizes from 6' to over 150' (1.8 m to 45.7 m). *Figure 31* shows a worker tied off to a retractable lanyard when working on a boom lift. Retractable lanyards are used where movement or close proximity with the ground will render standard lanyards ineffective or inefficient. Retractable lanyards can be longer than 6' (1.8 m) because when a fall occurs, they grab hold within 2' (0.6 m). This quick reaction also eliminates the need for a shock absorber.

> **WARNING!**
> Do not use a shock absorber in line with a retractable lanyard unless it is manufactured into the system because it may interfere with the response of the lanyard.

OSHA standards require PFAS to control falls for any working surface 6' (1.8 m) or more above the ground. If you are forced to attach below the D-ring on your harness, you will fall that distance plus the 6' (1.8 m) length of a standard lanyard. This could result in a fall of up to 12' (3.7 m). A PFAS is not designed to meet those forces and will transfer much higher impacts when it arrests the fall.

RETRACTABLE
LANYARD

Figure 31 Proper fall protection on a boom lift.

To combat this, the American National Standards Institute (ANSI) has changed the equipment manufacturing standards for fall protection. The new standard is commonly referred to as the 11' (3.3 m) lanyard. The lanyard is still only 6' (1.8 m) long, but the allowable deployment (extendable length) of the shock absorber has been increased. This lessens the force transferred to the body through the harness, reducing the likelihood of injury.

Before using a PFAS, you must examine the space below any potential fall point. Make sure it is clear of any hazards or obstructions. When assessing the required free length of a PFAS, you must account for the deployment of the shock absorber. In addition, PFAS manufacturers require up to a 3' (0.9 m) safety factor. Taking this safety factor into account, *Table 6* indicates the minimum distance required between any hazard/obstruction and the worker or attachment point.

As you can see, a typical single-story building may not provide enough height to use standard lanyards. Retractable lanyards that engage and stop falls in less than 2' (0.6 m) are usually the best choice for fall protection when used below 20' (6.1 m) or on any type of lift equipment.

ANSI has also addressed the locking mechanism (gate) on snap hooks. OSHA requires the gate to be rated for 350 pounds (158.8 kg) when the rest of the system is rated at 5,000 pounds (2,268 kg). This creates a weak link in the PFAS. ANSI issued a new standard increasing the strength of these gates to 3,600 pounds (1,632.9 kg).

Anchorage devices – Anchorage devices and points are the interface between the PFAS and the structure to which they are attached. This point must hold 5,000 pounds (2,268 kg). This is the equivalent of a full-size extended cab pickup truck.

A lanyard cannot be wrapped around an anchorage and then attached to itself unless it is specially designed with cross arm straps. Cross arm straps are made of webbing, 2" (50 mm) wide, of any necessary length, with two different size D-rings. The cross-arm straps are passed over whatever object you are going to attach them to and wrapped around to reduce the length of the lanyard. Beam clamps, wire hangers, trolleys, and other manufactured devices are also used for specific applications.

What's wrong with this picture?

Figure Credit: Mike Powers

Table 6 Required Free Lengths for Various Lanyards

Lanyard Type	Lanyard Length	Shock Absorber Length	Average Worker Height	Safety Factor	Required Free Length
Standard 6' lanyard with 3.5' shock absorber	6'	3.5'	6'	3'	18.5'
New ANSI 6' lanyard with 4' shock absorber	6'	4'	6'	3'	19'
New ANSI 11' lanyard with 5' shock absorber	6' (11' freefall)	5'	N/A (accounted for by lanyard)	3'	20'

Equipment inspection process – All PFAS must be inspected when received and before each use. Check the manufacturer's tag for the manufacturer's inspection date. If the fall equipment has no date, it should be disposed of immediately. Carefully look over the webbing. If you observe any burns, ripped stitches, color marker threads, distorted grommets, bent or cracked buckle tongues, distorted D-rings, or bent, cracked, or pulled fabrics, remove the PFAS from service. Retractable lanyards must undergo the same inspection process as other PFAS. In addition, you must also pull out the entire lanyard for inspection, let it back in, and then pull out 2' to 4' (0.6 to 1.2 m) of lanyard and give it a swift tug to see if it properly engages. If any of these inspections fail, the equipment must be destroyed immediately or tagged DO NOT USE and returned to the shop.

Rescue – Never use fall protection equipment to pull anyone up; always use ladders or equipment to rescue from below. If the standard equipment is not available to provide rescue, a plan must be created before work can proceed. Rescues must be accomplished from below using ladders, lifts, and/or scaffolds.

WARNING!
Unless a fallen person is in immediate danger, never attempt to lift the worker up by the lanyard. This could cause an additional drop for the fallen worker and/or injure the rescuers.

Immediately summon the fire department to assist in the rescue effort unless you can rescue the person without assistance. Rescue must take place as quickly as possible, as hanging from a harness presents additional hazards. If you fall, continue to move your limbs while awaiting rescue. This will help maintain circulation in your lower extremities.

WARNING!
All fall protection equipment that is involved in a fall must be taken out of service and destroyed. Any employees involved in a fall must receive medical attention, even if they do not feel they have been injured. Falls can cause internal injuries that are not readily apparent to the victim. After rescuing a victim, seek medical attention before removing a harness or laying a worker down to avoid reflow syndrome, a dangerous condition in which the oxygen-starved blood that has pooled in the extremities rapidly reenters the circulatory system, potentially resulting in organ damage or death. In order to slow the absorption of this stale blood, keep the victim's harness on, do not release the leg straps, and have the victim rest in a sitting position until medical help arrives.

PFAS selection – The type of system selected depends on the fall hazards associated with the work to be performed. First, a hazard analysis must be conducted by the job-site supervisor prior to the start of work. Based on the hazard analysis, the job-site supervisor and project manager, in consultation with the safety manager, will select the appropriate fall protection system. All employees must be instructed in the use of the fall protection system before starting work.

WARNING!
Per OSHA, any employee whose weight, including tools, exceeds 310 pounds (140.6 kg) cannot wear fall protection.

Putting It All Together

This module has described a professional approach to electrical safety. How does this professional outlook differ from an everyday attitude? What do you think are the key features of a professional philosophy of safety?

4.0.0 Section Review

1. Maintaining three-point contact on a ladder means that _____.
   a. you must either have both feet and one hand on the ladder or both hands and one foot as you climb
   b. at least three of the four rail points (two side rails and two feet) must be resting on a solid surface
   c. you may reach as far to the left and right as required as long as you have one hand and both feet on the ladder
   d. all splices must be attached at three points

2. Which of the following is true with regard to lifting?
   a. Step carefully over objects rather than going around them.
   b. To avoid foot injuries, stand as far away as possible when lifting a load.
   c. Lift with your back muscles.
   d. Lift by straightening your legs.

3. Which of the following is true with regard to hand tools?
   a. Point the blade of a sharp tool away from yourself to avoid injury if it slips during use.
   b. A dull tool is safer than a sharp one.
   c. The side of a wedge can be used as a hammer if necessary.
   d. Mushroomed heads on impact tools are preferred because they offer a larger surface area for striking.

4. Which of the following is considered a confined space?
   a. Narrow alleyway between buildings
   b. Shallow trench
   c. Grain hopper
   d. Horse pen

5. Fluorescent light fixtures may contain _____.
   a. acid
   b. lead
   c. CFCs
   d. PCBs

6. Fall protection is required when working at elevations of _____.
   a. 6' (1.8 m) or more
   b. 7' (2.1 m) or more
   c. 8' (2.4 m) or more
   d. 10' (3 m) or more

1. The most life-threatening hazards on a construction site typically include all of the following, except _____.

   a. falls
   b. electric shock
   c. being crushed or struck by falling or flying objects
   d. chemical burns

2. If a person's heart begins to fibrillate due to an electrical shock, the solution is to _____.

   a. leave the person alone until the fibrillation stops
   b. immerse the person in ice water
   c. use the Heimlich maneuver
   d. have a qualified person use a defibrillator

3. Low-voltage conductors _____.

   a. are not powerful enough to cause death from electrocution
   b. must exceed 480V to cause death
   c. are responsible for most electrocution deaths due to the frequency of contact
   d. are unlikely to cause injury if rubber-soled shoes are worn

4. A shock of 5mA can cause injury through _____.

   a. respiratory arrest
   b. suffocation
   c. involuntary movement
   d. electrical burns

5. Scheduling off-shift maintenance work is a type of _____.

   a. hazard elimination
   b. engineering control
   c. hazard substitution
   d. administrative control

6. A machine guard is a type of _____.

   a. hazard elimination
   b. engineering control
   c. hazard substitution
   d. administrative control

7. A work permit is a type of _____.

   a. hazard elimination
   b. engineering control
   c. hazard substitution
   d. administrative control

8. The least effective method of preventing risk is through the use of _____.

   a. elimination
   b. substitution
   c. engineering controls
   d. PPE

9. The type of hard hat that can withstand a voltage of 20,000VAC for three minutes is a _____.

   a. Class A
   b. Class C
   c. Class E
   d. Class G

10. Class 0 rubber gloves are used when working with voltages less than _____.

    a. 500 volts
    b. 1,000 volts
    c. 5,000 volts
    d. 7,500 volts

11. An important use of a hot stick is to _____.

    a. replace busbars
    b. test for voltage
    c. replace fuses
    d. test for continuity

12. Which of these statements correctly describes a double-insulated power tool?

    a. There is twice as much insulation on the power cord.
    b. It can safely be used in place of a grounded tool.
    c. It is made entirely of plastic or other non-conductive material.
    d. The entire tool is covered in rubber.

13. Which of the following applies in a lockout/ tagout procedure?

    a. Only the supervisor can install lockout/ tagout devices.
    b. If several employees are involved, the lockout/tagout equipment is applied only by the first employee to arrive at the disconnect.
    c. Lockout/tagout devices applied by one employee can be removed by another employee as long as it can be verified that the first employee has left for the day.
    d. Lockout/tagout devices are installed by every employee involved in the work.

14. The *NEC®* covers the _____.

    a. minimum requirements for the installation of electrical systems
    b. manufacturing specifications for electrical equipment
    c. workplace hazards associated with electrical equipment
    d. operating instructions for electrical equipment

15. What is the proper distance from the feet of a straight ladder to the wall?

    a. one-fourth the working height of the ladder
    b. one-half the height of the ladder
    c. a distance equal to the height of the first three rungs
    d. one-fourth of the square root of the height of the ladder

16. What are the minimum and maximum distances (in inches) that a scaffold plank can extend beyond its end support?

    a. 4; 8
    b. 6; 10
    c. 6; 12
    d. 8; 12

17. What happens to the permits used to enter a confined space?

    a. They are reviewed at the end of each shift and discarded at the completion of the job.
    b. They are recorded in the job journal at the end of each shift and retained for a year.
    c. They are recorded in the job journal at the end of each shift and submitted to OSHA.
    d. They are reviewed at the end of each shift and retained for three months.

18. The best way to protect yourself from solvent hazards is to _____.

    a. always wear vinyl gloves and a paper filter mask
    b. ask a co-worker for instructions
    c. ask the supplier
    d. read and follow all instructions on the product's SDS

19. It is safe to assume that asbestos may be found in any facility constructed before _____.

    a. 1980
    b. 1985
    c. 1990
    d. 1995

20. A PFAS anchorage point for a 6′ (1.8 m) lanyard must be able to hold _____.

    a. 250 pounds (113.4 kg)
    b. 500 pounds (226.8 kg)
    c. 1,000 pounds (453.6 kg)
    d. 5,000 pounds (2,268 kg)

Trade Terms Quiz

Fill in the blank with the correct term that you learned from your study of this module.

1. A life-threatening condition of the heart in which the muscle fibers contract irregularly is called _____.

2. Tools that have a case made of nonconductive material and have been constructed so that the case is insulated from electrical energy are called _____.

3. Chemicals known as _____ may be found in certain types of large transformers and capacitors.

4. A _____ will de-energize a circuit or a portion of it if the current to ground exceeds some predetermined value.

5. Situations that put a worker at risk due to the demands of the task, conditions, worker attitude, and/or environment are known as _____.

6. A _____ has a three-prong plug at the end of its power cord or some other means to ensure that stray current travels to ground without passing through the body of the operator.

7. One who has demonstrated the skills and knowledge related to the construction and operation of the electrical equipment and installations and has received safety training to identify and avoid the hazards involved is known as a _____.

8. The _____ is an approach limit at a distance from an exposed energized electrical conductor or circuit part within which there is an increased likelihood of electric shock.

9. The _____ is an approach limit at a distance from an exposed energized electrical conductor or circuit part within which a shock hazard exists.

10. The _____ is an approach limit at a distance from exposed energized electrical conductors or circuit parts within which a person could receive a second-degree burn if an electrical arc flash were to occur.

11. The amount of thermal energy impressed on a surface at a certain distance from the source of an electrical arc is known as the _____.

12. A first-year electrical apprentice is most likely to be an _____.

13. A garment's _____ is its maximum incident energy resistance demonstrated prior to material breakdown, or at the onset of a second-degree skin burn.

14. An _____ is used to study a worker's potential exposure to arc flash energy.

15. A temporary ground would be removed using a _____.

Trade Terms

Arc flash boundary (AFB)
Arc flash risk assessment
Arc rating
Double-insulated/ ungrounded tools
Error precursors

Fibrillation
Grounded tool
Ground fault circuit interrupter (GFCI)
Hot stick
Incident energy

Limited approach boundary
Polychlorinated biphenyls (PCBs)
Qualified person

Restricted approach boundary
Unqualified person

Supplemental Exercises

1. *CFR Part 1910* covers the

   _____

2. *CFR Part 1926* covers the

   _____

3. True or False? You should use a compressor for an air test on high-voltage gloves because it has more pressure and is much faster.

   _____

4. There are six classes of rubber protective equipment. List the six classes and their voltage ratings.

   _____

5. All conductors, buses, and connections should be considered _____ until proven otherwise.

6. The distance from the ladder feet to the base of the wall or support should be about _____ the vertical distance from the bottom to the top of the ladder.

7. As it is erected, each part of a scaffolding shall be carefully _____

8. List the general guidelines used in identifying a confined space.

   _____

9. Before determining the proper procedure for confined space entry, a(n) _____ must be performed.

10. Fall protection must be used when employees are on a walking or working surface that is _____ or more above a lower level and has an unprotected edge or side.

Michael J. Powers
Tri-City Electrical Contractors, Inc.

How did you choose a career in the electrical field?

My father was an electrician and after I "burned out" with a career in fast-food management, I decided to choose a completely different field.

Tell us about your apprenticeship experience.

It was excellent! I worked under several very knowledgeable electricians and had a pretty good selection of teachers. Over my four-year apprenticeship, I was able to work on a variety of jobs, from Photomats to kennels to colleges.

What positions have you held and how did they help you to get where you are now?

I have been an electrical apprentice, a licensed electrician, a job-site superintendent, a master electrician, and am currently a corporate safety and training director. The knowledge I acquired in electrical theory in apprenticeship school and preparing for my licensing exams, as well as the practical on-the-job experience over thirty years in the trade, were wonderful training for my current position.

I also serve on the authoring team for NCCER's Electrical curricula, which has provided me with not only the opportunity to share what I have learned, but is also a great way to keep current in other areas by meeting with electricians from a variety of disciplines (we have commercial, residential, and industrial electricians on the team, as well as instructors).

What would you say was the single greatest factor that contributed to your success?

Choosing a company that recognized and rewarded competent, hard workers and provided them with the support and guidance to allow them to develop and succeed in the industry.

What does your current job entail?

I am responsible for safe work practices and procedures through the job-site management team at Tri-City Electrical Contractors, Inc. I also assist in developing, delivering, and administering the training program, from apprenticeship to in-house to outsourced training.

What advice do you have for trainees?

Training in all its aspects is the key to your success and advancement in the industry. Any time you are given a training opportunity, take it, even if it might not appear relevant at the time. Eventually, all knowledge can be applied to some situation.

Most importantly, have fun! The construction industry is composed of good people. I firmly believe that construction workers, as a group, are much more honest and direct than any other comparable group. Wait—did I say comparable group? That's a misstatement—there is no comparable group. Construction workers build America!

Trade Terms Introduced in This Module

Arc flash boundary (AFB): An approach limit at a distance from exposed energized electrical conductors or circuit parts within which a person could receive a second-degree burn if an electrical arc flash were to occur.

Arc flash risk assessment: A study investigating a worker's potential exposure to arc flash energy, conducted for the purpose of injury prevention and the determination of safe work practices and appropriate levels of PPE.

Arc rating: The maximum incident energy resistance demonstrated by a material (or a layered system of materials) prior to material breakdown, or at the onset of a second-degree skin burn. Expressed in joules/cm^2 or calories/cm^2.

Double-insulated/ungrounded tools: Electrical tools that are constructed so that the case is insulated from electrical energy. The case is made of a nonconductive material.

Error precursors: Situations that put a worker at risk due to the demands of the task, conditions, worker attitude, and/or environment.

Fibrillation: Very rapid irregular contractions of the muscle fibers of the heart that result in the muscle being unable to contract and pump blood properly.

Grounded tool: An electrical tool with a three-prong plug at the end of its power cord or some other means to ensure that stray current travels to ground without passing through the body of the user. The ground plug is bonded to the conductive frame of the tool.

Ground fault circuit interrupter (GFCI): A protective device that functions to de-energize a circuit or portion thereof within an established period of time when a current to ground exceeds some predetermined value. This value is less than that required to operate the overcurrent protective device of the supply circuit.

Hot stick: An insulated tool designed for the manual operation of disconnecting switches, fuse removal and insertion, and the application and removal of temporary grounds.

Incident energy: The amount of thermal energy impressed on a surface at a certain distance from the source of an electrical arc. Incident energy is typically expressed in calories per square centimeter (cal/cm^2).

Limited approach boundary: An approach limit at a distance from an exposed energized electrical conductor or circuit part within which a shock hazard exists.

Polychlorinated biphenyls (PCBs): Toxic chemicals that may be contained in liquids used to cool certain types of large transformers and capacitors.

Qualified person: One who has demonstrated the skills and knowledge related to the construction and operation of the electrical equipment and installations and has received safety training to identify and avoid the hazards involved.

Restricted approach boundary: An approach limit at a distance from an exposed energized electrical conductor or circuit part within which there is an increased likelihood of electric shock.

Unqualified person: A person who is not a qualified person.

Additional Resources

This module presents thorough resources for task training. The following resource material is suggested for further study.

29 CFR Parts 1900–1910, Standards for General Industry. Occupational Safety and Health Administration, U.S. Department of Labor.

29 CFR Part 1926, Standards for the Construction Industry. Occupational Safety and Health Administration, U.S. Department of Labor.

Managing Electrical Hazards, Latest Edition. Upper Saddle River, NJ: Pearson Education, Inc.

National Electrical Code® Handbook, Latest Edition. Quincy, MA: National Fire Protection Association.

Standard for Electrical Safety in the Workplace (NFPA 70E®), Latest Edition. Quincy, MA: National Fire Protection Association.

Figure Credits

U.S. Department of Labor, Tables 1, 2, and 5

Fluke, Figure 3

Eaton, Figure 5

Mike Powers, Module Opener, Figures 7, 12, 17, 19, 23, 26–28, 30

Brady Corporation, Figure 11B

Reprinted with permission from NFPA 70E®-2018, *Standard for Electrical Safety in the Workplace®*, Copyright © 2017, National Fire Protection Association, Quincy, MA. This reprinted material is not the complete and official position of the NFPA on the referenced subject, which is represented only by the standard in its entirety which may be obtained through the NFPA website at *www.nfpa.org.*, Table 4

Honeywell | Salisbury, Figures 14, 15

National Field Services, Figure 16

©Stanley Black & Decker, Figure 21A

Greenlee / A Textron Company, Figure 21B, 21D

Klein Tools, Inc., Figure 21C

The Master Lock Company, Figure 24

Courtesy of RIDGID®: RIDGID® is the registered trademark of RIDGID, Inc., Figures 22, 25

Section Review Answer Key

Section 1.0.0

Answer	Section Reference	Objective
1. b	1.1.0	1a
2. c	1.2.0	1b
3. d	1.3.0	1c
4. b	1.4.3	1d

Section 2.0.0

Answer	Section Reference	Objective
1. b	2.1.0	2a
2. a	2.2.1	2b

Section 3.0.0

Answer	Section Reference	Objective
1. b	3.1.4	3a
2. c	3.2.0	3b

Section 4.0.0

Answer	Section Reference	Objective
1. a	4.1.1	4a
2. d	4.2.0	4b
3. a	4.3.1	4c
4. c	4.4.1	4d
5. d	4.5.5	4e
6. a	4.6.0	4f

NCCER CURRICULA — USER UPDATE

NCCER makes every effort to keep its textbooks up-to-date and free of technical errors. We appreciate your help in this process. If you find an error, a typographical mistake, or an inaccuracy in NCCER's curricula, please fill out this form (or a photocopy), or complete the online form at **www.nccer.org/olf**. Be sure to include the exact module ID number, page number, a detailed description, and your recommended correction. Your input will be brought to the attention of the Authoring Team. Thank you for your assistance.

Instructors – If you have an idea for improving this textbook, or have found that additional materials were necessary to teach this module effectively, please let us know so that we may present your suggestions to the Authoring Team.

NCCER Product Development and Revision

13614 Progress Blvd., Alachua, FL 32615

Email: curriculum@nccer.org
Online: www.nccer.org/olf

❏ Trainee Guide ❏ Lesson Plans ❏ Exam ❏ PowerPoints Other _____

Craft / Level: _____ Copyright Date: _____

Module ID Number / Title: _____

Section Number(s): _____

Description:
_____
_____
_____
_____
_____

Recommended Correction:
_____
_____
_____
_____

Your Name: _____

Address: _____
_____

Email: _____ Phone: _____

This page is intentionally left blank.

Introduction to Electrical Circuits

OVERVIEW

All kinds of instruments use electrical circuitry to function. This module discusses basic atomic theory and electrical theory, which are the fundamental concepts behind electricity in every setting. It also covers electrical units of measurement and explains how Ohm's law and the power equation can be used to determine unknown values. This module also includes electrical schematic diagrams.

Module 26103-20

Trainees with successful module completions may be eligible for credentialing through the NCCER Registry. To learn more, go to **www.nccer.org** or contact us at 1.888.622.3720. Our website, **www.nccer.org**, has information on the latest product releases and training.

Your feedback is welcome. You may email your comments to **curriculum@nccer.org**, send general comments and inquiries to **info@nccer.org**, or fill in the User Update form at the back of this module.

This information is general in nature and intended for training purposes only. Actual performance of activities described in this manual requires compliance with all applicable operating, service, maintenance, and safety procedures under the direction of qualified personnel. References in this manual to patented or proprietary devices do not constitute a recommendation of their use.

26103-20 V10.0

Objectives

When you have completed this module, you will be able to do the following:

1. Describe atomic structure as it relates to electricity.
   a. Identify the components of an atom.
   b. Compare the atomic structures of conductors and insulators.
   c. Identify the role of magnetism in electrical devices.
   d. Identify the basic components in a power distribution system.
2. Identify electrical units of measurement.
   a. Define current.
   b. Define voltage.
   c. Define resistance.
   d. Use Ohm's law to solve for unknown circuit values.
3. Read schematic diagrams.
   a. Identify the symbol for a resistor and determine its value based on color codes.
   b. Distinguish between series and parallel circuits.
   c. Identify the instruments used to measure circuit values.
   d. Calculate electrical power.

Performance Tasks

This is a knowledge-based module. There are no performance tasks.

Trade Terms

Ammeter	Electrons	Ohmmeter	Solenoids
Amperes (A)	Insulator	Ohm's law	Transformers
Atoms	Joule (J)	Power	Valence shell
Battery	Kilo	Proton	Volts (V)
Charge	Matter	Relays	Voltage
Circuit	Mega	Resistance	Voltage drop
Conductors	Neutrons	Resistors	Voltmeter
Coulomb	Nucleus	Schematic	Watts (W)
Current	Ohms (Ω)	Series circuit	

Industry Recognized Credentials

If you are training through an NCCER-accredited sponsor, you may be eligible for credentials from NCCER's Registry. The ID number for this module is 26103-20. Note that this module may have been used in other NCCER curricula and may apply to other level completions. Contact NCCER's Registry at 888.622.3720 or go to **www.nccer.org** for more information.

> **NOTE**
>
> NFPA 70®, *National Electrical Code®* and *NEC®* are registered trademarks of the National Fire Protection Association, Quincy, MA.

Contents

Figures and Tables

This page is intentionally left blank.

1.0.0 ATOMIC STRUCTURE AND ELECTRICITY

Objective

Describe atomic structure as it relates to electricity.

a. Identify the components of an atom.
b. Compare the atomic structures of conductors and insulators.
c. Identify the role of magnetism in electrical devices.
d. Identify the basic components in a power distribution system.

Trade Terms

Amperes (A): The basic unit of measurement for electrical current, represented by the letter A.

Atoms: The smallest particles to which an element may be divided and still retain the properties of the element.

Battery: A DC voltage source consisting of two or more cells that convert chemical energy into electrical energy.

Charge: A quantity of electricity that is either positive or negative.

Circuit: A complete path for current flow.

Conductors: Materials through which it is relatively easy to maintain an electric current.

Current: The movement, or flow, of electrons in a circuit. Current (I) is measured in amperes.

Electrons: Negatively charged particles that orbit the nucleus of an atom.

Insulator: A material through which it is difficult to conduct an electric current.

Matter: Any substance that has mass and occupies space.

Neutrons: Electrically neutral particles (neither positive nor negative) that have the same mass as a proton and are found in the nucleus of an atom.

Nucleus: The center of an atom. It contains the protons and neutrons of the atom.

Ohms (Ω): The basic unit of measurement for resistance, represented by the symbol Ω.

Power: The rate of doing work, or the rate at which energy is used or dissipated. Electrical power is measured in watts.

Proton: The smallest positively charged particle of an atom. Protons are contained in the nucleus of an atom.

Relays: Electromechanical devices consisting of a coil and one or more sets of contacts. Used as a switching device.

Resistance: An electrical property that opposes the flow of current through a circuit. Resistance (R) is measured in ohms.

Solenoids: Electromagnetic coils used to control a mechanical device such as a valve.

Transformers: Devices consisting of one or more coils of wire wrapped around a common core. Transformers are commonly used to step voltage up or down.

Valence shell: The outermost ring of electrons that orbit about the nucleus of an atom.

Volts (V): The unit of measurement for voltage, represented by the letter V. One volt is equivalent to the force required to produce a current of one ampere through a resistance of one ohm.

Voltage: The driving force that makes current flow in a circuit. Voltage, often represented by the letter E, is also referred to as *electromotive force* (*emf*), *difference of potential*, or *electrical pressure*.

Watts (W): The basic unit of measurement for electrical power, represented by the letter W.

Electricity is a form of energy that can be used by electrical devices such as motors, lights, TVs, heaters, and numerous other devices to perform work. Electricity is also used to control non-electrical devices that perform work. For example, although your car's engine runs on gasoline (rather than electricity), you wouldn't be able to start it or turn it off without the electrical system and battery. In order to work with electricity, you need to know how it is produced and how it acts in electrical circuits.

An electrical circuit contains, at minimum, a voltage source, a load, and conductors (wires) to carry the electrical current (*Figure 1*). The circuit should also have a means to stop and start the current, such as a switch.

Electricity is all about cause and effect. The presence of voltage (measured in volts) in a closed circuit will cause current (measured in amperes, or amps) to flow. Voltage can also be described as electrical pressure; the more pressure you apply, the more current will flow. However, the amount of current flow is also determined by how much resistance (measured in ohms) the load offers to

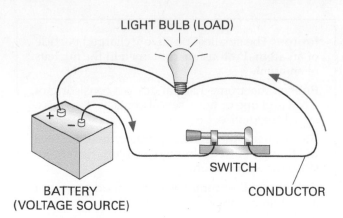

LIGHT BULB (LOAD)

SWITCH

**BATTERY
(VOLTAGE SOURCE)**

CONDUCTOR

Figure 1 Basic electrical circuit.

the flow of current. In order to convert electrical energy into work, the load consumes energy. The amount of energy a device consumes is called **power**, and is expressed in watts. **Volts (V)**, **amperes (A)**, **ohms (Ω)**, and **watts (W)** are related in such a way that if any one of them changes, the others are proportionally affected. This relationship can be seen using basic math principles that you will learn in this module. You will also learn how electricity is produced and how test instruments are used to measure electricity.

In order to understand electrical theory, you must first understand the basic concepts of atomic theory. Atomic theory explains the construction and behavior of **atoms**, including the transfer of **electrons** that results in current flow.

1.1.0 Components of an Atom

The atom is the smallest part of an element that enters into a chemical change, but it does so in the form of a charged particle. These charged particles are called ions and are of two types—positive and negative. A positive ion may be defined as an atom that has become positively charged. A negative ion may be defined as an atom that has become negatively charged. One of the properties of charged ions is that ions of the same charge tend to repel one another, whereas ions of unlike charge will attract one another. The term **charge** can be taken to mean a quantity of electricity that is either positive or negative.

The structure of an atom is best explained by a detailed analysis of the simplest of all atoms: the hydrogen atom. A hydrogen atom, shown in *Figure 2*, is composed of a **nucleus** containing one **proton** and a single orbiting electron. As the electron revolves around the nucleus, it is held in this orbit by two counteracting forces. One of these forces is called centrifugal force, which is the force that tends to cause the electron to fly

outward as it travels around its circular orbit. The second force acting on the electron is electrostatic force, which pulls the electron toward the nucleus. This is caused by the mutual attraction between the positively charged nucleus and the negatively charged electron. At some given radius (distance from the nucleus), the two forces will balance each other, providing a stable path for the electron.

Electrostatic force can be either a repelling force or an attracting force, depending on the polarity (positive or negative charge) of the components involved. The following rules apply:

- A proton (+) repels another proton (+).
- An electron (–) repels another electron (–).
- A proton (+) attracts an electron (–).

An atom is made up of three types of subatomic particles that each have unique electrical characteristics: electrons, protons, and **neutrons**. The protons and neutrons are located in the center, or nucleus, of the atom, and the electrons travel around the nucleus in orbits.

Because protons are relatively heavy, the repelling force they exert on one another in the nucleus of an atom has little effect.

The attracting and repelling forces on charged materials occur because of the electrostatic lines of force that exist around the charged materials. In a negatively charged object, the lines of force of the excess electrons combine to produce an electrostatic field that has lines of force coming into the object from all directions. In a positively charged object, the lines of force of the excess protons combine to produce an electrostatic field that has lines of force going out of the object in all directions. The electrostatic fields either aid or oppose each other to attract or repel.

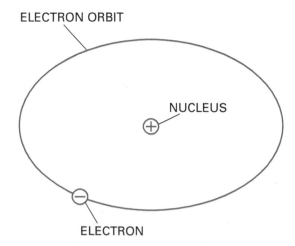

ELECTRON ORBIT

NUCLEUS

ELECTRON

Figure 2 Hydrogen atom.

Why Bother Learning Theory?

Many trainees wonder why they need to bother learning the theory behind how things operate. They figure, why should I learn how it works as long as I know how to install it? The answer is, if you only know how to install something (e.g., run wire, connect switches, etc.), that's all you are ever going to be able to do. For example, if you don't know how your car operates, how can you troubleshoot it? The answer is, you can't. You can only keep changing out the parts until you finally hit on what is causing the problem. (How many times have you seen people do this?) Remember, unless you understand not only how things work but why they work, you'll only be a parts changer. With theory behind you, there is no limit to what you can do.

1.1.1 The Nucleus

The nucleus is the central part of the atom. It is made up of heavy particles called protons and neutrons. The proton is a charged particle containing the smallest known unit of positive electricity. The neutron has no electrical charge. The number of protons in the nucleus determines how the atom of one element differs from the atom of another element.

Although a neutron is actually a unique particle, it is generally thought of as an electron and proton combined, and is electrically neutral. Because they are electrically neutral, neutrons are not considered relevant to the electrical nature of atoms.

1.1.2 Electrical Charges

The negative charge of an electron is equal to the positive charge of a proton. The charges of an electron and a proton are called *electrostatic charges*. The lines of force associated with each particle produce electrostatic fields. Because of the way these fields act together, charged particles can attract or repel one another. The Law of Electrical Charges states that particles with like charges repel each other and those with unlike charges attract each other. The Law of Electrical charges is illustrated in *Figure 3*.

1.2.0 Atomic Structures of Conductors and Insulators

The difference between atoms, with respect to chemical activity and stability, depends on the number and position of the electrons included within the atom. In general, the electrons reside in groups of orbits called *shells*. The shells are arranged in steps that correspond to fixed energy levels.

The outer shell of an atom is called the **valence shell**, and the electrons contained in this shell are called *valence electrons* (*Figure 4*). The number

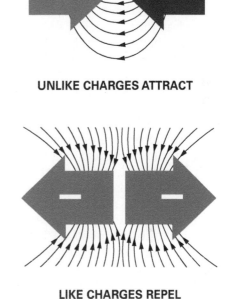

UNLIKE CHARGES ATTRACT

LIKE CHARGES REPEL

Figure 3 Law of Electrical Charges.

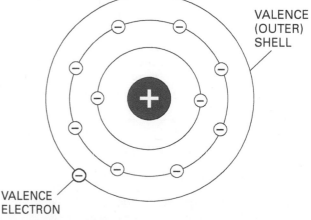

VALENCE (OUTER) SHELL

VALENCE ELECTRON

Figure 4 Valence shell and electrons.

of valence electrons determines an atom's ability to gain or lose an electron, which in turn determines the chemical and electrical properties of the atom. An atom that is lacking only one or two electrons from its outer shell will easily gain electrons to complete its shell, but a large amount of energy is required to free any of its electrons. An atom having a relatively small number of electrons in its outer shell in comparison to the number of electrons required to fill the shell will easily lose these valence electrons.

When it comes to electricity, the valence electrons are of the most concern. That is because they are easiest electrons to break loose from their parent atom. All of the elements that make up **matter** may be placed into one of three categories: conductors, insulators, and semiconductors. Usually, a conductor has three or fewer valence electrons, an **insulator** has five or more, and a semiconductor has four.

Conductors are elements that will readily conduct a flow of electricity. Because of their strong conducting abilities, they are formed into wire and used whenever it is desired to transfer electrical energy from one point to another. Copper and silver are examples of conductors.

In contrast, insulators are elements that do not conduct electricity to any great degree. These are used when it is desirable to prevent the flow of electricity. Porcelain and plastic are examples of good insulators.

Semiconductors are elements that are not good conductors but cannot be used as insulators either because their electrical characteristics fall in between those of conductors and those of insulators. Germanium and silicon are examples of semiconductors. As you will learn later in your training, semiconductors play a crucial role in electronic circuits.

1.3.0 Magnetism in Electrical Devices

The operation of many electrical components relies on the power of magnetism. Motors, **relays**, **transformers**, and **solenoids** are examples. Magnetized iron generates a magnetic field consisting of magnetic lines of force, also known as magnetic

Think About It
Electrical Charges

Think about the things you come in contact with every day. Where do you see or find examples of electrostatic attraction?

flux lines (*Figure 5*). Magnetic objects within the field will be attracted or repelled by the magnetic field. The more powerful the magnet, the more powerful the magnetic field around it. Each magnet has a north pole and a south pole. Opposing poles attract each other, and like poles repel each other.

Electricity also produces magnetism. Current flowing through a conductor produces a small magnetic field around the conductor. If the conductor is coiled around an iron bar, the result is an electromagnet (*Figure 6*) that attracts and repels other magnetic objects just like an iron magnet. This is the basis on which electric motors and other components operate.

1.4.0 Power Distribution

Electricity comes from electrical generating plants operated by utilities like your local power company. Steam from coal-burning or nuclear power plants is used to power huge generators called turbines, which generate electricity. There are also hydroelectric power plants, solar power generating plants, and wind-driven turbines. *Figure 7* illustrates how electricity is safely distributed from generating stations to industrial facilities, businesses, and homes.

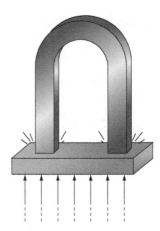

Figure 5 Magnetism.

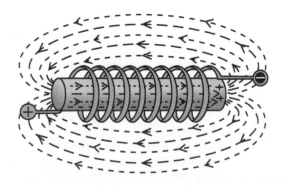

Figure 6 Electromagnet.

The electrical power that travels through long-distance transmission lines may be as high as 750,000V. Devices known as *transformers* are used to step the voltage down to lower levels as it reaches electrical substations and eventually homes, offices, and factories. The voltage you receive at home is usually about 240V. At the wall outlet where you plug in small appliances such as televisions and toasters, the voltage is about 120V (*Figure 8*). Electric stoves, clothes dryers, water heaters, and central air conditioning systems usually require the full 240V. Commercial buildings and factories may receive anywhere from 208V to 575V, depending on the amount of power their machines consume.

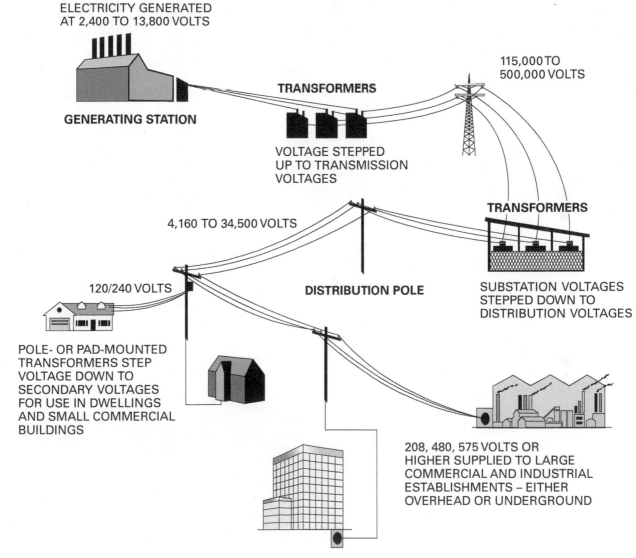

ELECTRICITY GENERATED
AT 2,400 TO 13,800 VOLTS

GENERATING STATION

TRANSFORMERS

VOLTAGE STEPPED
UP TO TRANSMISSION
VOLTAGES

115,000 TO
500,000 VOLTS

TRANSFORMERS

4,160 TO 34,500 VOLTS

DISTRIBUTION POLE

SUBSTATION VOLTAGES
STEPPED DOWN TO
DISTRIBUTION VOLTAGES

120/240 VOLTS

POLE- OR PAD-MOUNTED
TRANSFORMERS STEP
VOLTAGE DOWN TO
SECONDARY VOLTAGES
FOR USE IN DWELLINGS
AND SMALL COMMERCIAL
BUILDINGS

208, 480, 575 VOLTS OR
HIGHER SUPPLIED TO LARGE
COMMERCIAL AND INDUSTRIAL
ESTABLISHMENTS – EITHER
OVERHEAD OR UNDERGROUND

Figure 7 Electrical power distribution.

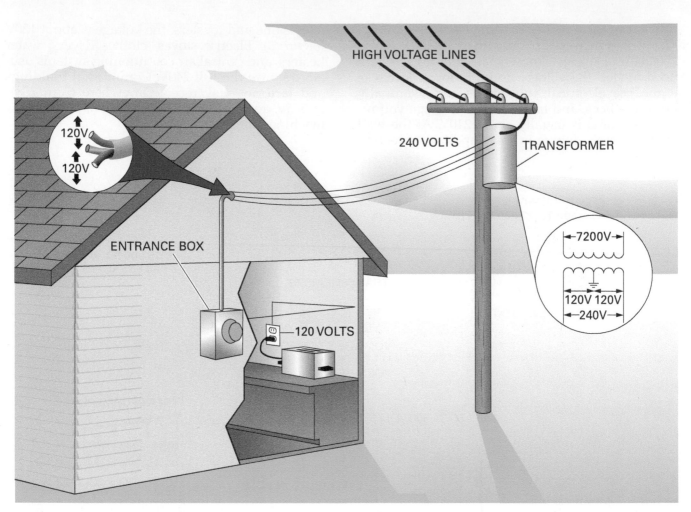

HIGH VOLTAGE LINES

120V
120V

240 VOLTS

TRANSFORMER

ENTRANCE BOX

120 VOLTS

7200V

120V 120V
240V

Figure 8   Power distribution within a home.

Transformers

Large distribution transformers at power substations step down the power to the level required for local distribution. Pole transformers like the one shown here step it down further to the voltages needed for homes and businesses.

Hydroelectric Plants

Hydroelectric plants use the power generated by water to drive turbines that produce electricity.

Figure Credit: US Army Corps of Engineers

1.0.0 Section Review

1. A proton repels a(n) _____.

    a. electron
    b. proton
    c. neutron
    d. negative ion

2. An atom with seven valence electrons is most likely a(n)_____.

    a. insulator
    b. conductor
    c. capacitor
    d. semiconductor

3. Current flowing through a conductor coiled around an iron bar produces a(n)_____.

    a. insulator
    b. capacitor
    c. electromagnet
    d. turbine

4. The voltage used by a typical television or toaster is _____.

    a. 60V
    b. 120V
    c. 180V
    d. 240V

SECTION TWO

2.0.0 ELECTRICAL UNITS OF MEASUREMENT

Objective

Identify electrical units of measurement.
  a. Define current.
  b. Define voltage.
  c. Define resistance.
  d. Use Ohm's law to solve for unknown circuit values.

Trade Terms

Coulomb: A unit of electrical charge equal to 6.25×10^{18} electrons (or 6.25 quintillion electrons). A coulomb is the common unit of quantity used for specifying the size of a given charge.

Joule (J): A unit of measurement for doing work, represented by the letter J. One joule is equal to one newton-meter (Nm).

Ohm's law: A statement of the relationships among current, voltage, and resistance in an electrical circuit: current (I) equals voltage (E) divided by resistance (R). Generally expressed as a mathematical formula: $I = E/R$.

Resistors: Any devices in a circuit that resist the flow of electrons.

Various circuit values, such as voltage, current, and resistance, can be measured to determine circuit characteristics. Even if only two of these values are known, the remaining unknown value can be calculated using Ohm's law.

2.1.0 Defining Current

The movement of the flow of electrons is called *current*. Electrical current is often represented by the letter I. The basic unit in which current is measured is the ampere (A), also called an amp. One ampere of current is defined as the movement of 1 coulomb of charge past any point of a conductor in a second. An electron has 1.6×10^{-19} coulombs of charge. Therefore, it takes 6.25×10^{18} electrons to make up one coulomb of charge, as shown in the following equation:

$$\frac{1}{1.6 \times 10^{-19}} = 6.25 \times 10^{18} \text{ electrons}$$

If two particles, one having charge Q_1 and the other charge Q_2, are a distance (d) apart, then the force between them is given by Coulomb's law, which states that the force is directly proportional to the product of the two charges and inversely proportional to the square of the distance between them:

$$\text{Force} = \frac{k \times Q_1 \times Q_2}{d^2}$$

If Q_1 and Q_2 are both positive or both negative, then the force is positive; it is repulsive (a repelling force). If Q_1 and Q_2 are of opposite charges, then the force is negative; it is attractive (an attracting force). The letter k equals a constant with a value of 10^9.

Electrical current is a rate, which can be defined as an equation:

$$I = \frac{Q}{T}$$

Where:

  I = current (amperes)
  Q = charge (coulombs)
  T = time (seconds)

Charge differs from current in that charge (Q) is an accumulation of charge, whereas current (I) measures the intensity of moving charges.

In a conductor, such as copper wire, the free electrons are charges that can be forced to move with relative ease by a potential difference. If a potential difference is connected across two ends of a copper wire, as shown in *Figure 9*, the applied voltage forces the free electrons to move. This current is a flow of electrons from the point of negative charge (–) at one end of the wire, moving through the wire to the positive charge (+) at the other end. The direction of the electron flow is from the negative side of the battery, through the wire, and back to the positive side of the battery. The direction of current flow is therefore from a point of negative potential to a point of positive potential.

Think About It

Current Flow

Why do you need two wires to use electrical devices? Why can't current simply move to a lamp and be released as light energy?

There are two theories of electron flow. Most electricians use *electron flow theory*, which describes current flow from negative to positive. However, electronics engineers frequently use *conventional electron flow*, which describes current flow in the opposite direction (from positive to negative). To many, this makes it easier to analyze complex electronics. This curriculum uses electron flow theory (from negative to positive). It is important to be aware of this to avoid communication issues and miscalculations.

2.2.0 Defining Voltage

An electric charge has the ability to do the work of moving another charge by attraction or repulsion. The ability of a charge to do work is called its *potential*. When one charge is different from another, there is a difference in potential between them. The sum of the difference of potential of all the charges in the electrostatic field is referred to as the *potential difference*, *electromotive force (emf)*, or *voltage*. Voltage is often represented by the letter E.

One volt is the potential difference between two points for which one coulomb of electricity will do one **joule (J)** of work. A battery is one of several means of creating voltage. It chemically creates a large reserve of free electrons at the negative (–) terminal. The positive (+) terminal has electrons chemically removed and will therefore accept electrons if an external path is provided from the negative (–) terminal. When a battery is no longer able to chemically deposit electrons at the negative (–) terminal, it is said to be dead, or in need of recharging. Batteries are usually rated in volts. Large batteries are also rated in ampere-hours, where one ampere-hour is a current of one amp supplied for one hour.

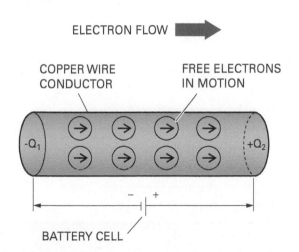

Figure 9 Potential difference causing electric current.

Law of Electrical Force

In the 18th century, a French physicist named Charles de Coulomb was concerned with how electric charges behaved. He watched the repelling forces exerted by opposite electric charges measuring the twist in a wire. An object's weight acted as a turning force to twist the wire, and the amount of twist was proportional to the object's weight. After many experiments with opposing forces, de Coulomb proposed the Inverse Square Law, later known as the Law of Electrical Force.

2.3.0 Defining Resistance

Resistance is directly related to the ability of a material to conduct electricity. All conductors have very low resistance; insulators have very high resistance. Resistance can be defined as the opposition to current flow. To add resistance to a circuit, electrical components called **resistors** are used. A resistor is a device whose resistance to current flow is a known, specified value. Resistance is measured in ohms and is represented by the symbol R in equations. One ohm is defined as the amount of resistance that will limit the current in a conductor to one ampere when the voltage applied to the conductor is one volt. The symbol for an ohm is Ω.

The resistance of a wire is proportional to the length of the wire, inversely proportional to the cross-sectional area of the wire, and dependent upon the kind of material of which the wire is made. The relationship for finding the resistance of a wire is:

$$R = \rho \frac{L}{A}$$

Where:

R = resistance (ohms)
L = length of wire (feet)
A = area of wire (circular mils, CM, or cm^2)
ρ = specific resistance (ohm-CM/ft or microhm-CM)

A mil equals 0.001 inch; a circular mil is the cross-sectional area of a wire one mil in diameter.

The specific resistance is a constant that depends on the material of which the wire is made. (The specific resistance of a material is also called its *resistivity*.) The resistance of copper building wire is 12.9 ohm-CM/ft, whereas that of aluminum is 21.3 ohm-CM/ft. Note that the resistance also varies by ambient temperature.

2.4.0 Calculating Circuit Values

Ohm's law defines the relationship between current, voltage, and resistance. If any two of these quantities are known, the third can be calculated using Ohm's law. (For example, if you know the current flowing to a lamp and how much voltage is applied, you can determine the resistance of the lamp.) There are three ways to express Ohm's law mathematically, each of which solves for a different circuit value:

- *Current* – The current in a circuit is equal to the voltage applied to the circuit divided by the resistance of the circuit:

$$I = \frac{E}{R}$$

- *Resistance* – The resistance of a circuit is equal to the voltage applied to the circuit divided by the current in the circuit:

$$R = \frac{E}{I}$$

- *Voltage* – The applied voltage to a circuit is equal to the product of the current and the resistance of the circuit:

$$E = I \times R \quad or \quad E = IR$$

Where:

I = current (amperes)
R = resistance (ohms)
E = voltage, or emf (volts)

If any two of the quantities E, I, or R are known, the third can be calculated. The Ohm's law equations can be memorized and practiced effectively by using an Ohm's law circle, as shown in *Figure 10*. To find the equation for E, I, or R when two quantities are known, cover the unknown third quantity. The other two quantities in the circle will indicate how the covered quantity may be found.

Example 1:

Find I when E = 120V and R = 30Ω.

$$I = \frac{E}{R}$$

$$I = \frac{120V}{30Ω}$$

$$I = 4A$$

This formula shows that in a DC circuit, current (I) is directly proportional to voltage (E) and inversely proportional to resistance (R).

The Visual Language of Electricity

Learning to read circuit diagrams is like learning to read a book—first you learn to read the letters, then you learn to read the words, and before you know it, you are reading without paying attention to the individual letters anymore. Circuits are the same way—you will struggle at first with the individual pieces, but before you know it you will be reading a circuit without even thinking about it. Studying the table below will help you to understand the fundamental language of electricity.

What's Measured	Unit of Measurement and Symbol		Ohm's Law Symbol
Amount of current	Amp	A	I
Electrical power	Watt	W	P
Force of current	Volt	V	E
Resistance to current	Ohm	Ω	R

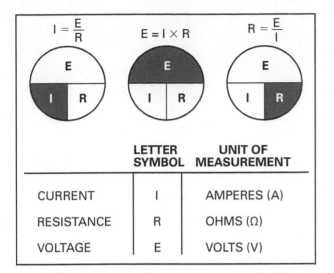

	LETTER SYMBOL	UNIT OF MEASUREMENT
CURRENT	I	AMPERES (A)
RESISTANCE	R	OHMS (Ω)
VOLTAGE	E	VOLTS (V)

Figure 10 Ohm's law circle.

Example 2:

Find R when E = 240V and I = 20A.

$$R = \frac{E}{I}$$

$$R = \frac{240V}{20A}$$

$$R = 12\Omega$$

Example 3:

Find E when I = 15A and R = 8Ω.

$$E = I \times R$$
$$E = 15A \times 8\Omega$$
$$E = 120V$$

Joule's Law

While other scientists of the 19th century were experimenting with batteries, cells, and circuits, James Joule was theorizing about the relationship between heat and energy. He discovered, contrary to popular belief, that work did not just move heat from one place to another; work, in fact, generated heat. Furthermore, he demonstrated that over time a relationship existed between the temperature of water and electric current. These ideas formed the basis for the concept of energy. In his honor, the modern unit of energy was named the joule.

Think About It

Voltage Matters

Standard household voltage is different around the world, from 100V in Japan to 600V in Bombay, India. Many countries have no standard voltage; for example, France varies from 110V to 360V. If you were to plug a 120V hair dryer into England's 240V, you would burn out the dryer. Use basic electric theory to explain exactly what would happen to destroy the hair dryer.

Units of Electricity and Volta

A disagreement with a fellow scientist over the twitching of a frog's leg eventually led 18th-century physicist Alessandro Volta to theorize that when certain objects and chemicals come into contact with each other, they produce an electric current. Believing that electricity came from contact between metals only, Volta coined the term *metallic electricity*. To demonstrate his theory, Volta placed two discs, one of silver and the other of zinc, into a weak acidic solution. When he linked the discs together with wire, electricity flowed through the wire. Thus, Volta introduced the world to the battery, also known as the Voltaic pile. Volta needed a term to measure the strength of the electric push or the flowing charge, which is now called a *volt*.

2.0.0 Section Review

1. Coulomb's law can be used to calculate the _____.

    a. direction of current flow
    b. force between two charges
    c. speed of electron movement
    d. loss of current

2. The negative terminal of a battery contains a large reserve of free _____.

    a. electrons
    b. protons
    c. neutrons
    d. acid

3. Which of the following does *not* have an effect on wire resistance?

    a. Temperature
    b. Type of power
    c. Wire length
    d. Material

4. Find the applied voltage when the current is 30A and the resistance is 4Ω.

    a. 30V
    b. 60V
    c. 120V
    d. 180V

3.0.0 READING SCHEMATIC DIAGRAMS

Objective

Read schematic diagrams.
  a. Identify the symbol for a resistor and determine its value based on color codes.
  b. Distinguish between series and parallel circuits.
  c. Identify the instruments used to measure circuit values.
  d. Calculate electrical power.

Trade Terms

Ammeter: An instrument for measuring electrical current.

Kilo: A prefix used to indicate one thousand (for example, one kilowatt is equal to one thousand watts).

Mega: A prefix used to indicate one million; for example, one megawatt is equal to one million watts.

Ohmmeter: An instrument used for measuring resistance.

Schematic: A type of drawing in which symbols are used to represent the components in a system.

Series circuit: A circuit with only one path for current flow.

Voltage drop: The change in voltage across a component that is caused by the current flowing through it and the amount of resistance opposing it.

Voltmeter: An instrument for measuring voltage. The resistance of the voltmeter is fixed. When the voltmeter is connected to a circuit, the current passing through the meter will be directly proportional to the voltage at the connection points.

The simple electric circuit shown in *Figure 1* appears again in *Figure 11* in both pictorial and **schematic** forms. The schematic diagram is a shorthand way to draw an electric circuit, and circuits are usually represented in this way. In addition to the connecting wire, three components are shown symbolically: the battery, the switch, and the lamp. Note the positive (+) and negative (–) markings in both the pictorial

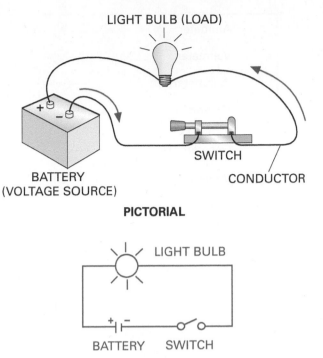

Figure 11 Electrical circuit.

and schematic representations of the battery. The schematic components represent the pictorial components in a simplified manner. A schematic diagram is one that shows, by means of graphic symbols, the electrical connections and functions of the different parts of a circuit.

The standard graphic symbols for commonly used electrical and electronic components are shown in *Figure 12*.

3.1.0 Resistors

The function of a resistor is to offer a particular resistance to current flow. For a given current and known resistance, the change in voltage across the component, or **voltage drop**, can be predicted using Ohm's law. Voltage drop refers to a specific amount of voltage used, or developed, by that component. An example is a very basic circuit of a 10V battery and a single resistor in a **series circuit**. The voltage drop across that resistor is 10V because it is the only component in the circuit and all voltage must be dropped across that resistor. Similarly, for a given applied voltage, the current that flows may be predetermined by selection of the resistor value. The required power dissipation largely dictates the construction and physical size of a resistor.

The two most common types of electronic resistors are wire-wound and carbon composition construction. A typical wire-wound resistor consists of a length of nickel wire wound on a

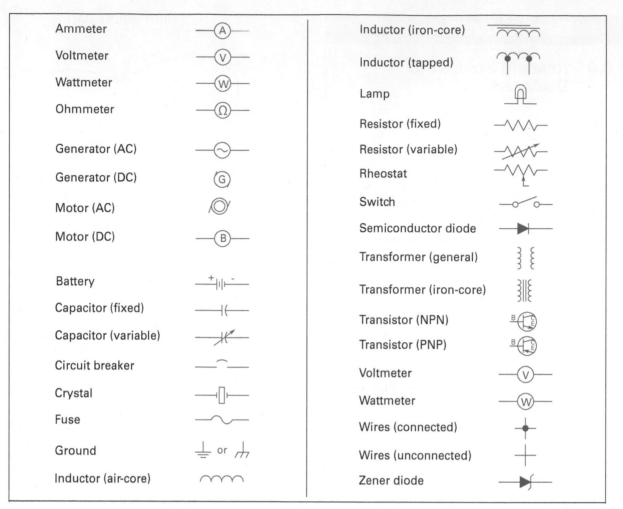

Ammeter	—Ⓐ—	Inductor (iron-core)	
Voltmeter	—Ⓥ—	Inductor (tapped)	
Wattmeter	—Ⓦ—	Lamp	
Ohmmeter	—Ω—	Resistor (fixed)	—／\／\—
Generator (AC)	—∿—	Resistor (variable)	
Generator (DC)	Ⓖ	Rheostat	
Motor (AC)	◎	Switch	—o⁄ o—
Motor (DC)	—Ⓑ—	Semiconductor diode	—▶⊢
Battery	—⁺‖⊢⁻	Transformer (general)	
Capacitor (fixed)	—⊣⊢	Transformer (iron-core)	
Capacitor (variable)		Transistor (NPN)	
Circuit breaker		Transistor (PNP)	
Crystal		Voltmeter	—Ⓥ—
Fuse		Wattmeter	—Ⓦ—
Ground	⏚ or	Wires (connected)	
Inductor (air-core)		Wires (unconnected)	
		Zener diode	—▶⊢

Figure 12 Standard schematic symbols.

ceramic tube and covered with porcelain. Low-resistance connecting wires are provided, and the resistance value is usually printed on the side of the component. *Figure 13* illustrates the construction of typical resistors. Carbon composition resistors are constructed by molding mixtures of powdered carbon and insulating materials into a cylindrical shape. An outer sheath of insulating material affords mechanical and electrical protection, and copper connecting wires are provided at each end. Carbon composition resistors are smaller and less expensive than the wire-wound type. However, the wire-wound type is the more rugged of the two and is able to survive much larger power dissipations than the carbon composition type.

Most resistors have standard fixed values, so they can be termed fixed resistors. Variable

Using Your Intuition

Learning the meanings of various electrical symbols may seem overwhelming, but if you take a moment to study *Figure 12*, you will see that most of them are intuitive—that is, they are shaped (in a symbolic way) to represent the actual object. For example, the battery shows + and –, just like an actual battery. The motor has two arms that suggest a spinning rotor. The transformer shows two coils. The resistor has a jagged edge to suggest pulling or resistance. Connected wires have a black dot that reminds you of solder. Unconnected wires simply cross. The fuse stretches out in both directions as though to provide extra slack in the line. The circuit breaker shows a line with a break in it. The capacitor shows a gap. The variable resistor has an arrow like a swinging compass needle. As you learn to read schematics, take the time to make mental connections between the symbol and the object it represents.

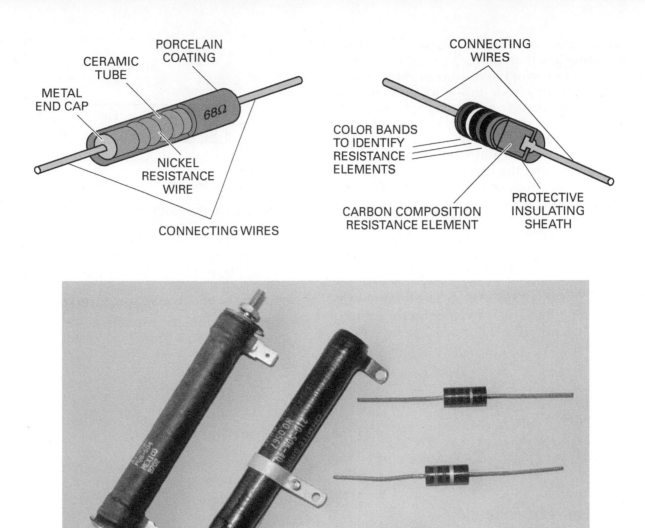

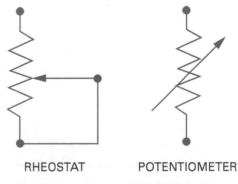

Figure 13 Common resistors.

resistors, also known as adjustable resistors, are used a great deal in electronics. Two common symbols for a variable resistor are shown in *Figure 14*.

A variable resistor consists of a coil of closely wound insulated resistance wire formed into a partial circle. The coil has a low-resistance terminal at each end, and a third terminal is connected to a movable contact with a shaft adjustment facility. The movable contact can be set to any point on a connecting track that extends over one (uninsulated) edge of the coil. Using the adjustable

contact, the resistance from either end terminal to the center terminal can be adjusted from zero to the maximum coil resistance. Another type of variable resistor is known as a decade resistance box—a laboratory component that contains precise values of switched series-connected resistors.

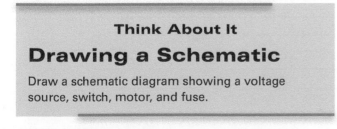

Think About It
Drawing a Schematic
Draw a schematic diagram showing a voltage source, switch, motor, and fuse.

RHEOSTAT   POTENTIOMETER

Figure 14 Symbols used for variable resistors.

Because carbon composition resistors are small in size—some less than 0.4 inch (1 cm) long—it is not convenient to print the resistance value on the side. Instead, a color code in the form of colored bands is used to identify the resistance value and tolerance. The color code is illustrated in *Figure 15*. Starting from one end of the resistor, the first two bands identify the first and second digits of the resistance value, and the third band indicates the number of zeros. An exception to this is when the third band is either silver or gold, which indicates a 0.01 or 0.1 multiplier, respectively. The fourth band is always either silver or gold, and in this position, silver indicates a ± 10% tolerance and gold indicates a ± 5% tolerance. Where no fourth band is present, the resistor tolerance is ± 20%.

You can put this information to practical use by determining the range of values for the carbon resistor in *Figure 16*. The color code for this resistor is as follows:

- Brown = 1, black = 0, red = 2, gold = a tolerance of ± 5%
- First digit = 1, second digit = 0, number of zeros (2) = 1,000Ω

Since this resistor has a value of 1,000Ω ± 5%, the resistor can range in value from 950Ω to 1,050Ω.

3.2.0 Series versus Parallel Circuits

You will often hear the terms *series circuit* and *parallel circuit* during your training. These terms refer to the way loads are connected in the circuit.

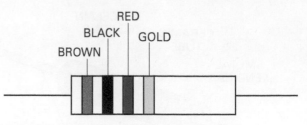

Figure 16 Sample color codes on a fixed resistor.

3.2.1 Series Circuits

A series circuit provides only one path for current flow and is a voltage divider. The total resistance (R_T) of a series circuit is equal to the sum of the individual resistances in the circuit. The 12V series circuit in *Figure 17* (*A*) has two 30Ω loads. Therefore, the total resistance is 60Ω. Using the total resistance and applied voltage with Ohm's law, you can determine that the amount of current flowing in the circuit is 0.2A. This is calculated as follows:

$$I = \frac{E}{R}$$

$$I = \frac{12V}{60Ω}$$

$$I = 0.2A$$

If there were five 30Ω loads, the total resistance would be 150Ω. The current flow is the same through all the loads. The voltage measured across any one of the loads (also called the *voltage drop* across a load) depends on the resistance of that load. The sum of all the voltage drops in a circuit is equal to the total voltage applied to the circuit.

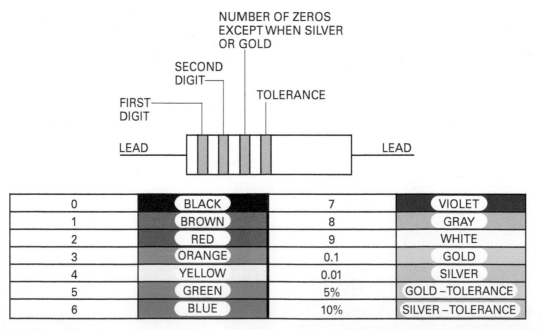

0	BLACK	7	VIOLET
1	BROWN	8	GRAY
2	RED	9	WHITE
3	ORANGE	0.1	GOLD
4	YELLOW	0.01	SILVER
5	GREEN	5%	GOLD – TOLERANCE
6	BLUE	10%	SILVER – TOLERANCE

Figure 15 Resistor color codes.

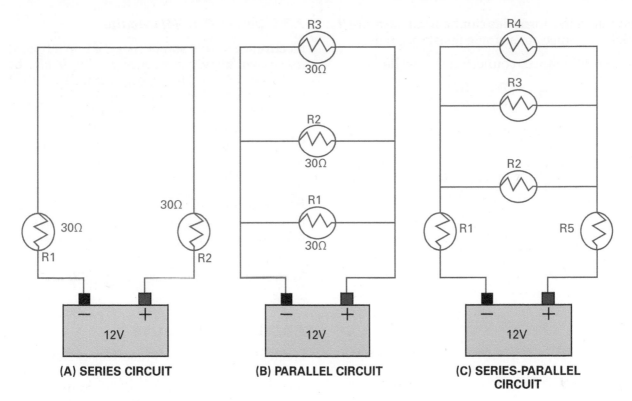

Figure 17 Types of circuits.

Circuits containing loads in series are uncommon. An important trait of a series circuit is that if the circuit is open at any point, no current will flow. For example, if you have five light bulbs connected in series and one of them blows, all five lights will go off.

3.2.2 Parallel Circuits

In a parallel circuit, each load is connected directly to the voltage source; therefore, the voltage drop through each of the loads is the same, and the current is divided between the loads. (The amount of current across each load depends on its resistance.) The source sees the circuit as two or more individual circuits containing one load each. In the parallel circuit in *Figure 17* (B), the source sees three circuits, each containing a 30Ω load. The current flow through any load is determined by the resistance of that load. Thus, the total current drawn by the circuit is the sum of the individual currents. The total resistance of a parallel circuit is calculated differently from that of a series circuit. In a parallel circuit, the total resistance is less than the smallest of the individual resistances.

For example, each of the 30Ω loads in *Figure 17* (B) draws 0.4A at 12V; therefore, the total current is 1.2A. This is calculated as follows:

$$I = \frac{E}{R} = \frac{12V}{30\Omega} = 0.4A \text{ per circuit}$$

$$0.4A \text{ per circuit} \times 3 \text{ circuits} = 1.2A$$

Now, Ohm's law can be used again to calculate the total resistance:

$$R = \frac{E}{I} = \frac{12V}{1.2A} = 10\Omega$$

This example was simple because all the resistances were the same value. The process is the same when the resistances are different, but the current calculation has to be done for each load. The individual currents are added to get the total current.

Unlike series circuits, parallel circuits continue working even if one circuit opens. Household circuits are wired in parallel. In fact, almost all the load circuits you encounter will be parallel circuits.

The following formulas can be used to convert parallel resistances to a single resistance value:

- For two resistances connected in parallel:

$$R_T = \frac{R1 \times R2}{R1 + R2}$$

- For three or more resistances connected in parallel:

$$R_T = \frac{1}{\dfrac{1}{R1} + \dfrac{1}{R2} + \dfrac{1}{R3}}$$

Example:

1. The total resistance of the parallel circuit below is 6Ω.

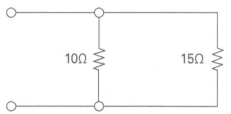

Total resistance (R_T) =

$$\frac{R1 \times R2}{R1 + R2} = \frac{10 \times 15}{10 + 15} = \frac{150}{25} = 6\Omega$$

2. The total resistance of the parallel circuit below is 4.76Ω.

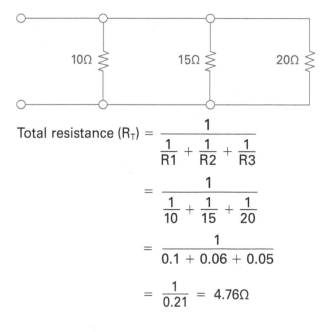

Total resistance (R_T) = $\dfrac{1}{\dfrac{1}{R1} + \dfrac{1}{R2} + \dfrac{1}{R3}}$

$$= \frac{1}{\dfrac{1}{10} + \dfrac{1}{15} + \dfrac{1}{20}}$$

$$= \frac{1}{0.1 + 0.06 + 0.05}$$

$$= \frac{1}{0.21} = 4.76\Omega$$

3.2.3 *Series-Parallel Circuits*

Electronic circuits sometimes contain a hybrid arrangement known as a *series-parallel circuit*, shown in *Figure 17* (C). However, you will rarely find loads connected in this arrangement.

To determine the total resistance of a series-parallel circuit, the parallel loads must be converted to their equivalent series resistance. The load resistances are then added to determine total circuit resistance.

> **NOTE**
>
> Because loads are rarely connected in a series-parallel arrangement, you will not have to determine characteristics of series-parallel circuits very often. However, these calculations are presented in detail in Module 26104-20 of this curriculum, "Electrical Theory."

3.3.0 Electrical Meters

Electricians frequently use test meters to measure voltage, current, and resistance. The most common test meter is the volt-ohm-milliammeter (VOM), also called a *multimeter*. (Multimeters usually measure voltage, resistance, and current.) *Figure 18* shows both digital and analog multimeters.

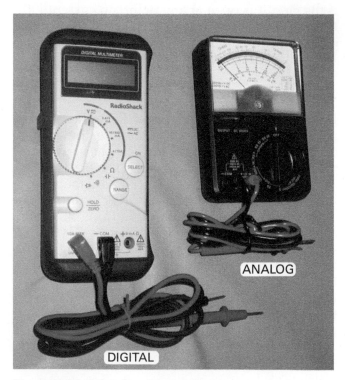

Figure 18 Digital and analog meters.

An analog meter is described as *analog* because the pointer moves in proportion to the value being measured. The person using the meter must then interpret the scale to determine the measured value, although digital meters display the result numerically on the screen.

Multimeters are commonly used to measure AC and DC voltage, DC current, and resistance. They can also be used to measure AC current in the milliamp range. For larger current values, it is usually necessary to use a clamp-on ammeter (*Figure 19*).

> **WARNING!**
>
> Only qualified individuals may use these meters. Consult your company's safety policy for applicable rules.

3.3.1 Measuring Current

A clamp-on ammeter is used to measure current. The jaws of the ammeter are placed around a single conductor (*Figure 20*), and current flowing through the wire creates a magnetic field, inducing a proportional current in the ammeter jaws. This current is read by the meter movement and appears as a direct readout or, on an analog meter, as a deflection of the meter needle.

In-line ammeters (*Figure 21*) are less common and must be connected in series with the circuit, which means that the circuit must be opened.

Aside from following good safety practices, there are a few things to remember when measuring current:

- If the ammeter jaws are dirty or misaligned, the meter will not read correctly.
- Meters are precision instruments and should be handled carefully to avoid damage.
- An analog meter can be damaged when the power of the current far exceeds the selected scale. When using an analog meter, always start at the highest range on the meter and work down.
- Do not clamp the meter jaws around two different conductors at the same time, or an inaccurate reading will result.

3.3.2 Measuring Voltage

A **voltmeter** must be connected in parallel with (across) the component or circuit to be tested (*Figure 22*). If a circuit function is not operating, the voltmeter can be used to determine if the correct voltage is available to the circuit. Voltage must be checked with power applied.

Is It a Series Circuit?

When the term *series circuit* is used, it refers to the way the loads are connected. The same is true for parallel and series-parallel circuits. You will rarely, if ever, find loads connected in series, or in a series-parallel arrangement. The simple circuit shown here illustrates this point. At first glance, you might think it is a series-parallel circuit. On closer examination, you can see that there are only two loads—the relay and the contactor—and they are connected in parallel. Therefore, it is a parallel circuit. The control devices are wired in series with the loads, but only the loads are considered in determining the type of circuit.

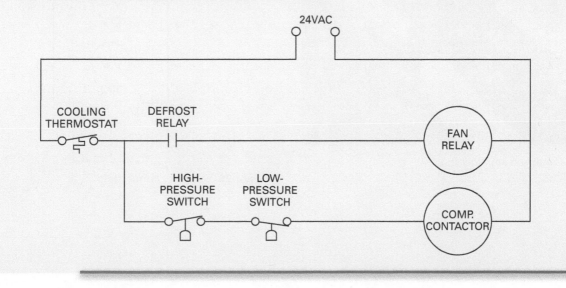

Figure 19 Clamp-on ammeter.

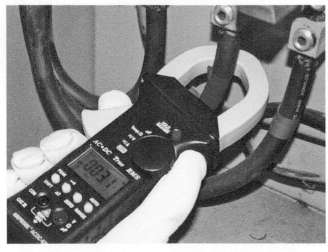

Figure 20 Clamp-on ammeter in use.

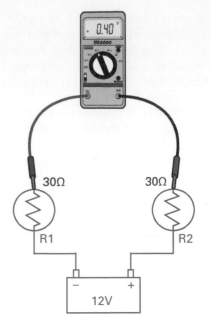

30Ω R1   30Ω R2

12V

SERIES CIRCUIT

Figure 21 In-line ammeter test setup.

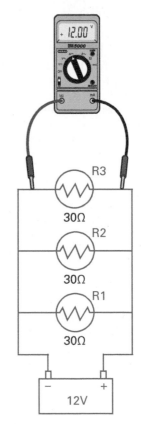

R3
30Ω

R2
30Ω

R1
30Ω

12V

Figure 22 Voltmeter connection.

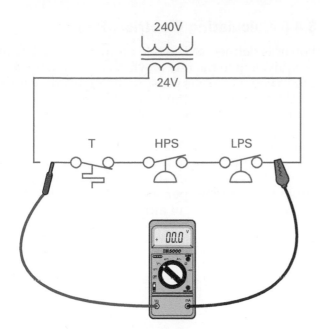

Figure 23 Ohmmeter connection for continuity testing.

3.3.3 Measuring Resistance

An ohmmeter contains an internal battery that acts as a voltage source. Therefore, resistance measurements are always made with the system power shut off. Sometimes, an ohmmeter is used to measure resistance in a load; motor windings are a good example. More often, an ohmmeter is used to check continuity in a circuit. A wire or closed switch offers negligible resistance. With the ohmmeter connected as in *Figure 23* and the three switches closed, the current produced by the ohmmeter battery will flow unopposed and the meter will show zero resistance. The circuit has continuity; that is, it is continuous. If a switch is open, however, there is no path for current and the meter will see infinite resistance (lack of continuity).

A continuity tester (*Figure 24*) is a simple device consisting primarily of a battery and either an audible or visual indicator. It can be used in place of an ohmmeter to test the continuity of a wire and to identify individual wires contained in a conduit or other raceway. To test the continuity of a wire, strip the insulation off the end of the wire to be tested at one end of the conduit run, then connect (short) the wire to the metal conduit. At the other end of the conduit run, clip the alligator clip lead of the tester to the conduit and touch the probe to the end of the wire under test. If the tester audible alarm sounds or the indicator light comes on, there is continuity. Note that this only indicates there is continuity between the two points being tested; it does not indicate the actual value of the resistance. If there is no indication, the wire is open.

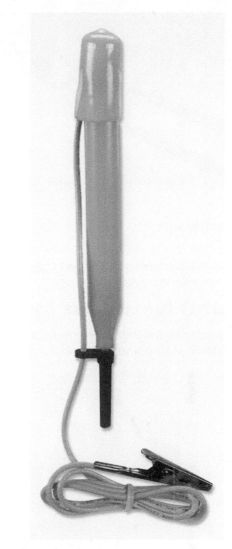

Figure 24 Continuity tester.

To identify individual wires in a conduit run, touch the tester probe to the wires in the conduit one at a time until the tester audible alarm sounds or the indicator lights. Then, put matching identification tags on both ends of the wire. Continue this procedure until all the wires have been identified.

3.3.4 Voltage Testers

Figure 25 shows one of the wide varieties of devices available for checking for the presence of voltage. It can be used as a troubleshooting tool and as a safety device to make sure the voltage is turned off before touching any terminals or conductors. When the probes are touched to the circuit, the light on the instrument will turn on if a voltage is present. Instruments like these are available in several voltage ranges, so it is important to know something about the circuit you are checking.

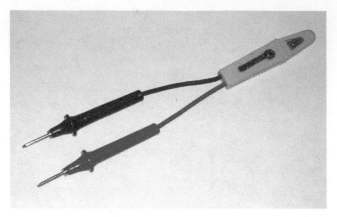
Figure 25 Voltage tester.

3.4.0 Calculating Electrical Power

Power is defined as the rate of doing work, which is equivalent to the rate at which energy is used or dissipated. Electrons passing through a resistance dissipate energy in the form of heat. In electrical circuits, power is measured in units called watts (W). The power in watts equals the rate of energy conversion. One watt of power is equal to the energy used by 1 volt to move electrical charge at a rate of 1 coulomb per second. Since 1 ampere is equal to 1 coulomb per second, power in watts is equal to the product of amperes and volts.

The work done in an electrical circuit can be useful work or wasted work. In both cases, the

Old and New Test Instruments

Early electricians used individual meters to test circuit parameters. Today, those instruments seem primitive given the availability of all-purpose instruments like the clamp-on ammeter with remote read capability shown here.

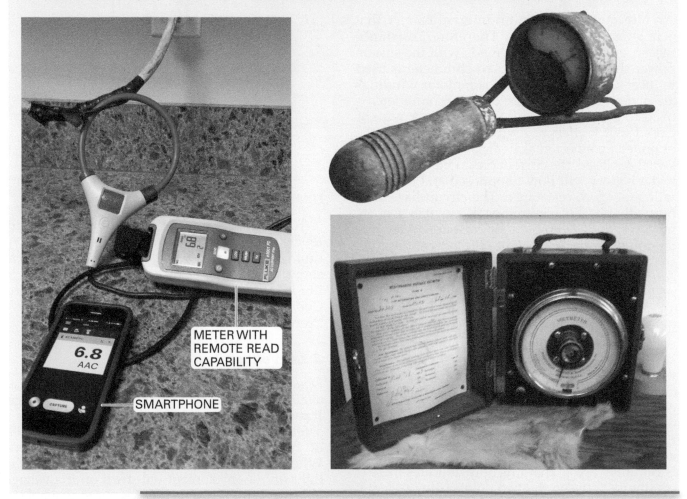

METER WITH REMOTE READ CAPABILITY

SMARTPHONE

rate at which the work is done is still measured in power. The turning of an electric motor is useful work. The heating of wires or resistors in a circuit is wasted work, since no useful function is performed by the heat.

The unit of electrical work is the joule. This is the amount of work done by one coulomb flowing through a potential difference of one volt. If five coulombs flow through a potential difference of one volt, five joules of work are done. The time it takes these coulombs to flow through the potential difference has no bearing on the amount of work done.

Amperes are a measurement of current, which is a rate of current flow. This rate measures how much electrical charge passes a point in a certain amount of time. As previously discussed, one ampere is equal to a rate of 1 coulomb per second. A joule is a measurement of work (or dissipated energy); one joule of work is done when 1 ampere moves through 1 volt in a second. This rate of one joule per second is the basic unit of power, and is called a *watt*. Therefore, a watt is the power used when one ampere of current flows through a potential difference of 1 volt, as shown in *Figure 26*.

Mechanical power is usually measured in units of horsepower (hp). To convert from horsepower to watts, multiply the number of horsepower by 746. To convert from watts to horsepower, divide the number of watts by 746. *Table 1* lists conversions for common units of power.

The kilowatt-hour (kWh) is commonly used for large amounts of electrical work or energy. (The prefix **kilo** means one thousand, so 1 kilowatt is equal to 1,000 watts.) The amount is calculated as the product of the power in kilowatts multiplied by the time in hours during which the power

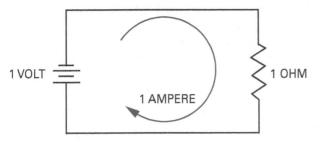

Figure 26 One watt.

Table 1 Conversion Table

1,000 watts (W) =	1 kilowatt (kW)
1,000,000 watts (W) =	1 megawatt (MW)
1,000 kilowatts (kW) =	1 megawatt (MW)
1 watt (W) =	0.00134 horsepower (hp)
1 horsepower (hp) =	746 watts (W)

is used. If a light bulb uses 300W or 0.3kW for 4 hours, the amount of energy is 0.3 × 4, which equals 1.2kWh.

Very large amounts of electrical work or energy are measured in megawatts (MW). (The prefix **mega** means one million, so 1 megawatt is equal to 1 million watts.)

3.4.1 Power Equation

When one ampere flows through a difference of two volts, two watts of power must be used. In other words, the number of watts used is equal to the number of amperes of current times the potential difference. This is called *the power equation*, and it is expressed as follows:

$$P = I \times E \quad or \quad P = IE$$

Where:

P = power used in watts
I = current in amperes
E = potential difference in volts

The equation is sometimes called Ohm's law for power, because it is similar to Ohm's law. This equation is used to find the power consumed in a circuit or load when the values of current and voltage are known.

As an example, suppose the total load current in the main line equals 20A. Then the power in watts from the 120V line is:

$$P = I \times E$$
$$P = 20A \times 120V$$
$$P = 2400W \quad or \quad 2.4kW$$

If this power is used for five hours, then the energy of work supplied in kilowatt-hours (kWh) equals:

$$2.4kW \times 5 \text{ hours} = 12kWh$$

The second form of the power equation is used to find the voltage when the power and current are known:

$$E = \frac{P}{I}$$

Power

We take electrical power for granted, never stopping to think how surprising it is that a flow of submicroscopic electrons can pump thousands of gallons of water or illuminate a skyscraper. Our lives now constantly rely on the ability of the electron to do work. Think about your day up to this moment. How has electrical power shaped your experience?

The third form of the power equation is used to find the current when the power and voltage are known:

$$I = \frac{P}{E}$$

Using these three equations, the power, voltage, or current in a circuit can be calculated whenever any two of the values are already known.

Example 1:

Calculate the power in a circuit where the source of 100V produces 2A in a 50Ω resistance.

P = IE

P = 2A × 100V

P = 200W

This means the source generates 200W of power while the resistance dissipates 200W in the form of heat.

Example 2:

Calculate the source voltage in a circuit that consumes 1,200W at a current of 5A.

$$E = \frac{P}{I}$$

$$E = \frac{1200W}{5A}$$

E = 240V

Example 3:

Calculate the current in a circuit that consumes 600W with a source voltage of 120V.

$$I = \frac{P}{E}$$

$$I = \frac{600W}{120V}$$

I = 5A

Components that use the power dissipated in their resistance are generally rated in terms of power. The power is rated at normal operating voltage, which is usually 120V. For instance, an appliance that draws 5A at 120V would dissipate 600W. The rating for the appliance would then be 600W/120V.

To calculate I or R for components rated in terms of power at a specified voltage, it may be convenient to use the power formula in different forms. There are three basic power formulas, but each can be rearranged into two other forms for a total of nine combinations, as shown in *Figure 27*. Note that all of these formulas are based on Ohm's law (E = IR) and the power formula (P = IE).

3.4.2 Power Rating of Resistors

If too much current flows through a resistor, the heat caused by the current will damage or destroy the resistor. This heat is caused by I^2R heating, which is power loss expressed in watts. This is expressed as the following equation:

$$P = I^2R$$

Every resistor is given a wattage, or power rating, to show how much I^2R heating it can take before it burns out. This means that a resistor with a power rating of 1W will burn out if it is used in a circuit where the current causes it to dissipate heat at a rate greater than 1W.

If the power rating of a resistor is known, the maximum current it can carry is found by using an equation derived from $P = I^2R$:

$$P = I^2R$$

Divide both sides of the equation by R:

$$\frac{P}{R} = \frac{I^2R}{R}$$

$$P/R = I^2$$

Take the square root of each side:

$$\sqrt{P/R} = \sqrt{I^2}$$

Transpose the equation; this is the equation used to find the current.

$$I = \sqrt{P/R}$$

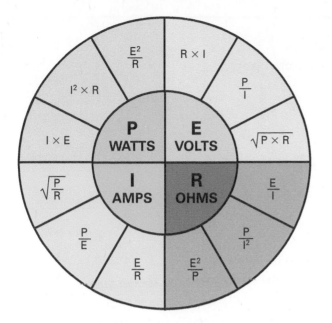

Figure 27 Expanded Ohm's law circle.

Using this equation, find the maximum current that can be carried by a 1Ω resistor with a power rating of 4W:

$$I = \sqrt{P/R} = \sqrt{4/1} = \sqrt{4} = 2A$$

If such a resistor conducts more than 2A, it will dissipate more than its rated power, causing it to burn out.

Power ratings assigned by resistor manufacturers are usually based on the resistors being mounted in an open location where there is free air circulation, and where the temperature is not higher than 104°F (40°C). If a resistor is mounted in a small, crowded, enclosed space, or where the temperature is higher than 104°F (40°C), there is a good chance it will burn out even before its power rating is exceeded. Also, some resistors are designed to be attached to a chassis or frame that will carry away the heat.

Think About It

Putting It All Together

Notice the common electrical devices in the building you're in. What is their wattage rating? How much current do they draw? How would you test their voltage or amperage?

3.0.0 Section Review

1. A resistor with a gold band in the fourth position has a tolerance of _____.

    a. ± 1%
    b. ± 5%
    c. ± 10%
    d. ± 20%

2. The total resistance in a 120V series circuit with three 10Ω resistors is _____.

    a. 1.2Ω
    b. 12Ω
    c. 30Ω
    d. 120Ω

3. Which of the following is true regarding the use of a clamp-on ammeter?

    a. Place the jaws around a single conductor at a time.
    b. Place the jaws around an even number of conductors only.
    c. Start at the lowest meter range and work up.
    d. Misaligned meter jaws can be corrected by tapping them with a hammer.

4. Calculate the power in a circuit where the source of 240V produces 10A.

    a. 12W
    b. 240W
    c. 1,200W
    d. 2,400W

1. An electrical circuit contains, at minimum, a(n) _____.
   a. voltage source, load, and overcurrent device
   b. ammeter, load, and voltage source
   c. voltage source, load, and conductors
   d. conductor, switch, and load

2. A type of subatomic particle with a positive charge is a(n) _____.
   a. proton
   b. neutron
   c. electron
   d. nucleus

3. Which of the following substances is considered an insulator?
   a. Gold
   b. Copper
   c. Silver
   d. Porcelain

4. The voltage commonly supplied to a residence by the local utility is _____.
   a. 120V
   b. 240V
   c. 480V
   d. 208V

5. Current is measured in units called _____.
   a. joules
   b. coulombs
   c. amperes
   d. volt-amperes

6. Joules are _____.
   a. units of work
   b. the potential difference between two points
   c. the difference between EMF and current
   d. the rate of current flow

7. Another term used for voltage is _____.
   a. emf
   b. coulomb
   c. current
   d. joule

8. All conductors have _____.
   a. current flow
   b. some resistance
   c. EMF
   d. voltage potential

9. In order to calculate the current flowing in a circuit _____.
   a. multiply voltage by resistance (E × R)
   b. divide resistance by power (R/P)
   c. multiply power by resistance (P × R)
   d. divide voltage by resistance (E/R)

10. The color band that represents tolerance on a resistor is the _____.
    a. 4th band
    b. 3rd band
    c. 2nd band
    d. 1st band

11. In a parallel circuit, the total resistance is _____ the smallest resistance.
    a. greater than
    b. equal to
    c. less than
    d. proportional to

12. Ammeters measure _____.
    a. voltage
    b. resistance
    c. current
    d. power

13. Circuit continuity is checked using the _____ function of a multimeter.
    a. ammeter
    b. voltmeter
    c. ohmmeter
    d. wattmeter

14. Resistance is measured in _____.

    a. ohms

    b. amperes

    c. volt-amperes

    d. coulombs

15. The power in a circuit with 120 volts and 5 amps is _____.

    a. 5 watts

    b. 120 watts

    c. 240 watts

    d. 600 watts

Trade Terms Quiz

Fill in the blank with the correct term that you learned from your study of this module.

1. A(n) _____ is an instrument for measuring electrical current.

2. Measured in amperes, _____ is the flow of electrons in a circuit.

3. _____ are the force required to produce a current of one ampere through a resistance of one ohm.

4. Voltage is measured with a(n) _____.

5. _____ are the basic unit of measurement for electrical current.

6. One volt is the potential difference between two points for which one coulomb of electricity will do one _____ of work.

7. A(n) _____ is a quantity of electricity that is either positive or negative.

8. A(n) _____ is the common unit used for specifying the size of a given charge.

9. _____ is the driving force that makes current flow in a circuit.

10. _____ are the basic unit of measurement for electrical power.

11. _____ are the smallest particles of an element that will still retain the properties of that element.

12. The _____ is the center of an atom.

13. Both found in the nucleus of an atom, a _____ is an electrically positive particle and _____ are electrically neutral particles.

14. The outermost ring of electrons orbiting the nucleus of an atom is known as the _____.

15. _____ are negatively charged particles that orbit the nucleus of an atom.

16. The definition of _____ is any substance that has mass and occupies space.

17. The prefix used to indicate one thousand is _____.

18. The prefix used to indicate one million is _____.

19. Consisting of two or more cells, a _____ converts chemical energy into electrical energy.

20. A(n) _____ is a complete path for current flow.

21. _____ are materials through which it is relatively easy to maintain an electric current.

22. A(n) _____ is a material through which it is difficult to conduct an electric current.

23. _____ are the basic unit of measurement for resistance.

24. The instrument that is used to measure resistance is called a(n) _____.

25. _____ is a statement of the relationship between current, voltage, and resistance in an electrical circuit.

26. _____ is the rate of doing work or the rate at which energy is used or dissipated.

27. Measured in ohms, _____ is the electrical property that opposes the flow of current through a circuit.

28. _____ are components that normally oppose current flow in a DC circuit.

29. A(n) _____ is a drawing in which symbols are used to represent the components in a system.

30. A(n) _____ has only one route for current flow.

31. The change in voltage across a component is called _____.

32. _____ are electromechanical components used as switching devices.

33. Devices containing one or more coils of wire wrapped around a common core are called _____.

34. Electromagnetic devices used to control a mechanical device such as a valve are called _____.

Trade Terms

Ammeter	Electrons	Ohmmeter	Solenoids
Amperes (A)	Insulator	Ohm's law	Transformers
Atoms	Joule (J)	Power	Valence shell
Battery	Kilo	Proton	Volts (V)
Charge	Matter	Relays	Voltage
Circuit	Mega	Resistance	Voltage drop
Conductors	Neutrons	Resistors	Voltmeter
Coulomb	Nucleus	Schematic	Watts (W)
Current	Ohms (Ω)	Series circuit	

1. An atom that is missing only one or two electrons from its outer shell will _____.

2. To find voltage when both current and resistance are known, use the formula _____.

3. A resistor with a color code of yellow, orange, red, and silver has a tolerance of _____.

4. True or False? When using an in-line ammeter, it is important to connect the meter in series.

5. If a toaster draws 6.2A of current at 120V, how many kilowatt-hours of energy will be used in 3.5 hours?

   _____.

6. If a battery sends a current of 10A through a circuit for one hour, how many coulombs will flow through the circuit?

   _____.

7. The charged particles of an atom are called _____.

8. Conductors have _____ or fewer valence electrons.

9. The sum of the difference in potential of all the charges in an electrostatic field is called

   _____.

10. Electric charge is measured in _____.

11. Find the resistance when the voltage is 120V and the current is 6A.

    _____.

12. Give the resistance value and tolerance of a resistor where the color bands are red, yellow, brown, and silver.

    _____.

13. What is the power in a 120V circuit with a current of 12.5A?

    _____.

14. What is the voltage in a 30Ω circuit with a power rating of 480W?

    _____.

15. An ohmmeter is used to measure resistance and check for

    _____.

E.L. Jarrell
Associated Builders and Contractors

Eurlin Layne (E.L.) Jarrell is a prime example of a master electrician giving back to the electrical community by teaching and mentoring.

After serving in the US Army, E.L. went to work for Cities Services, now known as CITGO. He stayed at CITGO for 38 years, retiring in 1995. It was during his employment at CITGO that he first received apprenticeship training in the electrical field.

While at CITGO, E.L. worked as a process unit operator before moving to the electrical department. While there, he worked as a trainee electrician for three years until he became a first-class electrician. A few years later, he was promoted to temporary supervisor, planning and scheduling shut-down maintenance. In 1983, he passed the Block Master Electrician test for the City of Lake Charles, Louisiana. In 1997, E.L. became involved with Associated Builders and Contractors (ABC).

E.L. is currently the Electrical Department Head for the ABC Training Center, where he works in the lab, overseeing students doing hands-on electrical work. During his first semester teaching at the ABC Training Center, it became clear to E.L. that many students simply had no time to study because they worked 10-hour days, drove over 100 miles to work, and had family obligations. In response, E.L. began an in-class study guide. He encouraged students to form study groups, and he gave students time to study in class.

E.L. was an instrumental member of NCCER's Technical Review Committee, which completely rewrote all four levels of NCCER's Electrical curriculum. In addition, E.L. is currently a member of both NCCER's National Skills Assessment Written Test Committee and the Performance Verification Packet for Industrial Electricians Committee.

E.L. has decided to give back to the electrical community with his expertise and mentoring. Many of E.L.'s students have become his personal friends. He says, "At this point in my life, I just want to continue being the best electrical instructor that I can be and share some of my knowledge and experience with my students and hope that I can make a difference in their lives and careers."

Trade Terms Introduced in This Module

Ammeter: An instrument for measuring electrical current.

Amperes (A): The basic unit of measurement for electrical current, represented by the letter A

Atoms: The smallest particles to which an element may be divided and still retain the properties of the element.

Battery: A DC voltage source consisting of two or more cells that convert chemical energy into electrical energy.

Charge: A quantity of electricity that is either positive or negative.

Circuit: A complete path for current flow.

Conductors: Materials through which it is relatively easy to maintain an electric current.

Coulomb: A unit of electrical charge equal to 6.25×10^{18} electrons (or 6.25 quintillion electrons). A coulomb is the common unit of quantity used for specifying the size of a given charge.

Current: The movement, or flow, of electrons in a circuit. Current (I) is measured in amperes.

Electrons: Negatively charged particles that orbit the nucleus of an atom.

Insulator: A material through which it is difficult to conduct an electric current.

Joule (J): A unit of measurement for doing work, represented by the letter J. One joule is equal to one newton-meter (Nm).

Kilo: A prefix used to indicate one thousand (for example, one kilowatt is equal to one thousand watts).

Matter: Any substance that has mass and occupies space.

Mega: A prefix used to indicate one million; for example, one megawatt is equal to one million watts.

Neutrons: Electrically neutral particles (neither positive nor negative) that have the same mass as a proton and are found in the nucleus of an atom.

Nucleus: The center of an atom. It contains the protons and neutrons of the atom.

Ohms (Ω): The basic unit of measurement for resistance, represented by the symbol Ω.

Ohmmeter: An instrument used for measuring resistance.

Ohm's law: A statement of the relationships among current, voltage, and resistance in an electrical circuit: current (I) equals voltage (E) divided by resistance (R). Generally expressed as a mathematical formula: $I = E/R$.

Power: The rate of doing work, or the rate at which energy is used or dissipated. Electrical power is measured in watts.

Proton: The smallest positively charged particle of an atom. Protons are contained in the nucleus of an atom.

Relays: Electromechanical devices consisting of a coil and one or more sets of contacts. Used as a switching device.

Resistance: An electrical property that opposes the flow of current through a circuit. Resistance (R) is measured in ohms.

Resistors: Any devices in a circuit that resist the flow of electrons.

Schematic: A type of drawing in which symbols are used to represent the components in a system.

Series circuit: A circuit with only one path for current flow.

Solenoids: Electromagnetic coils used to control a mechanical device such as a valve.

Transformers: Devices consisting of one or more coils of wire wrapped around a common core. Transformers are commonly used to step voltage up or down.

Valence shell: The outermost ring of electrons that orbit about the nucleus of an atom.

Volts (V): The unit of measurement for voltage, represented by the letter V. One volt is equivalent to the force required to produce a current of one ampere through a resistance of one ohm.

Voltage: The driving force that makes current flow in a circuit. Voltage, often represented by the letter E, is also referred to as electromotive force (emf), difference of potential, or electrical pressure.

Voltage drop: The change in voltage across a component that is caused by the current flowing through it and the amount of resistance opposing it.

Voltmeter: An instrument for measuring voltage. The resistance of the voltmeter is fixed. When the voltmeter is connected to a circuit, the current passing through the meter will be directly proportional to the voltage at the connection points.

Watts (W): The basic unit of measurement for electrical power, represented by the letter W.

Additional Resources

This module presents thorough resources for task training. The following reference material is recommended for further study.

Electronics Fundamentals: Circuits, Devices, and Applications, Thomas L. Floyd. New York, NY: Pearson Education, Inc.

Principles of Electric Circuits, Thomas L. Floyd. New York, NY: Pearson Education, Inc.

Figure Credits

Greenlee / A Textron Company, Module Opener

Courtesy of Extech Instruments, a FLIR Company, Figure 19

Tim Dean, Figure 20

Amprobe, Figure 24

SECTION 1.0.0

Answer	Section Reference	Objective
1. b	1.1.0	1a
2. a	1.2.0	1b
3. c	1.3.0	1c
4. b	1.4.0	1d

SECTION 2.0.0

Answer	Section Reference	Objective
1. b	2.1.0	2a
2. a	2.2.0	2b
3. b	2.3.0	2c
4. c	2.4.0	2d

SECTION 3.0.0

Answer	Section Reference	Objective
1. b	3.1.0	3a
2. c	3.2.1	3b
3. a	3.3.1	3c
4. d	3.4.1	3d

2.0.0 SECTION REVIEW

Question 4

Using Ohm's Law, solve for the voltage (E) using the current and resistance values:

$$E = I \times R$$
$$E = 30A \times 4\Omega$$
$$E = 120V$$

The applied voltage is **120V**.

3.0.0 SECTION REVIEW

Question 2

Find the sum of the resistances:

$$R_T = 10\Omega + 10\Omega + 10\Omega = 30\Omega$$

The total resistance is **30Ω**.

Question 4

Using the power equation, multiply the current by the voltage to find the power:

$$P = IE$$
$$P = 10A \times 240V$$
$$P = 2400W$$

The power is **2,400W**.

This page is intentionally left blank.

NCCER CURRICULA — USER UPDATE

NCCER makes every effort to keep its textbooks up-to-date and free of technical errors. We appreciate your help in this process. If you find an error, a typographical mistake, or an inaccuracy in NCCER's curricula, please fill out this form (or a photocopy), or complete the online form at **www.nccer.org/olf**. Be sure to include the exact module ID number, page number, a detailed description, and your recommended correction. Your input will be brought to the attention of the Authoring Team. Thank you for your assistance.

Instructors – If you have an idea for improving this textbook, or have found that additional materials were necessary to teach this module effectively, please let us know so that we may present your suggestions to the Authoring Team.

NCCER Product Development and Revision

13614 Progress Blvd., Alachua, FL 32615

Email: curriculum@nccer.org
Online: www.nccer.org/olf

❏ Trainee Guide   ❏ Lesson Plans   ❏ Exam   ❏ PowerPoints   Other _____

Craft / Level: _____   Copyright Date: _____

Module ID Number / Title: _____

Section Number(s): _____

Description: _____

_____

_____

_____

_____

Recommended Correction: _____

_____

_____

_____

_____

Your Name: _____

Address: _____

_____

Email: _____   Phone: _____

This page is intentionally left blank.

Electrical Theory

OVERVIEW

Knowledge of electrical circuits is essential in the electrical field. Sound understanding of basic circuits, as well as the methods for calculating the electrical energy within them, forms the foundation for utilizing these principles in practical applications. This module explains how to apply Ohm's law to series, parallel, and series-parallel circuits. It also covers Kirchhoff's voltage and current laws.

Module 26104-20

Trainees with successful module completions may be eligible for credentialing through the NCCER Registry. To learn more, go to **www.nccer.org** or contact us at 1.888.622.3720. Our website, **www.nccer.org**, has information on the latest product releases and training.

Your feedback is welcome. You may email your comments to **curriculum@nccer.org**, send general comments and inquiries to **info@nccer.org**, or fill in the User Update form at the back of this module.

This information is general in nature and intended for training purposes only. Actual performance of activities described in this manual requires compliance with all applicable operating, service, maintenance, and safety procedures under the direction of qualified personnel. References in this manual to patented or proprietary devices do not constitute a recommendation of their use.

26104-20 V10.0

Objectives

When you have completed this module, you will be able to do the following:

1. Calculate values in resistive circuits.
   a. Identify resistances in series.
   b. Identify resistances in parallel.
   c. Simplify series-parallel circuits.
   d. Apply Ohm's law to various types of circuits.
2. Apply Kirchhoff's laws to various types of circuits.
   a. Use Kirchhoff's current law.
   b. Use Kirchhoff's voltage law.

Performance Tasks

This is a knowledge-based module. There are no performance tasks.

Trade Terms

Kirchhoff's current law
Kirchhoff's voltage law
Parallel circuits
Series circuit
Series-parallel circuits

Industry Recognized Credentials

If you are training through an NCCER-accredited sponsor, you may be eligible for credentials from NCCER's Registry. The ID number for this module is 26104-20. Note that this module may have been used in other NCCER curricula and may apply to other level completions. Contact NCCER's Registry at 888.622.3720 or go to **www.nccer.org** for more information.

NFPA 70®, *National Electrical Code*® and *NEC*® are registered trademarks of the National Fire Protection Association, Quincy, MA.

Contents

Figures

1.0.0 RESISTIVE CIRCUITS

Objective

Calculate values in resistive circuits.
   a. Identify resistances in series.
   b. Identify resistances in parallel.
   c. Simplify series-parallel circuits.
   d. Apply Ohm's law to various types of circuits.

Trade Terms

Parallel circuits: Circuits containing two or more parallel paths through which current can flow.

Series circuit: A circuit that has only one path for current flow.

Series-parallel circuits: Circuits that contain both series and parallel current paths.

Resistance, which is measured in ohms (Ω), is calculated in different ways depending on the type of circuit. Different equations are used for series and parallel circuits. Resistance calculations are often used to determine other circuit characteristics, such as voltage and current.

> **NOTE**
>
> There are two theories of electron flow. Most electricians use *electron flow theory*, which describes current flow from negative to positive. However, electronics engineers frequently use *conventional electron flow*, which describes current flow in the opposite direction (from positive to negative). To many, this makes it easier to analyze complex electronics. This curriculum uses electron flow theory (from negative to positive). It is important to be aware of this to avoid communication issues and miscalculations.

1.1.0 Resistances in Series

A **series circuit** is a circuit that has only one path for current flow. In the series circuit shown in *Figure 1*, the current (I) is the same in all parts of the circuit. This means that the current flowing through R_1 is the same as the current flowing through R_2 and R_3, and it is also the same as the current supplied by the battery. Unknown values can be found using any variation of Ohm's law

or the power equation. The three forms of Ohm's law (used to find voltage, resistance, or current) are as follows:

To find voltage:
$$E = I \times R$$

To find resistance:
$$R = E \div I$$

To find current:
$$I = E \div R$$

Where:

   E = voltage in volts (V)
   I = current in amperes, or amps (A)
   R = resistance in ohms (Ω)

The three forms of the power equation (used to find power, current, or voltage) are as follows:

To find power:
$$P = I \times E$$

To find voltage:
$$E = P \div I$$

To find current:
$$I = P \div E$$

Where:

   P = power in watts (W)
   I = current in amperes
   E = voltage in volts

When resistances are connected in series, like the example shown in *Figure 1*, the total resistance in the circuit is equal to the sum of all the resistances in the circuit:

$$R_T = R_1 + R_2 + R_3 + R_N$$

Where:

   R_T = total resistance
   R_1, R_2, R_3, R_n = each of the resistances in series
   (There can be any number of these.)

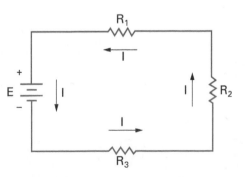

Figure 1 Series circuit.

Example 1:

The circuit shown in *Figure 2 (A)* has 50Ω, 75Ω, and 100Ω resistors in series. Find the total resistance of the circuit.

Add the values of the three resistors in series:

$$R_T = R_1 + R_2 + R_3$$
$$R_T = 50Ω + 75Ω + 100Ω$$
$$R_T = 225Ω$$

The total resistance is 225Ω.

Example 2:

The circuit shown in *Figure 2 (B)* has three lamps connected in series with the resistances shown. Find the total resistance of the circuit.

Add the values of the three lamp resistances in series:

$$R_T = R_1 + R_2 + R_3$$
$$R_T = 20Ω + 40Ω + 60Ω$$
$$R_T = 120Ω$$

The total resistance is 120Ω.

1.2.0 Resistances in Parallel

Circuits that contain two or more parallel paths through which current can flow are called **parallel circuits**. The total resistance in a parallel circuit is given by the formula:

$$R_T = \cfrac{1}{\cfrac{1}{R_1} + \cfrac{1}{R_2} + \cfrac{1}{R_3}}$$

Where:

R_T = total resistance in parallel
R_1, R_2, and R_3 = each of the branch resistances
(There can be any number of these.)

Example 1:

Find the total resistance of the 2Ω, 4Ω, and 8Ω resistors in parallel shown in *Figure 3*.

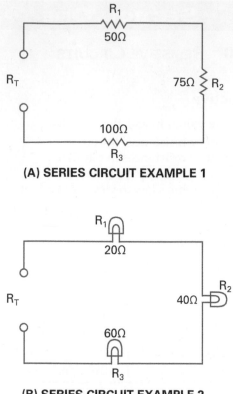

(A) SERIES CIRCUIT EXAMPLE 1

(B) SERIES CIRCUIT EXAMPLE 2

Figure 2 Example series circuits.

To find the total resistance, use the following steps:

Step 1 Write the formula for the three resistances in parallel:

$$R_T = \cfrac{1}{\cfrac{1}{R_1} + \cfrac{1}{R_2} + \cfrac{1}{R_3}}$$

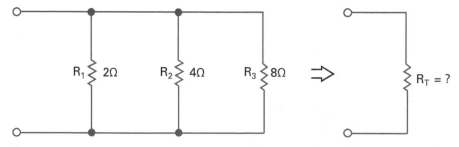

Figure 3 Parallel branch.

Series Circuits

Simple series circuits are often used as voltage dividers, but are seldom encountered in practical wiring. The only simple series circuit you may recognize is an older strand of Christmas lights, in which the entire string went dead when one lamp burned out. Think about what the actual wiring of a series circuit would look like in household receptacles. How would the circuit physically be wired? What kind of illumination would you get if you wired your household receptacles in series and plugged half a dozen lamps into those receptacles?

Step 2 Substitute the resistance values:

$$R_T = \frac{1}{\frac{1}{2} + \frac{1}{4} + \frac{1}{8}}$$

$$R_T = \frac{1}{0.5 + 0.25 + 0.125}$$

$$R_T = \frac{1}{0.875} = 1.14\Omega$$

Note that when resistances are connected in parallel, the total resistance is always less than the resistance of any single branch. In this example:

$$R_T = 1.14\Omega \; < \; R_1 = 2\Omega$$
$$R_T = 1.14\Omega \; < \; R_2 = 4\Omega$$
$$R_T = 1.14\Omega \; < \; R_3 = 8\Omega$$

Example 2:

Add a fourth parallel resistor of 2Ω to the circuit in *Figure 3*. What is the new total resistance, and what is the net effect of adding another resistance in parallel?

To find the total resistance, use the following steps:

Step 1 Write the formula for four resistances in parallel:

$$R_T = \frac{1}{\frac{1}{R_1} + \frac{1}{R_2} + \frac{1}{R_3} + \frac{1}{R_4}}$$

Step 2 Substitute the resistance values:

$$R_T = \frac{1}{\frac{1}{2} + \frac{1}{4} + \frac{1}{8} + \frac{1}{2}}$$

$$R_T = \frac{1}{0.5 + 0.25 + 0.125 + 0.5}$$

$$R_T = \frac{1}{1.375} = 0.73\Omega$$

The net effect of adding another resistance in parallel is a reduction of the total resistance from 1.14Ω to 0.73Ω.

1.2.1 Simplified Formulas

When there are only two resistors in a parallel circuit, it is often easier to calculate the total resistance by multiplying the two resistances and dividing the product by the sum of the resistances. This is sometimes called the *product-over-sum method*, and it is shown in the following equation (this formula only works if there are two resistors in the circuit):

$$R_T = \frac{R_1 \times R_2}{R_1 + R_2}$$

Where:

R_T = total resistance of unequal resistors in parallel

R_1, R_2 = two unequal resistors in parallel

When resistors in parallel are all equal to one another, their total resistance is equal to the resistance of one resistor divided by the number of resistors in the circuit, as shown in the following equation:

$$R_T = \frac{R}{N}$$

Where:

R_T = total resistance of equal resistors in parallel
R = resistance of one of the equal resistors
N = number of equal resistors

If two resistors with the same resistance are connected in parallel, their equivalent resistance is equal to one resistor of half of that value, as shown in *Figure 4*.

The two 200Ω resistors in parallel are the equivalent of one 100Ω resistor; the two 100Ω resistors are the equivalent of one 50Ω resistor; and the two 50Ω resistors are the equivalent of one 25Ω resistor.

Example 1:

What is the total resistance of a 6Ω (R_1) resistor and an 18Ω (R_2) resistor in parallel?

Because there are only 2 resistors, the product-over-sum method can be used:

$$R_T = \frac{R_1 \times R_2}{R_1 + R_2}$$

$$R_T = \frac{6 \times 18}{6 + 18}$$

$$R_T = \frac{108}{24} = 4.5\Omega$$

Example 2:

Find the total resistance of a 100Ω (R_1) resistor and a 150Ω (R_2) resistor in parallel.

Because there are only 2 resistors, the product-over-sum method can be used:

$$R_T = \frac{R_1 \times R_2}{R_1 + R_2}$$

$$R_T = \frac{100 \times 150}{100 + 150}$$

$$R_T = \frac{15,000}{250} = 60\Omega$$

1.3.0 Series-Parallel Circuits

Finding current, voltage, and resistance in series circuits and parallel circuits is fairly easy. When working with either type of arrangement (series or parallel), use only the rules that apply to that type. In **series-parallel circuits**, some parts of the circuit are connected in series, and other parts are connected in parallel. Thus, in some parts the rules for series circuits apply, and in other parts, the rules for parallel circuits apply.

To analyze or solve a problem involving a series-parallel circuit, it is necessary to recognize which parts of the circuit are series connected and which parts are parallel connected. This can be obvious if the circuit is simple. However, more complex circuits must be redrawn, putting it into a form that is easier to recognize.

In a series circuit, the current is the same at all points. A parallel circuit has one or more points where the current divides and flows in separate

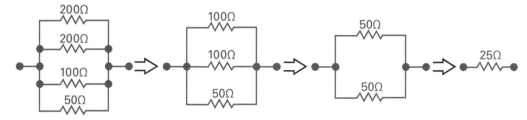

Figure 4 Equal resistances in a parallel circuit.

Think About It

Parallel Circuits

An interesting fact about circuits is the drop in resistance in a parallel circuit as more resistors are added. But this fact does not mean that you can add an endless number of devices, such as lamps, in a parallel circuit. Why not?

branches. A series-parallel circuit has both separate branches and series loads. The easiest way to find out whether a circuit is a series, parallel, or series-parallel circuit is to start at the negative terminal of the power source and trace the path of current through the circuit back to the positive terminal of the power source. If the current does not divide anywhere, it is a series circuit. If the current divides into separate branches, but there are no series loads, it is a parallel circuit. If the current divides into separate branches and there are also series loads, it is a series-parallel circuit. *Figure 5* shows electric lamps connected in series, parallel, and series-parallel circuits.

After determining that a circuit is series-parallel, redraw the circuit so that the branches and the series loads are more easily recognized. This is especially helpful when computing the total resistance of the circuit. *Figure 6 (A)* shows resistors connected in a series-parallel circuit, and *Figure 6 (B)* shows the equivalent circuit redrawn in a simpler form.

1.3.1 Reducing Series-Parallel Circuits

Most of the time, all that is known about a series-parallel circuit is the applied voltage and the values of the individual resistances. To find the voltage drop across any of the loads or the current in any of the branches, the total circuit current must also be known. To find the total current, the total resistance of the circuit must be known.

To find the total resistance, reduce the circuit to its simplest form. Usually, this means simplifying the values of all the resistors into a single,

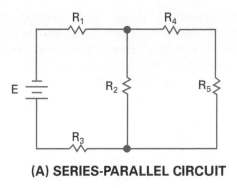

(A) SERIES-PARALLEL CIRCUIT

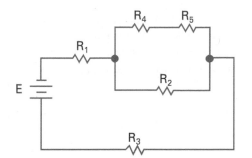

(B) REDRAWN VERSION OF THE CIRCUIT

Figure 6 Redrawing a series-parallel circuit.

Think About It

Series-Parallel Circuits

Explain *Figure 6.* Which resistors are in series and which are in parallel?

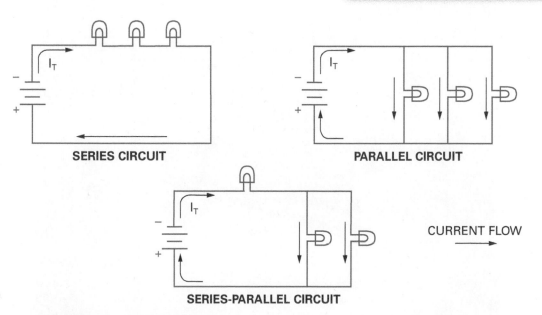

Figure 5 Series, parallel, and series-parallel circuits.

equivalent (effective) value. This simple series circuit has the equivalent resistance of the series-parallel circuit it was derived from, and also has the same total current. There are four basic steps in reducing a series-parallel circuit:

Step 1 If necessary, redraw the circuit so that all parallel combinations of resistances and series resistances are easily recognized.

Step 2 For each parallel combination of resistances, calculate their effective resistance (their equivalent series resistance).

Step 3 Replace each of the parallel combinations with one resistance whose value is equal to the effective resistance of that combination. This provides a circuit with all series loads.

Step 4 Find the total resistance of this circuit by adding the resistances of all the series loads.

The following are four general steps for finding voltage drops across individual resistors; this step-by-step process will be further explained and illustrated in the following sections:

Step 1 Reduce the circuit into a simpler form by converting any parallel resistances into an equivalent series resistance value.

Step 2 Find the total resistance by adding all of the series resistances in the simplified circuit.

Step 3 Now that you have the total resistance (R_T), you can use this along with the applied voltage (E_T) to find the total current (I_T). Do this by using Ohm's law to solve for I_T ($I_T = E_T \div R_T$).

Step 4 Use Ohm's law with the individual resistances and total current in order to find the voltage drops across each resistor.

Examine the series-parallel circuit shown in *Figure 7* and reduce it to an equivalent series circuit.

Parallel Circuits

Most practical circuits are wired in parallel, like the pole lights shown here.

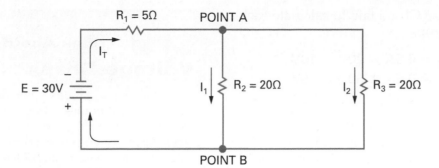

Figure 7 Reducing a series-parallel circuit.

In this circuit, resistors R_2 and R_3 are connected in parallel, but resistor R_1 is in series with both the battery (applied voltage) and the parallel combination of R_2 and R_3. The current I_T leaving the negative terminal of the voltage source travels through resistor R_1 before it is divided at the junction of resistors R_1, R_2, and R_3 (Point A). It then passes through the two branches formed by resistors R_2 and R_3.

Given the information in *Figure 7*, you can calculate the resistance of R_2 and R_3 in parallel, and then add it to the series resistance R_1 to find the total resistance of the circuit, R_T.

The total resistance of the circuit is the sum of R_1 and the equivalent resistance of R_2 and R_3 in parallel. To find R_T, first find the resistance of R_2 and R_3 in parallel. Because the two resistances have the same value of 20Ω, the resulting equivalent resistance is 10Ω. (As previously discussed, two equal resistances in parallel are equivalent to half of one of the resistances.) Therefore, the total resistance is 15Ω ($5\Omega + 10\Omega$).

1.4.0 Ohm's Law

Ohm's law is one of the primary tools used to find unknown values in resistive circuits. As previously discussed, unknown circuit parameters (voltage, current, and resistance) can be found by using Ohm's law and the techniques for determining equivalent resistance.

1.4.1 Applying Ohm's Law in Series Circuits

In resistive circuits, you can find unknown circuit parameters by using Ohm's law combined with the techniques for determining equivalent resistance. Ohm's law may be applied to an entire series circuit, and it may also be applied to the individual parts of the circuit. When it is used on a particular part of a circuit, the voltage across that part is equal to the current in that part multiplied by the resistance of that part.

For example, given the information in *Figure 8*, calculate the total resistance (R_T) and the total current (I_T).

Find R_T:

$$R_T = R_1 + R_2 + R_3$$
$$R_T = 20 + 50 + 120$$
$$R_T = 190\Omega$$

Find I_T using Ohm's law:

$$I_T = \frac{E_T}{R_T}$$

$$I_T = \frac{95}{190}$$

$$I_T = 0.5A$$

Now find the voltage drop across each individual resistor. In a series circuit, the current is the same through each resistor; that is, $I = 0.5A$ through each resistor. Now that you know the current and resistance for each individual

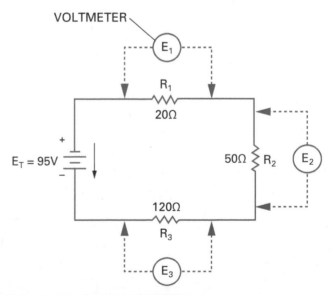

Figure 8 Calculating voltage drops.

resistor, you can use Ohm's law to calculate each resistor's voltage drop:

$$E_1 = I_1 \times R_1 = 0.5A \times 20\Omega = 10V$$
$$E_2 = I_2 \times R_2 = 0.5A \times 50\Omega = 25V$$
$$E_3 = I_3 \times R_3 = 0.5A \times 120\Omega = 60V$$

As we have just calculated, the voltage drops E_1, E_2, and E_3 in *Figure 8* are 10V, 25V, and 60V, respectively. Voltage drops are also known as *IR drops*. Their effect is to reduce the voltage that is available to be applied across the rest of the components in the circuit. The sum of the voltage drops in any series circuit is always equal to the voltage that is applied to the circuit. (The total voltage, E_T, is the same as the applied voltage.) In this example, $E_T = 10 + 25 + 60 = 95V$.

1.4.2 Applying Ohm's Law in Parallel Circuits

A parallel circuit is a circuit in which two or more components branch off of the same voltage source. The branches are like their own individual series circuits, but they all connected to the same battery (voltage source), as illustrated in *Figure 9*. The resistors R_1, R_2, and R_3 are in parallel with each other and with the battery. Each parallel branch has its own individual current. When the total current I_T leaves the voltage source E_T, part I_1 of the current the total current will flow through R_1, part I_2 will flow through R_2, and the remainder I_3 will flow through R_3. The branch currents (I_1, I_2, and I_3) can be different. However, if a voltmeter is connected across R_1, R_2, and R_3, the voltage drops E_1, E_2, and E_3 will each be equal to the source voltage E_T.

The total current I_T is equal to the sum of all branch currents. This formula applies for any number of parallel branches, whether the resistances are equal or unequal.

Using Ohm's law, the current for an individual branch equals the voltage for that branch divided by the total resistance of that branch. Since the voltage drop across each branch of a parallel

Think About It

Voltage Drops

Calculating voltage drops is not just a schoolroom exercise. It is important to know the voltage drop when sizing circuit components. What would happen if you sized a component without accounting for a substantial voltage drop in the circuit?

circuit is equal to the applied voltage, you do not have to calculate individual voltage drops for each branch. The following equations illustrate this for the three branches in *Figure 9*:

$$\text{Branch 1: } I_1 = \frac{E_1}{R_1} = \frac{E_T}{R_1}$$

$$\text{Branch 2: } I_2 = \frac{E_2}{R_2} = \frac{E_T}{R_2}$$

$$\text{Branch 3: } I_3 = \frac{E_3}{R_3} = \frac{E_T}{R_3}$$

With the same applied voltage, any branch that has less resistance allows more current through it than a branch with higher resistance. The sum of the branch currents in a parallel circuit is equal to the total current (I_T).

Example 1:

The two branches R_1 and R_2, shown in *Figure 10* (*A*), draw a total line current of 20A across a 110V power line. (R_1 and R_2 are in parallel.) Branch R_1 takes 12A. What is the current I_2 in branch R_2?

Since we know that the total current (I_T) is equal to the sum of each branch current, we can rearrange the formula to find I_2, and then substitute given values, as follows:

$$I_T = I_1 + I_2$$
$$I_2 = I_T - I_1$$
$$I_2 = 20 - 12 = 8A$$

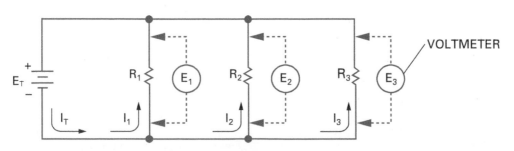

Figure 9 Parallel circuit.

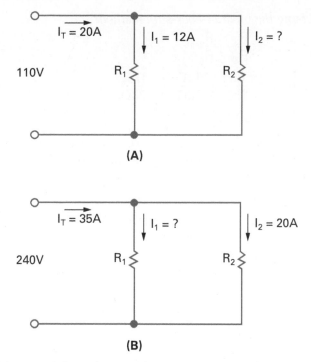

(A)

(B)

Figure 10 Solving for an unknown current.

Example 2:

As shown in *Figure 10* (B), the two branches R_1 and R_2 across a 240V power line draw a total line current of 35A. Branch R_2 takes 20A. What is the current I_1 in branch R_1?

Rearrange the formula to find I_1, and then substitute given values:

$$I_T = I_1 + I_2$$
$$I_1 = I_T - I_2$$
$$I_1 = 35 - 20 = 15A$$

1.4.3 Applying Ohm's Law in Series-Parallel Circuits

Series-parallel circuits combine the elements and characteristics of both the series and parallel configurations. By properly applying the equations and methods previously discussed, you can determine the values of individual components of the circuit. *Figure 11* shows a simple series-parallel circuit with a 1.5V battery.

The following steps outline the procedure for determining voltage drops across and current flows through all resistors in a series-parallel circuit:

> **NOTE**
>
> How these steps are applied to a particular circuit will depend strictly upon the arrangement of its components.

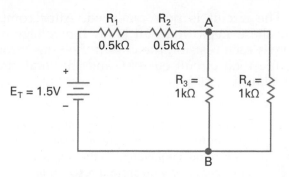

Figure 11 Series-parallel circuit.

Step 1 If necessary, redraw the circuit to make the series and parallel relationships of its components more obvious. Be sure to maintain the values and relationships of the original circuit. Unless careful consideration is taken during this step, it is easy to alter the circuit into a different one by accident. (Label this drawing "Diagram 1".)

Step 2 Reduce all groups of resistors in parallel to a single equivalent series resistor (using any of the previously discussed methods). Do this for each parallel group.

Step 3 Redraw the circuit using the equivalent series resistors in place of the parallel groups. The circuit should now be a simple series circuit. (Label this drawing "Diagram 2".)

Step 4 Determine the total circuit resistance (R_T) of Diagram 2 by adding all of the resistances.

Step 5 Determine the total circuit (I_T) by using the power supply voltage (E_T), the total circuit resistance (I_T), and the correct form of Ohm's law ($I_T = E_T \div R_T$).

Step 6 Determine the voltage drop (E_{Rn}) across each resistance in Diagram 2 by using the total current (I_T), the resistor's value (R_n), and the correct form of Ohm's law ($E_{Rn} = I_T \times R_n$).

Step 7 Referring back to Diagram 1, determine the branch current of each parallel resistor. The voltage drop across each parallel resistor group will be equal to the calculated voltage drop across its equivalent resistance (calculated in Step 6).

Step 8 Determine the branch current (I_{Rn}) of each individual parallel resistor using the voltage drop across the resistor (E_{Rn}), the branch resistance (R_n), and the correct form of Ohm's law ($I_{Rn} = E_{Rn} \div R_n$).

The circuit is now analyzed. After completing these steps, you will know the voltage drop across each resistor, the current flowing through it, the total circuit current, and the total circuit resistance.

The current and voltage associated with each component can be determined by first simplifying the circuit to find the total current, and then working across the individual components.

The circuit in *Figure 11* can be broken into two components: the series resistances R_1 and R_2, and the parallel resistances R_3 and R_4.

R_1 and R_2 can be added together to form the equivalent series resistance R_{1+2}:

$$R_{1+2} = R_1 + R_2$$
$$R_{1+2} = 0.5k\Omega + 0.5k\Omega$$
$$R_{1+2} = 1k\Omega$$

R_3 and R_4 can be totaled using either the general reciprocal formula we have been using, or, since there are two resistances in parallel, the product-over-sum method. Both methods are shown as follows:

Using the reciprocal formula:

$$R_{3+4} = \cfrac{1}{\cfrac{1}{R_3} + \cfrac{1}{R_4}}$$

$$R_{3+4} = \cfrac{1}{\cfrac{1}{1k\Omega} + \cfrac{1}{1k\Omega}}$$

$$R_{3+4} = \cfrac{1}{\cfrac{1}{1000\Omega} + \cfrac{1}{1000\Omega}}$$

$$R_{3+4} = \cfrac{1}{0.001 + 0.001} = \cfrac{1}{0.002}$$

$$R_{3+4} = 500\Omega \ or \ 0.5k\Omega$$

Using the product-over-sum method:

$$R_{3+4} = \frac{R_3 \times R_4}{R_3 + R_4}$$

$$R_{3+4} = \frac{1k\Omega \times 1k\Omega}{1k\Omega + 1k\Omega} = \frac{1k\Omega}{2k\Omega}$$

$$R_{3+4} = 500\Omega \ or \ 0.5k\Omega$$

The equivalent circuit containing the R_{1+2} resistance of 1kΩ and the R_{3+4} resistance of 0.5kΩ is shown in *Figure 12*.

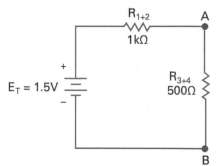

Figure 12 Simplified series-parallel circuit.

Using the Ohm's law relationship that total current equals voltage divided by circuit resistance, the circuit current can be determined. First, however, total circuit resistance must be found. Because the simplified circuit consists of two resistances in series, they are simply added together to obtain total resistance.

$$R_T = R_{1+2} + R_{3+4}$$
$$R_T = 1k\Omega + 0.5k\Omega$$
$$R_T = 1.5k\Omega$$

Applying this to the current/voltage equation:

$$I_T = \frac{E_T}{R_T} = \frac{1.5V}{1.5k\Omega} = 1mA \ or \ 0.001A$$

Now that the total current is known, voltage drops across individual components can be determined:

$$E = I \times R$$
$$E_1 = I_T \times R_1 = 1mA \times 0.5k\Omega = 0.5V$$
$$E_2 = I_T \times R_2 = 1mA \times 0.5k\Omega = 0.5V$$

Because the total voltage equals the sum of all voltage drops, the voltage drop from A to B (E_{AB}) can be determined by subtracting the E_{R1} and E_{R2} voltage drops from the total voltage, as follows:

$$E_T = E_{R1} + E_{R2} + E_{AB}$$
$$E_{AB} = E_T - E_{R1} - E_{R2}$$
$$E_{AB} = 1.5V - 0.5V - 0.5V$$
$$E_{AB} = 0.5V$$

Since R_3 and R_4 are in parallel, some of the total current must pass through each resistor. R_3 and R_4 are equal, so the same current should flow through each branch using the relationship:

$$I = \frac{E}{R}$$

$$I_{R3} = \frac{E_{R3}}{R_3}$$

$$I_{R3} = \frac{0.5V}{1k\Omega} \quad or \quad \frac{0.5V}{1000\Omega}$$

$$I_{R3} = 0.5mA \quad or \quad 0.0005A$$

Therefore, the total current for the circuit passes through R_1 and R_2 and is evenly divided between R_3 and R_4.

1.0.0 Section Review

1. Which of the following is true regarding a series circuit?

   a. It has more than one path for current flow.
   b. The current flow is the same through all resistors.
   c. The current flow varies depending on the strength of the resistor.
   d. The current flow through the resistors is always lower than the source current.

2. Which of the following is true regarding a parallel circuit?

   a. It has more than one path for current flow.
   b. The resistance is the same through all circuit components.
   c. The total resistance is equal to the sum of the individual resistances.
   d. The current flow through each resistor is always less than the total resistance.

3. Which of the following is true regarding a series-parallel circuit?

   a. Once reduced to an equivalent series circuit, the total resistance is found by adding the reciprocal of each resistance and then dividing by 1.
   b. The resistance is the same through all circuit components.
   c. If a circuit is reduced to an equivalent series circuit, the total resistance is found by adding the loads.
   d. The current flow through each resistor is always less than the total resistance.

4. Two branches of a circuit (R_1 and R_2) across a 120V power line draw a total line current of 30A. Branch R_1 takes 15A. What is the current I_2 in branch R_2?

   a. 12A
   b. 15A
   c. 17A
   d. 30A

2.0.0 KIRCHHOFF'S LAWS

Objective

Apply Kirchhoff's laws to various types of circuits.

    a. Use Kirchhoff's current law.
    b. Use Kirchhoff's voltage law.

Trade Terms

Kirchhoff's current law: The statement that the total amount of current flowing through a parallel circuit is equal to the sum of the amounts of current flowing through each current path.

Kirchhoff's voltage law: The statement that the sum of all the voltage drops in a circuit is equal to the source voltage of the circuit.

Kirchhoff's laws provide a simple, practical method of solving for unknown parameters in a circuit. Kirchhoff developed laws that applied to both voltage and current.

2.1.0 Kirchhoff's Current Law

In its most general form, **Kirchhoff's current law** states at any point in a circuit, the total current entering that point must equal the total current leaving that point. For parallel circuits, this implies that the current in a parallel circuit is equal to the sum of the currents in each branch.

When using Kirchhoff's laws to solve circuits, it is necessary to adopt conventions that determine the algebraic signs for current and voltage terms. A convenient system for current is to consider all current flowing into a branch point as positive, and all current directed away from that point as negative.

For example, in *Figure 13*, the currents can be written as:

$$I_A + I_B - I_C = 0 \quad or \quad 5A + 3A - 8A = 0$$

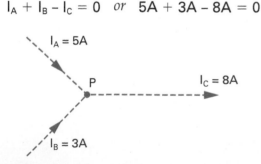

Figure 13 Kirchhoff's current law.

Currents I_A and I_B are positive terms because these currents flow into P, but I_C, directed out of P, is negative.

For a circuit application, refer to Point C at the top of the diagram in *Figure 14*. The total current I_T into Point C is 6A. It then divides into two currents, the I_3 (2A) and I_{4+5} (4A), which are both directed out of point C. Note that I_{4+5} is the current through R_4 and R_5. Write the algebraic equation again:

$$I_T - I_3 - I_{4+5} = 0$$

Then substitute the values for each current:

$$6A - 2A - 4A = 0$$

For the opposite direction, refer to Point D at the bottom of *Figure 14*. Here, the branch currents into Point D combine to equal the mainline current I_T returning to the voltage source. Now, I_T is directed out from Point D, with I_3 and I_{4+5} directed in. The algebraic equation is:

$$I_3 + I_{4+5} - I_T = 0$$
$$2A + 4A - 6A = 0$$

Note that at either Point C or Point D, the sum of the 2A and 4A branch currents must equal the 6A total line current. The amount of current coming into any point on a circuit must be equal to the amount of current leaving the circuit. Therefore, Kirchhoff's current law can also be stated as:

$$I_{IN} = I_{OUT}$$

For *Figure 14*, the equations for current can be written as follows.

At Point C:

$$6A = 2A + 4A$$

At Point D:

$$2A + 4A = 6A$$

Kirchhoff's current law is the basis for the practical rule in parallel circuits that the total line current must equal the sum of the branch currents.

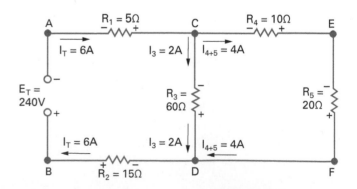

Figure 14 Application of Kirchhoff's current law.

2.2.0 Kirchhoff's Voltage Law

Kirchhoff's voltage law states that the algebraic sum of the voltages around any closed path is zero.

Referring to *Figure 15*, the sum of the voltage drops around the circuit must equal the voltage applied to the circuit:

$$E_A = E_1 + E_2 + E_3$$

Where:

E_A = voltage applied to the circuit
E_1, E_2, and E_3 = voltage drops in the circuit

Another way of stating this law is the algebraic sum of the voltage rises and voltage drops must be equal to zero. In the *Figure 15* example, the battery created the rise and the resistors created the drops. A voltage source is considered a voltage rise; a voltage across a resistor is a voltage drop. (For convenience in labeling, letter subscripts are shown for voltage sources and numerical subscripts are used for voltage drops.) This form of the law can be written by transposing the right members to the left side:

$$\text{Voltage applied} - \text{sum of voltage drops} = 0$$
$$E_A - E_1 - E_2 - E_3 = 0$$
$$E_A - (E_1 + E_2 + E_3) = 0$$

2.2.1 Loop Equations

Kirchhoff's voltage law may be applied more generally through something called a *loop equation*. Any closed circuit path for current flow is called a *loop*. A loop equation specifies the voltages around the loop.

Before writing loop equations, however, it is essential to develop a convention for deciding if a component in the loop is a voltage rise (+) or voltage drop (–). Do not assume that all sources (like batteries) are rises and all resistors are drops, as this convention can sometimes be faulty. Also, it doesn't work if you have a loop with nothing but resistors and no sources.

The following convention is simple and easy to apply. Work your way around the loop in whichever direction you wish. Whenever you encounter a component, note the polarity (positive or negative sign) that you encounter first. If the sign is positive, treat the component as a voltage rise. If the sign is negative, treat the component as a voltage drop. Examine *Figure 16*.

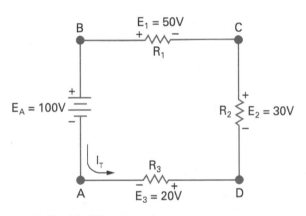

Figure 15 Kirchhoff's voltage law.

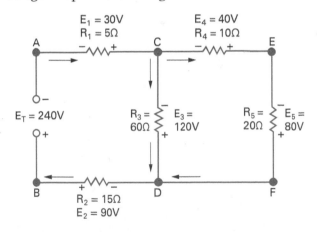

Figure 16 Loop equation.

Consider the inside loop through A, C, D, and B. This includes the voltage drops E_1, E_3, and E_2, and the source E_T. In a clockwise direction, starting at Point A, the algebraic sum of the voltages is:

$$- E_1 - E_3 - E_2 + E_T = 0$$
or
$$- 30V - 120V - 90V + 240V = 0$$

Voltages E_1, E_3, and E_2 have a negative value, because there is a decrease in voltage seen across each of the resistors in a clockwise direction. However, the source E_T is a positive term because an increase in voltage is seen in that same direction.

For the opposite direction, going counterclockwise in the same loop from Point B, E_T is negative while E_1, E_2, and E_3 have positive values. Therefore:

$$- E_T + E_2 + E_3 + E_1 = 0$$
or
$$- 240V + 90V + 120V + 30V = 0$$

When the negative term is transposed, the equation becomes:

$$240V = 90V + 120V + 30V$$

In this form, the loop equation shows that Kirchhoff's voltage law is really the basis for the practical rule in series circuits that the sum of the voltage drops must equal the applied voltage.

For example, determine the voltage E_B for the circuit shown in *Figure 17*. The direction of the current flow is shown by the arrow. First, mark the polarity of the voltage drops across the resistors and trace the circuit in the direction of the current flow starting at Point A. Then write the loop equation around the circuit:

$$- E_3 - E_B - E_2 - E_1 + E_A = 0$$

Solve for E_B:

$$E_B = E_A - E_3 - E_2 - E_1$$
$$E_B = 15V - 2V - 6V - 3V$$
$$E_B = 4V$$

Since E_B was found to be positive, the assumed direction of current is in fact the actual direction of current.

In its most general form, Kirchhoff's voltage law states that the algebraic sum of all the potential differences in a closed loop is equal to zero. A closed loop means any completely closed path consisting of wire, resistors, batteries, or other components. For series circuits, this implies that the sum of the voltage drops around the circuit is equal to the applied voltage. For parallel circuits, this implies that the voltage drops across all branches are equal.

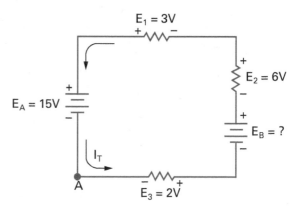

Figure 17 Applying Kirchhoff's voltage law.

2.0.0 Section Review

1. Who stated that the current entering a point must equal the total current leaving that point?

    a. Kirchhoff
    b. Joule
    c. Pascal
    d. Ohm

2. In a circuit with an applied voltage of 10V and voltage drops of $E_1 = 2V$, $E_2 = 5V$, and $E_3 = ?$, the value of E_3 must be _____.

    a. 3V
    b. 7V
    c. 12V
    d. 17V

1. True or False? You can use Ohm's law to find the value of each branch current in a parallel circuit.

   a. True
   b. False

2. Find the total resistance in a series circuit with three resistances of 10Ω, 20Ω, and 30Ω.

   a. 1Ω
   b. 15Ω
   c. 20Ω
   d. 60Ω

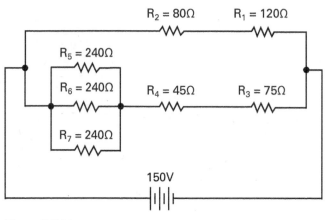

Figure RQ01

3. The total resistance in *Figure RQ01* is _____.

   a. 100Ω
   b. 129Ω
   c. 157Ω
   d. $1,040\Omega$

4. In a parallel circuit, the voltage across each path is equal to the _____.

   a. total circuit resistance times path current
   b. source voltage minus path voltage
   c. path resistance times total current
   d. applied voltage

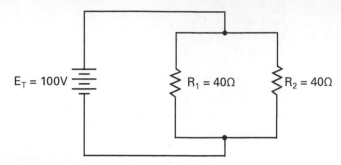

Figure RQ02

5. The value for total current in *Figure RQ02* is

   _____.

   a. 1.25A
   b. 2.50A
   c. 5A
   d. 10A

6. A resistor of 32Ω is in parallel with a resistor of 36Ω, and a 54Ω resistor is in series with the pair. When 350V is applied to the combination, the current through the 54Ω resistor is

   _____.

   a. 2.87A
   b. 3.26A
   c. 4.93A
   d. 5.86A

7. A 242Ω resistor is in parallel with a 180Ω resistor, and a 420Ω resistor is in series with the combination in a 27V circuit. A current of 22mA flows through the 242Ω resistor. The current through the 180Ω resistor is _____.

   a. 29.6mA
   b. 36.4mA
   c. 59.4mA
   d. 60.3mA

8. Which statement is *not* correct relative to Kirchhoff's current law?

   a. $I_A + I_B - I_C = 0$.
   b. The total current entering a point must equal total current leaving a point.
   c. The total line current must equal the sum of the branch currents.
   d. Kirchhoff's laws apply to current only.

9. Kirchhoff's voltage law states that the algebraic sum of the voltages around any closed current path is _____.

   a. infinity
   b. zero
   c. twice the current
   d. always less than the individual voltages duc to voltage drop

10. Two 24Ω resistors are in parallel, and a 42Ω resistor is in series with the combination. When 78V is applied to the three resistors, the voltage drop across the 42Ω resistor is about _____.

    a. 49.8V
    b. 55.8V
    c. 60.5V
    d. 65.3V

Trade Terms Quiz

Fill in the blank with the correct term that you learned from your study of this module.

1. _____ states that the total amount of current flowing through a parallel circuit is equal to the sum of the amounts of current flowing through each current path.

2. _____ states that the sum of all the voltage drops in a circuit is equal to the source voltage of the circuit.

3. _____ contain both series and parallel current paths.

4. _____ contain two or more parallel paths through which current can flow.

5. A _____ contains only one path for current flow.

Trade Terms

Kirchhoff's current law
Kirchhoff's voltage law
Parallel circuits

Series circuit
Series-parallel circuits

1. Which formula is used for series circuits?

    a. $R_T = \dfrac{R}{N}$

    b. $R_T = \dfrac{1}{\dfrac{1}{R_1} + \dfrac{1}{R_2} + \dfrac{1}{R_3}}$

    c. $R_T = R_1 + R_2 + R_3$

    d. $R_T = \dfrac{R_1 \times R_2}{R_1 + R_2}$

2. If two or more resistors are connected in parallel, the _____.

    a. total resistance is higher than any single resistor

    b. total resistance is lower than any single resistor

    c. current flow varies based on voltage fluctuation

    d. resistance depends on the power rating of the circuit

3. The formula for calculating the total resistance in a series circuit with three resistors is _____.

    a. $R_T = R_1 + R_2 + R_3$
    b. $R_T = R_1 - R_2 - R_3$
    c. $R_T = R_1 \times R_2 \times R_3$

    d. $R_T = \dfrac{1}{\dfrac{1}{R_1} + \dfrac{1}{R_2} + \dfrac{1}{R_3}}$

4. True or False? Parallel circuits are current dividers.

5. The formula for calculating the total resistance in a parallel circuit with three resistors is _____.

    a. $R_T = R_1 + R_2 + R_3$
    b. $R_T = R_1 - R_2 - R_3$
    c. $R_T = R_1 \times R_2 \times R_3$

    d. $R_T = \dfrac{1}{\dfrac{1}{R_1} + \dfrac{1}{R_2} + \dfrac{1}{R_3}}$

6. True or False? In a series circuit, E_T must equal the sum of the individual voltage drops.

7. When calculating series-parallel combination circuits, it is important to _____.

    a. reduce to series loads, then add the loads

    b. replace parallel combinations with one value

    c. reduce to simpler circuits where possible

    d. All of the above.

8. True or False? Kirchhoff's voltage law can be used to determine voltage drops.

9. Two resistors are connected in parallel; R_1 is 90 ohms, R_2 is 45 ohms. What is the total resistance of R_1 and R_2 in parallel?

    _____

10. Two resistors are connected in parallel; R_2 is 90 ohms, R_3 is 45 ohms. A third resistor (R_1, which is 20 ohms) is connected in series with R_2 and R_3.

    a. What is the total resistance of this circuit?

    _____

    b. If the voltage source of this circuit is 150V, what is the total current flow?

    _____

    (Hint: sketch the circuit before solving.)

James Mitchem

JEM Electrical Consulting Services

Jim Mitchem owns his own electrical consulting firm. During his career in the electrical industry, he worked his way up from apprentice to technical services manager, and is now a consultant on large commercial and industrial installations.

How did you become an electrician?

Quite by accident. A couple of years after college, I was working as a relief operator in a plant when the lead electrician retired, creating a vacancy. I liked the idea that electricians were expected to use their knowledge and initiative to keep the place running. I applied and was accepted as a trainee.

How did you get your training?

I took an electrical apprenticeship course by correspondence, and I was fortunate enough to work with good people who helped me along. I worked in an environment that exposed me to a variety of equipment and applications, and just about everyone I've ever worked with has taught me something. Now I'm passing my knowledge on to others.

What kinds of work have you done in your career?

I've worked as an apprentice, journeyman, instrument and controls technician, instrument fitter, foreman, general foreman, superintendent, and startup engineer. Each of these positions required that I learn new skills, both technical and managerial. My experience in many disciplines and types of projects has given me a high level of credibility with my clients.

Now I act as a technical resource and troubleshooter in functions such as safety, quality assurance, and training. I visit job sites to help solve problems and help out with commissioning and startup.

What factor or factors have contributed the most to your success?

There are several factors. Two very important ones have been a desire to learn and a willingness to do whatever is asked of me. I also keep an eye on the big picture. When I'm on a job, I'm not just pulling wire, I'm building a power plant or whatever the project is.

Any advice for apprentices just beginning their careers?

Keep learning! And don't depend on others to train you. Take the initiative to buy or borrow books and trade journals. Take licensing tests and do whatever is necessary to keep your licenses current. Finally, make sure you know your own personal and professional values and work with a company that shares those values.

Trade Terms Introduced in This Module

Kirchhoff's current law: The statement that the total amount of current flowing through a parallel circuit is equal to the sum of the amounts of current flowing through each current path.

Kirchhoff's voltage law: The statement that the sum of all the voltage drops in a circuit is equal to the source voltage of the circuit.

Parallel circuits: Circuits containing two or more parallel paths through which current can flow.

Series circuit: A circuit that has only one path for current flow.

Series-parallel circuits: Circuits that contain both series and parallel current paths.

Additional Resources

This module presents thorough resources for task training. The following resource material is recommended for further study.

Electronics Fundamentals: Circuits, Devices, and Applications, Thomas L. Floyd. New York, NY: Pearson Education, Inc.
Principles of Electric Circuits, Thomas L. Floyd. New York, NY: Pearson Ecustion, Inc.

Figure Credit

Greenlee / A Textron Company, Module Opener

SECTION 1.0.0

Answer	Section Reference	Objective
1. b	1.1.0	1a
2. a	1.2.0	1b
3. c	1.3.1	1c
4. b	1.4.2	1d

SECTION 2.0.0

Answer	Section Reference	Objective
1. a	2.1.0	2a
2. a	2.2.0	2b

1.0.0 SECTION REVIEW

Question 4

Since the total current (I_T) is equal to the sum of the branch currents, you can use this equation to find the unknown branch current (I_2), as follows:

$$I_T = I_1 + I_2$$
$$I_2 = I_T - I_1$$
$$I_2 = 30 - 15 = 15A$$

The current I_2 in branch R_2 is **15A**.

2.0.0 SECTION REVIEW

Question 2

Use Kirchhoff's voltage law, and use the principles of algebra (rearrange the equation) to solve for E_3:

$$E_A = E_1 + E_2 + E_3$$
$$E_3 = E_A - (E_1 + E_2)$$
$$E_3 = 10V - (2V + 5V)$$
$$E_3 = 10V - 7V$$
$$E_3 = 3V$$

The value of E_3 is **3V**.

NCCER CURRICULA — USER UPDATE

NCCER makes every effort to keep its textbooks up-to-date and free of technical errors. We appreciate your help in this process. If you find an error, a typographical mistake, or an inaccuracy in NCCER's curricula, please fill out this form (or a photocopy), or complete the online form at **www.nccer.org/olf**. Be sure to include the exact module ID number, page number, a detailed description, and your recommended correction. Your input will be brought to the attention of the Authoring Team. Thank you for your assistance.

Instructors – If you have an idea for improving this textbook, or have found that additional materials were necessary to teach this module effectively, please let us know so that we may present your suggestions to the Authoring Team.

NCCER Product Development and Revision

13614 Progress Blvd., Alachua, FL 32615

Email: curriculum@nccer.org
Online: www.nccer.org/olf

❑ Trainee Guide   ❑ Lesson Plans   ❑ Exam   ❑ PowerPoints   Other _____

Craft / Level: _____   Copyright Date: _____

Module ID Number / Title: _____

Section Number(s): _____

Description: _____

_____

_____

_____

_____

Recommended Correction: _____

_____

_____

_____

_____

Your Name: _____

Address: _____

_____

Email: _____   Phone: _____

This page is intentionally left blank.

Introduction to the
National Electrical Code®

OVERVIEW

The *NEC®* is one of the most important tools for electricians. When used together with the applicable electrical code for your local area, the *NEC®* provides the minimum requirements for the installation of electrical systems. This module describes the purpose of the *NEC®* and explains how to use it to find the installation requirements for various electrical devices and wiring methods. It also provides an overview of the National Electrical Manufacturers Association and Nationally Recognized Testing Laboratories.

Module 26105-20

Trainees with successful module completions may be eligible for credentialing through the NCCER Registry. To learn more, go to **www.nccer.org** or contact us at 1.888.622.3720. Our website, **www.nccer.org**, has information on the latest product releases and training.

Your feedback is welcome. You may email your comments to **curriculum@nccer.org**, send general comments and inquiries to **info@nccer.org**, or fill in the User Update form at the back of this module.

This information is general in nature and intended for training purposes only. Actual performance of activities described in this manual requires compliance with all applicable operating, service, maintenance, and safety procedures under the direction of qualified personnel. References in this manual to patented or proprietary devices do not constitute a recommendation of their use.

26105-20 V10.0

INTRODUCTION TO THE NATIONAL ELECTRICAL CODE®

Objectives

When you have completed this module, you will be able to do the following:

1. Explain the purpose and history of the *NEC*®.
   a. Identify key dates in the history of the *NEC*®.
   b. Describe how changes are made to the *NEC*®.
   c. Identify the other organizations that produce standards for the manufacture and use of electrical products.
2. Navigate the *NEC*®.
   a. Explain the layout of the *NEC*®.
   b. Use the *NEC*® to find specific installation requirements.

Performance Tasks

Under the supervision of the instructor, you should be able to do the following:

1. Use *NEC Article 90* to determine the scope of the *NEC*®. State what is covered and what is not covered by the *NEC*®.
2. Find the definition of the term *feeder* in the *NEC*®.
3. Look up the *NEC*® requirements needed to install an outlet near a swimming pool.
4. Find the minimum wire bending space required for two No. 1/0 AWG conductors installed in a junction box or cabinet and entering opposite the terminal.

Trade Terms

Articles
Chapters
Exceptions
Informational Note
Institute for Electrical and Electronics Engineers (IEEE)
International Electrotechnical Commission (IEC)

National Electrical Manufacturers Association (NEMA)
National Fire Protection Association (NFPA)
Nationally Recognized Testing Laboratories (NRTLs)
Parts
Sections

Industry Recognized Credentials

If you are training through an NCCER-accredited sponsor, you may be eligible for credentials from NCCER's Registry. The ID number for this module is 26105-20. Note that this module may have been used in other NCCER curricula and may apply to other level completions. Contact NCCER's Registry at 888.622.3720 or go to **www.nccer.org** for more information.

Contents

Figures

1.0.0 PURPOSE AND HISTORY OF THE NEC®

Objective

Explain the purpose and history of the NEC®.

a. Identify key dates in the history of the NEC®.
b. Describe how changes are made to the NEC®.
c. Identify the other organizations that produce standards for the manufacture and use of electrical products.

Trade Terms

Articles: The articles are the main topics of the NEC®, beginning with *NEC Article 90, Introduction*, and ending with *NEC Article 840, Premises-Powered Broadband Communications Systems*.

Institute for Electrical and Electronics Engineers (IEEE): A professional organization that develops international standards impacting electronics, telecommunications, information technology, and power generation products and services.

International Electrotechnical Commission (IEC): An international organization that develops consensus standards for all electrical and electronic technologies.

National Electrical Manufacturers Association (NEMA): The association that maintains and improves the quality and reliability of electrical products.

National Fire Protection Association (NFPA): The publisher of the NEC®. The NFPA develops codes and standards to minimize the possibility and effects of fire.

Nationally Recognized Testing Laboratories (NRTLs): Product safety certification laboratories that are responsible for testing and certifying electrical equipment.

NFPA 70®, also known as the *National Electrical Code® (NEC®)*, is published by the National Fire Protection Association (NFPA). The NEC® is one of the most important tools for the electrician. When used together with the applicable electrical code for your local area, the NEC® provides the minimum requirements for the installation of electrical systems. Unless otherwise specified, always use the latest edition of the NEC® as your on-the-job reference. It specifies the minimum provisions necessary for protecting people and property from electrical hazards. In some areas, however, local laws may specify different editions of the NEC®, so be sure to use the edition specified by your employer. Also, bear in mind that the NEC® specifies only minimum requirements, so local or job requirements may be more stringent.

The primary purpose of the NEC® is the practical safeguarding of persons and property from hazards arising from the use of electricity [*NEC Section 90.1(A)*]. A thorough knowledge of the NEC® is one of the first requirements for becoming a trained electrician. The NEC® is probably the most widely used and generally accepted code in the world. It has been translated into several languages. It is used as an electrical installation, safety, and reference guide in the United States. Compliance with NEC® requirements increases the safety of electrical installations—the reason the NEC® is so widely used.

Although *NEC Section 90.1(A)* states, "This *Code* is not intended as a design specification or an instruction manual for untrained persons," it does provide a sound basis for the study of electrical installation procedures—under the proper guidance. The NEC® has become the standard reference in the electrical construction industry. Anyone involved in electrical work should obtain the latest edition and refer to it frequently.

All electrical work must comply with the currently adopted NEC® and all local ordinances. Like most codes, the NEC® gets easier to work with once you understand the terminology and how the information is structured.

> **NOTE**
> This module is not a substitute for the NEC®. You must acquire a copy of the most recent edition and keep it handy at all times. The more you know about the NEC®, the better an electrician you will become.

Think About It

The NEC®

Why do you think it's necessary to have a standard set of procedures for electrical installations? Find out who does the electrical inspection in your area. Who determines what will be inspected, when it will be inspected, and who will do the inspection?

1.1.0 Evolution of the *NEC*®

The *NEC*® has a rich history, as shown in the timeline in *Figure 1*. From its conceptual beginnings in the late 19th century to its wide adoption today, the *NEC*® has become the primary driver for ensuring the safe installation of electrical systems. In 1881, the National Association of Fire Engineers met in Richmond, Virginia. From this meeting came the idea to draft the first *National Electrical Code*®. The first nationally recommended electrical code was published by the National Board of Fire Underwriters (now the American Insurance Association) in 1895.

In 1896, the National Electric Light Association (NELA) was working to make the requirements of the fire insurance organizations and electrical utilities fit together. NELA succeeded in promoting a conference that would result in producing a standard national code. The NELA code would serve the interests of the insurance industry, operating concerns, manufacturing, and industry.

The conference produced a set of requirements that was unanimously accepted. In 1897, the first edition of the *NEC*® was published, and the *NEC*® became the first cooperatively produced national code. The organization that produced the *NEC*® was known as the *National Conference on Standard Electrical Rules*. This group became a permanent organization, and its job was to develop the *NEC*®.

In 1911, the NFPA took over administration and control of the *NEC*®. However, the National Board of Fire Underwriters continued to publish the *NEC*® until 1962. From 1911 until now, the *NEC*® has experienced several major changes, as well as regular three-year updates. In 1923, the *NEC*® was rearranged and rewritten, and in 1937, it was editorially revised.

In 1949, the NFPA reorganized the *NEC*® into its present structure. The present structure consists of a Correlating Committee and Code-Making Panels (CMPs). The Correlating Committee consists of principal voting members and alternates. The principal function of the Correlating Committee is to ensure that:

- No conflict of requirements exists
- Correlation has been achieved
- NFPA regulations governing committee projects have been followed
- A practical schedule of revision and publication is established and maintained

Each of the CMPs has members who are experts on particular subjects and have been assigned certain articles to supervise and revise as required. Members of the CMPs represent special interest groups such as trade associations, electrical contractors, electrical designers and engineers, electrical inspectors, electrical manufacturers and suppliers, electrical testing laboratories, and insurance organizations. NFPA membership is drawn from the fields listed above.

Each panel is structured so that not more than one-third of its members are from a single interest group. The members of the *NEC*® CMPs create or revise requirements for the *NEC*® through researching, debating, analyzing, weighing, and reviewing new input. These proposed changes are then reviewed by the Correlating Committee, which ensures that the changes are consistent between articles of the *NEC*® as well as other NFPA codes and standards.

In addition to publishing the *NEC*®, the duties of the NFPA include the following:

- Developing, publishing, and distributing standards that are intended to minimize the possibility and effects of fire and explosion
- Conducting fire safety education programs for the general public
- Providing information on fire protection, prevention, and suppression
- Compiling annual statistics on causes and occupancies of fires, large-loss fires (over one million dollars), fire deaths, and firefighter casualties
- Providing field service by specialists on electricity, flammable liquids and gases, and marine fire concerns
- Conducting research projects that apply statistical methods and operations research to develop computer models and data management systems

In 1959, the *NEC*® adopted a new numbering system. In 1962, the NFPA took over publishing the *NEC*® from the NBFU.

1.2.0 *NEC*® Revision Process

Anyone can contribute to the *NEC*® development and change process, regardless of whether they are on a formal NFPA or *NEC*® committee. Proposals are reviewed by technical committees made up of representatives of major interest groups, individuals with specific expertise, or panel members who represent particular industries.

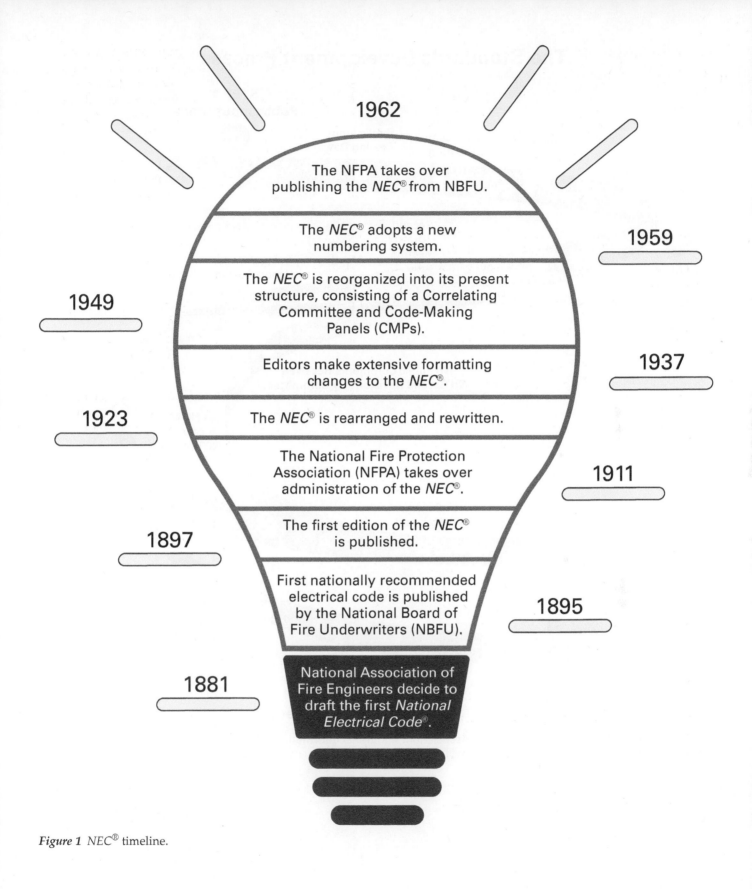

1962

The NFPA takes over publishing the *NEC*® from NBFU.

The *NEC*® adopts a new numbering system.

1959

The *NEC*® is reorganized into its present structure, consisting of a Correlating Committee and Code-Making Panels (CMPs).

1949

1937

Editors make extensive formatting changes to the *NEC*®.

The *NEC*® is rearranged and rewritten.

1923

The National Fire Protection Association (NFPA) takes over administration of the *NEC*®.

1911

The first edition of the *NEC*® is published.

1897

First nationally recommended electrical code is published by the National Board of Fire Underwriters (NBFU).

1895

National Association of Fire Engineers decide to draft the first *National Electrical Code*®.

1881

Figure 1 *NEC*® timeline.

The Standards Development Process

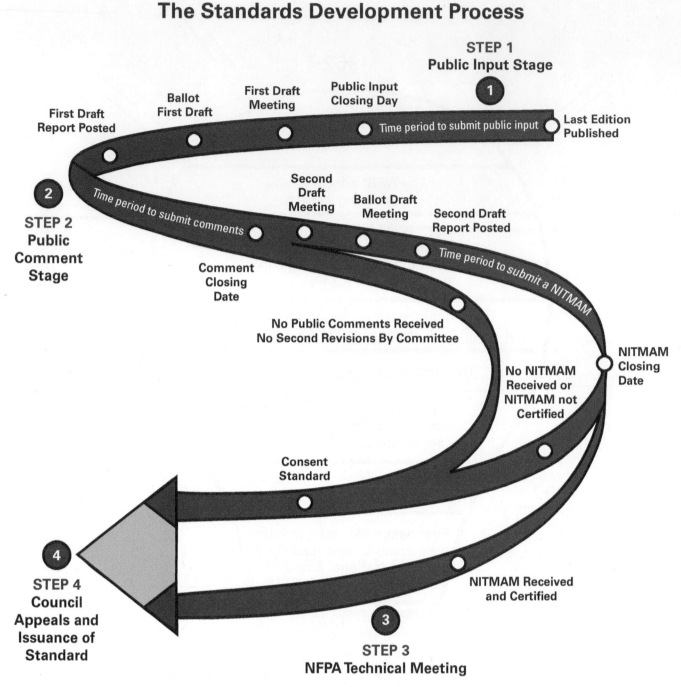

STEP 1
Public Input Stage
①

Last Edition Published

Time period to submit public input

Public Input Closing Day

First Draft Meeting

Ballot First Draft

First Draft Report Posted

STEP 2
Public Comment Stage
②

Time period to submit comments

Comment Closing Date

Second Draft Meeting

Ballot Draft Meeting

Second Draft Report Posted

Time period to submit a NITMAM

No Public Comments Received
No Second Revisions By Committee

NITMAM Closing Date

No NITMAM Received or NITMAM not Certified

Consent Standard

NITMAM Received and Certified

STEP 4
Council Appeals and Issuance of Standard
④

STEP 3
NFPA Technical Meeting
③

Copyright © 2019, National Fire Protection Association. Additional information on the NFPA standards development process can be obtained through the NFPA web site at *www.nfpa.org*.

Figure 2 NEC® standards development process.

By using the NFPA website, individuals can submit their proposal. Each proposal follows the path shown in *Figure 2*. The four steps of the NFPA standards development process are:

Step 1 *Public Input* – NFPA accepts input through its website at *www.nfpa.org/X* (where X = the standard number). For example, the internet address to submit input for the *National Electrical Code®* (NFPA 70®) is *www.nfpa.org/70*.

Step 2 *Public Comment* – Comments related to the first draft are submitted during this step. These comments are reviewed by the Technical Committee during its Second Draft Meeting. After this period, amending motions to any previously submitted comments can be made using the Notice of Intent to Make a Motion (NITMAM) procedure.

Step 3 *NFPA Technical Meeting* – After the Public Input and Public Comment stages, additional discussion about the standard is held at the NFPA Technical Meeting, which occurs at the annual NFPA Conference and Expo.

Step 4 *Standards Council Action* – When the council convenes, it considers any appeals and determines whether to issue the standard. If the decision is made to proceed, the standard takes effect 20 days after issuance.

1.3.0 Other Organizations and Laboratories

In addition to the NFPA, there are many other organizations that produce standards for the manufacture and/or use of electrical products.

What's wrong with these pictures?

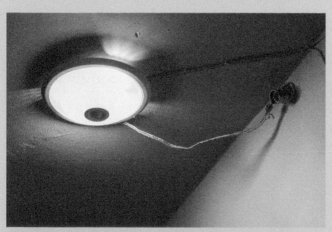

Figure Credit: iStock@Veni vidi...shoot

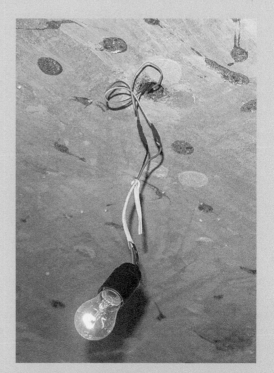

Figure Credit: 123RF@Aleksandr Proshkin

Figure Credit: 123RF@Thanayu Jongwattanasilkul

1.3.1 Nationally Recognized Testing Laboratories

Nationally Recognized Testing Laboratories (NRTLs) are product safety certification laboratories. These laboratories perform extensive testing of new products to make sure they are built to established standards for electrical and fire safety. NRTLs establish and operate product safety certification programs to make sure that items produced under the service are safeguarded against reasonably foreseeable risks. NRTLs maintain a worldwide network of field representatives who make unannounced visits to factories to check products bearing their safety marks.

OSHA publishes a list of Nationally Recognized Testing Laboratories. The following organizations are currently recognized by OSHA as NRTLs:

- Bay Area Compliance Laboratories
- Canadian Standards Association (CSA) (also known as *CSA International*)
- Communication Certification Laboratory, Inc. (CCL)
- Curtis-Straus LLC (CSL)
- Eurofins MET Labs (MET) (formerly MET Laboratories, Inc.)
- FM Approvals LLC
- International Association of Plumbing and Mechanical Officials EGS (IAPMO)
- Intertek Testing Services NA, Inc. (ITSNA) (formerly ETL)
- Nemko USA, Nemko North America, Inc. (NNA)
- NSF International (NSF)
- QAI Laboratories, LTD (QAI)
- QPS Evaluation Services, Inc. (QPS)
- SGS U.S. Testing Company, Inc. (SGSUS) (formerly UST-CA)
- Southwest Research Institute (SWRI)
- TÜV Rheinland of North America (TÜV)
- TÜV Rheinland PTL, LLC (TÜVPTL)
- TÜV SÜD America (TÜVAM)
- TÜV SÜD Product Services GmbH (TÜVPSG)
- Underwriters Laboratories (UL)

1.3.2 National Electrical Manufacturers Association

The National Electrical Manufacturers Association (NEMA) was founded in 1926. It is made up of companies that manufacture equipment used for generation, transmission, distribution, control, and utilization of electric power. The objectives of NEMA are to maintain and improve the quality and reliability of products, to ensure safety standards in the manufacture and use of products, and to develop product standards covering such matters as naming, ratings, performance, testing, and dimensions. NEMA participates in developing the *NEC®* and advocates its acceptance by state and local authorities.

Think About It

Code Changes

This photograph shows the first code book and the current code book. Code changes occur every three years. Who can suggest changes to the *National Electrical Code®*? What might be reasons for submitting changes?

1.3.3 Standards and Globalization

An important but often misunderstood role of standards is their effect on world markets. Two organizations that play critical roles in worldwide electrical standards are the International Electrotechnical Commission (IEC) and the Institute for Electrical and Electronics Engineers (IEEE). IEC's focus is preparing and publishing international standards for electrical and electronic technologies. IEEE is a professional organization that develops international standards impacting electronics, telecommunications, information technology, and power-generation products and services. Both organizations, along with the assistance of other standards groups, have begun working toward the globalization of electronics standards. This is important because many stages of the electronics manufacturing process involve the manufacture, distribution, and assembly of parts from multiple countries. This has exposed the need for private and public organizations across the globe to create international standards that protect consumer safety while creating a framework that shares knowledge, promotes trade, and increases economic progress for all participating nations.

1.0.0 Section Review

1. Today, the *NEC*® is administered by the _____.
   a. DOT
   b. OSHA
   c. NFPA
   d. DOE

2. The internet address to submit input for the *National Electrical Code*® is _____.
   a. **www.nfpa.org/70**
   b. **www.nfpa.org/72**
   c. **www.nfpa.org/77**
   d. **www.nfpa.org/80**

3. Two organizations working toward the globalization of electronics standards are _____.
   a. IEC and IEEE
   b. IEEE and NELA
   c. NFPA and OSHA
   d. IEC and NBFU

2.0.0 NAVIGATING THE *NEC*®

Objective

Navigate the *NEC*®.
a. Explain the layout of the *NEC*®.
b. Use the *NEC*® to find specific installation requirements.

Performance Tasks

1. Use *NEC Article 90* to determine the scope of the *NEC*®. State what is covered and what is not covered by the *NEC*®.
2. Find the definition of the term *feeder* in the *NEC*®.
3. Look up the *NEC*® requirements needed to install an outlet near a swimming pool.
4. Find the minimum wire bending space required for two No. 1/0 AWG conductors installed in a junction box or cabinet and entering opposite the terminal.

Trade Terms

Chapters: Chapters contain a group of articles related to a broad category. Nine chapters form the broad structure of the *NEC*®.

Exceptions: Exceptions follow the applicable sections of the *NEC*® and allow alternative methods to be used under specific conditions.

Informational Note: Explanatory material that follows specific *NEC*® sections.

Parts: Certain articles in the *NEC*® are subdivided into parts that cover a specific topic. Parts have Roman numeral designations (e.g., *NEC Article 250, Part IX, Instruments, Meters, and Relays*).

Sections: Parts and articles are subdivided into sections. Sections have numeric designations that follow the article number and are preceded by a period (e.g., *NEC Section 501.4*).

In order to make the best use of the *NEC*®, you must first become familiar with its overall structure (see *Figure 3*). The *NEC*® is organized into nine **chapters**, followed by informative annexes, the index, a schedule for the next code cycle, and information on submitting proposed changes.

The main body of the *NEC*® begins with *NEC Article 90, Introduction*. This introduction provides an overview of the *NEC*®. Topics covered in this article include the following:

- Purpose of the *NEC*®
- Scope of the code book
- Code arrangement
- Enforcement
- Mandatory rules, permissive rules, and explanatory material
- Formal interpretation
- Examination of equipment for safety
- Wiring planning
- Units of measurement

NEC Chapters 1 through 8 are also subdivided into articles. Each chapter focuses on a general category of electrical application, such as *NEC Chapter 2, Wiring and Protection*. Articles are subdivided into **parts**, such as *NEC Article 210, Part I, General Provisions*. Each part contains one or more **sections**. Each section gives examples of a specific application of the *NEC*®, such as *NEC Section 210.4, Multiwire Branch Circuits*.

NEC Chapter 9 contains tables referenced by any of the articles in *NEC Chapters 1 through 8. Informative Annexes A through J* provide additional information for applying *NEC*® requirements.

The *NEC*® uses several types of text or typography. Here is an explanation of each of them:

- *Bold black letters* – Headings for each *NEC*® application are written in **bold** black letters.
- Exceptions – Explain the circumstances under which a specific part of the *NEC*® does not apply. Exceptions are written in *italics* under the part of the *NEC*® to which they pertain.

NEC® Layout

Remember, chapters contain a group of articles relating to a broad category. An article is a specific subject within that category, such as *NEC Article 250, Grounding and Bonding*, which is in *NEC Chapter 2* relating to wiring and protection. When an article applies to different installations in the same category, it will be divided into parts using Roman numerals. Any specific requirements in any of the articles may also have exception(s) to the main rules.

- **Informational Note** – Explains something in an application, suggests other sections to read about the application, or provides tips about the application. It is defined in the text by the term *Informational Note* before an indented paragraph.
- *Figures* – May be included with explanations to show how an application might appear.
- *Tables* – Often included to provide more detailed application requirements, such as for different wire sizes, box sizes, or motor types.

In addition to different types of text, the *NEC®* contains two types of rules, as defined in *NEC Section 90.5*: mandatory rules and permissive rules, which are defined as follows:

- *Mandatory rules* – Mandatory rules contain the words "shall" or "shall not" and must be obeyed.
- *Permissive rules* – Permissive rules identify actions that are allowed but not required and typically cover options or alternative methods. Permissive rules are indicated by the phrases "shall be permitted" or "shall not be required."

> **NOTE**
> Be aware that local ordinances may amend requirements of the *NEC®*. This means that a state, city, or county may have additional requirements, exceptions, or prohibitions that must be followed in that jurisdiction.

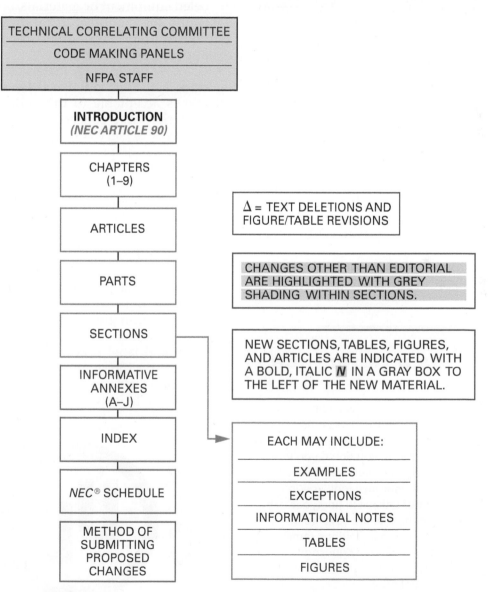

Figure 3 The layout of the *NEC®*.

2.1.0 Layout of the *NEC®*

The chapters of the *NEC®* are organized into the following major categories:

- *NEC Chapters 1*, *2*, *3*, *and 4* – The first four chapters present the rules for the design and installation of electrical systems. They generally apply to all electrical installations.
- *NEC Chapters 5*, *6*, *and 7* – These chapters are concerned with special occupancies, equipment, and conditions. Rules in these chapters may modify or amend those in any of the first seven chapters. See *NEC Figure 90.3*, *Code Arrangement*.
- *NEC Chapter 8* – This chapter covers various types of communications systems. It is not subject to the requirements of *Chapters 1 through 7* unless the requirements are specifically referenced in *Chapter 8*.
- *NEC Chapter 9* – This chapter contains tables that are applicable when referenced by other chapters in the *NEC®*.
- *Informative Annexes A through J* – These annexes contain helpful information that is not mandatory.

2.1.1 *NEC Chapter 1, General*

NEC Article 100 contains a list of common definitions used in the *NEC®*. Definitions specific to a single article are included in the second section of each article. For example, *NEC Section 240.2* contains definitions specific to overcurrent protection.

Two definitions from *NEC Article 100* you should become familiar with are:

- *Labeled* – "Equipment or materials to which has been attached a label, symbol, or other identifying mark of an organization that is acceptable to the authority having jurisdiction and concerned with product evaluation, that maintains periodic inspection of production of labeled equipment or materials, and by whose labeling the manufacturer indicates compliance with appropriate standards or performance in a specified manner."
- *Listed* – "Equipment, materials, or services included in a list published by an organization that is acceptable to the authority having jurisdiction and concerned with evaluation of products or services, that maintains periodic inspection of production of listed equipment or materials, or periodic evaluation of services, and whose listing states that either the equipment, material, or service meets appropriate designated standards or has been tested and found suitable for a specified purpose."

The definitions for labeled and listed are important because in addition to the installation rules, you must pay close attention to the type and quality of materials used in electrical wiring systems. Nationally recognized testing laboratories are product safety certification laboratories that list and label products to indicate that they have met specific safety standards. Underwriters Laboratories, also called *UL*, is one such laboratory. The UL label is shown in *Figure 4*.

Figure 4 Underwriters Laboratories label.

NEC Article 110 provides the general requirements for electrical installations. It contains information about marking requirements, required working space around electrical equipment, and more. It is important for you to become familiar with this information. One section that you will often hear referenced is *NEC Section 110.12*, which requires that all electrical equipment be installed in a neat and workmanlike manner. In other words, an untidy job site is not only a safety hazard, it is also a violation of the *NEC®*.

2.1.2 NEC Chapter 2, Wiring and Protection

NEC Chapter 2 discusses wiring design and protection, the information electricians need most often. It covers the use and identification of grounded conductors, branch circuits, feeders, calculations, services, overcurrent protection, grounding, and surge protective devices. This is essential information for all types of electrical systems. If you encounter a problem related to the design or installation of a conventional electrical system, this chapter should provide the solution.

NEC Article 200 contains important information on the use and identification of grounded conductors.

NEC Articles 210 and 215 cover the provisions and requirements for branch circuits, their ratings and required outlets, and the installation and overcurrent protection requirements for feeders.

Underwriters Laboratories

The Chicago World's Fair was opened in 1893 and, thanks to Edison's introduction of the electric light bulb, included a light display named the *Palace of Electricity*. But all was not perfect—wires soon sputtered and crackled, and, ironically, the Palace of Electricity caught fire. The fair's insurance company brought in a troubleshooting engineer named William Henry Merrill, who found faulty and brittle insulation, worn-out and deteriorated wiring, bare wires, and overloaded circuits. Merrill called for standards in the electrical industry and then set up a testing laboratory above a Chicago firehouse to do just that. As a result, an independent testing organization, Underwriters Electrical Bureau, was born. Now known as *Underwriters Laboratories (UL)*, this organization is now an internationally recognized authority on product safety testing, safety certification, and standards development.

NEC Article 220 contains the requirements for calculating branch circuit, feeder, and service loads, including the calculation methods for farm loads.

NEC Articles 225 and 230 cover outside branch circuit, feeder, and service installation requirements.

NEC Article 240 covers the requirements for overcurrent protection and overcurrent protective devices, including the standard ratings for fuses and fixed-trip circuit breakers. *NEC Article 240, Part IX* covers overcurrent protection for installations over 1,000 volts (V).

NEC Article 242 covers the requirements for surge protective devices (1,000V or less) and surge arrestors (over 1,000V).

NEC Article 250 covers the requirements for grounding and bonding electrical systems. It lists the systems required, permitted, and not permitted to be grounded and provides the requirements for grounding connection locations. It also covers accepted methods of grounding and bonding, including types and sizes of grounding and bonding conductors and electrodes.

2.1.3 NEC Chapter 3, Wiring Methods and Materials

NEC Chapter 3 lists the rules on wiring methods and materials. The materials and procedures to use on a particular system depend on the type of building construction, type of occupancy, location of the wiring in the building, type of atmosphere in the building or in the area surrounding the building, and mechanical factors. Note that the general requirements for conductors and wiring methods that form an integral part of manufactured equipment are not included in the requirements of the code per *NEC Section 300.1(B)*.

NEC Article 300 provides the general requirements for all wiring methods and materials, including information such as minimum cover requirements and permitted wiring methods for areas above suspended ceilings.

NEC Article 310 contains a description of acceptable conductors for the wiring methods contained in *NEC Chapter 3*.

NEC Articles 312 and 314 give rules for boxes, cabinets, conduit bodies, meter socket enclosures, and raceway fittings. Outlet boxes vary in size and shape, depending on their use, the size of the raceway, the number of conductors entering the box, the type of building construction, and the atmospheric conditions of the area. These articles should answer most questions on the selection and use of these items.

The *NEC®* does not describe in detail all types and sizes of outlet boxes. However, the manufacturers of outlet boxes provide excellent catalogs showing their products. Collect these catalogs, since these items are essential to your work.

NEC Articles 320 through 340 cover cables of one or more conductors, such as nonmetallic-sheathed and metal-clad cable.

NEC Articles 342 through 356 cover conduit wiring systems, such as rigid and flexible metal and nonmetallic conduit.

NEC Articles 358 through 362 cover tubing wiring methods, such as electrical metallic and nonmetallic tubing.

NEC Articles 366 through 390 cover other wiring methods, such as busways and wireways. *NEC Article 392* covers cable trays. *NEC Article 393* covers low-voltage suspended ceiling power distribution systems.

Think About It

Junction Boxes

Find the rule in the *NEC®* that explains whether a junction box without devices can be supported solely by two or more lengths of rigid metal conduit (RMC). Explain the technical terminology in everyday language.

2.1.4 NEC Chapter 4, Equipment for General Use

NEC Article 400 covers the use and installation of flexible cords and cables, including the trade name, type letter, wire size, number of conductors, conductor insulation, outer covering, and use of each type. *NEC Article 402* covers fixture wires, again giving the trade name, type letter, and other important details.

NEC Article 404 covers the requirements for the uses and installation of switches, switching devices, and circuit breakers where used as switches.

NEC Article 406 covers the requirements for the installation of receptacles, cord connectors, and attachment plugs (cord caps).

NEC Article 408 covers the requirements for switchboards, switchgear, and panelboards used to control light and power circuits.

NEC Article 410 provides the installation requirements for luminaires in various locations.

NEC Article 422 covers the use of electric appliances in any occupancy. It includes kitchen appliances, heating appliances, cord-and-plug connected equipment, and others.

NEC Article 424 covers fixed electric space-heating equipment. It includes heating cable, unit heaters, boilers, central systems, and other approved equipment.

NEC Article 430 covers electric motors, including electrical connections, motor controls, and overload protection.

NEC Articles 440 through 460 cover air conditioning and refrigerating equipment, generators, transformers, phase converters, and capacitors.

NEC Article 480 provides requirements related to battery-operated electrical systems. Storage batteries are seldom thought of as part of a conventional electrical system, but they often provide standby emergency lighting service. They may also supply power to security systems that are separate from the main AC electrical system.

NEC Chapter 4 also covers industrial control panels, low-voltage lighting, industrial process

Think About It

Disconnects

How would you proceed to find the *NEC®* rule for the maximum number of disconnects permitted for a service?

and pipeline heating equipment, ice melting, resistors and reactors, and equipment over 1,000V.

2.1.5 NEC Chapter 5, Special Occupancies

NEC Chapter 5 covers special occupancy areas. These are areas where sparks or heat generated by electrical equipment may cause an explosion or fire. The hazard may be due to the atmosphere of the area or the presence of a volatile material in the area. Commercial garages, aircraft hangars, and service stations are typical special occupancy locations.

NEC Article 500 covers the different types of special occupancy atmospheres where an explosion is possible. The atmospheric groups listed in this article were established to make it easy to test and approve equipment for various types of uses.

NEC Articles 501, 502, and 503 cover the installation of explosion-proof wiring in hazardous (classified) locations. An explosion-proof system is designed to prevent the ignition of a surrounding explosive atmosphere when arcing occurs within the electrical system. Classes include:

- *Class I* – Areas containing flammable gases or vapors in the air. Class I areas include paint spray booths, dyeing plants where hazardous liquids are used, and gas generator rooms (*NEC Article 501*).
- *Class II* – Areas where combustible dust is present, such as grain-handling and storage plants, dust and stock collector areas, and sugar-pulverizing plants (*NEC Article 502*). These are areas where, under normal operating conditions, there may be enough combustible dust in the air to produce explosive or ignitable mixtures.
- *Class III* – Areas that are hazardous because of the presence of easily ignitable fibers or other particles in the air, although not in large enough quantities to produce ignitable mixtures (*NEC Article 503*). Class III locations include cotton mills, rayon mills, and clothing manufacturing plants.

Each of the three classes can be further categorized into Division 1 and Division 2. Division 1 includes normal operating conditions; Division 2 relates to abnormal operating conditions.

NEC Articles 511 and 514 regulate garages and fuel dispensing locations where volatile or flammable liquids are used.

NEC Article 520 regulates theaters and similar occupancies where fire and panic can create hazards to life and property. Projection rooms and adjacent areas must be properly ventilated and wired for the protection of operating personnel and others using the area.

NEC Article 590 provides guidance for all temporary construction installations.

NEC Chapter 5 also covers floating buildings, marinas, agricultural buildings, service stations, bulk storage plants, health care facilities, mobile homes and parks, and temporary installations.

2.1.6 NEC Chapter 6, Special Equipment

NEC Chapter 6 provides the requirements for various types of special equipment not covered in other chapters of the *NEC®*.

NEC Article 600 covers electric signs and outline lighting. *NEC Article 604* covers manufactured wiring systems. *NEC Article 610* applies to cranes and hoists.

NEC Article 620 covers elevators, dumbwaiters, escalators, and moving walks. The manufacturer is responsible for most of this work. The electrician usually just furnishes a feeder terminating in a disconnect means in the elevator equipment room. The electrician may also be responsible for a lighting circuit to a junction box midway in the elevator shaft for connecting the elevator cage lighting cable and exhaust fans.

NEC Article 630 regulates electric welding equipment. It is normally treated as a piece of industrial power equipment requiring a special power outlet, but there are special conditions that apply to the circuits supplying welding equipment that are outlined in this article.

NEC Article 640 covers wiring for sound recording and similar equipment. This type of equipment normally requires low-voltage wiring. Special outlet boxes or cabinets are usually provided with the equipment, but some items may be mounted in or on standard outlet boxes. Some sound recording systems require direct current. It is supplied from rectifying equipment, batteries, or motor generators. Low-voltage alternating current comes from relatively small transformers connected on the primary side to a 120V circuit within the building.

Other items covered in *NEC Chapter 6* include electric vehicle charging (*NEC Article 625*), X-ray equipment (*NEC Article 660*), induction and dielectric heat-generating equipment (*NEC Article 665*), industrial machinery (*NEC Article 670*), swimming pools and fountains (*NEC Article 680*), solar photovoltaic (PV) systems (*NEC Article 690*), and wind electric systems (*NEC Article 694*).

2.1.7 NEC Chapter 7, Special Conditions

In most commercial buildings, the *NEC®* and local ordinances require a means of lighting public rooms, halls, stairways, and entrances. There must be enough light to allow the occupants to exit from the building if the general building lighting is interrupted. Exit doors must be clearly indicated by illuminated exit signs. *NEC Chapter 7* covers the installation of emergency and legally required standby systems.

Emergency and standby circuits are arranged so that they can automatically transfer to an alternate source of current, usually storage batteries or gasoline-driven generators. An alternative in some occupancies is to connect them to the supply side of the main service. In doing so, disconnecting the main service switch will not disconnect the emergency circuits. *NEC Chapter 7* also covers fire alarms and a variety of other equipment, systems, and conditions that are not easily categorized elsewhere in the *NEC®*.

2.1.8 NEC Chapter 8, Communications Systems

NEC Chapter 8 is a special category for wiring associated with electronic communications systems including telephone, radio and TV, satellite dish, network-powered broadband systems, and community antenna television and radio distribution systems.

2.1.9 NEC Chapter 9, Tables

NEC Chapter 9 contains detailed tables along with explanatory notes. These tables cover topics such as conduit bends, raceway fill, and conductor dimensions. You will become very familiar with these tables as you proceed through your electrical training.

2.1.10 Informative Annexes A through J

Informative Annexes A through J provide informational material and examples that are helpful when applying *NEC®* requirements.

- *Informative Annex A* contains a list of product safety standards. These standards provide further references for requirements that are in addition to the *NEC®* requirements for the electrical components mentioned.
- *Informative Annex B* contains information for determining ampacities of conductors under engineering supervision.
- *Informative Annex C* contains the conduit fill tables for multiple conductors of the same size and type within the accepted raceways.
- *Informative Annex D* contains examples of calculations for branch circuits, feeders, and services, as well as load calculations.
- *Informative Annex E* contains information on types of building construction.
- *Informative Annex F* provides information on critical operations power systems.
- *Informative Annex G* is for informational purposes; it covers supervisory control and data acquisition (SCADA) systems.
- *Informative Annex H* contains suggested requirements for administration and enforcement of the *NEC®*. The text of this annex is a "model law" for a local jurisdiction to use.
- *Informative Annex I* contains the recommended torque tables from *UL Standard 486A–B*.
- *Informative Annex J* contains the ADA requirements for accessible design.

2.2.0 Finding Specific Installation Requirements

To locate information for a specific installation, use the following steps:

Step 1 Familiarize yourself with *NEC Articles 90,100, and 110* to gain an understanding of the material covered in the *NEC®* and the definitions used in it.

Step 2 Turn to the *Table of Contents* at the beginning of the *NEC®*.

Step 3 Locate the chapter that focuses on the desired category.

Step 4 Find the article pertaining to your specific application.

Step 5 Turn to the page indicated. Each application will begin with a bold heading.

> **NOTE**
>
> An index is provided at the end of the *NEC®*. The index lists specific topics and provides a reference to the location of the material within the *NEC®*. The index is helpful when you are looking for a specific topic rather than a general category.

After becoming familiar with *NEC Articles 90, 100, and 110*, move on to the rest of the *NEC®*. There are several key sections frequently used in servicing electrical systems.

Using the 2020 *NEC®*, follow along with these sample scenarios to familiarize yourself with the layout of the *NEC®*.

2.2.1 Installing Type SE Cable

Suppose you are installing Type SE (service-entrance) cable on the side of a home. You know the cable must be secured, but you are not sure of the spacing between cable clamps. To locate this information, use the following procedure:

Step 1 Look in the *NEC®* *Table of Contents* and follow down the list until you find an appropriate category. (Or you can use the index at the end of the book.)

Step 2 *NEC Article 230, Services*, will probably catch your eye first, so turn to the page where it begins.

Step 3 Scan down through the section numbers until you come to *NEC Section 230.51, Mounting Supports*. Upon reading this section, you will find in paragraph *(A) Service-Entrance Cables* that "service-entrance cables shall be supported by straps or other approved means within 300 mm (12 in.) of every service head, gooseneck, or connection to a raceway or enclosure and at intervals not exceeding 750 mm (30 in.)."

After reading this section, you will know that a cable strap is required within 12" (300 mm) of the service head and within 12" (300 mm) of the meter base. Furthermore, the cable must be secured in between these two termination points at intervals not exceeding 30" (750 mm).

2.2.2 Installing Track Lighting

It is important for electricians to determine if they are violating the *NEC®*, regardless of the type of installation. For example, assume that you are installing track lighting in a residential occupancy. The owners want the track located behind the curtain of their sliding glass patio doors. To determine if this is an *NEC®* violation, follow these steps:

Step 1 Start with *NEC Article 100* to become familiar with the definition of track lighting found under *Lighting Track*.

Step 2 Look in the *NEC® Table of Contents* and find the chapter that contains information about the general application on which you are working. *NEC Chapter 4, Equipment for General Use*, covers track lighting.

Step 3 Now look for the article that fits the specific category on which you are working. In this case, *NEC Article 410* covers luminaires, lampholders, and lamps.

Step 4 Locate the part within *NEC Article 410* that deals with the specific application. For this example, refer to *Part XIV, Lighting Track*.

Step 5 Turn to the page listed.

Step 6 Go to *NEC Section 410.151* and read the information contained therein. Note that paragraph *(C) Locations Not Permitted* under *NEC Section 410.151* states the following: "Lighting track shall not be installed in the following locations: (1) Where likely to be subjected to physical damage; (2) In wet or damp locations; (3) Where subject to corrosive vapors; (4) In storage battery rooms; (5) In hazardous (classified) locations; (6) Where concealed; (7) Where extended through walls or partitions; (8) Less than 1.5 m (5 ft) above the finished floor except where protected from physical damage or track operating at less than 30 volts rms open-circuit voltage; (9) Where prohibited by 410.10(D)."

Step 7 Read *NEC Section 410.151(C)* carefully. Do you see any conditions that would violate any *NEC®* requirements if the track lighting was installed in the area specified? In checking these items, you will probably note condition (6), "Where concealed." Since the track lighting is to be installed behind a curtain, this sounds like an *NEC®* violation. You need to check further.

Step 8 To determine the *NEC®* definition of *concealed*, turn to *NEC Article 100, Definitions* and find the main term *concealed*. It reads: "**Concealed.** Rendered inaccessible by the structure or finish of the building." Although the track lighting may be out of sight if the curtain is drawn, it will still be readily accessible for maintenance. Consequently, the track lighting is really not concealed according to the *NEC®* definition.

When using the *NEC®* to determine electrical installation requirements, remember that you will usually need to refer to more than one section. Sometimes the *NEC®* itself refers the reader to other articles and sections. Eventually, you will become familiar enough with the *NEC®* to know which other sections pertain to the installation at hand. It can be confusing, but time and experience using the *NEC®* will make it much easier. A pictorial road map of some *NEC®* topics is shown in *Figure 5*.

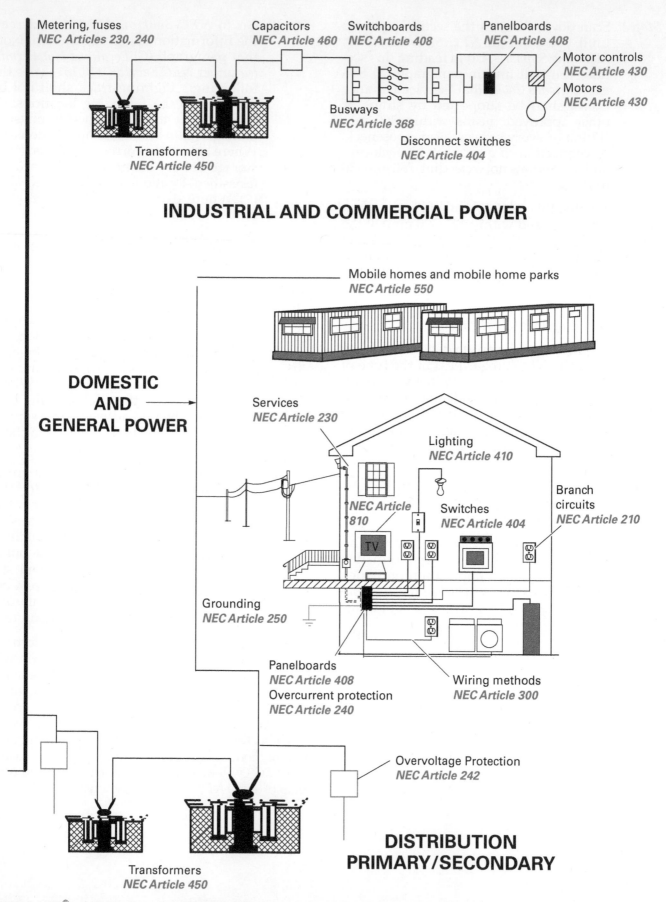

Metering, fuses
NEC Articles 230, 240

Capacitors
NEC Article 460

Switchboards
NEC Article 408

Panelboards
NEC Article 408

Motor controls
NEC Article 430

Motors
NEC Article 430

Busways
NEC Article 368

Transformers
NEC Article 450

Disconnect switches
NEC Article 404

INDUSTRIAL AND COMMERCIAL POWER

Mobile homes and mobile home parks
NEC Article 550

DOMESTIC AND GENERAL POWER

Services
NEC Article 230

Lighting
NEC Article 410

NEC Article 810

Switches
NEC Article 404

Branch circuits
NEC Article 210

Grounding
NEC Article 250

Panelboards
NEC Article 408

Overcurrent protection
NEC Article 240

Wiring methods
NEC Article 300

Overvoltage Protection
NEC Article 242

DISTRIBUTION PRIMARY/SECONDARY

Transformers
NEC Article 450

Figure 5 NEC® references for industrial, commercial, and residential power.

Other NFPA Codes

In addition to the *NEC®*, NFPA also publishes many other codes. Two that are of interest to the electrician are NFPA 70E®, *Standard for Electrical Safety in the Workplace* and NFPA 70B, *Recommended Practice for Electrical Equipment Maintenance*. NFPA 70E® provides direction on the safe installation, operation, and maintenance of electrical equipment. NFPA 70B provides direction for performing electrical tests, inspection, and maintenance procedures. Both documents are excellent resources for the electrical professional.

Think About It

Conformance and Electrical Equipment

Which other resources are available for finding information about the use of electrical equipment and materials?

Think About It

Putting It All Together

Examine the electrical components and products used in your home or classroom and consider the quality of the work. Do you see any components or products that have not been listed or labeled? If so, how might these devices put you in harm's way? Do you see any code violations?

2.0.0 Section Review

1. Information on calculating branch circuit loads can be found in _____.

   a. *NEC Article 110*
   b. *NEC Article 220*
   c. *NEC Article 342*
   d. *NEC Article 404*

2. Where in *NEC Article 406* can you find information on installing receptacles in damp or wet locations?

   a. *NEC Section 406.3*
   b. *NEC Section 406.9*
   c. *NEC Section 406.5*
   d. *NEC Section 406.12*

---

1. Which word or phrase best describes the *NEC®* requirements for the installation of electrical systems?

   a. Minimum
   b. Most stringent
   c. Design specification
   d. Complete

2. All of the following groups are usually represented on the Code-Making Panels, except _____.

   a. trade associations
   b. electrical inspectors
   c. insurance organizations
   d. government lobbyists

3. Mandatory and permissive rules are defined in _____.

   a. *NEC Article 90*
   b. *NEC Article 100*
   c. *NEC Article 110*
   d. *NEC Article 200*

4. The general design and installation of electrical systems is covered in _____.

   a. *NEC Chapters 1, 2, and 7*
   b. *NEC Chapters 1, 2, 3, and 4*
   c. *NEC Chapters 6, 7, and 8*
   d. *NEC Chapters 5, 6, 7, and 9*

5. Devices such as radios, televisions, and telephones are covered in _____.

   a. *NEC Chapter 8*
   b. *NEC Chapter 7*
   c. *NEC Chapter 6*
   d. *NEC Chapter 5*

6. *NEC Article 110* covers _____.

   a. branch circuits
   b. definitions
   c. general requirements for electrical installations
   d. wiring design and protection

7. Cable trays are covered in _____.

   a. *NEC Article 330*
   b. *NEC Article 342*
   c. *NEC Article 368*
   d. *NEC Article 392*

8. Installation procedures for luminaires are provided in _____.

   a. *NEC Article 410*
   b. *NEC Article 408*
   c. *NEC Article 366*
   d. *NEC Article 460*

9. Theaters are covered in _____.

   a. *NEC Article 338*
   b. *NEC Article 110*
   c. *NEC Article 430*
   d. *NEC Article 520*

10. *NEC Article 600* covers _____.

    a. track lighting
    b. electric signs and outline lighting
    c. X-ray equipment
    d. emergency lighting systems

11. Examples of branch circuit calculations can be found in _____.

    a. *NEC Informative Annex A*
    b. *NEC Informative Annex C*
    c. *NEC Informative Annex D*
    d. *NEC Informative Annex G*

12. Information on conductors for general wiring can be found in _____.

    a. *NEC Article 280*
    b. *NEC Article 310*
    c. *NEC Article 404*
    d. *NEC Article 340*

13. Within *NEC Article 310*, locate information on stranded conductors. Which size conductors must be stranded when installed in raceways?

    a. 14 AWG or larger
    b. 12 AWG or larger
    c. 10 AWG or larger
    d. 8 AWG or larger

14. Information on switchboards and panel-boards can be found in _____.

    a. *NEC Article 285*
    b. *NEC Article 300*
    c. *NEC Article 408*
    d. *NEC Article 540*

15. Within *NEC Article 408*, find the section that covers switchboard clearances. Assuming the switchboard does not have a noncombustible shield, what is the minimum clearance between the top of the switchboard and a combustible ceiling?

    a. 12" (300 mm)
    b. 18" (400 mm)
    c. 24" (600 mm)
    d. 36" (900 mm)

Trade Terms Quiz

Fill in the blank with the correct term that you learned from your study of this module.

1. _____ are the main topics of the *NEC*®.

2. Nine _____ form the broad structure of the *NEC*®.

3. Certain articles in the *NEC*® are subdivided into _____ with Roman numeral designations.

4. Parts and articles are subdivided into numbered _____.

5. Although they follow the applicable sections of the *NEC*®, _____ allow alternative methods to be used under specific conditions.

6. A(n) _____ is explanatory material that follows specific *NEC*® sections.

7. The _____ is an organization that maintains and improves the quality and reliability of electrical products.

8. The _____ publishes the *NEC*®; it also develops other codes and standards to minimize the possibility and effects of fire.

9. _____ are organizations that are responsible for testing and certifying electrical equipment.

10. A professional organization that develops international standards impacting electronics, telecommunications, information technology, and power-generation products and services is the _____.

11. The _____ is an international organization that develops consensus standards for all electrical and electronic technologies.

Trade Terms

Articles
Chapters
Exceptions
Informational Note
Institute for Electrical and
　Electronics Engineers (IEEE)

International Electrotechnical
　Commission (IEC)
National Electrical Manufacturers
　Association (NEMA)
National Fire Protection Association
　(NFPA)

Nationally Recognized Testing
　Laboratories (NRTLs)
Parts
Sections

1. The *NEC*® is made up of _____ chapters.

2. The *NEC*® provides the _____ requirements for the installation of electrical systems.

3. *NEC Chapter 2* covers _____.
   a. wiring and protection
   b. lighting
   c. occupancy
   d. emergency

4. Which of the following covers electric motors?
   a. *NEC Article 230*
   b. *NEC Article 330*
   c. *NEC Article 430*
   d. *NEC Article 130*

5. *NEC Article 520* covers _____ and similar occupancies where fire and panic can cause hazards to life and property.

6. To size grounding conductors, refer to _____.
   a. *NEC Chapter 1*
   b. *NEC Chapter 2*
   c. *NEC Chapter 3*
   d. *NEC Chapter 4*

7. Wiring methods are covered in _____.
   a. *NEC Chapter 1*
   b. *NEC Chapter 2*
   c. *NEC Chapter 3*
   d. *NEC Chapter 4*

8. Gas stations are covered in _____.
   a. *NEC Chapter 5*
   b. *NEC Chapter 6*
   c. *NEC Chapter 7*
   d. *NEC Chapter 8*

9. Emergency systems are covered in _____.
   a. *NEC Chapter 6*
   b. *NEC Chapter 7*
   c. *NEC Chapter 8*
   d. *NEC Chapter 9*

10. Fire alarm systems are covered in _____.
    a. *NEC Chapter 6*
    b. *NEC Chapter 7*
    c. *NEC Chapter 8*
    d. *NEC Chapter 9*

11. Definitions are covered in _____.

12. Units of measurement are covered in _____.

13. X-ray equipment is covered in _____.
    a. *NEC Article 665*
    b. *NEC Article 660*
    c. *NEC Article 670*
    d. *NEC Article 650*

14. _____ contains tables referenced by other chapters.
    a. *NEC Chapter 8*
    b. *NEC Chapter 9*
    c. *NEC Chapter 7*
    d. *NEC Chapter 6*

15. Electric signs are covered in _____.
    a. *NEC Chapter 8*
    b. *NEC Chapter 7*
    c. *NEC Chapter 6*
    d. *NEC Chapter 9*

Steven Gene Newton
SET
National Field Services

How did you choose a career in the electrical field?

After spending 13 years in the Navy on nuclear submarines, I planned to work for the Martin Marietta space station, but it was suspended. I needed to find a job so I could take care of my family, so I decided to work for an electrical testing company since I knew electrical power and electronics.

How important is education and training in construction?

Education and training are not only important in construction, but also in life. You should always be in the process of continually training and educating yourself.

How important are NCCER credentials to your career?

NCCER credentials—like any credential—prove that someone else has ensured you are qualified for a position or to perform a task. Organizations like the National Institute for Certification in Engineering Technologies (NICET), National Society of Professional Engineers (NSPE), National Center for Construction Education and Research (NCCER), and the InterNational Electrical Testing Association (NETA) are third parties who validate your knowledge and capabilities. The credentials they provide are essential for identifying individuals in industry who are competent at various levels.

What kinds of work have you done in your career?

Everything from construction supervision to project management and technical work. I have worked from the beginning of the design phase, through the construction phase, and to the final commissioning and in-service phases. My work in each of these phases dealt with electrical, electronics, and instrumentation duties.

Would you recommend construction as a career?

I would absolutely suggest a career in construction, or in another construction-related field like startup testing. This kind of career gives someone the flexibility of traveling overseas or staying in one place. For the younger person, this career allows you the flexibility of traveling when you are young, and ultimately, of deciding where you want to end up as you develop in your career.

What advice would you give to those new to the field?

Study while you are young. Put as much effort into your career as possible before you are married or while your children are young. This approach will allow you to be there for your family when they need you the most. Put yourself in a position to make decisions about your own destiny and to choose your career path

James Westfall
Instructor
ABC of Western Pennsylvania

How did you choose a career in the electrical field?

I knew from an early age that working with my hands would be an integral part of my life. Most of my family has a fear of electricity, so I chose to face that fear and was determined to succeed.

Who inspired you to enter the industry?

My father and grandfather instilled in me the importance of learning a trade. My high school vocational teacher, Mr. Rick Bell, inspired me to go as far as possible in the trade and to give back to others.

What types of training have you received?

My training began in high school in a two-year Electrical Program at the Jefferson County Joint Vocational School in Bloomingdale, Ohio. I continued my education by earning an associate degree in Electrical Engineering from Jefferson Community College in Steubenville, Ohio. I am currently working toward my Vocational Instructor Certificate with plans to earn a BA in Vocational Education through Indiana University of Pennsylvania.

How important is education and training in construction?

Changes in technology occur daily and revisions to the *National Electrical Code*® and safety procedures are constantly changing as well, so it is vital that electricians continue to learn. Staying on top of these through training determines an electrician's employability and success in the trade.

How important are NCCER credentials to your career?

Qualified electricians are in high demand. NCCER offers credentials that matter to employers who are looking to fill open positions. The NCCER curriculum offers challenging coursework and portability nationwide.

How has training/construction impacted your life?

The construction field has offered me the opportunity to provide for a family, buy new cars, and give back to younger electricians. Taking pride in workmanship cannot be overstated. When I drive by a home, office, or retail store that I helped build, it gives me a sense of pride that cannot be taken away or replaced.

Would you recommend construction as a career?

I am a proponent of the construction trades as a whole, so it is important to me to educate people about the opportunities that the trades provide. The building trades are an integral part of our country's success as a nation and it is exciting to see the final product of your hard work. The people you meet and work with become a part of your family and you make friends that last a lifetime. Plus, the financial benefits are felt immediately. You don't have to wait years to begin your career, you earn as you work.

What advice would you give to those new to the field?

Become a student of the trade. Never quit studying. Ask questions and be willing to learn.

Trade Terms Introduced in This Module

Articles: The articles are the main topics of the *NEC®*, beginning with *NEC Article 90, Introduction*, and ending with *NEC Article 840, Premises-Powered Broadband Communications Systems*.

Chapters: Chapters contain a group of articles related to a broad category. Nine chapters form the broad structure of the *NEC®*.

Exceptions: Exceptions follow the applicable sections of the *NEC®* and allow alternative methods to be used under specific conditions.

Informational Note: Explanatory material that follows specific *NEC®* sections.

Institute for Electrical and Electronics Engineers (IEEE): A professional organization that develops international standards impacting electronics, telecommunications, information technology, and power-generation products and services.

International Electrotechnical Commission (IEC): An international organization that develops consensus standards for all electrical and electronic technologies.

National Electrical Manufacturers Association (NEMA): The association that maintains and improves the quality and reliability of electrical products.

National Fire Protection Association (NFPA): The publishers of the *NEC®*. The NFPA develops codes and standards to minimize the possibility and effects of fire.

Nationally Recognized Testing Laboratories (NRTLs): Product safety certification laboratories that are responsible for testing and certifying electrical equipment.

Parts: Certain articles in the *NEC®* are subdivided into parts that cover a specific topic. Parts have Roman numeral designations (e.g., *NEC Article 250*, *Part IX*, *Instruments, Meters, and Relays*).

Sections: Parts and articles are subdivided into sections. Sections have numeric designations that follow the article number and are preceded by a period (e.g., *NEC Section 501.4*).

Additional Resources

This module presents thorough resources for task training. The following reference material is recommended for further study.

National Electrical Code® Handbook, Latest Edition. Quincy, MA: National Fire Protection Association.

NFPA 70B®, *Recommended Practice for Electrical Maintenance*, Latest Edition. Quincy, MA: National Fire Protection Association.

NFPA 70E®, *Standard for Electrical Safety in the Workplace*, Latest Edition. Quincy, MA: National Fire Protection Association.

Figure Credits

UL and the UL logo are trademarks of UL LLC, Figure 4

John Traister, Figure 5

Section Review Answer Key

SECTION 1.0.0

Answer	Section Reference	Objective
1. c	1.1.0	1a
2. a	1.2.0	1b
3. a	1.3.3	1c

SECTION 2.0.0

Answer	Section Reference	Objective
1. b	2.1.2	2a
2. b	2.2.0	2b

NCCER CURRICULA — USER UPDATE

NCCER makes every effort to keep its textbooks up-to-date and free of technical errors. We appreciate your help in this process. If you find an error, a typographical mistake, or an inaccuracy in NCCER's curricula, please fill out this form (or a photocopy), or complete the online form at **www.nccer.org/olf**. Be sure to include the exact module ID number, page number, a detailed description, and your recommended correction. Your input will be brought to the attention of the Authoring Team. Thank you for your assistance.

Instructors – If you have an idea for improving this textbook, or have found that additional materials were necessary to teach this module effectively, please let us know so that we may present your suggestions to the Authoring Team.

NCCER Product Development and Revision
13614 Progress Blvd., Alachua, FL 32615

Email: curriculum@nccer.org
Online: www.nccer.org/olf

❏ Trainee Guide   ❏ Lesson Plans   ❏ Exam   ❏ PowerPoints   Other _____

Craft / Level: _____   Copyright Date: _____

Module ID Number / Title: _____

Section Number(s): _____

Description: _____

_____

_____

_____

_____

Recommended Correction: _____

_____

_____

_____

_____

Your Name: _____

Address: _____

_____

Email: _____   Phone: _____

This page is intentionally left blank.

Device Boxes

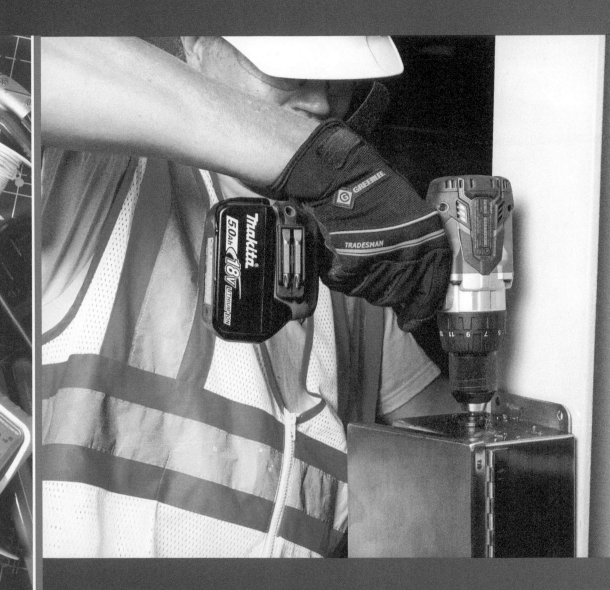

OVERVIEW

Electricians work with device boxes almost every day on every project, making a thorough understanding of the types of boxes available and their applications essential. This module describes the various types of boxes and explains how to calculate the *NEC®* fill requirements for outlet and junction boxes under 100 cubic inches (1,650 cubic centimeters).

Module 26106-20

Trainees with successful module completions may be eligible for credentialing through the NCCER Registry. To learn more, go to **www.nccer.org** or contact us at 1.888.622.3720. Our website, **www.nccer.org**, has information on the latest product releases and training..

Your feedback is welcome. You may email your comments to **curriculum@nccer.org**, send general comments and inquiries to **info@nccer.org**, or fill in the User Update form at the back of this module.

This information is general in nature and intended for training purposes only. Actual performance of activities described in this manual requires compliance with all applicable operating, service, maintenance, and safety procedures under the direction of qualified personnel. References in this manual to patented or proprietary devices do not constitute a recommendation of their use.

26106-20 V10.0

Objectives

When you have completed this module, you will be able to do the following:

1. Size and install outlet boxes.
   a. Identify boxes and their applications.
   b. Size outlet boxes.
   c. Install outlet boxes.
2. Size and install pull and junction boxes.
   a. Size pull and junction boxes.
   b. Install pull and junction boxes.

Performance Tasks

Under the supervision of the instructor, you should be able to do the following:

1. Identify the appropriate box type and size for a given application.
2. Select the minimum size pull or junction box for the following applications:
   - Conduit entering and exiting for a straight pull.
   - Conduit entering and exiting at an angle.

Trade Terms

Connector	Pull box
Explosion-proof	Raintight
Handy box	Watertight
Junction box	Weatherproof
Outlet box	

Industry Recognized Credentials

If you are training through an NCCER-accredited sponsor, you may be eligible for credentials from NCCER's Registry. The ID number for this module is 26106-20. Note that this module may have been used in other NCCER curricula and may apply to other level completions. Contact NCCER's Registry at 888.622.3720 or go to **www.nccer.org** for more information.

> **NOTE**
>
> NFPA 70®, *National Electrical Code*® and *NEC*® are registered trademarks of the National Fire Protection Association, Quincy, MA.

Contents

Figures

1.0.0 OUTLET BOXES

Objective

Size and install outlet boxes.
  a. Identify boxes and their applications.
  b. Size outlet boxes.
  c. Install outlet boxes.

Performance Task

  1. Identify the appropriate box type and size for a given application.

Trade Terms

Connector: Device used to physically connect conduit or cable to an outlet box, cabinet, or other enclosure.

Explosion-proof: Designed and constructed to withstand an internal explosion without creating an external explosion or fire.

Handy box: Single-gang outlet box used for surface mounting to enclose receptacles or wall switches on concrete or concrete block construction of industrial and commercial buildings; nongangable; also made for recessed mounting; also known as a utility box.

Outlet box: A metallic or nonmetallic box installed in an electrical wiring system from which current is taken to supply some apparatus or device.

Raintight: Constructed or protected so that exposure to a beating rain will not result in the entrance of water under specified test conditions.

Watertight: Constructed so that moisture will not enter the enclosure under specified test conditions.

Weatherproof: Constructed or protected so that exposure to the weather will not interfere with successful operation.

On every job, boxes are required. All of these must be sized, installed, and supported to meet current *NEC®* requirements. Because the *NEC®* limits the number of conductors, fittings, and devices allowed in each outlet or switch box according to its size, you must install boxes that are large enough to accommodate the number of conductors that must be spliced in the box or fed through it. Therefore, a knowledge of the various types of boxes and the volume of each is essential.

1.1.0 Identifying Boxes and Their Applications

Besides being able to calculate the required box sizes, you must also know how to select the proper type of box for any given application. For example, metallic boxes used in concrete deck pours are different from those used as device boxes in residential or commercial buildings. Boxes used for the support of lighting fixtures or for securing devices in outdoor installations will be different from the two types just mentioned. Boxes for use in certain hazardous locations will further differ in construction; many must be rated as being explosion-proof.

You must also know what fittings are available for terminating the various wiring methods in these boxes.

Electrical drawings rarely indicate the exact type of outlet box to be used in a given area, with the possible exception of boxes used in hazardous locations. Electricians who lack practical on-the-job experience may not always choose the best box for a given application. The use of improper boxes and other ill-adapted materials will cause excessive time to be taken on the job. For example, outlet boxes for use with only Type AC cable should contain built-in clamps; many times boxes will be ordered with knockouts only. This latter case requires extra connector usage. Each connection may require only a few additional seconds, but when these are added up over the period of a large project, much additional time is wasted. Because the cost of labor is an expensive item, any excess labor required will more than offset any savings gained from the use of inadequate materials.

Labor is often wasted due to obstructions from debris that enter raceways during the general construction work. Most of these obstructions can be avoided by plugging all raceway openings with capped bushings or similar means of protection (*Figure 1*). Care should also be taken to thoroughly tighten all fittings, couplings, and so on. All openings in boxes must be closed as specified by *NEC Section 314.17(A)*. Knockout closures are used for this purpose (*Figure 1*).

Outlet boxes normally fall into three categories:

- Pressed steel boxes with knockouts of various sizes for raceway or cable entrances
- Cast iron, aluminum, or brass boxes with threaded hubs of various sizes and locations for raceway entrances
- Nonmetallic boxes

Pressed steel boxes also fall into two categories:

- Boxes with conduit, electric metallic tubing, and cable
- Boxes designed for use with specific types of surface metal raceways

Outlet boxes vary in size and shape depending upon their use, the size of the raceway, the number of conductors entering the box, the type of building construction, the atmospheric conditions of the area, and special requirements.

Outlet box covers are usually required to adapt the box to the particular use it is to serve. For example, a 4" (100 mm) square box is adapted to one-gang or two-gang switches or receptacles by the use of either one-gang or two-gang flush device covers. A one-gang cast hub box can be adapted to provide a vapor-proof switch or a vapor-proof receptacle cover. Special outlet box hangers are available to facilitate their installation, particularly in frame building construction.

> **NOTE**
>
> Metric conversions vary depending on rounding and whether the dimension is matched to a standard metric size for a particular box, conductor, raceway, or other product. All metric conversions in this module are taken from the *NEC*®.

The types of enclosures used as outlet and device boxes for the support of fixtures, or for securing devices such as switches, receptacles, or other equipment on the same yoke or strap, are available in various sizes and shapes. These enclosures may be used in the one-gang, two-gang, three-gang, or four-gang types. Ceiling outlet boxes are available in various shapes. Device boxes are those that are usually installed to support receptacles and switches.

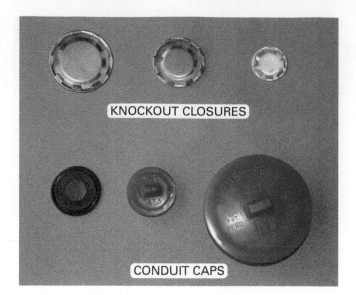

Figure 1 Conduit caps and knockout closures.

Boxes installed for the support of luminaires (lighting fixtures) are required to be listed for the purpose. Most device boxes are typically not designed or listed for use to support luminaires. The use of device boxes to support luminaires is addressed in *NEC Section 314.27(A)*.

A floor box that is listed specifically for installation in a floor is required where receptacles or junction boxes are installed in a floor. Listed floor boxes are provided with covers and gaskets to exclude surface water and cleaning compounds.

A box used at fan outlets is not permitted to be used as the sole support for ceiling (paddle) fans, unless it is listed for the application as the sole means of support. Where a ceiling fan does not exceed 70 lbs (32 kg) in weight, it is permitted to be supported by outlet boxes listed and identified for such use. Boxes designed to support more than 35 lbs (16 kilograms) must be marked with the maximum weight to be supported. See *NEC Section 314.27(C)*. These boxes must be rigidly supported from a structural member of the building. A paddle fan box and its related accessories are shown in *Figure 2*.

NEC Section 314.27(D) states that boxes used for support of utilization equipment other than ceiling-suspended (paddle) fans shall meet the requirements of *NEC Section 314.27(A)* for support of a luminaire that is the same size and weight.

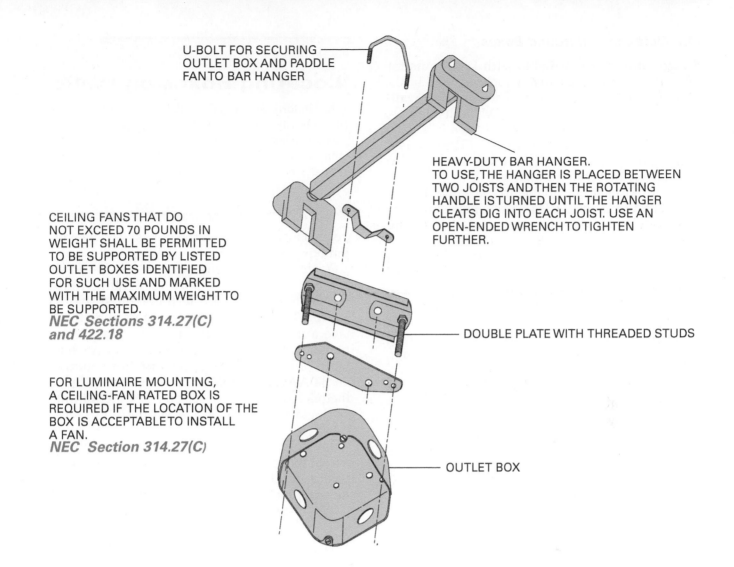

U-BOLT FOR SECURING OUTLET BOX AND PADDLE FAN TO BAR HANGER

HEAVY-DUTY BAR HANGER. TO USE, THE HANGER IS PLACED BETWEEN TWO JOISTS AND THEN THE ROTATING HANDLE IS TURNED UNTIL THE HANGER CLEATS DIG INTO EACH JOIST. USE AN OPEN-ENDED WRENCH TO TIGHTEN FURTHER.

CEILING FANS THAT DO NOT EXCEED 70 POUNDS IN WEIGHT SHALL BE PERMITTED TO BE SUPPORTED BY LISTED OUTLET BOXES IDENTIFIED FOR SUCH USE AND MARKED WITH THE MAXIMUM WEIGHT TO BE SUPPORTED. *NEC Sections 314.27(C) and 422.18*

FOR LUMINAIRE MOUNTING, A CEILING-FAN RATED BOX IS REQUIRED IF THE LOCATION OF THE BOX IS ACCEPTABLE TO INSTALL A FAN. *NEC Section 314.27(C)*

DOUBLE PLATE WITH THREADED STUDS

OUTLET BOX

Figure 2 Typical fan box for installation in an existing ceiling.

Special Considerations

Before you can effectively plan and route conductors to their termination points and then select and install the correct types and sizes of boxes, you will need to have the following information:

- Length and number of conductors
- Conductor ampacity
- Allowances for voltage drops
- Environment

Outlet Boxes in Poured Concrete

When installing outlet boxes in poured concrete structures, stuff the inside of the boxes with newspaper and cover the openings with duct tape or use capped bushings to keep concrete out of the raceway system.

1.1.1 Octagon and Round Boxes

Octagon boxes are available with knockouts for use with either conduit (using locknuts and bushings) or cable box connectors. They are also available with both Type AC and NM cable clamps. The standard width of octagon boxes is 4" (100 mm), with depths available in 1¼" (32 mm), 1½" (38 mm), or 2⅛" (54 mm). Extension rings are also available for increasing the depth.

Round boxes are available in the same dimensions, but *NEC Section 314.2* prevents using such boxes where conduit or connectors—requiring locknuts and bushings—are connected to the side of the box. The conduit or box connector must terminate in the top of such boxes. Round boxes with cable clamps, however, are permitted for use with Type AC and NM cables that may terminate in either the side or top of the box. Nonmetallic round boxes are permitted only with open wiring on insulators, concealed knob-and-tube wiring, Type NM cable, and nonmetallic raceways.

Figure 3 shows typical metallic octagon boxes and an octagon extension ring. *Figure 3 (A)* shows a box with concentric knockouts for conduit or box connectors, while *Figure 3 (B)* shows a box that utilizes cable clamps. *Figure 3 (C)* shows the extension ring. *Figure 4* shows a nonmetallic round box and fixture ring with a bar hanger for mounting between studs or joists.

Octagon and round boxes are used for wall-mounted lighting fixtures (luminaires). However, covers are available for octagon boxes that will support receptacles and switches. Blank covers

Locating Boxes on Walls

The Uniform Building Code (UBC) requires boxes on opposite sides of fire-rated walls between occupancies to be separated by at least 24" (600 mm) to maintain the fire rating. (See *NEC Section 300.21, Informational Note.*)

are also available when the box is used as a junction box.

Figure 5 shows a cross section of a round shallow box used for supporting lighting fixtures that have integral wire termination space. *NEC Section 314.24(A)* requires that this and all boxes have a minimum depth of ½" (12.7 mm). Boxes intended to enclose flush devices must not have a depth of less than ¹⁵⁄₁₆" (23.8 mm) for conductors No. 14 AWG and smaller, 1³⁄₁₆" (30.2 mm) for conductors No. 12 or 10 AWG, and 2¹⁄₁₆" (52.4 mm) for conductors No. 8, 6, or 4 AWG per *NEC Section 314.24(B)*.

Boxes enclosing equipment or devices that project more than 1⅞" (48 mm) into the box shall have a depth that is no less than the depth of the equipment plus ⅛" (6 mm) per *NEC Section 314.24(B)(1)*.

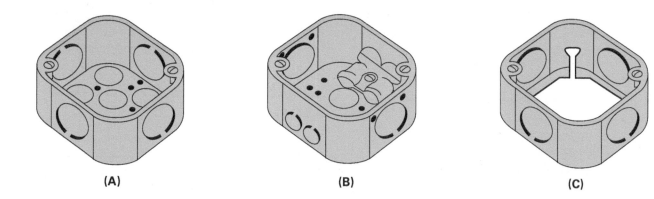

(A) (B) (C)

Figure 3 Typical octagon boxes and octagon extension ring.

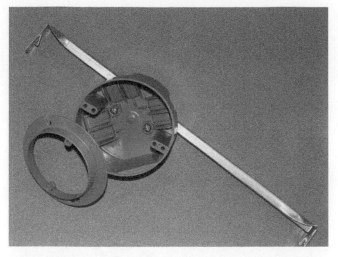

Figure 4 Nonmetallic round box and fixture ring with bar hanger.

1.1.2 Square Boxes

Square boxes are typically available in 4" (100 mm) and $4^{11}/_{16}$" (120 mm) square sizes. Both are available in depths of $1^{1}/_{4}$" (32 mm), $1^{1}/_{2}$" (38 mm), and $2^{1}/_{8}$" (54 mm). Extension rings are also available to further increase the depth. These boxes are available with or without mounting brackets for fastening to structural members. Boxes designed for use with cable may have either Type AC or NM clamps for securing the cable at the box entrance points. Square boxes may be used with a single or two-gang device ring (e.g., plaster ring or tile ring) for mounting receptacles or switches. A ring with a round opening is also available for mounting lighting fixtures. Blank covers are available when the boxes are used as junction boxes.

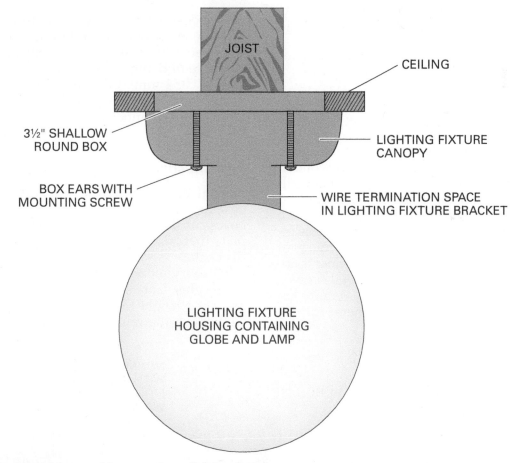

Figure 5 Shallow round box used for mounting a lighting fixture.

Identifying Boxes

After installing all the boxes in a complex installation, you can save yourself much time and trouble later on if you label each box location according to type. Using a can of spray paint, paint small dots on the floor below each box. Use one dot for a plug receptacle, two dots for a switch box, and three dots for a wall light. After the drywall installers are finished, you'll be able to easily identify any wall box.

Figure 6 Square box with extension ring.

Figure 6 shows a square box with a rectangular extension ring that is used to bring the box to the finished surface. Square boxes can be used as junction boxes, and when the number of conductors warrants more capacity than is available in other types of boxes.

1.1.3 Device Boxes

Device boxes house switches and receptacles. They are designed for flush mounting mainly in residential and some commercial applications. Device boxes are available with or without cable clamps and brackets for mounting to wooden structural members. This type of box is also available with plaster ears for installation in finished wall partitions. Three types of device boxes are shown in *Figure 7*. As you can see, some boxes include integral nails for direct installation into wall studs. Boxes used in metal stud environments are installed using self-tapping sheet metal screws, while those installed in concrete may require the use of special powder-actuated fasteners.

A special single-gang box with the mounting ears on the inside of the box is called a **handy box** (also known as a utility box). Such boxes are available in depths of $1\frac{1}{2}$" (38 mm), $1\frac{7}{8}$" (48 mm), and $2\frac{1}{8}$" (54 mm). Care must be exercised when using these boxes because their limited volume restricts the number of conductors permitted in the box.

1.1.4 Masonry Boxes

Special boxes known as masonry or concrete boxes are used in flat-slab construction jobs. These boxes consist of a sleeve with external ears and a plate that is attached after the sleeve is nailed to the deck.

Masonry boxes are manufactured in different heights. Care should be taken to use boxes of sufficient height to allow the knockouts to come well above the reinforcing rods. This eliminates

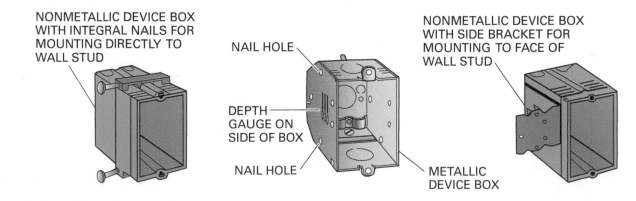

Figure 7 Typical device boxes.

Use of Round Boxes is Limited

Round boxes are not to be used in cases where conduit must enter from the side of the box because it is difficult to make a good connection with a locknut or bushing on a rounded surface.

Knockouts

When completing building renovations, it may be difficult to remove a knockout from a previously installed box without dislodging the box. One way to do it is to drill into the knockout and partially insert a self-tapping screw. Then use diagonal or side-cut pliers to pull the knockout from the box.

the need for offsets in the conduit where it enters a box. *Figure 8* shows a practical application of a masonry box.

1.1.5 Boxes for Wet and Damp Locations

In damp or wet locations, boxes and fittings must be placed or equipped to prevent moisture or water from entering and accumulating within the box or fitting. It is recommended that approved boxes of nonconductive material be used with nonmetallic sheathed cable or approved nonmetallic conduit when the cable or conduit is used in locations where there is likely to be occasional moisture present. Boxes installed in wet locations must be listed for such use as stated in *NEC Section 314.15*.

A wet location is any location subject to saturation with water or other liquids, such as locations exposed to weather or water, washrooms, garages, and interiors that might be hosed down. Underground installations or those in concrete slabs or masonry in direct contact with the earth must be considered wet locations. **Raintight** or **watertight** equipment (including fittings) may satisfy the requirements for **weatherproof** equipment. Boxes with threaded conduit hubs and gasketed covers will normally prevent water from entering the box except for condensation within the box.

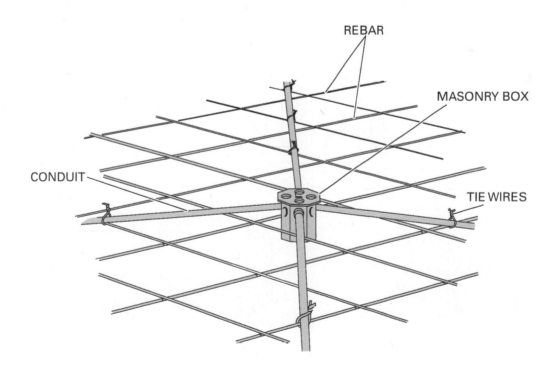

Figure 8 Masonry box.

A damp location is a location subject to some degree of moisture. Such locations include partially protected outdoor locations—such as under canopies, marquees, and roofed open porches. It also includes interior locations subject to moderate degrees of moisture—such as some basements, some barns, and cold storage warehouses.

Weatherproof covers for outdoor receptacles must be chosen with care. All 15A and 20A 125–250V receptacles in wet locations require a cover that is weatherproof whether or not the receptacle is in use. Other receptacles may or may not require these covers, depending on the rating. *NEC Sections 406.9(A) and (B)* cover installation of receptacles in damp or wet locations.

1.2.0 Sizing Outlet Boxes

In general, the maximum number of conductors permitted in standard outlet boxes is listed in *NEC Table 314.16(A)*. These figures apply where no fittings or devices such as fixture studs, cable clamps, switches, or receptacles are contained in the box and where no grounding conductors are part of the wiring within the box. Obviously, in all modern residential wiring systems there will be one or more of these items contained in the outlet box. Therefore, where one or more of the above-mentioned items are present, the number of conductors is reduced by one less than that shown in the table for each type of fitting and

by two for each device strap. For example, a deduction of two conductors must be made for each strap containing a device such as a switch or duplex receptacle; a further deduction of one conductor shall be made for up to four grounding conductors entering the box. For example, a 3" × 2" × 3½" (75 mm × 50 mm × 90 mm) box is listed in the table as containing a maximum number of eight No. 12 wires. If the box contains a cable clamp and a duplex receptacle, three wires will have to be deducted from the total of eight providing for only five No. 12 wires. If a ground wire is used, only four No. 12 wires may be used, which might be the case when a three-wire cable with ground is used to feed a three-way wall switch. Also, each looped conductor over 12" (300 mm) used in the box counts as two conductors.

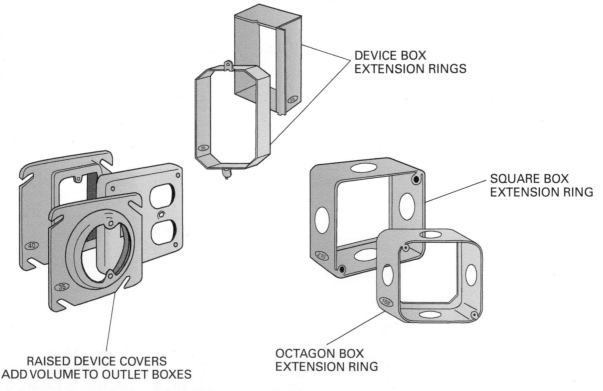

DEVICE BOX EXTENSION RINGS

SQUARE BOX EXTENSION RING

OCTAGON BOX EXTENSION RING

RAISED DEVICE COVERS ADD VOLUME TO OUTLET BOXES

Figure 9 Devices or components that add to outlet box capacity.

A pictorial definition of stipulated conditions as they apply to *NEC Section 314.16(A)* is shown in *Figure 9* through *Figure 11*. *Figure 9* illustrates an assortment of raised covers and outlet box extensions. These components, when combined with the appropriate outlet boxes, serve to increase the usable space. Each type is marked with its capacity, which may be added to the figures in *NEC Table 314.16(A)* to calculate the increased number of conductors allowed.

Figure 10 shows typical wiring configurations and devices that must be counted as conductors when calculating the total capacity of outlet boxes. A wire passing through the box without a splice or tap is counted as one conductor. Therefore, a cable containing two wires that passes in and out of an outlet box without a splice or tap is counted as two conductors. However, a wire that enters a box and is either spliced or connected to a terminal, and then exits again,

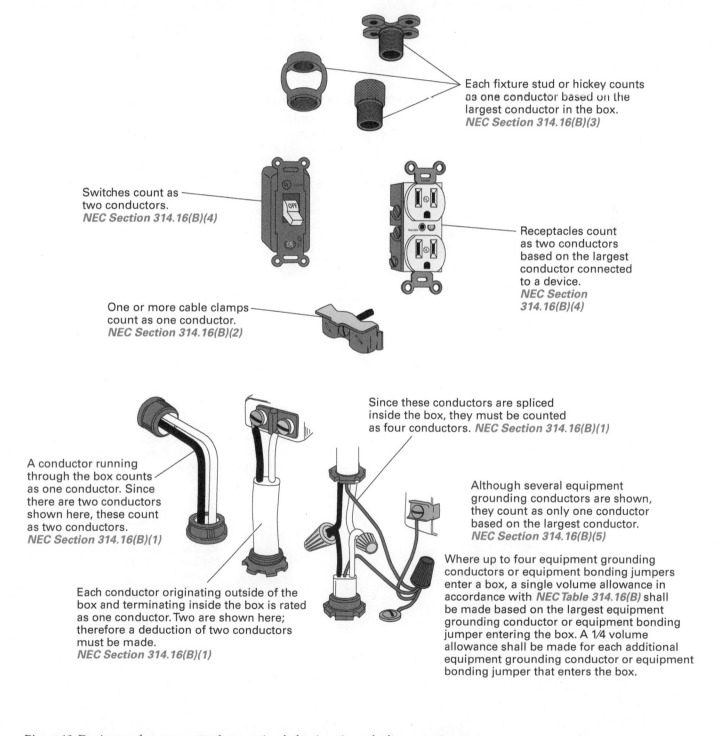

Each fixture stud or hickey counts as one conductor based on the largest conductor in the box.
NEC Section 314.16(B)(3)

Switches count as two conductors.
NEC Section 314.16(B)(4)

Receptacles count as two conductors based on the largest conductor connected to a device.
NEC Section 314.16(B)(4)

One or more cable clamps count as one conductor.
NEC Section 314.16(B)(2)

Since these conductors are spliced inside the box, they must be counted as four conductors. *NEC Section 314.16(B)(1)*

A conductor running through the box counts as one conductor. Since there are two conductors shown here, these count as two conductors.
NEC Section 314.16(B)(1)

Although several equipment grounding conductors are shown, they count as only one conductor based on the largest conductor.
NEC Section 314.16(B)(5)

Each conductor originating outside of the box and terminating inside the box is rated as one conductor. Two are shown here; therefore a deduction of two conductors must be made.
NEC Section 314.16(B)(1)

Where up to four equipment grounding conductors or equipment bonding jumpers enter a box, a single volume allowance in accordance with *NEC Table 314.16(B)* shall be made based on the largest equipment grounding conductor or equipment bonding jumper entering the box. A 1/4 volume allowance shall be made for each additional equipment grounding conductor or equipment bonding jumper that enters the box.

Figure 10 Devices and components that require deductions in outlet box capacity.

is counted as two conductors. In the case of two cables that each have two wires, the total conductors counted will be four. Wires that enter and terminate in the same box are counted as individual conductors and in this case, the total count would be two conductors. Remember, when one or more grounding wires enter the box and are joined, a deduction of only one is required, regardless of their number.

Further components that require deduction adjustments from those specified in *NEC Table 314.16(A)* include fixture studs and hickeys [*NEC Section 314.16(B)(3)*]. One conductor must be deducted from the total for each type of fitting used. Two conductors must be deducted for each strap-mounted device, such as duplex receptacles and wall switches; a deduction of one conductor is made when one or more internally mounted cable clamps are used [*NEC Section 314.16(B)(2) and*

(4)]. When mixed conductor sizes are installed on a yoke, the deduction is based on the largest wire size in the box per *NEC Table 314.16(B)*.

Figure 11 shows components that may be used in outlet boxes without affecting the total number of conductors. Such items include grounding clips and screws, wire nuts, and cable connectors when the latter are inserted through knockout holes in the outlet box and secured with locknuts. Prewired fixture wires are not counted against the total number of allowable conductors in an outlet box; neither are conductors originating and terminating in the box, such as pigtails.

To better understand how outlet boxes are sized, use the following as an example: Two No. 12 AWG conductors are installed in trade size ½" EMT (MD 12) and terminate into a metallic outlet box containing one duplex receptacle. What size outlet box will meet *NEC®* requirements?

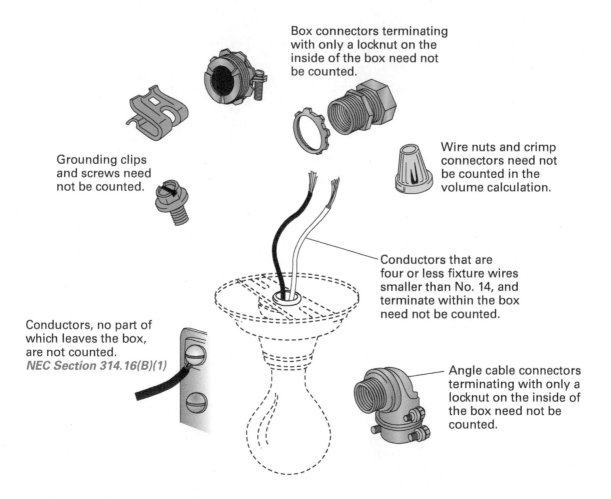

Box connectors terminating with only a locknut on the inside of the box need not be counted.

Grounding clips and screws need not be counted.

Wire nuts and crimp connectors need not be counted in the volume calculation.

Conductors that are four or less fixture wires smaller than No. 14, and terminate within the box need not be counted.

Conductors, no part of which leaves the box, are not counted. *NEC Section 314.16(B)(1)*

Angle cable connectors terminating with only a locknut on the inside of the box need not be counted.

Figure 11 Items that may be disregarded when calculating outlet box capacity.

The first step is to count the total number of conductors and equivalents that will be used in the box (*NEC Section 314.16*). The following steps describe how to determine the size of the outlet box:

Step 1 Calculate the total number of conductors and their equivalents:

One receptacle = 2
+ Two #12 conductors = 2
————————————————
Total #12 conductors = 4

Step 2 Determine the amount of space required for each conductor. *NEC Table 314.16(B)* gives the box volume required for each conductor:

US Measure:
No. 12 AWG = 2.25 in^3

Metric:
No. 12 AWG = 36.9 cm^3

Step 3 Calculate the outlet box space required by multiplying the volume required for each conductor by the number of conductors found in Step 1 above.

US Measure:
4 × 2.25 in^3 = 9.00 in^3

Metric:
4 × 36.9 cm^3 = 147.6 cm^3

Step 4 Once you have determined the required box capacity, refer to *NEC Table 314.16(A)* and note that a 3" x 2" x 2" (75 mm x 50 mm x 50 mm) box comes closest to our requirements. This box size is rated for 10.0 cubic inches (164 cubic centimeters).

For another example, if four No. 12 conductors enter the box, two additional No. 12 conductors must be added to our previous count for a total of six conductors.

US Measure:
6 × 2.25 in^3 = 13.5 in^3

Metric:
6 × 36.9 cm^3 = 221 cm^3

Again, refer to *NEC Table 314.16(A)* and note that a 3" × 2" × 2¾" (75 mm × 50 mm × 70 mm) device box with a rated capacity of 14.0 cubic inches (230 cubic centimeters) is the closest device box that meets *NEC* requirements. Of course, any box with a larger capacity is permitted.

Allocating Space in a Box

Here is a summary of the *NEC®* rules for determining the capacity of a box. Add the number of conductor equivalents indicated for each device:

Each conductor	1
Each looped conductor over 12" (300 mm)	2
Each strap-mounted device	2
Each fixture stud or hickey	1
Up to four grounding conductors	1
One or more cable clamps	1

Also add $\frac{1}{4}$ volume allowance for each additional grounding conductor or equipment bonding jumper (over 4) that enters the box.

1.3.0 Installing Outlet Boxes

In addition to box fill, there are a number of other considerations when installing boxes. These include additional *NEC®* requirements and making the actual wiring connections.

Some of the general *NEC®* requirements are as follows:

- The box selected must be listed for the given application (for example, a box used in a wet location must be listed for use in that location).
- As discussed previously, the box must have sufficient volume, as listed in *NEC Table 314.16(A)*, and must allow sufficient free space for conductors, as listed in *NEC Table 314.16(B)*.
- Conductors entering boxes and fittings must be protected from abrasion.
- Boxes must be installed and supported properly, and the finished installation must be accessible for later repair or maintenance.

You must also consider the type of box cover or canopy to be used, as well as the type of box. Refer to *NEC Sections 314.15 through 314.30* for additional details.

1.3.1 NEC® Requirements for Receptacle Locations

NEC Section 210.52 states the minimum requirements for the location of receptacles in dwelling units. It specifies that in each kitchen, family room, dining room, living room, parlor, library, den, sunroom, bedroom, recreation room, or similar room or area, receptacle outlets shall be installed so that no point along the floor line in any wall space is more than 6' (1.8 m), measured horizontally, from an outlet in that space, including any wall space 2' (600 mm) or more in width and the wall space occupied by fixed panels in walls, but excluding sliding panels (*Figure 12*). When spaced in this manner, a 6' (1.8 m) extension cord will reach a receptacle from any point along the wall line. Receptacle outlets shall, insofar as practicable, be spaced equal distances apart. Receptacle outlets in floors shall not be counted as part of the required number of receptacle outlets unless located within 18" (450 mm) of the wall.

The *NEC®* defines wall space as a wall that is unbroken along the floor line by doorways, fireplaces, or similar openings. Each wall space that is 2' (600 mm) or more in width must be treated individually and separately from other wall spaces within the room. This minimizes the use of cords across doorways, fireplaces, and similar openings.

At least one receptacle is required in each laundry area, within 3' (900 mm) of bathroom basins, on the outside of the building at the front and back (GFCI protected), on any balcony, deck, or porch accessible from the inside of the house, in each basement, in each attached and detached garage or vehicle bay, in each hallway 10' (3 m) or more in length, in each foyer larger than 60 sq. ft (5.6 sq. m), and at an accessible location for servicing any HVAC equipment.

Although no actual *NEC®* requirements exist for mounting heights and positioning of receptacles, other than the prohibition against mounting receptacles face up on countertops and similar work surfaces, there are certain *NEC®* requirements regarding receptacle placement. For example, *NEC Section 210.52(C)(3)* states that receptacles shall be not more than 20" (500 mm) above a kitchen countertop. Also, where allowed, receptacles may not be located more than 12" (300 mm) below the countertop surface or located where the countertop extends more than 6" (150 mm) beyond its support base. In addition to these *NEC®* guidelines, certain installation methods have become standard in the electrical industry. *Figure 13* shows common mounting heights of duplex receptacles used on conventional residential and small commercial installations. However, these dimensions are frequently varied to suit the building structure. For example, ceramic tile might be placed above a kitchen or bathroom countertop. If the dimensions in *Figure 13* put the receptacle part of the way out of the tile, the mounting height should be adjusted to place the

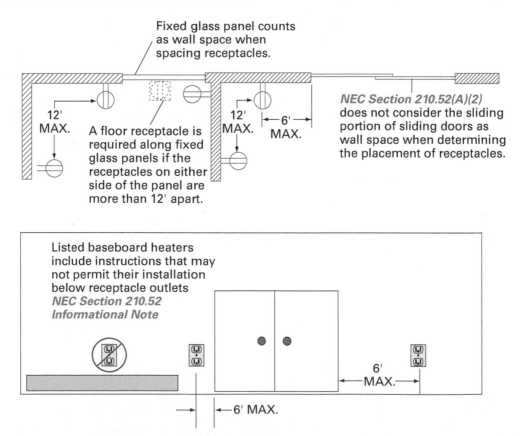

Figure 12 *NEC®* requirements for receptacle locations in dwelling units.

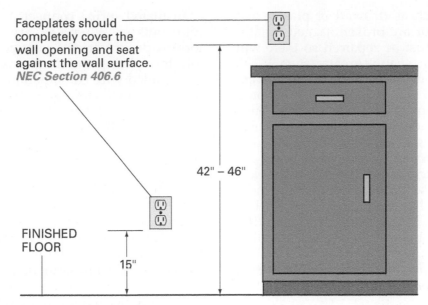

Figure 13 Mounting heights of duplex receptacles.

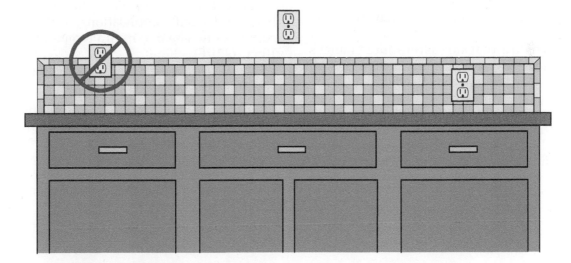

Figure 14 Adjusting mounting heights.

receptacle either completely in the tile or completely out of the tile, as shown in *Figure 14*.

Refer again to *Figure 13* and note that the mounting heights are given to the bottom of the outlet box. Many dimensions on electrical drawings are given to the center of the outlet box or receptacle. However, during the actual installation, workers installing the outlet boxes can mount them more accurately (and in less time) by lining up the bottom of the box with a chalk mark rather than trying to eyeball this mark to the center of the box.

A decade or so ago, most electricians mounted receptacle outlets 12" (300 mm) from the finished floor to the center of the outlet box. However, a survey of more than 500 homeowners shows that they prefer a mounting height of 15" (380 mm) from the finished floor to the bottom of the outlet

box. It is easier to plug and unplug the cord assemblies at this height. However, always check the working drawings, written specifications, details of construction, and local codes for measurements that may affect the mounting height of a particular receptacle outlet. For example, persons using wheelchairs may require more specific receptacle height locations to fit their individual needs.

NEC Section 314.20 requires all outlet boxes installed in walls or ceilings of concrete, tile, or other noncombustible material, such as plaster or drywall, to be installed in such a manner that the front edge of the box or fitting is not set back from the finished surface by more than ¼" (6 mm). Where walls and ceilings are constructed of wood or other combustible materials, outlet boxes

Wall surfaces such as drywall or plaster that contain wide gaps or are broken, jagged, or otherwise damaged, must be repaired so there will be no gaps or open spaces greater than ⅛" (3 mm) between the outlet box and the wall material (*NEC Section 314.21*). These repairs should be made prior to installing the faceplate. Such repairs are best made using a noncombustible caulking or spackling compound. See *Figure 15*.

Many industrial/commercial outlet boxes are surface-mounted and connect directly to the electrical raceway. These boxes are often mounted to the building steel or to a support channel or strut that is welded or clamped to the steel. The installation of outlet boxes or raceways should be coordinated so as not to interfere with piping, ductwork, fireproofing, and other items that may or may not already be in place.

1.3.2 Making Connections

Before any installation begins, study the electrical floor plan and consult with the builder or architect to ensure that no last-minute changes have been made in the electrical system. Install all boxes in accordance with the electrical drawings. The spacing should be as even as possible. The box center is the midpoint on the vertical dimension of the box. Make sure that you check the door swing direction so that the switches are not installed behind a door. Measure the height of the switch boxes from the floor so that they will be at the proper height when installed. After the boxes are installed, the wires must be spliced.

Insulated spring connectors, commonly called wire nuts or Wirenuts®, are solderless connectors made in various color-coded sizes that allow for splicing the hundreds of different solid or stranded wire combinations typically encountered in branch-circuit and fixture splicing applications (*Figure 16*). Several varieties of wire nuts are available, but the following are the ones used most often:

- Those for use on wiring systems 300V and under
- Those for use on wiring systems 600V and under (1,000V in lighting fixtures and signs)

Some types of wire nuts have thin wings on each side of the connector to facilitate their installation. Wire nuts are normally made in sizes to accommodate conductors as small as No. 22 AWG up to as large as No. 10 AWG, with practically any combination of those sizes in between.

Always check the listing requirements on the package prior to use. Some wire nuts are only listed for specific applications, such as copper to copper only. The maximum temperature rating is 105°C (221°F).

The general procedure for splicing wires with wire nuts is as follows:

Step 1 Select the proper size wire nut to accommodate the wires being spliced. Wire nut packages contain charts that list the allowable combinations of wires by size. Refer to the label on the wire nut box or container for this information.

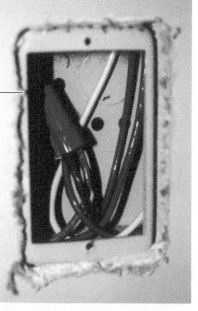

Gaps or openings around outlet box must not be greater than ⅛"; repair if necessary.
NEC Section 314.21

Figure 15 Gaps or openings around outlet boxes must be repaired.

Figure 16 Wire nuts.

Step 2 Select the appropriate tool (*Figure 17*), and then strip the insulation from the ends of the wires to be spliced. The length of insulation stripped off is typically about ½" (13 mm) (*Figure 18*), but may vary depending on the wire size and the wire nut being used. Follow the manufacturer's directions given on the wire nut package.

Step 3 Stick the ends of the wires into the wire nut and turn clockwise until tight. The wire nut draws the conductors and insulation into the body of the connector. See *Figure 19*.

Step 4 After making the connections within a box, tuck the wires neatly into the back of the box. That way, when the painters and plasterers come along to finish the walls, the wires will not be covered in drywall compound or paint.

NOTE

Some manufacturers of wire nuts require that the wires be pre-twisted before screwing on the nut. Also, some manufacturers recommend using a nut driver to tighten the wire nut. Always follow the manufacturer's instructions.

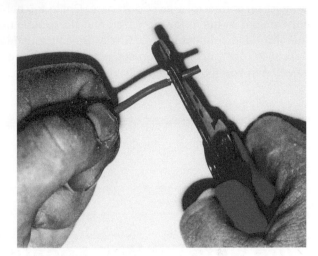

Figure 18 Stripping the insulation.

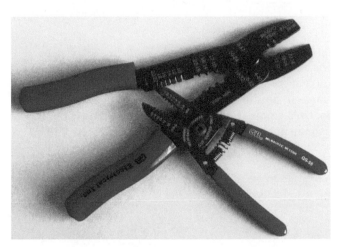

Figure 17 Stripping tools.

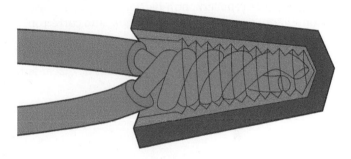

Figure 19 Wires installed in wire nut.

Making Connections

Conductor connections can be made using either push-in connectors or twist-on connectors, as shown here. Get in the habit of making your connections in the following order: first the grounding conductors, then the grounded conductors, and finally, the ungrounded conductors. If you approach each job in a systematic manner, it will soon become second nature to you and you'll be less likely to make mistakes.

(A) NO. 14/2 NM-B CABLE WITH PUSH-IN WIRE CONNECTORS

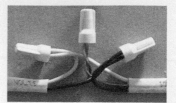

(B) NO. 12/2 NM-B CABLE WITH TWIST-ON WIRE NUTS

Wire Nuts for Wet Locations

Specially designed wire nuts are made for use in wet locations and/or direct burial applications. These wire nuts have a water repellent, non-hardening sealant inside the body that completely seals out moisture to protect the conductors against moisture, fungus, and corrosion. The sealant remains in a gel state and will not melt or run out of the wire nut body throughout the life of the connection. Unlike other types of wire nuts, this type can be used one time only. The wire nut can be backed off, eliminating the need to cut the wires for future or retrofit applications, but once removed, it must be discarded.

Think About It

Putting It All Together

Turn off the power in one area of your home and then remove some of the switch and receptacle plates. Examine the wiring inside each box. Is the box adequately sized for the number of wires and devices?

1.0.0 Section Review

1. A receptacle is usually housed in a(n) _____.

   a. pull box
   b. round box
   c. junction box
   d. device box

2. If a box can hold eight No. 12 conductors and contains a cable clamp, a switch, and a looped conductor that is 13" (330 mm) long, the final number of No. 12 conductors allowed in this box is _____.

   a. two
   b. three
   c. four
   d. five

3. A box installed in a wet location must be _____.

   a. installed upside down
   b. plastic
   c. sealed with caulk
   d. listed for the application

2.0.0 PULL AND JUNCTION BOXES

Objective

Size and install pull and junction boxes.
a. Size pull and junction boxes.
b. Install pull and junction boxes.

Performance Task

2. Select the minimum size pull or junction box for the following applications:
   • Conduit entering and exiting for a straight pull.
   • Conduit entering and exiting at an angle.

Trade Terms

Junction box: An enclosure where one or more raceways or cables enter, and in which electrical conductors can be, or are, spliced.

Pull box: A sheet metal box-like enclosure used in conduit runs to facilitate the pulling of cables from point to point in long runs, or to provide for the installation of conduit support bushings needed to support the weight of long riser cables, or to provide for turns in multiple conduit runs.

A pull box or junction box is provided in an electrical installation to facilitate the installation of conductors, or to provide a junction point for the connection of conductors, or both. In some instances, the location and size of pull boxes are designated on the drawings. In most cases, however, the electricians on the job must determine the proper number, location, and sizes of pull or junction boxes to facilitate conductor installation.

2.1.0 Sizing Pull and Junction Boxes

Pull boxes should be as large as possible. Workers need space within the box for both hands and in the case of the larger wire sizes, workers will need room for their arms to feed the wire. *NEC Sections 314.28(A) through (E)* specify that pull and junction boxes must provide adequate space and dimensions for the installation of conductors. For raceways containing conductors of No. 4 or larger, and for cables containing conductors of No. 4 or larger, the minimum dimensions of pull or junction boxes installed in a raceway or cable run shall comply with the following:

- In straight pulls, the length of the box shall not be less than eight times the trade diameter or metric designator (MD) of the largest raceway. (Note that the MD is not a direct conversion between inches and millimeters, but is very close to the trade sizes.)
- Where angle or U pulls are made, the distance between each raceway entry inside the box and the opposite wall of the box shall not be less than six times the trade diameter or MD of the largest raceway in a row. This distance shall be increased for additional entries by the amount of the sum of the trade diameters/MDs of all other raceway entries in the same row on the same wall of the box. Each row shall be calculated individually, and the single row that provides the maximum distance shall be used.
- Also where angle or U pulls are made, the distance between raceway entries enclosing the same conductor shall not be less than six times the trade diameter/MD of the larger raceway.
- When transposing cable size into raceway size, the minimum trade size/MD raceway required for the number and size of conductors in the cable shall be used.

Figure 20 shows a junction box with four runs of conduit. This is a straight pull, contains conductors sized No. 4 or larger, and trade size 4" (MD 103) conduit is the largest size in the group. The minimum length required for the box can be determined using the following formula:

$$\frac{\text{Trade size of conduit} \times 8 \ [\text{per } \textit{NEC Section 314.28(A)(1)}]}{= \text{minimum length of box}}$$

When calculating box size, always use the largest conduit size in the group. In the example shown in *Figure 20*, the calculation is as follows:

US Measure:
Trade size 4" × 8 = 32"
Metric:
MD 103 × 8 = 824 mm

Therefore, this particular pull box must be at least 32" (824 mm) in length. The width of the box, however, need only be of sufficient size to enable locknuts and bushings to be installed on all the conduits or connectors entering the enclosure.

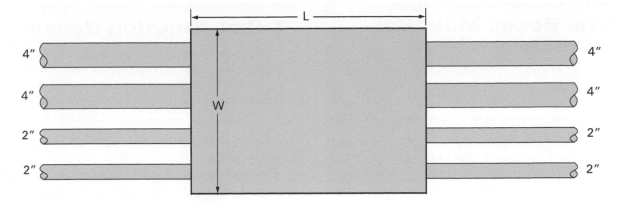

Figure 20 Pull box with straight conduit runs.

Junction or pull boxes containing conductors sized No. 4 or larger in which the conductors are pulled at an angle (*Figure 21*) must have a distance of not less than six times the trade diameter of the largest conduit [*NEC Section 314.28(A)(2)*]. The distance must be increased for additional conduit entries by the amount of the sum of the diameter of all other conduits entering the box on the same side. The distance between raceway entries enclosing the same conductors must not be less than six times the trade diameter of the largest conduit.

Since the 4" (MD 103) conduit is the largest size in this case:

US Measure:
$$L_1 = 6 \times 4" + (3" + 2") = 29$$

Metric:
$$L_1 = 6 \times MD\ 103 + (MD\ 78 + MD\ 53) = 749\ mm$$

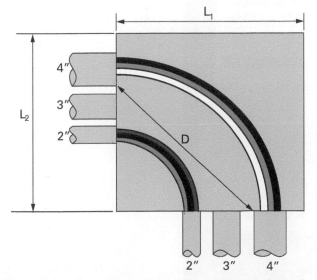

Figure 21 Pull box with conduit runs entering at right angles.

Because the same conduit runs are located on the adjacent wall of the box, L_2 is calculated in the same way; therefore, $L_2 = 29"$ (749 mm).

The distance (D) = 6 × 4" or 24" (6 × MD 103 = 618 mm). This is the minimum distance permitted between conduit entries enclosing the same conductor.

The depth of the box need only be of sufficient size to permit locknuts and bushings to be properly installed. In this case, a 6"-deep (152 mm-deep) box would suffice.

2.2.0 Installing Pull and Junction Boxes

Long runs of conductors should not be made in one pull. Pull boxes, installed at convenient intervals, will relieve much of the strain on the conductors. The length of the pull, in many cases, is left to the judgment of the workers or their supervisor, and the condition under which the work is installed.

The installation of pull boxes may seem to cause extra work and expense, but they save a considerable amount of time and hard work when pulling conductors. Properly placed, they eliminate bends and elbows and do away with the necessity of fishing from both ends of a conduit run.

If possible, pull boxes should be installed in a location that allows electricians to work easily and conveniently. For example, in an installation

Using Pull Boxes

Pull boxes make it easier to install conductors. They may also be installed to avoid having more than 360° worth of bends in a single run. (Remember, if a pull box is used, it is considered the end of the run for the purposes of the *NEC*® 360° rule.)

Metal Boxes Must Be Grounded

Metal boxes are good conductors. Therefore, when metal boxes are used, they must be grounded to the circuit grounding system.

Label Junction Boxes

It's a good idea to label every junction cover with the circuit number, the panel it came from, and its destination. The next person to service the installation will be grateful for this extra help.

where the conduit comes up a corner of a wall and changes direction at the ceiling, a pull box that is installed too high will force the electrician to stand on a ladder when feeding conductors, and will allow no room for supporting the weight of the wire loop or for the cable-pulling tools.

Unless the contract drawings or project engineer state otherwise, it is just as easy for the pull boxes to be placed at chest height or a similar location that allows workers to stand on the floor with sufficient room for both wire loop and tools.

In some electrical installations, a number of junction boxes must be installed to route the conduit in the shortest, most economical way. The *NEC®* requires all junction boxes to be readily accessible. This means that a person must be able to get to the conductors inside the box without

removing plaster, wall covering, or any other part of the building.

Junction boxes or pull boxes must be securely fastened in place on walls or ceilings or adequately suspended. All unused openings, such as knockouts, must be securely covered.

While certain sizes of factory-constructed boxes are available with concentric knockouts, in many instances it will be necessary to have them custom built to meet the job requirements. When it is not possible to accurately anticipate the raceway entrance requirements, it will be necessary to cut the required knockouts on the job.

In the case of large pull boxes and troughs, shop drawings should be prepared prior to the construction of these items with all required knockouts accurately indicated in relation to the conduit run requirements.

Knockout Punch Kit

Often, conduit must enter boxes, cabinets, or panels that do not have precut knockouts. In these cases, a knockout punch can be used to make a hole for the conduit connection.

2.0.0 Section Review

1. A straight pull contains two raceways. One of the raceways has a trade size of 3" and one has a trade size of 2". The length of the box must be _____.

   a. 16"
   b. 24"
   c. 26"
   d. 32"

2. If possible, pull boxes should be installed _____.

   a. high on the wall for security
   b. at a height/location for ease when pulling conductors
   c. behind wall coverings for a neater appearance
   d. with extra knockouts left open for ventilation

1. The maximum weight allowed by the *NEC®* when ceiling fans are mounted directly to an approved, unmarked outlet box is _____.

   a. 25 pounds
   b. 35 pounds
   c. 45 pounds
   d. 55 pounds

2. Square metal boxes are available in sizes of 4" and _____.

   a. $4^{11}/_{16}$"
   b. 5"
   c. $5^1/_4$"
   d. 6"

3. A box installed under a roofed open porch is considered a _____.

   a. dry location
   b. wet location
   c. damp location
   d. weatherproof location

4. How many conductor(s) must be deducted for each strap-mounted device in a device box?

   a. One
   b. Two
   c. Three
   d. Four

5. The capacity of an outlet box can be increased by using _____.

   a. fixture studs
   b. strap-mounted devices
   c. wire nuts
   d. raised device covers

6. A conductor under 12" (300 mm) long running through a box counts as the conductor equivalent of _____.

   a. 0
   b. 1
   c. 2
   d. 12

7. If there are four equipment grounding conductors in a box, you must deduct _____ conductor(s) from the total allowed.

   a. 1
   b. 2
   c. 3
   d. 4

8. Box installations are covered in _____.

   a. *NEC Sections 314.1 through 314.4*
   b. *NEC Sections 314.15 through 314.30*
   c. *NEC Sections 314.40 through 314.44*
   d. *NEC Sections 314.70 through 314.72*

9. If the largest trade diameter of a raceway entering a pull box is 3", and it is a straight pull, the minimum size box allowed is _____.

   a. 20"
   b. 24"
   c. 30"
   d. 36"

10. The depth of a pull box with conduit runs entering at right angles _____.

   a. need only be deep enough to permit the proper installation of locknuts and bushings
   b. must be eight times the diameter of the smallest conduit
   c. must be four times the diameter of the largest conduit
   d. must be six times the diameter of the smallest conduit

Trade Terms Quiz

Fill in the blank with the correct term that you learned from your study of this module.

1. _____ means constructed or protected so that exposure to beating rain will not result in the entrance of water under specified test conditions.

2. A box designed and constructed to withstand an internal explosion without creating an external explosion or fire is called _____.

3. A box constructed so that moisture will not enter the enclosure under specified test conditions is _____.

4. A(n) _____ is an enclosure used to facilitate the installation of cables from point to point in long runs.

5. A box constructed or protected so that exposure to the weather will not interfere with successful operation is referred to as _____.

6. A(n) _____ is a metallic or nonmetallic box installed in an electrical wiring system from which current is taken to supply some apparatus or device.

7. A single-gang outlet box used for surface mounting to enclose receptacles or wall switches on concrete or concrete block construction is called a(n) _____.

8. A(n) _____ is an enclosure where one or more raceways or cables enter, and in which electrical conductors may be spliced.

9. A(n) _____ is a device used to physically connect conduit or cable to an outlet box, cabinet, or other enclosure.

Trade Terms

Connector	Junction box	Raintight
Explosion-proof	Outlet box	Watertight
Handy box	Pull box	Weatherproof

1. All boxes must be sized, installed, and supported to meet the current _____ requirements.

2. Blank covers are used when boxes serve as _____.

3. Each conductor originating outside of the box and terminating inside of the box counts as _____ conductor(s).

4. When sizing outlet boxes, deduct _____ conductor(s) for each switch.

5. For a straight pull, the minimum length of the box must be at least _____ times the trade diameter of the largest raceway.

6. The minimum length for a junction box in which the conductors are pulled at an angle and that contains two trade size 3" conduits and one trade size 4" conduit entering each side is _____.

7. Octagon and round boxes are used mostly for

   _____.

8. When sizing outlet boxes, _____ conductor(s) must be deducted for each strap-mounted device.

9. An extension ring is used to increase the _____ of the box.

10. What is the maximum distance between receptacles along the floor line in any wall space?

Gary Edgington
Baker Electric

What made you decide to become an electrician?

I had an early fascination with electricity and used to work on motorized cars, lights, or about anything else that would plug into the wall or run from a battery. My interest continued through high school. After watching an electrician wire a house, I made up my mind to be in the electrical field somewhere.

How did you learn the trade?

After graduating high school in 1972, I called an electrical contractor, Kinsey Electric, on a regular basis until he hired me. Harry was a seasoned electrical contractor, who did residential, commercial, and agricultural wiring. Because it was a one-man operation, I had an advantage in working with the owner and getting some good hands-on training along with acquiring his good work ethics. Harry was a well-respected person in the community, with a reputation of being the person to call if you wanted it done right. He was my first role model. Along with the electrical training, Harry taught me that if you're going to do a job, be the best at it and take pride in what you do.

What factor or factors have contributed most to your success?

After working with Harry for a few years, I went to work at a manufacturing facility as an industrial electrician. I was fortunate to meet two more role models who greatly influenced me in my career. One was a retired Air Force electrician, who taught me the importance of knowledge and the need to be able to find answers to what you don't know. The second was an electrician from Manchester, England. He helped me recognize my abilities and inspired me to acquire more. Above everything, he gave me confidence in myself which has helped me throughout my life.

What kind of jobs did you hold on the way to your current position?

In addition to my job as an electrician helper/ electrician for Kinsey and as an industrial electrician, I became a licensed electrical contractor in 1978. After being in the electrical contracting industry for 22 years, and employing up to fifty people at times, I sold my business and became an electrical consultant. As the owner of the business I have found myself doing everything from electrical work in the field to estimating to project and business management. During this time and since then I have taught electrical apprentice classes and *National Electrical Code®* classes for thirteen years.

What does an electrical consultant do?

As a consultant I have taught electrical classes and code updates, been involved in the electrical design process, done code review on projects, and have been an expert in court cases.

What advice would you give to someone entering the electrical field?

My advice to someone entering the field is to embrace every opportunity to learn something new, to develop good work practices and ethics, and to think safety all of the time.

What advice do you have for trainees?

As a trainee in this field you need to practice at your profession. Be accurate in your work and pursue knowledge. You will find the ability and the confidence you need to succeed in this field.

What would you say was the single greatest factor that contributed to your success?

The single greatest factor for me has been getting close to a professional in the field and trying to follow in their footsteps.

Trade Terms Introduced in This Module

Connector: Device used to physically connect conduit or cable to an outlet box, cabinet, or other enclosure.

Explosion-proof: Designed and constructed to withstand an internal explosion without creating an external explosion or fire.

Handy box: Single-gang outlet box used for surface mounting to enclose receptacles or wall switches on concrete or concrete block construction of industrial and commercial buildings; nongangable; also made for recessed mounting; also known as a utility box.

Junction box: An enclosure where one or more raceways or cables enter, and in which electrical conductors can be, or are, spliced.

Outlet box: A metallic or nonmetallic box installed in an electrical wiring system from which current is taken to supply some apparatus or device.

Pull box: A sheet metal box-like enclosure used in conduit runs to facilitate the pulling of cables from point to point in long runs, or to provide for the installation of conduit support bushings needed to support the weight of long riser cables, or to provide for turns in multiple conduit runs.

Raintight: Constructed or protected so that exposure to a beating rain will not result in the entrance of water under specified test conditions.

Watertight: Constructed so that moisture will not enter the enclosure under specified test conditions.

Weatherproof: Constructed or protected so that exposure to the weather will not interfere with successful operation.

Additional Resources

This module presents thorough resources for task training. The following resource material is suggested for further study:

National Electrical Code® Handbook, Latest Edition. Quincy, MA: National Fire Protection Association.

Figure Credits

Greenlee / A Textron Company, Module Opener
John Traister, Figures 9–12, 20

Section Review Answer Key

Section 1.0.0

Answer	Section Reference	Objective
1. d	1.1.3	1a
2. b	1.2.0	1b
3. d	1.3.0	1c

Section 2.0.0

Answer	Section Reference	Objective
1. b	2.1.0	2a
2. b	2.2.0	2b

1.0.0 SECTION REVIEW

Question 2

Per *NEC Section 314.16*, subtract values for the box's contents to find the remaining space:

Total box capacity – cable clamp – switch – 13" looped conductor =
8 wires – 1 – 2 – 2 = 3 wires

A maximum of three additional wires can fit in the box.

2.0.0 SECTION REVIEW

Question 1

Use the formula for determining box size:

$$\frac{\text{Trade size of conduit} \times 8 \text{ [per } \textit{NEC Section 314.28(A)(1)}]}{= \text{minimum length of box}}$$

Use the largest trade size to determine the minimum length of the box:

3" × 8 = 24"

The length of the box must be **24"**.

This page is intentionally left blank.

NCCER CURRICULA — USER UPDATE

NCCER makes every effort to keep its textbooks up-to-date and free of technical errors. We appreciate your help in this process. If you find an error, a typographical mistake, or an inaccuracy in NCCER's curricula, please fill out this form (or a photocopy), or complete the online form at **www.nccer.org/olf**. Be sure to include the exact module ID number, page number, a detailed description, and your recommended correction. Your input will be brought to the attention of the Authoring Team. Thank you for your assistance.

Instructors – If you have an idea for improving this textbook, or have found that additional materials were necessary to teach this module effectively, please let us know so that we may present your suggestions to the Authoring Team.

NCCER Product Development and Revision

13614 Progress Blvd., Alachua, FL 32615

Email: curriculum@nccer.org
Online: www.nccer.org/olf

❏ Trainee Guide   ❏ Lesson Plans   ❏ Exam   ❏ PowerPoints   Other _____

Craft / Level: _____ Copyright Date: _____

Module ID Number / Title: _____

Section Number(s): _____

Description:
_____

_____

_____

_____

_____

Recommended Correction:
_____

_____

_____

_____

_____

Your Name: _____

Address: _____

_____

Email: _____ Phone: _____

This page is intentionally left blank.

Hand Bending

OVERVIEW

The art of conduit bending is dependent upon the skills of the electrician and requires a working knowledge of basic terms and proven procedures. Practice, knowledge, and training will help you gain the skills necessary for proper conduit bending and installation. This module describes methods for hand bending conduit. It covers 90-degree bends, back-to-back bends, offsets, and saddle bends. It also describes how to cut, ream, and thread conduit.

Module 26107-20

26107-20 V10.0

26107-20
Hand Bending

Objectives

When you have completed this module, you will be able to do the following:

1. Select and use hand bending equipment.
   a. Use geometry to make a bend.
   b. Make 90° bends.
   c. Make offset bends.
2. Cut, ream, and thread conduit.
   a. Cut conduit using a hacksaw.
   b. Cut conduit using a pipe cutter.
   c. Ream conduit.
   d. Thread conduit.
   e. Cut and join PVC conduit.

Performance Tasks

Under the supervision of the instructor, you should be able to do the following:

1. Make 90° bends, back-to-back bends, offsets, and saddle bends using a hand bender.
2. Cut, ream, and thread conduit.

Trade Terms

90° bend
Back-to-back bend
Concentric bends
Developed length
Gain

Offset
Rise
Segment bend
Stub-up

Industry Recognized Credentials

If you are training through an NCCER-accredited sponsor, you may be eligible for credentials from NCCER's Registry. The ID number for this module is 26107-20. Note that this module may have been used in other NCCER curricula and may apply to other level completions. Contact NCCER's Registry at 888.622.3720 or go to **www.nccer.org** for more information.

> **NOTE**
> NFPA 70®, *National Electrical Code*® and *NEC*® are registered trademarks of the National Fire Protection Association, Quincy, MA.

Contents

Figures and Tables

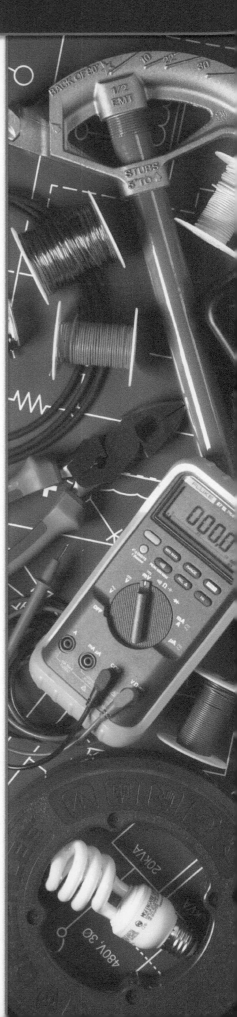

This page is intentionally left blank.

1.0.0 SELECTING AND USING HAND BENDING EQUIPMENT

Objective

Select and use hand bending equipment.
a. Use geometry to make a bend.
b. Make 90° bends.
c. Make offset bends.

Performance Task

1. Make 90° bends, back-to-back bends, offsets, and saddle bends using a hand bender.

Trade Terms

90° bend: A bend that changes the direction of the conduit by 90°.

Back-to-back bend: Any bend formed by two 90° bends with a straight section of conduit between the bends.

Concentric bends: 90° bends made in two or more parallel runs of conduit with the radius of each bend increasing from the inside of the run toward the outside.

Developed length: The actual length of the conduit that will be bent.

Gain: Because a conduit bends in a radius and not at right angles, the length of conduit needed for a bend will not equal the total determined length. Gain is the distance saved by the arc of a 90° bend.

Offset: An offset is two bends placed in a piece of conduit to change elevation to go over or under obstructions or for proper entry into boxes, cabinets, etc.

Rise: The length of the bent section of conduit measured from the bottom, centerline, or top of the straight section to the end of the conduit being bent.

Segment bend: A large bend formed by multiple short bends or shots.

Stub-up: Another name for the rise in a section of conduit at 90°. Also, a term used for conduit penetrating a slab or the ground.

The art of conduit bending depends on the skills of the electrician and requires a working knowledge of basic terms and proven procedures. Practice, knowledge, and training will help you develop the skills necessary for proper conduit bending and installation. *Figure 1* shows hand benders. Hand benders are convenient to use on the job because they are portable and no electrical power is required. Hand benders have a shape that supports the walls of the conduit being bent.

These benders are used to make various bends in smaller conduit (trade sizes $\frac{1}{2}$" to $1\frac{1}{4}$" or Metric Designator sizes MD 16 to MD 35). Most hand benders are sized to bend rigid conduit and electrical metallic tubing (EMT) of corresponding sizes. For example, a single hand bender can bend either $\frac{3}{4}$" EMT (MD 21) or $\frac{1}{2}$" rigid conduit (MD 16). The next larger size of hand bender will bend either 1" EMT (MD 27) or $\frac{3}{4}$" rigid conduit (MD 21). This is because the corresponding sizes of conduit have nearly equal outside diameters.

The first step in making a good bend is familiarizing yourself with the bender. The manufacturer of the bender will typically provide documentation indicating starting points, distance between each offset, and other important values associated with that particular bender. There is no substitute for taking the time to review this information. It will make the job go faster and result in better bends.

> **CAUTION**
>
> When making bends, be sure you have a firm grip on the handle to avoid slippage and possible injury.

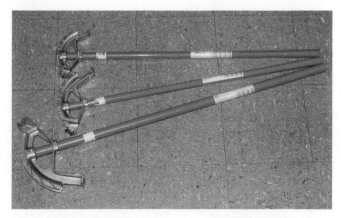

Figure 1 Hand benders.

When performing a bend, it is important to keep the conduit on a stable, firm, flat surface for the entire duration of the bend. Hand benders are designed to have force applied using one foot and the hands (*Figure 2*). It is important to use constant foot pressure as well as force on the handle to achieve uniform bends. Allowing the conduit to rise up or performing the bend on soft ground can result in distorting the conduit outside the bender.

A hickey (*Figure 3*) should not be confused with a hand bender. The hickey, which is used for RMC and IMC only, functions quite differently.

When you use a hickey to bend conduit, you are forming the bend as well as the radius. When using a hickey, be careful not to flatten or kink the conduit. Hickeys should be used with only RMC and IMC because very little support is given to the walls of the conduit being bent.

A hickey is a segment bending device. First, make a small bend of about 10°. Then, move the hickey to a new position and make another small bend. Continue this process until the bend is completed. Use a hickey for conduit **stub-ups** in slabs and decks.

Bend polyvinyl chloride (PVC) conduit using a heating unit (*Figure 4*). You must rotate the PVC regularly while it is in the heater so that it heats evenly. After heating the PVC, remove it and bend the PVC by hand. Some units use an electric heating element, while others use liquid propane (LP). After bending, a damp sponge or cloth is often used so that the PVC sets up faster.

When bending PVC that is 2" (MD 53) or larger in diameter, there is a risk of wrinkling or flattening the bend. A plug set eliminates this problem (*Figure 5*). Insert a plug into each end of the piece of PVC being bent. Then, use a hand pump to pressurize the conduit before bending it. The pressure is about 3 to 5 psi (7 to 21 kilopascals) before heat is applied.

1.1.0 Using Geometry to Make a Bend

Bending conduit requires that you use some basic geometry. You may already be familiar with most of the concepts needed; however, here is a review of the concepts directly related to this task.

A right triangle is defined as any triangle with a 90° angle. The side directly opposite the 90° angle

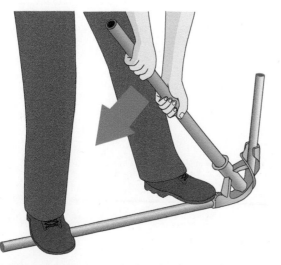

Figure 2 Pushing down on the bender to complete the bend.

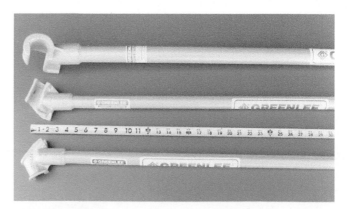

Figure 3 Hickeys.

is called the hypotenuse, and the side on which the triangle sits is the base. The vertical side is called the height. On the job, you will apply the relationships in a right triangle when making an offset bend. The offset forms the hypotenuse of a right triangle (*Figure 6*).

NOTE

There are reference tables for sizing offset bends based on these relationships (see *Appendix A*).

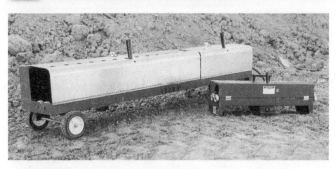

Figure 4 Typical PVC heating units.

Figure 5 Typical plug set.

A circle is defined as a closed curved line whose points are all the same distance from its center. The distance from the center point to the edge of the circle is called the *radius*. The length from one edge of the circle to the other edge through the center point (the full width of the circle) is called the *diameter*. (The diameter is always twice the length of the radius.) The distance around the circle is called the *circumference*. A circle can be divided into four equal quadrants. Each quadrant accounts for 90°, making a total of 360°. When you make a **90° bend**, you will use ¼ of a circle, or one quadrant (see *Figure 7*).

Concentric circles are circles that have a common center but different radii. The concept of concentric circles can be applied to **concentric bends** in conduit. The angle of each bend is 90°. Such bends have the same center point, but the radius of each is different.

To calculate the circumference of a circle, use the following formula:

$$C = \pi \times D \quad or \quad C = \pi D$$

Where:

C = circumference
π = 3.14
D = diameter

Another way of stating the formula for circumference is $C = 2\pi R$, where R equals the radius or ½ the diameter. When determining the arc of a quadrant, use the following formula:

$$\text{Length of arc} = 0.25 \times 2\pi R = 1.57R$$

For this formula, the arc of a quadrant equals ¼ the circumference of the circle or 1.57 times the radius.

A bending radius table is included in *Appendix B*.

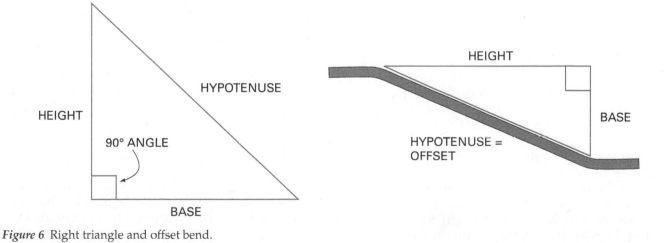

Figure 6 Right triangle and offset bend.

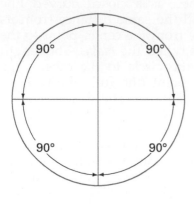

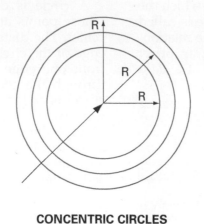

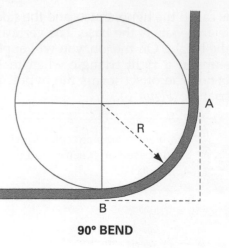

| CIRCLE | CONCENTRIC CIRCLES | 90° BEND |

Figure 7 Circles and 90° bends.

What's wrong with this picture?

1.2.0 Making 90° Bends

The 90° stub bend is probably the most basic bend of all. The stub bend is used much of the time, regardless of the type of conduit being installed. Before beginning to make the bend, you need to know two measurements:

- Desired rise or stub-up
- Take-up distance of the bender

The desired rise is the height of the stub-up. The take-up is the amount of conduit the bender will use to form the bend. Take-up distances are usually listed in the manufacturer's instruction manual. Typical bender take-up distances are shown in *Table 1*.

When you have determined the take-up, subtract it from the stub-up height. Mark that distance on the conduit (all the way around) at that distance from the end. The mark will indicate the point at which you will begin to bend the conduit. Line up the starting point on the conduit

Table 1 Typical Bender Take-Up Distances

EMT	Rigid/IMC	Take-Up
1/2"	—	5"
3/4"	1/2"	6"
1"	3/4"	8"
1 1/4"	1"	11"

with the starting point on the bender. Most benders have a mark, such as an arrow, to indicate this point. *Figure 8* shows the take-up required to achieve the desired stub-up length on a piece of 1/2" EMT (MD 16).

When you have lined up the bender, use one foot to hold the conduit steady. Keep your heel on the floor for balance. Apply constant pressure on the bender foot pedal with your other foot. Make sure you hold the bender handle perpendicular to the floor, and as far up as possible, to get maximum leverage. Then, bend the conduit in one smooth motion, pulling as steadily as possible. Avoid overstretching.

> **NOTE**
>
> When bending conduit, pay attention to which side of the mark you put the bender on. Putting the bender on the wrong side of the mark will result in gross error.

After finishing the bend, check to make sure you have the correct angle and measurement. Use the following steps to check a 90° bend:

Step 1 With the back of the bend on the floor, measure to the end of the conduit stub-up to make sure it is the right length.

Step 2 Check the 90° angle of the bend with a square or at the angle formed by the floor and a wall. A torpedo level may also be used.

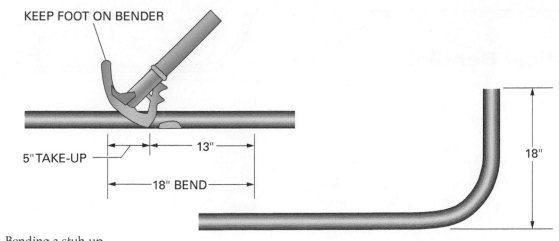

Figure 8 Bending a stub-up.

The above procedure will produce a 90° one-shot bend. That means that it took a single bend to form the conduit bend. A **segment bend** is any bend that is formed by a series of bends of a few degrees each, rather than a single one-shot bend. A shot is actually one bend in a segment bend. Segment or sweep bends must conform to the provisions of the *NEC*®.

1.2.1 Determining the Gain

In order to properly bend conduit, you must know how to determine the **gain** of the bender in use. The gain is the distance saved by the arc of a 90° bend. Knowing the gain can help you to precut, ream, and prethread both ends of the conduit before you bend it. This will make your work go more quickly because it is easier to work with conduit while it is straight. *Figure 9* shows that

Proper Bends

Kinks are created by bending too small a radius using a hickey.

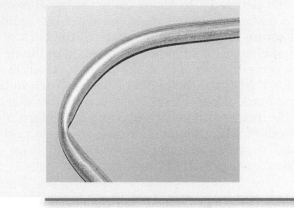

the overall **developed length** of a piece of conduit with a 90° bend is less than the sum of the horizontal and vertical distances when measured square to the corner. This is shown by the following equation:

Developed length = (A + B) − gain

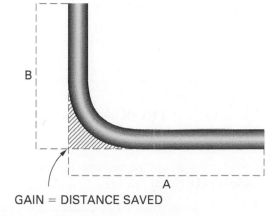

GAIN = DISTANCE SAVED

Figure 9 Gain.

Conduit Size	*NEC*® Radius	90° Gain
½"	4"	2⅝"
¾"	5"	3¼"
1"	6"	4"
1¼"	8"	5⅝"

TYPICAL GAIN TABLE

Practical Bending

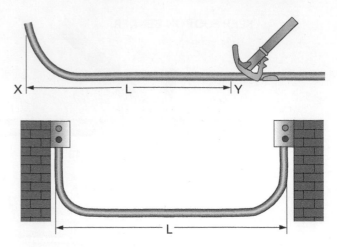

Figure 10 Back-to-back bends.

An example of a manufacturer's gain table is also shown in *Figure 9*. These tables are used to determine the gain for a certain size conduit.

1.2.2 Making Back-to-Back 90° Bends

A back-to-back bend consists of two 90° bends made on the same piece of conduit and placed back-to-back (*Figure 10*).

To make a back-to-back bend, make the first bend (labeled X in *Figure 10*) in the usual manner. To make the second bend, measure the required distance between the bends from the back of the first bend. This distance is labeled L in the figure. Reverse the bender on the conduit, as shown in *Figure 10*. Place the bender's back-to-back indicating mark at point Y on the conduit. Note that outside measurements from point X to point Y are used. Holding the bender in the reverse position and properly aligned, apply foot pressure and complete the second bend.

1.3.0 Making Offset Bends

Many situations require that the conduit be bent so that it can pass over objects such as beams and other conduits, or enter meter cabinets and junction boxes. Bends used for this purpose are called offsets. To produce an offset, two equal bends of less than 90° are required, a specified distance apart, as shown in *Figure 11*.

Offsets are a trade-off between space and the effort it will take to pull the wire. The larger the degree of bend, the harder it will be to pull the wire. The smaller the degree of bend, the easier it will be to pull the wire. Use the shallowest degree

of bend that will still allow the conduit to bypass the obstruction and fit in the given space.

When conduit is offset, some of the conduit length is used. If the offset is made into the area, an allowance must be made for this shrinkage. If the offset angle is away from the obstruction, the shrinkage can be ignored. *Table 2* shows the amount of shrinkage per inch of rise for common offset angles.

The formula for figuring the distance between bends is as follows:

Distance between bends =
depth of offset × multiplier

The distance between the offset bends can generally be found in the manufacturer's documentation for the bender. *Table 3* shows the distance between bends for the most common offset angles.

Calculations related to offsets are derived from the branch of mathematics known as trigonometry, which deals with triangles. The multipliers shown in *Table 2* represent the cosecant (CSC) of the related offset angle. Determine the multiplier by dividing the hypotenuse of the triangle created by the offset by the depth of the offset (*Figure 11*).

Basic trigonometry (trig) functions are briefly covered in *Appendix A*. As you will see in the next section, the tangent (TAN) of the offset angle is also used in calculating parallel offsets.

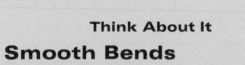

Think About It
Smooth Bends

Why are smooth bends so important?

Think About It
Gain

What is the difference between the gain and the take-up of a bend?

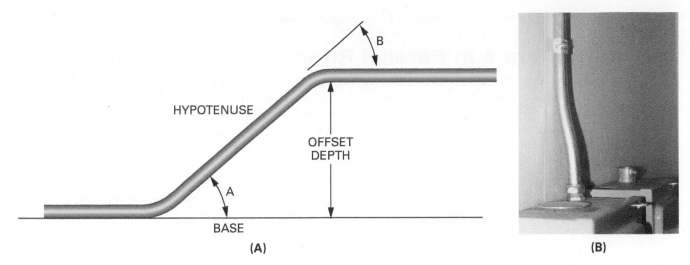

Figure 11 Offsets.

Table 2 Shrinkage Calculation

Offset Angle	Multiplier	Shrinkage (per inch of rise)
10° × 10°	6.0	$^1/_{16}$"
$22^1/_2$° × $22^1/_2$°	2.6	$^3/_{16}$"
30° × 30°	2.0	$^1/_4$"
45° × 45°	1.4	$^3/_8$"
60° × 60°	1.2	$^1/_2$"

Take-Up Method

When bending conduit using the take-up method, always place the bender on the conduit and make the bend facing the end of the conduit from which the measurements were taken. It helps to make a narrow mark with a soft lead pencil or marker completely around the conduit. This is called girdling.

Table 3 Common Offset Factors

Offset Depth	$22^1/_2$° Between Bends	Shrinkage	30° Between Bends	Shrinkage	45° Between Bends	Shrinkage	60° Beteween Bends	Shrinkage
2	$5^1/_4$	$^3/_8$	—	—	—	—	—	—
3	$7^3/_4$	$^9/_{16}$	6	$^3/_4$	—	—	—	—
4	$10^1/_2$	$^3/_4$	8	1	—	—	—	—
5	13	$^{15}/_{16}$	10	$1^1/_4$	7	$1^7/_8$	—	—
6	$15^1/_2$	$1^1/_8$	12	$1^1/_2$	$8^1/_2$	$2^1/_4$	$7^1/_4$	3
7	$18^1/_4$	$1^5/_{16}$	14	$1^3/_4$	$9^3/_4$	$2^5/_8$	$8^3/_8$	$3^1/_2$
8	$20^3/_4$	$1^1/_2$	16	2	$11^1/_4$	3	$9^5/_8$	4
9	$23^1/_2$	$1^3/_4$	18	$2^1/_4$	$12^1/_2$	$3^3/_8$	$10^7/_8$	$4^1/_2$
10	26	$1^7/_8$	20	$2^1/_2$	14	$3^3/_4$	12	5

Understanding trig functions will help you understand how offsets are determined. If you have a scientific calculator and understand these functions, you can calculate offset angles when you know the dimensions of the triangle created by the offset and the obstacle.

1.3.1 Making Parallel Offsets

Often, you must bend multiple pieces of conduit around a common obstruction. In this case, parallel offsets are made. Because the bends are laid out along a common radius, make an adjustment to ensure that the ends do not come out uneven, as shown in *Figure 12*.

Matching Bends in Parallel Runs

Suppose you are running 1" (MD 27) rigid conduit along with a 2" (MD 53) rigid conduit in a rack and you come to a 90° bend. If you used a 1" (MD 27) shoe, the radius would not match that of the 2" (MD 53) conduit bend. Instead, put the smaller conduit in the larger shoe to bend it. Both bends will now have the same radius. This trick will only work on rigid conduit. If done with EMT, it will flatten the pipe. If you have more than two parallel runs, you may have to make concentric bends. This will work for more, but the mark will change each time.

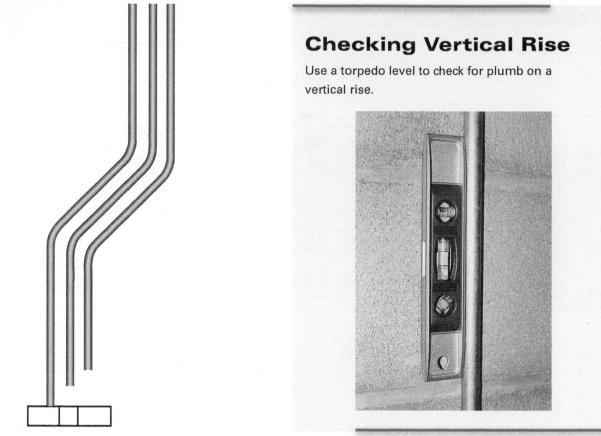

Figure 12 Incorrect parallel offsets.

Checking Vertical Rise

Use a torpedo level to check for plumb on a vertical rise.

First, find the center of the first bend of the innermost conduit, as shown in *Figure 13*. Each successive conduit must have its centerline moved farther away from the end of the pipe, as shown in *Figure 14*. Calculate the amount to add as follows:

Amount added =
center-to-center spacing ×
tangent (TAN) of ½ offset angle

Tangents can be found using the trig tables provided in *Appendix A*.

For example, *Figure 15* shows three pipes laid out as parallel and offset. The angle of the offset is 30°. The center-to-center spacing is 3". The start of the innermost pipe's first bend is 12".

The starting point of the second pipe will be:

12" + [center-to-center spacing ×
TAN (½ offset angle)] =

12" + (3" × TAN 15°) =

12" + (3" × 0.2679) =

12" + 0.8037" = 12.8037"

This is approximately $12^{13}/_{16}$".

The starting point for the outermost pipe is:

$$12^{13}/_{16}" + {}^{13}/_{16}" = 13^{5}/_{8}"$$

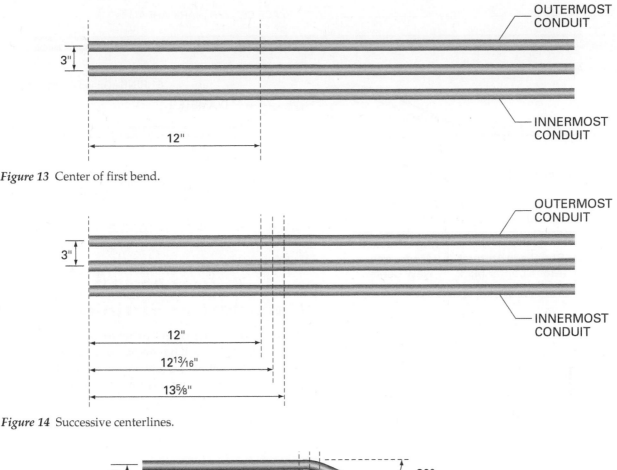

Figure 13 Center of first bend.

Figure 14 Successive centerlines.

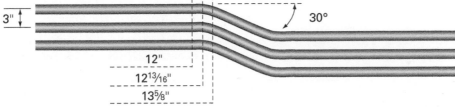

Figure 15 Parallel offset pipes.

1.3.2 Making Saddle Bends

Use a saddle bend to go around obstructions. *Figure 16* illustrates an example of a saddle bend that is required to clear a pipe obstruction. Making a saddle bend will cause the center of the saddle to shorten by $^3/_{16}$" (4.8 mm) for every inch of saddle depth (*Table 4*). For example, if the pipe diameter is 2" (MD 53), this would cause a $^3/_8$" (9.5 mm) shortening of the conduit on each side of the bend. When making saddle bends, the following steps should apply:

Step 1 Locate the center mark A on the conduit by using the size of the obstruction (i.e., pipe diameter) and calculate the shrink rate of the obstruction. For example, if the pipe diameter is 2", then $^3/_8$" of conduit will be lost on each side of the bend for a total shrinkage of $^3/_4$". This figure

will be added to the measurement from the end of the conduit to the centerline of the obstruction (for example, if the distance measured from the conduit end to the obstruction centerline was 15", the distance to A would be $15^3/_8$".

Step 2 Locate marks B and C on the conduit by measuring $2^1/_2$" for every 1" of saddle depth from the A mark (i.e., for the saddle depth of 2", the B mark would be 5" before the A mark and the C mark would be 5" after the A mark). Refer to the measurement locations shown in *Figure 17*.

Step 3 Refer to *Figure 18* and make a 45° bend at point A, make a $22^1/_2$° bend at point B, and make a $22^1/_2$° bend at point C. (Be sure to check the manufacturer's specifications.)

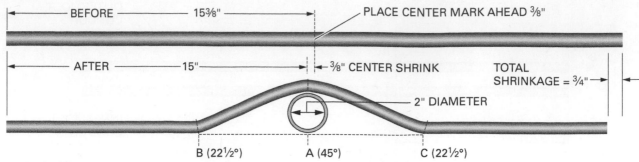

Figure 16 Saddle measurement.

Table 4 Shrinkage Chart for Saddle Bends with a 45°
Center Bend and Two 22½° Bends

Obstruction Depth	Shrinkage Amount (Move Center Mark Forward)	Make Outside Marks from New Center Mark
1	$^3/_{16}$"	$2^1/_2$"
2	$^3/_8$"	5"
3	$^9/_{16}$"	$7^1/_2$"
4	$^3/_4$"	10"
5	$^{15}/_{16}$"	$12^1/_2$"
6	$1^1/_8$"	15"
For each additional inch, add	$^3/_{16}$"	$2^1/_2$"

Think About It

Calculating Shrinkage

You're making a 30° by 30° offset to clear a 6" obstruction. What will be the distance between bends? What will be the developed length shrink? Make the same calculations for a 10" offset with 45° bends.

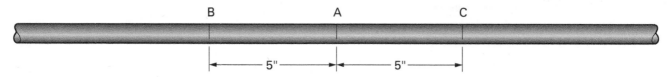

Figure 17 Measurement locations.

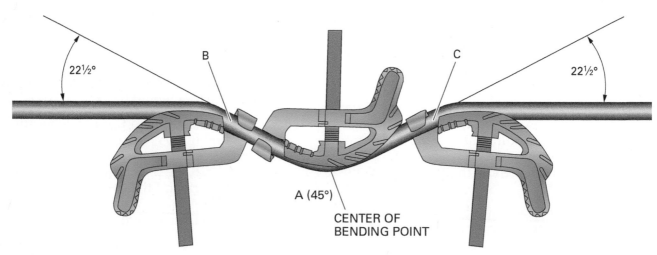

Figure 18 Location of bends.

Planning Bends

The more bends you make between pull points, the more difficult it is to pull the wires through the conduit. Therefore, plan your bends in advance, avoid sharp bends, and make as few bends as possible. The *NEC*® allows the bends in a single run of conduit to total 360° between pull points; however, 360° is not as much as you might think. For example, if you bend the conduit 90° for two corners of a room, with two 45° offsets where the conduit connects to a panelboard and junction box, you've used up your 360°.

1.3.3 Making Four-Bend Saddles

Four-bend saddles can be difficult. The reason is that four bends must be aligned exactly on the same plane. Extra time spent laying it out and performing the bends will pay off in not having to scrap the whole piece and start over.

Figure 19 illustrates that the four-bend saddle is really two offsets formed back-to-back. Working left to right, the procedure for forming this saddle is as follows:

Step 1 Determine the height of the offset.

Step 2 Determine the correct spacing for the first offset and mark the conduit.

Working with Conduit

Unprotected electrical cable is susceptible to physical damage; therefore, protect the wiring with conduit.

Step 3 Bend the first offset.

Step 4 Mark the start point for the second offset at the trailing edge of the obstruction.

Step 5 Mark the spacing for the second offset.

Step 6 Bend the second offset.

Figure 19 Typical four-bend saddle.

Bending Conduit

A good way to practice bending conduit is to use a piece of No. 10 or No. 12 solid wire and bend it to resemble the bends you need. This gives you some perspective on how to bend the conduit, and it will also help you to anticipate any problems with the bends.

Think About It

Equal Angles

Why is it important that the angles be identical when making an offset bend?

Using *Figure 20* as an example, a four-bend saddle using $\frac{1}{2}$" EMT (MD 16) is laid out as follows:

- Height of the box = 6"
- Width of the box = 8"
- Distance to the obstruction = 36"

Two 30° offsets will be used to form the saddle. It is created as follows:

Step 1 See *Figure 21*. Working from left to right, calculate the start point for the first bend. The distance to the obstruction is 36", the offset is 6", and the 30° multiplier from *Table 2* is 2.0:

Distance to first bend =
Distance to obstruction −
(offset × constant for angle) + shrinkage
36" − (6" × 2.0) + 1½" = 25½"

Step 2 Determine where the second bend will end to ensure the conduit clears the obstruction. See *Figure 22*.

Total length of first offset =
distance to first bend +
distance to second bend + shrinkage
25½" + 12" + 1½" = 39"

Step 3 Determine the start point of the second offset. The width of the box is 8"; therefore, the start point of the second offset should be 8" beyond the end of the first offset:

8" + 39" = 47"

Step 4 Determine the spacing for the second offset. Since the first and second offsets have the same rise and angle, the distance between bends will be the same, or 12".

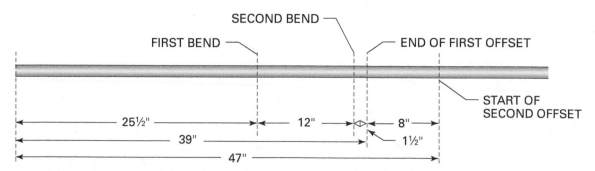

Figure 20 Four-bend saddle.

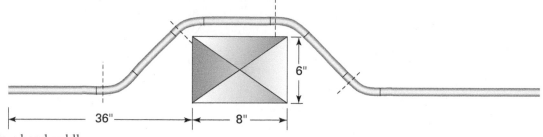

Figure 21 Four-bend saddle measurements.

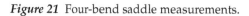

Figure 22 Bend and offset measurements.

Offset Benders

Use this offset bender for wall-mounted boxes with exposed conduit. It automatically matches the offset to the box knockout position and is a great timesaver when making multiple offsets.

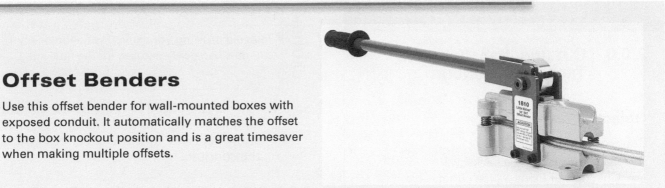

Figure Credit: Greenlee / A Textron Company

Think About It

Calculating Parallel Offsets

You're making parallel offsets of 45°, and the lengths of conduit are spaced 4" center to center. If the offset starts 12" down the pipe, what is the starting point for the bend on the second pipe?

1.0.0 Section Review

1. The first bend made with a hickey should be about _____.

    a. 10°
    b. 20°
    c. 30°
    d. 45°

2. A 90° bend is also known as a(n) _____.

    a. offset
    b. stub
    c. kick
    d. saddle

3. Making a saddle bend around a 4" obstruction will cause the center of the saddle to shorten by _____.

    a. $\frac{3}{16}$"
    b. $\frac{3}{8}$"
    c. $\frac{9}{16}$"
    d. $\frac{3}{4}$"

Section Two

2.0.0 Cutting, Reaming, and Threading Conduit

Objective

Cut, ream, and thread conduit.
  a. Cut conduit using a hacksaw.
  b. Cut conduit using a pipe cutter.
  c. Ream conduit.
  d. Thread conduit.
  e. Cut and join PVC conduit.

Performance Task

  2. Cut, ream, and thread conduit.

R MC, IMC, and EMT are available in standard 10' (3 m) lengths. When installing conduit, it is cut to fit the job requirements.

2.1.0 Cutting Conduit Using a Hacksaw

Conduit is normally cut using a hacksaw. To cut conduit with a hacksaw, proceed as follows:

> **WARNING!**
> Conduit edges are sharp and the cutting process produces metal debris that can damage eyes. Always wear safety goggles and gloves when cutting conduit.

Step 1 Inspect the blade of the hacksaw and replace it, if needed. A blade with 18, 24, or 32 cutting teeth per inch is recommended for conduit. Use a higher tooth count for EMT and a lower tooth count for rigid conduit and IMC. If the blade needs to be replaced, point the teeth toward the front of the saw when installing the new blade.

Step 2 Secure the conduit in a pipe vise.

Step 3 Rest the middle of the hacksaw blade on the conduit where the cut is to be made. Position the saw so the end of the blade is pointing slightly down and the handle is pointing slightly up. Push forward gently until the cut is started. Make even strokes until the cut is finished.

> **CAUTION**
> To avoid bruising your knuckles on the newly cut pipe, use gentle strokes for the final cut.

Step 4 Check the cut. The end of the conduit should be straight and smooth. *Figure 23* shows correct and incorrect cuts. Ream the conduit.

2.2.0 Cutting Conduit Using a Pipe Cutter

A pipe cutter can also be used to cut RMC and IMC. To use a pipe cutter, proceed as follows:

> **CAUTION**
> Conduit edges are sharp and the cutting process produces metal debris that can damage eyes. Always wear safety goggles and gloves when cutting conduit.

Step 1 Secure the conduit in a pipe vise and mark a place for the cut.

Step 2 Open the cutter and place it over the conduit with the cutter wheel on the mark.

Step 3 Tighten the cutter by rotating the screw handle.

> **CAUTION**
> Do not overtighten the cutter. Overtightening can break the cutter wheel and distort the wall of the conduit.

Step 4 Rotate the cutter to start the cut. It is important to remember that the rotation must pull the cutting blade into the conduit, as shown in *Figure 24*.

Step 5 Tighten the cutter handle $\frac{1}{4}$ turn for each full turn around the conduit. Again, make sure that you do not over-tighten it.

Step 6 Add a few drops of cutting oil to the groove and continue cutting. Avoid skin contact with the oil.

Step 7 When the cut is almost finished, stop cutting and snap the conduit to finish the cut. This reduces the ridge that can be formed on the inside of the conduit.

Step 8 Clean the conduit and cutter with a shop towel rag.

Step 9 Ream the conduit.

INCORRECT

CORRECT

Figure 23 Conduit ends after cutting.

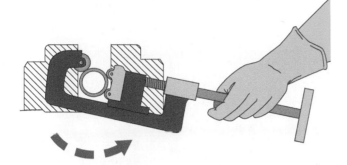

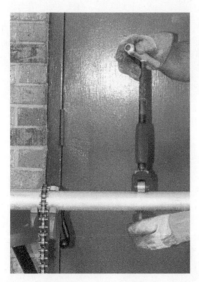

Figure 24 Cutter rotation.

2.3.0 Reaming Conduit

When the conduit is cut, the inside edge is sharp. This edge will damage the insulation of the wire when it is pulled through. To avoid this damage, use a reamer to smooth or ream the inside edge (*Figure 25*).

To ream the inside edge of a piece of conduit using a hand reamer, proceed as follows:

> **CAUTION**
>
> Conduit edges are sharp and the reaming process produces metal debris that can damage eyes. Always wear safety goggles and gloves when reaming conduit.

Step 1 Place the conduit in a pipe vise.

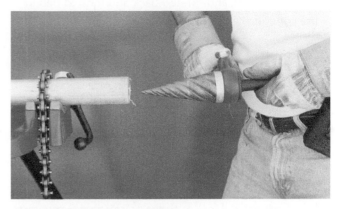

Figure 25 Rigid conduit reamer.

Step 2 Insert the reamer tip in the conduit.

Step 3 Apply light forward pressure and start rotating the reamer. *Figure 26* shows the proper way to rotate the reamer. Rotate the reamer using a downward motion. You could damage the reamer if you rotate it in the wrong direction. The reamer should bite as soon as you apply the proper pressure.

Step 4 Remove the reamer by pulling back on it while continuing to rotate it. Check the progress and then reinsert the reamer. Rotate the reamer until the inside edge is smooth. You should stop when all burrs have been removed.

> **NOTE**
>
> If a conduit reamer is not available, use a half-round file (the tang of the file must have a handle attached). You may use the nose of diagonal cutters or small hand reamers to ream EMT.

2.4.0 Threading Conduit

After cutting and reaming conduit, you will usually thread the conduit so it can be properly joined. Only RMC and IMC have walls thick enough for threading.

Use a tool used called a die to cut threads in conduit. Conduit dies are made to cut a taper of

EMT Reaming Tools

There are specialty tools available for reaming EMT. One example is shown here. This tool slips over the end of a square-shank screwdriver and is secured in place with setscrews. The tool is inserted into the end of the EMT and rotated back and forth to deburr the conduit.

Figure Credit: Klein Tools, Inc.

$^3\!/_4$ inch per foot. The number of threads per inch varies from 8 to 18, depending upon the diameter of the conduit. Use a thread gauge to measure how many threads per inch are cut.

A die head contains the threading dies. Use the die head with a hand-operated ratchet threader (*Figure 27*) or with a portable power drive.

To thread conduit using a hand-operated threader, proceed as follows:

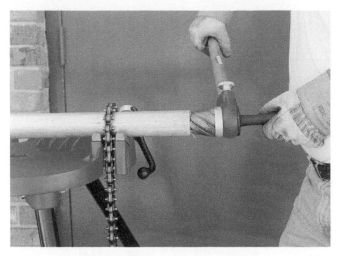

Figure 26 Reamer rotation.

> **CAUTION**
>
> Conduit edges are sharp and the threading process produces metal debris that can damage eyes. Always wear safety goggles and gloves when threading conduit.

Step 1 Insert the conduit in a pipe vise. Make sure the vise is fastened to a strong surface. Place supports, if necessary, to help secure the conduit.

Step 2 Determine the correct die and head. Inspect the die for damage, such as broken teeth. Never use a damaged die.

Step 3 Insert the die securely in the head. Make sure the proper die is in the appropriately numbered slot on the head.

Step 4 Determine the correct thread length to cut for the conduit size used (match the manufacturer's thread length).

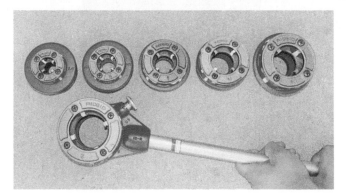

Figure 27 Hand-operated ratchet threader.

Step 5 Lubricate the die with cutting oil at the beginning and throughout the threading operation. Avoid skin contact with the oil.

Step 6 Cut threads to the proper length. Make sure that the conduit enters the tapered side of the die. Apply pressure and start turning the head. You should back off the head each quarter-turn to clear away chips.

Using Your Bender Head to Secure Conduit

To secure conduit while cutting, insert the conduit into the bender head, brace your foot against the bender to secure it, then proceed to cut the conduit.

Figure Credit: Tim Ely

Step 7 Remove the die when the proper cut is made. Threads should be cut only to the length of the die. Overcutting will leave the threads exposed to corrosion.

Step 8 Inspect the threads to make sure they are clean, sharp, and properly made. Use a thread gauge to measure the threads. The finished end should allow for a wrench-tight fit with one or two threads exposed.

> **NOTE**
> Ream the conduit again after threading to remove any burrs and edges. You must swab the cutting oil from the inside and outside of the conduit. Use a sandbox or drip pan under the threader to collect drips and shavings.

Using Tape

Use a piece of tape as a guide for marking your cutting lines around the conduit. This ensures a straight cut.

Oiling the Threader

For smoother operation, oil the threader often while threading the conduit.

You can also use die heads with portable power drives. Follow the same steps when using a portable power drive. Threading machines are often used on larger conduit and where frequent threading is required. Threading machines hold and rotate the conduit while the die is fed onto the conduit for cutting. When using a threading machine, make sure you secure the legs properly and follow the manufacturer's instructions.

Threading Conduit

The key to threading conduit is to start with a square cut. If you don't get it right, the conduit won't thread properly.

2.5.0 Cutting and Joining PVC Conduit

You can easily cut PVC conduit with a handsaw. Using a miter box or similar device when cutting 2" (MD 53) and larger PVC will ensure square cuts. You can deburr the cut ends using a pocket knife. Use a PVC cutter for smaller diameter PVC conduit, up to $1\frac{1}{2}$" (MD 41).

Use the following steps to join PVC conduit sections or attachments to plastic boxes:

> **WARNING!**
>
> Solvents and cements used with PVC are hazardous. Wear gloves and eye protection, and always follow the product instructions. Ensure that the area is well ventilated.

Step 1   Wipe all the contacting surfaces clean and dry.

Step 2   Apply a coat of cement (a brush or aerosol can is recommended) on both pieces to be joined.

Step 3   Press the conduit and fitting together and rotate about a half-turn to evenly distribute the cement.

Think About It

Putting It All Together

This module has stressed the precision necessary for creating accurate and uniform bends. Why is this important? What practical problems can result from sloppy or inaccurate bends?

> **NOTE**
>
> You must cement the PVC quickly. The PVC manufacturer usually provides aerosol spray cans of cement or the cement/brush combination. Make sure you use the recommended cement.

Forming PVC in the field requires a special tool called a hot box or other specialized methods. PVC may not be threaded when it is used for electrical applications.

PVC Cutters

Use a nylon string to cut PVC in place in awkward locations. However, it is best to use a PVC cutter to cut smaller trade sizes of PVC.

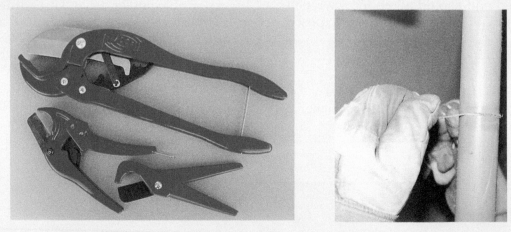

2.0.0 Section Review

1. Which of the following blades should you select when using a hacksaw to cut EMT?
   a. 10 teeth/inch
   b. 18 teeth/inch
   c. 22 teeth/inch
   d. 32 teeth/inch

2. When using a pipe cutter, tighten the cutter handle _____.
   a. $\frac{1}{4}$ turn for each full turn around the conduit
   b. $\frac{1}{2}$ turn for each full turn around the conduit
   c. $\frac{3}{4}$ turn for each full turn around the conduit
   d. a full turn for each full turn around the conduit

3. Smooth the cut metal conduit edges using _____.
   a. steel wool
   b. pliers
   c. a reamer
   d. a pocketknife

4. Conduit dies cut a taper of _____.
   a. $\frac{1}{4}$ inch/foot
   b. $\frac{1}{2}$ inch/foot
   c. $\frac{3}{4}$ inch/foot
   d. 1 inch/foot

5. When cutting and joining PVC, _____.
   a. use a PVC cutter for trade sizes 2" (MD 53) and larger
   b. use PVC cement on the inside and outside surfaces of the fitting only
   c. use PVC cement on the ends of both pieces to be joined
   d. press the conduit and fitting together and rotate about two turns to distribute the cement

1. The field bending of PVC requires a _____.
   a. hickey
   b. heating unit
   c. segmented bender
   d. one-shot bender

2. You can use a hickey to bend _____.
   a. RMC
   b. EMT
   c. PVC
   d. HDPE

3. What is the key to accurate bending with a hand bender?
   a. Correct size and length of handle
   b. Constant foot pressure on the back piece as well as force on the handle
   c. Using only the correct brand of bender
   d. Correct inverting of the conduit bender

4. In a right triangle, the side directly opposite the 90° angle is called the _____.
   a. right side
   b. hypotenuse
   c. altitude
   d. base

5. Prior to making a 90° bend, which two measurements must you know?
   a. Length of conduit and size of conduit
   b. Desired rise and length of conduit
   c. Size of bender and size of conduit
   d. Stub-up distance and take-up distance

6. A back-to-back bend is _____.
   a. a two-shot 90° bend
   b. two 90° bends made back-to-back on a single piece of conduit
   c. an offset with four bends back-to-back
   d. a segmented bend

7. To prevent the ends of the conduit from being uneven, what additional information must you use when making parallel offset bends?
   a. Center-to-center spacing and tangent of $\frac{1}{2}$ the offset angle
   b. Length of conduit and size of conduit
   c. Stub-up distance and take-up distance
   d. Offset angle and length of conduit

8. When making a saddle bend, the center of the saddle will cause the conduit to shrink _____ for every inch of saddle depth.
   a. $\frac{3}{8}$"
   b. $\frac{3}{16}$"
   c. $\frac{3}{4}$"
   d. $\frac{3}{32}$"

9. When using a pipe cutter, start the cut by rotating the cutter _____.
   a. blade $\frac{1}{2}$ turn tighter
   b. with the grain
   c. so the blade cuts into the conduit
   d. against the grain

10. Which of the following is typically threaded using a ratchet threader?
    a. EMT
    b. RMC
    c. LFNC
    d. FMC

Trade Terms Quiz

Fill in the blank with the correct term that you learned from your study of this module.

1. A right-angle bend is also called a(n) _____.

2. The rise in a section of conduit at 90° is called a(n) _____.

3. The _____ is the actual length of the conduit that will be bent.

4. A(n) _____ is two bends placed in a piece of conduit in order to navigate around obstructions.

5. A _____ is a large bend that is formed by multiple short bends or shots.

6. Two 90° bends with a straight section of conduit between them constitute a(n) _____.

7. _____ are 90° bends made in two or more parallel sections of conduit, where the radius of each bend in conduit after the inside bend is respectively increased.

8. _____ is the distance that is saved by the arc of a 90° bend.

9. _____ is the length of the bent section of conduit measured from the bottom, centerline, or top of the straight section to the end of the conduit being bent.

Trade Terms

90° bend
Back-to-back bend
Concentric bends
Developed length
Gain

Offset
Rise
Segment bend
Stub-up

1. The sizes of conduit that can be bent using a hand bender are _____.
    a. $\frac{1}{2}$ inch through $\frac{3}{4}$ inch
    b. $\frac{1}{2}$ inch through $1\frac{1}{4}$ inch
    c. $\frac{1}{2}$ inch through $1\frac{1}{2}$ inch
    d. $\frac{1}{2}$ inch through 2 inch

2. If a bender can be used to bend $\frac{3}{4}$-inch RMC, then it can also be used to bend _____ EMT.
    a. 1-inch
    b. 2-inch
    c. $2\frac{1}{2}$-inch
    d. 3-inch

3. The take-up on 1-inch EMT is _____.

4. The take-up on $\frac{1}{2}$-inch RMC is _____.

5. The typical gain on $\frac{1}{2}$-inch RMC bent at 90° is _____.

6. On an offset using 30° bends and a depth of 6", the conduit shrink is _____.
    a. $\frac{1}{16}$"
    b. $1\frac{1}{2}$"
    c. 1"
    d. $2\frac{1}{4}$"

7. On an offset using 30° bends and a depth of 6", the distance between bends is _____.
    a. 6"
    b. 7"
    c. 10"
    d. 12"

8. The conduit shrink is _____ per inch of offset when using 30° bends.
    a. $\frac{1}{16}$"
    b. $\frac{1}{8}$"
    c. $\frac{1}{4}$"
    d. $\frac{1}{2}$"

9. The multiplier for determining the distance between bends is _____ when bending offsets using 30° bends.
    a. 1.4
    b. 2.0
    c. 2.6
    d. 6.0

10. The multiplier for determining the distance between bends is _____ when bending offsets using 45° bends.
    a. 1.2
    b. 1.4
    c. 2.6
    d. 2.8

Timothy Ely

Beacon Electric Company

Tim Ely is a man who believes in giving something back to the industry that nurtured his successful career. Despite working in a demanding executive position, he serves on many industry committees and was instrumental in the development of the NCCER Electrical Program.

What made you decide to become an electrician?

During my last two years of high school, I worked for a do-it-all construction company. We laid concrete, installed roofs, hung drywall, installed plumbing, and did electrical work. I liked the electrical work the best.

How did you learn the trade?

I learned through on-the-job training, hard work, and studying on my own. I had good teachers who were patient with me and took the time to help me succeed.

What kinds of jobs did you hold on the way to your current position?

I started out wiring houses and did that for the first two years. Then I switched over to commercial and industrial work, and I worked as an apprentice in that area for two more years before becoming a journeyman. From there, I served as a lead electrician, then foreman, then city superintendent, then finally general superintendent before being promoted to my current job as vice president of construction.

What factor or factors have contributed most to your success?

Hard work helps a lot. I also try to bring a positive attitude to work with me every day. My family and friends have supported me throughout my career.

What does a vice president of construction do in your company?

In my job, I have responsibility for all the job sites, as well as the warehouse and service trucks. I also have responsibility for employee hiring, safety training, job planning and scheduling, quality control, and licensing. I personally hold 28 different state and city licenses, and I firmly believe that getting the training to obtain your licenses and then doing the in-service training to keep your licenses current are important factors in an electrician's success. For example, an electrical contractor can bid on jobs in a wide geographical area. Electricians working for that contractor can work on projects in different cities, even different states. Every place you go will require you to have a valid license.

What advice would you give to someone entering the electrical trade?

Work hard, treat people with respect, and keep an open mind. Be careful how you deal with people. Someone you offend today may wind up being your boss or a potential customer tomorrow.

Trade Terms Introduced in This Module

90° bend: A bend that changes the direction of the conduit by 90°.

Back-to-back bend: Any bend formed by two 90° bends with a straight section of conduit between the bends.

Concentric bends: 90° bends made in two or more parallel runs of conduit with the radius of each bend increasing from the inside of the run toward the outside.

Developed length: The actual length of the conduit that will be bent.

Gain: Because a conduit bends in a radius and not at right angles, the length of conduit needed for a bend will not equal the total determined length. Gain is the distance saved by the arc of a 90° bend.

Offset: An offset is two bends placed in a piece of conduit to change elevation to go over or under obstructions or for proper entry into boxes, cabinets, etc.

Rise: The length of the bent section of conduit measured from the bottom, centerline, or top of the straight section to the end of the conduit being bent.

Segment bend: A large bend formed by multiple short bends or shots.

Stub-up: Another name for the rise in a section of conduit at 90°. Also, a term used for conduit penetrating a slab or the ground.

Using Trigonometry to Determine Offset Angles and Multipliers

You do not have to be a mathematician to use trigonometry. Understanding the basic trig functions and how to use them can help you calculate unknown distances or angles. Assume that the right triangle below represents a conduit offset. If you know the length of one side and the angle, you can calculate the length of the other sides, or if you know the length of any two of the sides of the triangle, you can then find the offset angle using one or more of these trig functions. You can use a trig table such as that shown on the following pages or a scientific calculator to determine the offset angle. For example, if the cosecant of angle A is 2.6, the trig table tells you that the offset angle is $22\frac{1}{2}°$.

BENDING TABLE RADIUS

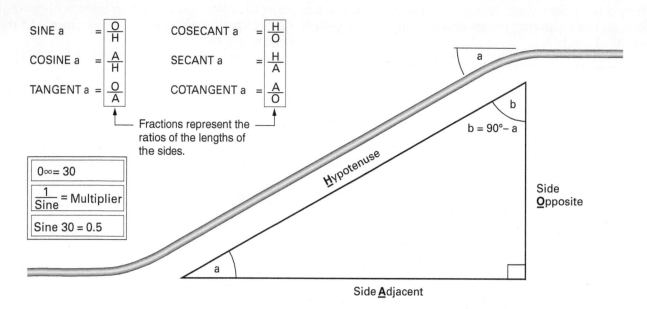

SINE a $= \left| \dfrac{O}{H} \right|$ COSECANT a $= \left| \dfrac{H}{O} \right|$

COSINE a $= \left| \dfrac{A}{H} \right|$ SECANT a $= \left| \dfrac{H}{A} \right|$

TANGENT a $= \left| \dfrac{O}{A} \right|$ COTANGENT a $= \left| \dfrac{A}{O} \right|$

Fractions represent the ratios of the lengths of the sides.

$0\infty = 30$

$\dfrac{1}{Sine} = Multiplier$

$Sine\ 30 = 0.5$

b = 90° − a

Hypotenuse

Side **O**pposite

Side **A**djacent

To determine the multiplier for the distance between bends in an offset:

1. Determine the angle of the offset: 30°

2. Find the sine of the angle: 0.5

3. Find the inverse (reciprocal) of the sine: $\dfrac{1}{0.5} = 2$. This is also listed in trig tables as the cosecant of the angle.

4. This number multiplied by the height of the offset gives the hypotenuse of the triangle, which is equal to the distance between bends.

ANGLE	SINE	COSINE	TANGENT	COTANGENT	COSECANT
1	0.0175	0.9998	0.0175	57.3000	57.3065
2	0.0349	0.9994	0.0349	28.6000	28.6532
3	0.0523	0.9986	0.0524	19.1000	19.1058
4	0.0698	0.9976	0.0699	14.3000	14.3348
5	0.0872	0.9962	0.0875	11.4000	11.4731
6	0.1045	0.9945	0.1051	9.5100	9.5666
7	0.1219	0.9925	0.1228	8.1400	8.2054
8	0.1392	0.9903	0.1405	7.1200	7.1854
9	0.1564	0.9877	0.1584	6.3100	6.3926
10	0.1736	0.9848	0.1763	5.6700	5.7587
11	0.1908	0.9816	0.1944	5.1400	5.2408
12	0.2079	0.9781	0.2126	4.7000	4.8097
13	0.2250	0.9744	0.2309	4.3300	4.4454
14	0.2419	0.9703	0.2493	4.0100	4.1335
15	0.2588	0.9659	0.2679	3.7300	3.8636
16	0.2756	0.9613	0.2867	3.4900	3.5915
17	0.2924	0.9563	0.3057	3.2700	3.4203
18	0.3090	0.9511	0.3249	3.0800	3.2360
19	0.3256	0.9455	0.3443	2.9000	3.0715
20	0.3420	0.9397	0.3640	2.7500	2.9238
21	0.3584	0.9336	0.3839	2.6100	2.7904
22	0.3746	0.9272	0.4040	2.4800	2.6694
23	0.3907	0.9205	0.4245	2.3600	2.5593
24	0.4067	0.9135	0.4452	2.2500	2.4585
25	0.4226	0.9063	0.4663	2.1400	2.3661
26	0.4384	0.8988	0.4877	2.0500	2.2811
27	0.4540	0.8910	0.5095	1.9600	2.2026
28	0.4695	0.8829	0.5317	1.8800	2.1300
29	0.4848	0.8746	0.5543	1.8000	2.0626
30	0.5000	0.8660	0.5774	1.7300	2.0000
31	0.5150	0.8572	0.6009	1.6600	1.9415
32	0.5299	0.8480	0.6249	1.6000	1.8870
33	0.5446	0.8387	0.6494	1.5400	1.8360
34	0.5592	0.8290	0.6745	1.4800	1.7883
35	0.5736	0.8192	0.7002	1.4300	1.7434
36	0.5878	0.8090	0.7265	1.3800	1.7012
37	0.6018	0.7986	0.7536	1.3300	1.6616
38	0.6157	0.7880	0.7813	1.2800	1.6242
39	0.6293	0.7771	0.8098	1.2300	1.5890
40	0.6428	0.7660	0.8391	1.1900	1.5557
41	0.6561	0.7547	0.8693	1.1500	1.5242
42	0.6691	0.7431	0.9004	1.1100	1.4944
43	0.6820	0.7314	0.9325	1.0700	1.4662
44	0.6947	0.7193	0.9657	1.0400	1.4395
45	0.7071	0.7071	1.0000	1.0000	1.4142

ANGLE	SINE	COSINE	TANGENT	COTANGENT	COSECANT
46°	0.7193	0.6947	1.0355	0.9660	1.4395
47°	0.7314	0.6820	1.0724	0.9330	1.3673
48°	0.7431	0.6691	1.1106	0.9000	1.3456
49°	0.7547	0.6561	1.1504	0.8690	1.3250
50°	0.7660	0.6428	1.1918	0.8390	1.3054
51°	0.7771	0.6293	1.2349	0.8100	1.2867
52°	0.7880	0.6157	1.2799	0.7810	1.2690
53°	0.7986	0.6018	1.3270	0.7540	1.2521
54°	0.8090	0.5878	1.3764	0.7270	1.2360
55°	0.8192	0.5736	1.4281	0.7000	1.2207
56°	0.8290	0.5592	1.4826	0.6750	1.2062
57°	0.8387	0.5446	1.5399	0.6490	1.1923
58°	0.8480	0.5299	1.6003	0.6250	1.1791
59°	0.8572	0.5150	1.6643	0.6010	1.1666
60°	0.8660	0.5000	1.7321	0.5770	1.1547
61°	0.8746	0.4848	1.8040	0.5540	1.1433
62°	0.8829	0.4695	1.8807	0.5320	1.1325
63°	0.8910	0.4540	1.9626	0.5100	1.1223
64°	0.8988	0.4384	2.0503	0.4880	1.1126
65°	0.9063	0.4226	2.1445	0.4660	1.1033
66°	0.9135	0.4067	2.2460	0.4450	1.0946
67°	0.9205	0.3907	2.3559	0.4240	1.0863
68°	0.9272	0.3746	2.4751	0.4040	1.0785
69°	0.9336	0.3584	2.6051	0.3840	1.0711
70°	0.9397	0.3420	2.7475	0.3640	1.0641
71°	0.9455	0.3256	2.9042	0.3440	1.0576
72°	0.9511	0.3090	3.0777	0.3250	1.0514
73°	0.9563	0.2924	3.2709	0.3060	1.0456
74°	0.9613	0.2756	3.4874	0.2870	1.0402
75°	0.9659	0.2588	3.7321	0.2680	1.0352
76°	0.9703	0.2419	4.0108	0.2490	1.0306
77°	0.9744	0.2250	4.3315	0.2310	1.0263
78°	0.9781	0.2079	4.7046	0.2130	1.0223
79°	0.9816	0.1908	5.1446	0.1940	1.0187
80°	0.9848	0.1736	5.6713	0.1760	1.0154
81°	0.9877	0.1564	6.3138	0.1580	1.0124
82°	0.9903	0.1392	7.1154	0.1410	1.0098
83°	0.9925	0.1219	8.1443	0.1230	1.0075
84°	0.9945	0.1045	9.5144	0.1050	1.0055
85°	0.9962	0.0872	11.4301	0.0880	1.0038
86°	0.9976	0.0698	14.3007	0.0700	1.0024
87°	0.9986	0.0523	19.0811	0.0520	1.0013
88°	0.9994	0.0349	28.6363	0.0350	1.0006
89°	0.9998	0.0175	57.2900	0.0180	1.0001
90°	1.0000	0.0000	—	0.0000	1.0000

Radius (Inches)	Radius Increments (Inches)									
	0	1	2	3	4	5	6	7	8	9
0	0.00	1.57	3.14	4.71	6.28	7.85	9.42	10.99	12.56	14.13
10	15.70	17.27	18.84	20.41	21.98	23.85	25.12	26.69	28.26	29.83
20	31.40	32.97	34.54	36.11	37.68	39.25	40.82	42.39	43.96	45.83
30	47.10	48.67	50.24	51.81	53.38	54.95	56.52	58.09	59.66	61.23
40	62.80	64.37	65.94	67.50	69.03	70.65	72.22	73.79	75.36	76.93
50	87.50	80.07	81.64	83.21	84.78	86.35	87.92	89.49	91.06	92.63
60	94.20	95.77	97.34	98.91	100.48	102.05	103.62	105.19	106.76	108.33
70	109.90	111.47	113.04	114.61	116.18	117.75	119.32	120.89	122.46	124.03
80	125.60	127.17	128.74	130.31	131.88	133.45	135.02	136.59	138.16	139.73
90	141.30	142.87	144.44	146.01	147.58	149.15	150.72	–	–	–

To find the developed length for the following angles, use a fraction of the 90° chart.

For	15°	22½°	30°	45°	60°	67½°	75°	90°
Take	⅙	¼	⅓	½	⅔	¾	⅚	See Chart

For any other degrees: Developed length = 0.01744 × radius × degrees.

Additional Resources

This module presents thorough resources for task training. The following reference material is recommended for further study.

Benfield Conduit Bending Manual, 2nd Edition. Overland Park, KS: EC&M Books.

National Electrical Code® Handbook, Latest Edition. Quincy, MA: National Fire Protection Association.

Tom Henry's Conduit Bending Package (includes DVD, book, and bending chart). Winter Park, FL: Code Electrical Classes, Inc.

Figure Credits

Greenlee / A Textron Company, Module Opener

John Traister, Table 1, Figures 8–10

Section Review Answer Key

Section 1.0.0

Answer	Section Reference	Objective
1. a	1.0.0	1a
2. b	1.2.0	1b
3. d	1.3.2; Table 4	1c

Section 2.0.0

Answer	Section Reference	Objective
1. d	2.1.0	2a
2. a	2.2.0	2b
3. c	2.3.0	2c
4. c	2.4.0	2d
5. c	2.5.0	2e

1.0.0 SECTION REVIEW

Question 3

Since making a saddle bend causes the center of the saddle to shorten $\frac{3}{16}$" for every inch of saddle depth:

$$\frac{3}{16}" \times 4" = \frac{12}{16}" = \frac{3}{4}"$$

The center of the saddle will shorten by $\frac{3}{4}$".

NCCER CURRICULA — USER UPDATE

NCCER makes every effort to keep its textbooks up-to-date and free of technical errors. We appreciate your help in this process. If you find an error, a typographical mistake, or an inaccuracy in NCCER's curricula, please fill out this form (or a photocopy), or complete the online form at **www.nccer.org/olf**. Be sure to include the exact module ID number, page number, a detailed description, and your recommended correction. Your input will be brought to the attention of the Authoring Team. Thank you for your assistance.

Instructors – If you have an idea for improving this textbook, or have found that additional materials were necessary to teach this module effectively, please let us know so that we may present your suggestions to the Authoring Team.

NCCER Product Development and Revision
13614 Progress Blvd., Alachua, FL 32615

Email: curriculum@nccer.org
Online: www.nccer.org/olf

❏ Trainee Guide   ❏ Lesson Plans   ❏ Exam   ❏ PowerPoints   Other _____

Craft / Level: _____   Copyright Date: _____

Module ID Number / Title: _____

Section Number(s): _____

Description: _____

_____

_____

_____

_____

Recommended Correction: _____

_____

_____

_____

_____

Your Name: _____

Address: _____

_____

Email: _____   Phone: _____

This page is intentionally left blank.

Wireways, Raceways, and Fittings

OVERVIEW

Electrical raceways present challenges and requirements involving proper installation techniques, general understanding of raceway systems, and applications of the *NEC®* to raceway systems. Acquiring quality installation skills for raceway systems requires practice, knowledge, and training. This module describes various types of raceway systems, along with their installation and *NEC®* requirements. It also describes the use of various conduit bodies.

Module 26108-20

Trainees with successful module completions may be eligible for credentialing through the NCCER Registry. To learn more, go to **www.nccer.org** or contact us at 1.888.622.3720. Our website, **www.nccer.org**, has information on the latest product releases and training.

Your feedback is welcome. You may email your comments to **curriculum@nccer.org**, send general comments and inquiries to **info@nccer.org**, or fill in the User Update form at the back of this module.

This information is general in nature and intended for training purposes only. Actual performance of activities described in this manual requires compliance with all applicable operating, service, maintenance, and safety procedures under the direction of qualified personnel. References in this manual to patented or proprietary devices do not constitute a recommendation of their use.

26108-20 V10.0

Objectives

When you have completed this module, you will be able to do the following:

1. Select and install raceway systems.
   a. Identify types of conduit and their applications.
   b. Properly bond conduit for use as a ground path.
   c. Install metal conduit fittings.
   d. Make conduit-to-box connections.
   e. Identify raceway supports.
   f. Identify installation requirements for various construction methods.
2. Select fasteners and anchors for the installation of raceway systems.
   a. Select and install tie wraps.
   b. Select and install screws.
   c. Select and install hammer-driven pins and studs.
   d. Identify the safety requirements for stud-type guns.
   e. Select and install masonry anchors.
   f. Select and install hollow-wall anchors.
   g. Select and install epoxy anchoring systems.
3. Select and install wireways and other specialty raceways.
   a. Identify types of wireways and their components.
   b. Install wireway supports.
   c. Identify and install specialty raceways.
4. Select and install cable trays.
   a. Identify cable tray types and fittings.
   b. Install cable tray supports.
5. Handle and store raceways.
   a. Handle raceways.
   b. Store raceways.

Performance Tasks

Under the supervision of the instructor, you should be able to do the following:

1. Identify the appropriate conduit body for a given application.
2. Identify and select various types and sizes of raceways, fittings, and fasteners for a given application.
3. Demonstrate how to install a raceway system.
4. Terminate a selected raceway system.

Trade Terms

Accessible	Raceways
Approved	Splice
Bonding wire	Tap
Cable trays	Trough
Conduit	Underwriters Laboratories, Inc. (UL)
Exposed location	Wireways
Kick	

Industry Recognized Credentials

If you are training through an NCCER-accredited sponsor, you may be eligible for credentials from NCCER's Registry. The ID number for this module is 26108-20. Note that this module may have been used in other NCCER curricula and may apply to other level completions. Contact NCCER's Registry at 888.622.3720 or go to **www.nccer.org** for more information.

> **NOTE**
>
> NFPA 70®, *National Electrical Code*® and *NEC*® are registered trademarks of the National Fire Protection Association, Quincy, MA.

Contents

Contents (continued)

Figures and Tables

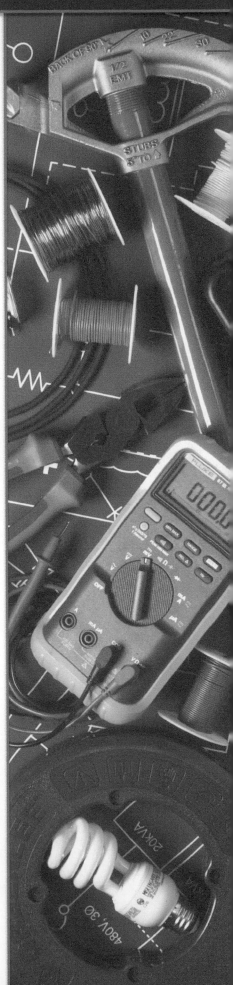

Figures and Tables (continued)

SECTION ONE

1.0.0 RACEWAY SYSTEMS

Objective

Select and install raceway systems.
  a. Identify types of conduit and their applications.
  b. Properly bond conduit for use as a ground path.
  c. Install metal conduit fittings.
  d. Make conduit-to-box connections.
  e. Identify raceway supports.
  f. Identify installation requirements for various construction methods.

Performance Task

1. Identify the appropriate conduit body for a given application.

Trade Terms

Accessible: Able to be reached, as for service or repair.

Approved: Meeting the requirements of an appropriate regulatory agency.

Bonding wire: A wire used to make a continuous grounding path between equipment and ground.

Conduit: A round raceway, similar to pipe, that houses conductors.

Exposed location: Not permanently closed in by the structure or finish of a building; able to be installed or removed without damage to the structure.

Kick: A bend in a piece of conduit, usually less than 45°, made to change the direction of the conduit.

Raceways: Enclosed channels designed expressly for holding wires, cables, or busbars, with additional functions as permitted in the NEC®.

Splice: Connection of two or more conductors.

Tap: Intermediate point on a main circuit where another wire is connected to supply electrical current to another circuit.

Underwriters Laboratories, Inc. (UL): An agency that evaluates and approves electrical components and equipment.

Wireways: Steel troughs designed to carry electrical wire and cable.

The term **raceways** refers to a wide range of circular and rectangular enclosed channels used to house electrical wiring. Raceways can be metallic or nonmetallic and come in different shapes. Depending on the particular purpose for which they are intended, raceways include enclosures such as underfloor raceways, flexible metal conduit, tubing, **wireways**, surface metal raceways, surface nonmetallic raceways, and support systems such as cable trays.

1.1.0 Types of Conduit and Their Applications

Conduit is a raceway with a circular cross section, similar to pipe, that contains wires or cables. Conduit is used to provide protection for conductors and route them from one place to another. Metal conduit also provides a permanent electrical path to ground. This equipment must be listed in accordance with the *NEC*®. There are many types of conduit used in the construction industry. The size of conduit to be used is determined by engineering specifications, local codes, and the *NEC*®. Refer to *NEC Chapter 9, Tables 1 and 4* and *Informative Annex C* for conduit fill with various conductors.

1.1.1 Electrical Metallic Tubing

Electrical metallic tubing (EMT) is the lightest duty tubing available for enclosing and protecting electrical wiring. EMT is widely used for residential, commercial, and industrial wiring systems. It is lightweight, easily bent and/or cut to shape, and is the least costly type of metallic conduit. Because the wall thickness of EMT is less than that of rigid conduit, it is often referred to as thinwall conduit. A comparison of inside and outside diameters of EMT to rigid metal conduit (RMC) and intermediate metal conduit (IMC) is shown in *Figure 1*.

NEC Section 358.10(A) permits the installation of EMT for either exposed or concealed work. Per *NEC Section 358.10(B)(1)*, galvanized steel and stainless steel EMT, elbows, and fittings shall be permitted to be installed in concrete, in direct contact with the earth, or in areas subject to severe corrosive influences where protected by corrosion protection and approved as suitable for the condition.

According to *NEC Section 358.10(D)*, where EMT is installed in wet locations, all supports, bolts, straps, screws, and so forth shall be of corrosion-resistant materials or protected against corrosion by corrosion-resistant materials.

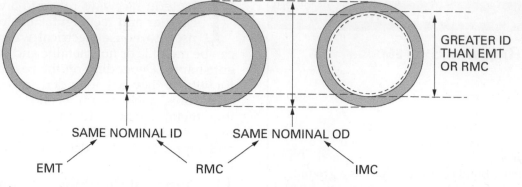

GREATER ID
THAN EMT
OR RMC

SAME NOMINAL ID   SAME NOMINAL OD

EMT   RMC   IMC

Figure 1 Conduit comparison.

According to *NEC Section 358.12*, EMT shall not be used where subject to severe physical damage or for the support of luminaires or other equipment, except conduit bodies no larger than the largest trade size of the tubing.

In a wet area, EMT and other metallic conduit must be installed to prevent water from entering the conduit system. In locations where walls are subject to regular wash-down [see *NEC Section 300.6(D)*], the entire conduit system must be installed to provide a $\frac{1}{4}$" (6 mm) air space between it and the wall or supporting surface. The entire conduit system is considered to include conduit, boxes, and fittings. To ensure resistance to corrosion caused by wet environments, EMT is galvanized. The term galvanized is used to describe the procedure in which the interior and exterior of the conduit are coated with a corrosion-resistant zinc compound.

EMT, being a good conductor of electricity, may be used as an equipment grounding conductor [see *NEC Section 250.118(4)*]. The conduit system must be tightly connected at each joint and provide a continuous grounding path from each electrical load to the service equipment. The connectors used in an EMT system ensure electrical and mechanical continuity throughout the system (see *NEC Sections 250.96, 300.10, and 358.42*).

> **NOTE**
> Support requirements for EMT are also covered in *NEC Section 358.30*. The types of supports will be discussed later in this module.

Because EMT is too thin for threads, fittings listed for EMT must be used. For wet or damp locations, listed compression fittings such as those shown in *Figure 2* are used. These fittings contain a compression ring made of metal that forms a raintight seal.

Think About It

EMT Use

Where would you use EMT? Are there any circumstances where EMT cannot be run through a suspended ceiling? What are some differences between EMT and rigid conduit?

When EMT compression couplings are used, they must be securely tightened, and when installed in masonry or concrete, they must be of the concrete-tight type. If installed in a wet location, they must be the raintight type. Refer to *NEC Section 358.42*.

EMT fittings for dry locations can be the setscrew type, the compression type, or the push-on type. Push-on fittings do not need tightening and are removable through the use of a special tool. To use the setscrew type, the ends of the EMT are inserted into the sleeve and the setscrews are tightened to make the connection. Various types of setscrew fittings are shown in *Figure 3*.

EMT sizes of $2\frac{1}{2}$" or metric designator (MD) 63 and larger have the same outside diameter as corresponding sizes of galvanized RMC. RMC threadless connectors may be used to connect EMT.

> **NOTE**
> EMT connectors smaller than $2\frac{1}{2}$" (MD 63), although they are the same size as RMC threadless connectors, may not be used to connect RMC.

Both setscrew and compression couplings are available in die-cast or steel construction. Steel couplings are stronger than die-cast types.

Support requirements for EMT are presented in *NEC Section 358.30*. As with most other metal conduit, EMT must be supported at intervals

---

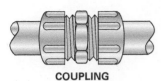
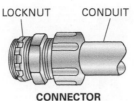

COUPLING   CONNECTOR

Figure 2 Compression fittings.

Figure 3 Setscrew fittings.

not to exceed 10' (3 m) and within 3' (900 mm) of each outlet box, junction box, cabinet, fitting, or terminating end of the conduit. An exception to *NEC Section 358.30(A), Exception 1* allows the fastening of unbroken lengths of EMT to be increased to a distance of 5' (1.5 m) where structural members do not readily permit fastening within 3' (900 mm).

Electrical nonmetallic tubing (ENT) is also available. It provides an economical alternative to EMT, but it can only be used in certain applications. See *NEC Article 362*.

1.1.2 Rigid Metal Conduit

Rigid metal conduit (RMC) is conduit that is constructed of metal of sufficient thickness to permit the cutting of pipe threads at each end. Specific information on RMC may be found in *NEC Article 344*. RMC provides the best physical protection for conductors of any of the various types of conduit. RMC is normally supplied in 10' (3 m) lengths, including a threaded coupling on one end.

RMC may be made from steel or aluminum. Rigid metal steel conduit may be stainless, galvanized, or enamel-coated inside and out. Because of its threaded fittings, RMC provides an excellent equipment grounding conductor as defined in *NEC Section 250.118(2)*. A piece of RMC is shown in *Figure 4 (A)*. The support requirements for RMC are presented in *NEC Section 344.30 and NEC Table 344.30(B)(2)*.

RMC is mostly used in industrial applications. RMC is heavier than EMT and IMC. It is more

EMT Installation

EMT is easily cut to size using a conduit cutter, such as the one shown here.

Figure Credit: Greenlee / A Textron Company

difficult to cut and bend, usually requires threading of cut ends, and has a higher purchase price than EMT and IMC. As a result, the cost of installing RMC is generally higher than the cost of installing EMT and IMC.

1.1.3 Plastic-Coated RMC

Plastic-coated RMC has a thin coating of polyvinyl chloride (PVC) over the RMC. See *Figure 4 (B)*. This combination is useful when an environment calls for the ruggedness of RMC along with the corrosion resistance of rigid nonmetallic conduit. Plastic-coated RMC requires special threading and bending techniques. Typical installations where plastic-coated RMC may be required are:

- Chemical plants
- Food plants
- Refineries
- Fertilizer plants
- Paper mills
- Wastewater treatment plants

1.1.4 Aluminum Conduit

Aluminum conduit has several characteristics that distinguish it from steel conduit. Because it has better corrosion resistance in wet environments and some chemical environments, aluminum conduit generally requires less maintenance in installations such as sewage treatment plants.

NEC Section 300.6(B) states that aluminum conduit used in concrete or in direct contact with soil requires supplementary corrosion protection. According to Underwriters Laboratories *Electrical Construction Equipment Directory* (UL Green Book), examples of supplementary protection are paints approved for the purpose (such as bitumastic

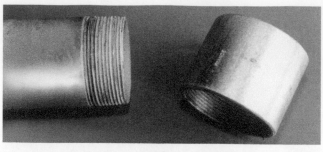

(A) RIGID METAL CONDUIT (RMC)

**THREADED PIPE
IDENTIFICATION**

BLUE REPRESENTS AN EVEN-
SIZED PIPE (1", 2", ETC.)

RED REPRESENTS A QUARTER-
INCH SIZE (¾" or 1¼")

BLACK REPRESENTS HALF-
INCH SIZES (1½", 2½", ETC.)

(B) PLASTIC-COATED RMC

Figure 4 Types of rigid metal conduit (RMC).

paint), tape wraps approved for the purpose, or PVC-coated conduit.

> **NOTE**
>
> Caution must be exercised to avoid burial of aluminum conduit in soil or concrete that contains calcium chloride. Calcium chloride may interfere with the corrosion resistance of aluminum conduit. Calcium chloride and similar materials are often added to concrete to speed concrete setting. It is important to determine if chlorides are to be used in the concrete prior to installing aluminum conduit. If chlorides are to be used, aluminum conduit must be avoided. Check with local authorities regarding this type of usage.

1.1.5 Black Enamel Steel Conduit

Rigid black enamel steel conduit (often called black conduit) is steel conduit that is coated with a black enamel. In the past, this type of conduit was used exclusively for indoor wiring. Black enamel steel conduit is no longer manufactured for sale in the United States. It is mentioned only because it may still be found in existing installations.

1.1.6 Intermediate Metal Conduit

Intermediate metal conduit (IMC) is a type of rigid steel conduit. It has a wall thickness that is less than that of RMC but greater than that of EMT. The weight of IMC is approximately ⅔ that of RMC. Because of its lower purchase price, lighter weight, and thinner walls, IMC installations are generally less expensive than comparable RMC installations. However, IMC installations still have high strength ratings.

> **NOTE**
>
> Additional information on IMC may be found in *NEC Article 342*.

The outside diameter of a given size of IMC is the same as that of the comparable size of RMC. Therefore, RMC fittings may be used with IMC. Because the threads on IMC and RMC are the same size, no special threading tools are needed to thread IMC. Some electricians feel that threading IMC is more difficult than threading RMC because IMC is somewhat harder.

The internal diameter of a given size of IMC is somewhat larger than the internal diameter of the same size of RMC because of the difference in wall thickness. Bending IMC requires the use of special shoes to support the conduit so it does not collapse.

The *NEC*® requires that IMC be identified along its length at 5' (1.5 m) intervals with the letters IMC. *NEC Sections 110.21(A)(1) and 342.120* describe this marking requirement.

Like RMC, IMC is permitted to act as an equipment grounding conductor, as defined in *NEC Section 250.118(3)*. The use of IMC may be restricted in some jurisdictions. It is important to investigate the requirements of each jurisdiction before selecting any materials.

1.1.7 Rigid Polyvinyl Chloride Conduit

The most common type of rigid nonmetallic conduit is manufactured from polyvinyl chloride (PVC). See *NEC Article 352*. Because PVC is noncorrosive, chemically inert, and non-aging, it is often used for installation in wet or corrosive environments. Corrosion problems found with steel and aluminum RMC do not occur with PVC. However, PVC may deteriorate under some conditions, such as extreme sunlight, unless marked sunlight resistant.

All PVC is marked according to standards established by the National Electrical Manufacturers Association (NEMA) or Underwriters Laboratories, Inc. (UL). A section of PVC is shown in *Figure 5*.

Since PVC is lighter than steel or aluminum rigid conduit, IMC, or EMT, it is considered easier to handle. PVC can usually be installed much faster than other types of conduit because the joints are made up with cement and require no threading.

PVC contains no metal. This characteristic reduces the voltage drop of conductors carrying alternating current in PVC compared to identical conductors in steel conduit.

Because PVC is nonconducting, it cannot be used as an equipment grounding conductor. An

---

Use of Aluminum Conduit

Aluminum conduit is used for special purposes such as high-cycle lines (400 cycles or above); around cooling towers, food service areas, and other applications in which corrosion is a factor; or where magnetic induction is a concern, such as near magnetic resonance imaging (MRI) equipment in hospitals.

---

equipment grounding conductor sized in accordance with *NEC Table 250.122* must be pulled in each PVC conductor run (except for underground service-entrance conductors).

PVC is available in a variety of lengths. However, some jurisdictions require it to be cut to 10' (3 m) prior to installation. PVC is subject to expansion and contraction directly related to the difference in temperature, plus any radiating effects on the conduit. In moderate climates, even a 10' (3 m) installation of PVC would require an expansion joint per the *NEC*®. Each straight section of conduit run must be treated independently from other sections when connected by elbows. To avoid damage to PVC caused by temperature changes, expansion couplings are used. *Figure 6* shows various PVC fittings. The inside of the coupling is sealed with one or more O-rings. This type of coupling may allow up to 6" (150 mm) of movement. Check the requirements of the local jurisdiction prior to installing PVC.

PVC is manufactured in the following two types:

- *Type EB* – Thin wall for underground use only when encased in concrete. Also referred to as Type I.
- *Type DB* – Thick wall for underground use without encasement in concrete. Also referred to as Type II.

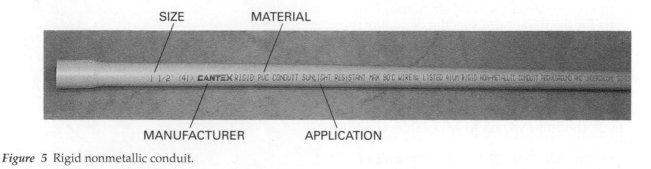

SIZE   MATERIAL

MANUFACTURER   APPLICATION

Figure 5 Rigid nonmetallic conduit.

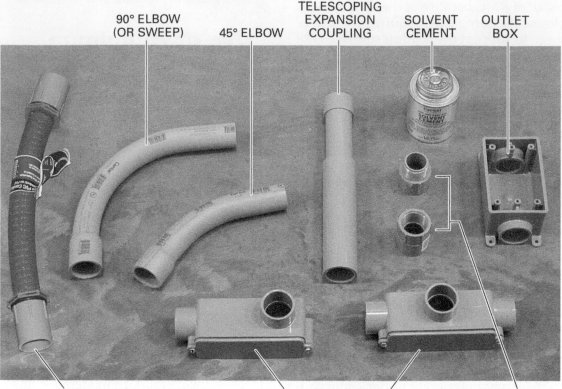

90° ELBOW (OR SWEEP)   45° ELBOW   TELESCOPING EXPANSION COUPLING   SOLVENT CEMENT   OUTLET BOX

0–90° FLEXIBLE ELBOW   CONDUIT BODIES   MALE AND FEMALE ADAPTERS

Figure 6 PVC fittings.

Type DB is available in the following two wall thicknesses:

- Schedule 40 is heavy wall for direct burial in the earth and aboveground installations.
- Schedule 80 is extra heavy wall for direct burial in the earth, aboveground installations for general applications, and installations where the conduit is subject to physical damage.

PVC is affected by higher-than-usual ambient temperatures. Support requirements for PVC are found in *NEC Section 352.30(B) and Table 352.30*. As with other conduit, it must be supported within 3' (900 mm) of each device or outlet box, junction box, conduit body, or other termination, but the maximum spacing between supports depends upon the size of the conduit. Some of the regulations for the maximum spacing of supports are:

- $\frac{1}{2}$" to 1" conduit (MD 16 to MD 27): every 3' (900 mm)
- $1\frac{1}{4}$" to 2" conduit (MD 35 to MD 53): every 5' (1.5 m)
- $2\frac{1}{2}$" to 3" conduit (MD 63 to MD 78): every 6' (1.8 m)
- $3\frac{1}{2}$" to 5" conduit (MD 91 to MD 129): every 7' (2.1 m)
- 6" conduit (MD 155): every 8' (2.5 m)

Liquidtight Conduit

Liquidtight conduit protects conductors from vapors, liquids, and solids. Liquidtight conduit that includes an inner metal core is widely used in commercial and industrial construction.

| STRAIGHT CONNECTOR | 45° CONNECTOR | 90° CONNECTOR |

Figure 7 Liquidtight flex connectors.

1.1.8 High-Density Polyethylene Conduit

High-density polyethylene conduit (HDPE) is a rigid nonmetallic conduit listed for underground installations. [It is not listed for aboveground use unless encased in 2" (50 mm) of concrete.] See *NEC Article 353*. It is suitable for direct burial or where encased in concrete. In many signaling and communications applications, it is provided on reels with conductors pre-installed and may be laid in a trench or plowed into the earth.

1.1.9 Liquidtight Flexible Nonmetallic Conduit

Liquidtight flexible nonmetallic conduit (LFNC) was developed as a raceway for industrial equipment where flexibility was required and protection of conductors from liquids was also necessary. LFNC is covered in *NEC Article 356*. Usage of LFNC has been expanded from industrial applications to outside and direct burial usage where listed.

Several varieties of LFNC have been introduced. The first product (LFNC-A) is commonly referred to as hose. It consists of an inner and outer layer of neoprene with a nylon reinforcing web between the layers. A second-generation product (LFNC-B), and most widely used, consists of a smooth wall, flexible PVC with a rigid PVC integral reinforcement rod. The third product (LFNC-C) is a nylon corrugated shape without any integral reinforcements. These three permitted LFNC raceway designs must be flame resistant with fittings approved for installation of electrical conductors. Nonmetallic connectors are listed for use and some liquidtight metallic flexible conduit connectors are dual-listed for both metallic and nonmetallic liquidtight flexible conduit.

LFNC is sunlight-resistant and suitable for use at conduit temperatures of 80°C dry and 60°C wet. It is available in $\frac{3}{8}$" (MD 12) through 4" (MD 103) sizes. *NEC Section 356.12* states that LFNC cannot be used where subject to physical damage. LFNC is also limited in length to no longer than 6' (1.8 m), except where properly secured, where flexibility is required, or as permitted by *NEC*

Section 356.10. Also, it cannot be used in any hazardous (classified) locations except as specified in other articles of the *NEC®*.

Liquidtight flexible metal conduit is a raceway of circular cross section having an outer liquidtight, nonmetallic, sunlight-resistant jacket over an inner flexible metal core with associated couplings and connectors covered by *NEC Article 350*.

Compression connectors are used to connect liquidtight flexible conduit to boxes or equipment. They are available in straight, 45°, and 90° configurations (*Figure 7*).

1.1.10 Flexible Metal Conduit

Flexible metal conduit (FMC), also called flex, may be used for many kinds of wiring systems. FMC is covered in *NEC Article 348*. Flexible metal conduit is made from a single strip of steel or aluminum, wound and interlocked. It is typically available in diameters from $\frac{3}{8}$" (MD 12) through 4" (MD 103). An illustration of flexible metal conduit is shown in *Figure 8*.

Flexible metal conduit is often used to connect equipment or machines that vibrate or move slightly during operation. Also, final connection to equipment having an electrical connection point that is marginally accessible is often accomplished with flexible metal conduit.

Flexible metal conduit is easily bent, but the minimum bending radius is the same as for other types of conduit. It should not be bent more than the equivalent of four quarter bends (360° total) between pull points (e.g., conduit bodies and boxes). It can be connected to boxes with a flexible conduit connector and to rigid conduit or EMT by using a combination coupling. A flexible-to-rigid combination coupling is shown in *Figure 9*.

Figure 8 Flexible metal conduit.

Figure 9 Combination coupling.

Flexible metal conduit is generally available in two types: nonliquidtight and liquidtight. *NEC Articles 348 and 350* cover the uses of flexible metal conduit.

Liquidtight flexible metal conduit has an outer covering of liquidtight, sunlight-resistant flexible material that acts as a moisture seal. It is intended for use in wet locations. It is used primarily for equipment and motor connections when movement of the equipment is likely to occur. The number of bends, size, and support requirements for liquidtight conduit are the same as for all flexible conduit. Fittings used with liquidtight conduit must also be of the liquidtight type.

Support requirements for flexible metal conduit are found in *NEC Sections 348.30 and 350.30*. Straps or other means of securing the flexible metal conduit must be spaced every $4^1/_2'$ (1.4 m) and within 12" (300 mm) of each end. (This spacing is closer together than for rigid conduit.) However, at terminals where flexibility is necessary, longer lengths are permitted per *NEC Sections 348.30 and 350.30, Exceptions*. For example, where flexibility is required for conduit sizes $^1/_2$" through $1^1/_4$" (MD 16 through MD 35), lengths of up to 36" (900 mm) without support are permitted. Failure to provide proper support for flexible conduit can make pulling conductors difficult.

1.2.0 Bonding Conduit

For safety reasons, most equipment that receives electrical power and has a metallic frame is bonded. In order to bond the equipment, an electrical connection must be made to connect the metal frame of the electrically powered equipment to the grounding point at the service-entrance equipment. This is usually done in one or both of the following ways:

- The frame of the equipment is connected to a wire (equipment grounding conductor), which is directly connected to the ground point at the grounding terminal.
- The frame of the equipment is connected (bonded) to a metal conduit or other type of raceway system, which provides an uninterrupted and low-impedance circuit to the ground point at the service-entrance equipment. The metal raceway or conduit acts as the equipment grounding conductor.

According to *NEC Section 250.96*, metal raceways, cable trays, cable armor, cable sheath, enclosures, frames, fittings, and other metal noncurrent-carrying parts that are to serve as equipment grounding conductors, with or without the use of supplementary equipment grounding conductors, shall be bonded where necessary to ensure electrical continuity and safely conduct any fault current likely to be imposed on them. Any nonconductive paint, enamel, or similar coating shall be removed at threads, contact points, and contact surfaces or be connected by means of fittings designed so as to make such removal unnecessary.

The purpose of the equipment grounding conductor is to provide a low-resistance path to ground for all equipment that receives power. This is done so that if an ungrounded conductor comes in contact with the frame of a piece of equipment, the circuit overcurrent device immediately acts to open the circuit. It also reduces the voltage to ground that would be present on the faulted equipment if a person came in contact with the equipment frame.

1.3.0 Metal Conduit Fittings

A large variety of conduit fittings are available to do electrical work. Manufacturers design and construct fittings to permit a multitude of applications. The type of conduit fitting used in a particular application depends upon the size and type of conduit, the type of fitting needed for the application, the location of the fitting, and the installation method. The requirements and proper applications of boxes, conduit bodies, or fittings are found in *NEC Section 300.15*. Some of the more common types of fittings are examined in the following sections.

> **NOTE**
>
> When using a combination coupling, be sure the flexible conduit is pushed as far as possible into the coupling. This covers the end and protects the conductors from damage.

1.3.1 Couplings

Couplings are sleeve-like fittings that are threaded inside to join two male threaded pieces of rigid conduit or IMC. A piece of conduit with a coupling is shown in *Figure 10*.

Other types of couplings may be used depending upon the location and type of conduit. Several types are shown in *Figure 11*.

COUPLING CONDUIT

Figure 10 Conduit and coupling.

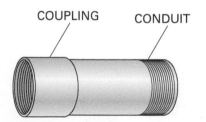

THREE-PIECE COUPLING   HINGED COUPLING

CONCRETE-TIGHT
SETSCREW   EMT TO RIGID

Figure 11 Metal conduit couplings.

1.3.2 Insulating Bushings

An insulating bushing is either nonmetallic or has an insulated throat. Insulating bushings are installed on the threaded end of conduit that enters an enclosure. Bushings can be grounding or nongrounding.

The purpose of a nongrounding insulating bushing is to protect the conductors from being damaged by the sharp edges of the threaded conduit end. *NEC Section 300.15(C)* states that where a conduit enters a box, fitting, enclosure, or conduit termination, a fitting must be provided to protect the wire from abrasion. *NEC Section 312.6(C)* references *NEC Section 300.4(G)*, which states that where ungrounded conductors of No. 4 AWG or larger enter a raceway in a cabinet or box enclosure, the conductors shall be protected by an integral threaded hub or boss or an identified/listed metal fitting providing a smoothly rounded insulating surface, unless the conductors are separated from the raceway fitting by substantial insulating material securely fastened in place. An exception is where threaded hubs or bosses that are an integral part of a cabinet, box, enclosure, or raceway provide a smoothly rounded or flared entry for conductors. Insulating bushings are shown in *Figure 12*.

Figure 12 Insulating bushings.

Installation of Conduit Bodies

It will be much easier to identify conduit bodies once you begin to see them in use. Here you see liquidtight nonmetallic conduit entering a Type T conduit body (A) and a Type LB conduit body in an outdoor commercial application (B).

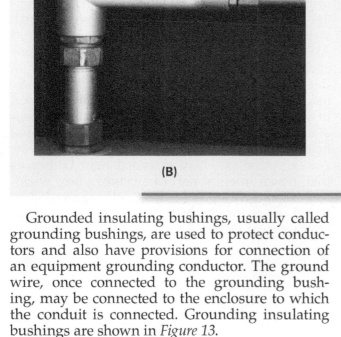

(B)

(A)

Grounded insulating bushings, usually called grounding bushings, are used to protect conductors and also have provisions for connection of an equipment grounding conductor. The ground wire, once connected to the grounding bushing, may be connected to the enclosure to which the conduit is connected. Grounding insulating bushings are shown in *Figure 13*.

1.3.3 Threaded Weatherproof Hubs

Threaded weatherproof hubs, also known as *Myers hubs*, are used for conduit entering a box in a wet location. *Figure 14* shows typical threaded weatherproof hubs.

1.3.4 Offset Nipples

Offset nipples are used to connect two pieces of electrical equipment in close proximity where a slight offset is required. They come in sizes ranging from $\frac{1}{2}$" to 2" (MD 16 to MD 53) in diameter. See *Figure 15*.

1.3.5 Conduit Bodies

Conduit bodies, also called condulets, are a separate portion of a conduit or tubing system that provide access through a removable cover(s) to the interior of the system at a junction of two or more sections of the system, a pull point, or at a terminal point of the system. They are

Figure 13 Grounding insulating bushings.

Figure 14 Threaded weatherproof hubs.

usually cast and are significantly higher in cost than the stamped steel boxes permitted with EMT. However, there are situations in which conduit bodies are preferable, such as in outdoor

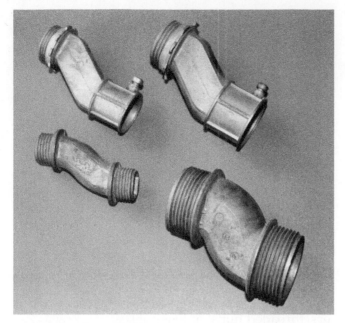

Figure 15 Offset nipples.

Table 1 Volume Required per Conductor [Data from *NEC Table 314.16(B)*]

Size of Conductor (AWG)	Free Space Within Box for Each Conductor
No. 18	1.5 cu in
No. 16	1.75 cu in
No. 14	2.0 cu in
No. 12	2.25 cu in
No. 10	2.5 cu in
No. 8	3.0 cu in
No. 6	5.0 cu in

Reprinted with permission from NFPA 70-2020, *National Electrical Code*®, Copyright © 2019, National Fire Protection Association, Quincy, MA. This reprinted material is not the complete and official position of the NFPA on the referenced subject, which is represented only by the standard in its entirety which may be obtained through the NFPA website at **www.nfpa.org**.

locations, for appearance's sake in an **exposed location**, or to change types or sizes of raceways. Also, conduit bodies do not have to be supported, as do stamped steel pull and junction boxes. They are also used when elbows or bends would not be appropriate.

NEC Section 314.16(C)(2) states that conduit bodies cannot contain a **splice**, **tap**, or device unless they are durably and legibly marked by the manufacturer with their volume capacity. The maximum number of conductors permitted in a conduit body is found using *NEC Table 314.16(B)*. (This information is shown in *Table 1*.)

Type C conduit bodies may be used to provide a pull point in a long conduit run or a conduit run that has bends totaling more than 360°. A Type C conduit body is shown in *Figure 16*.

When referring to conduit bodies, the letter L represents an elbow. A Type L conduit body is used as a pulling point for conduit that requires a 90° change in direction. The cover is removed, then the wire is pulled out, coiled on the ground or floor, reinserted into the other conduit body's opening, and pulled. The cover and its associated gasket are then replaced. Type L conduit bodies are available with the cover on the back (Type LB), on the sides (Type LL or LR), or on both sides (Type LRL). Type L conduit bodies are shown in *Figure 17*.

> **NOTE**
>
> The cover and gasket must be ordered separately. Do not assume that these parts come with conduit bodies when they are ordered.

Figure 16 Type C conduit body.

To identify Type L conduit bodies, use the following method:

Step 1 Hold the body like a pistol.

Step 2 Locate the opening on the body:

- If the opening is to the left, it is a Type LL.
- If the opening is to the right, it is a Type LR.
- If the opening is on top (back), it is a Type LB.
- If there are openings on both the left and the right, it is a Type LRL.

Type T conduit bodies are used to provide a junction point for three intersecting conduits and are used extensively in conduit systems. A Type T conduit body is shown in *Figure 18*.

Type X conduit bodies are used to provide a junction point for four intersecting conduits. The removable cover provides access to the interior of the X so that wire pulling and splicing may be performed. A Type X conduit body is shown in *Figure 19*.

1.3.6 Sealing Fittings

Hazardous locations in manufacturing plants and other industrial facilities involve a wide variety

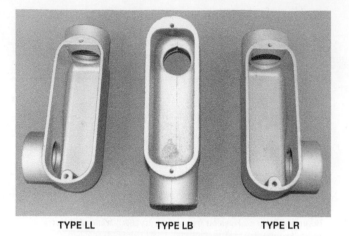

TYPE LL   **TYPE LB**   **TYPE LR**

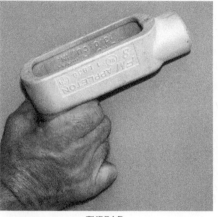

TYPE LB

Figure 17 Type L conduit bodies and how to identify them.

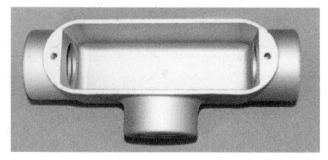

Figure 18 Type T conduit body.

Figure 19 Type X conduit body.

of flammable gases and vapors and ignitable dusts. These hazardous substances have widely different flash points, ignition temperatures, and flammable limits requiring fittings that can be sealed. Sealing fittings are installed in conduit runs to minimize the passage of gases, vapors, or flames through the conduit and reduce the accumulation of moisture. They are required by *NEC Sections 501.15 and 502.15* in hazardous locations where explosions may occur. They are also required where conduit passes from a hazardous location of one classification to another or to an unclassified location. Several types of sealing fittings are shown in *Figure 20*.

1.4.0 Making a Conduit-to-Box Connection

Conduit is joined to boxes by connectors, adapters, threaded hubs, or locknuts.

Bushings protect the wires from the sharp edges of the conduit. As previously discussed, bushings are usually made of plastic or metal. Some metal bushings have a grounding screw to permit a bonding wire to be installed.

Locknuts (*Figure 21*) are used on the inside and outside walls of the box to which the conduit is connected. A grounding locknut may be needed if a bonding wire is to be installed. Special sealing locknuts are also used in wet locations.

A means must be provided in each metal box for the connection of an equipment grounding conductor. The means shall be permitted to be a tapped hole or equivalent per *NEC Section 314.40(D)*. A proper conduit-to-box connection is shown in *Figure 22*.

In order to make a good connection, use the following procedure:

Step 1   Thread the external locknut onto the conduit. Run the locknut to the bottom of the threads.

Step 2   Insert the conduit into the box opening.

Step 3   If an inside locknut or grounding locknut is required, screw it onto the conduit inside the box opening.

Step 4   Screw the bushing onto the threads projecting into the box opening. Make sure the bushing is tightened as much as possible.

Step 5   Tighten the external locknut to secure the conduit to the box.

It is important that the bushings and locknuts fit tightly. For this reason, the conduit must enter

---

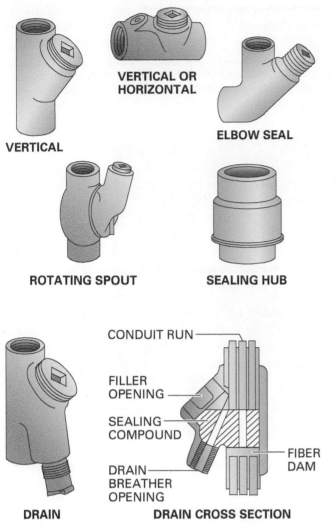

VERTICAL OR HORIZONTAL

VERTICAL

ELBOW SEAL

ROTATING SPOUT

SEALING HUB

CONDUIT RUN

FILLER OPENING

SEALING COMPOUND

DRAIN BREATHER OPENING

FIBER DAM

DRAIN

DRAIN CROSS SECTION

Figure 20 Sealing fittings.

SEALING LOCKNUT

STANDARD LOCKNUT

STANDARD LOCKNUT

GROUNDING LOCKNUT

Figure 21 Locknuts.

straight into the box. This may require that a box offset or **kick** be made in the conduit.

1.5.0 Raceway Supports

Raceway supports are available in many types and configurations. This section discusses common conduit supports found in electrical installations. *NEC Section 300.11(B)* discusses the requirements for branch circuit wiring that is supported from above suspended ceilings. Electrical equipment and raceways must have their own supporting methods and may not be supported by the supporting hardware of a fire-rated roof/ceiling assembly.

1.5.1 Straps

Straps are used to support conduit to a surface (see *Figure 23*). The spacing of these supports must conform to the minimum support spacing requirements for each type of conduit. One- and two-hole straps are used for all types of conduit: EMT, RMC, IMC, PVC, and flex. The straps can be flexible or rigid. Two-part straps are used to secure conduit to electrical framing channels (struts). Parallel and right angle beam clamps are also used to support conduit from structural members.

Clamp back straps can also be used with a backplate to maintain the $\frac{1}{4}$" (6 mm) spacing from the surface required for installations in wet locations.

1.5.2 Standoff Supports

The standoff support, often referred to as a Minerallac® (the name of a manufacturer of this type of support), is used to support conduit away from the supporting structure. In the case of the one-hole and two-hole straps, the conduit must be offset wherever a fitting occurs. If standoff supports are used, the conduit is held away from the supporting surface, and no offsets are required in the conduit at the fittings. Standoff supports may be used to support all types of conduit including RMC, IMC, EMT, PVC, and flex, as well as tubing installations. A standoff support is shown in *Figure 24*.

1.5.3 Electrical Framing Channels

Electrical framing channels or other similar framing materials are used together with Unistrut®-type conduit clamps to support conduit (*Figure 25*). They may be attached to a ceiling, wall, or other surface or be supported from a trapeze hanger.

1.5.4 Beam Clamps

Beam clamps are used with suspended hangers. The raceway is attached to or laid in the hanger. The hanger is suspended by a threaded rod. One end of the threaded rod is attached to the hanger and the other end is attached to a beam clamp.

Installing Sealing Fittings

These fittings must be sealed after the wires are pulled. A fiber dam is first packed into the base of the fitting between and around the conductors, then the liquid sealing compound is poured into the fitting. Speed sealing materials are also available that eliminate the need to insert a fiber dam.

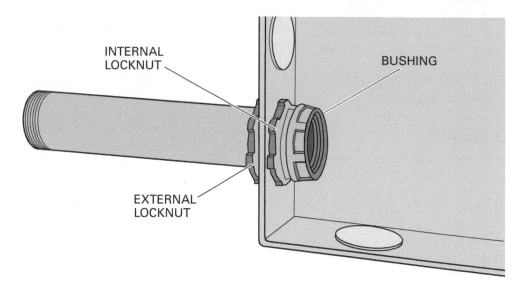

Figure 22   Conduit-to-box connection.

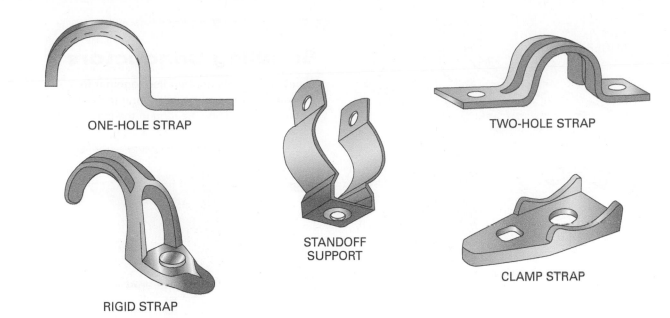

ONE-HOLE STRAP

TWO-HOLE STRAP

STANDOFF
SUPPORT

CLAMP STRAP

RIGID STRAP

Figure 23 Straps.

Figure 24 Standoff support.

Wall-Mounted Supports

This wall-mounted support has been fabricated to hold the conduit away from the metal building. While it is shown with only one raceway, additional raceways can be added to the framing channel.

Figure Credit: Tim Dean

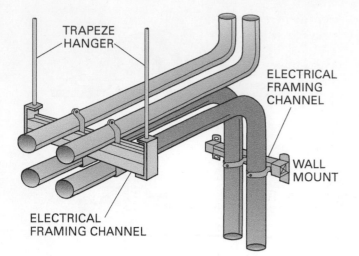

Figure 25 Electrical framing channels.

The beam clamp is then attached to a beam. A beam clamp with wireway support assembly is shown in *Figure 26*.

1.6.0 Installation Requirements for Various Construction Methods

Conduit and box installation varies with the type of construction. This section discusses some special requirements for masonry and concrete, metal framing, wood, and structural steel construction.

1.6.1 *Masonry and Concrete Flush-Mount Construction*

In a reinforced concrete construction environment, the conduit and boxes must be embedded in the concrete to achieve a flush surface. Ordinary boxes may be used, but special concrete boxes are preferred and are available in depths up to 6" (150 mm). These boxes have special breakaway ears by which they are nailed to the wooden forms for the concrete. When installing them, stuff the boxes tightly with paper to prevent concrete from seeping in. *Figure 27* shows an installed box.

Flush construction can also be done on existing concrete walls, but this requires chiseling a channel and box opening, anchoring the box and conduit, and then resealing the wall.

To achieve flush construction with masonry walls, the most acceptable method is for the electrician to work closely with the mason laying the blocks. When the construction blocks reach the convenience outlet elevation, boxes are made up as shown in *Figure 28*. The figure shows a raised tile ring or box device cover.

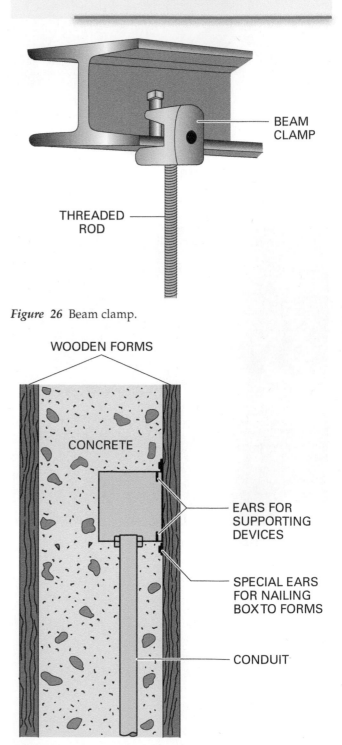

Bundling Conductors

When conductors are bundled together in a wireway their magnetic fields tend to cancel, thus minimizing inductive heating in the conductors.

Figure 26 Beam clamp.

Figure 27 Concrete flush-mount installation.

Figure 29 shows a masonry box that needs no extension or deep plaster ring to bring it to the surface.

Sections of conduit are then coupled in short (approximately 3' to 5' or 900 mm to 1.5 m) lengths. This is done because it is impractical for the mason to maneuver blocks over 10' (3 m) sections of conduit.

> **NOTE**
>
> The electrician must work with the mason to ensure the box is properly grouted and sealed.

1.6.2 Metal Stud Environment

Metal stud walls are a popular method of construction for the interior walls of commercial buildings. Metal stud framing consists of rela-

Figure 28 Box with raised ring.

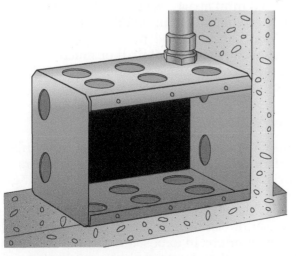

Figure 29 Three-gang concrete box.

tively thin metal channel studs, usually constructed of galvanized steel and with an overall dimension the same as standard wooden studs. Wiring in this type of construction is relatively easy when compared to masonry.

EMT conduit and MC cable are the most common type of wiring methods for metal stud environments. Metal studs usually have some number of pre-punched holes that can be used to route the conduit. If a pre-punched hole is not located where it needs to be, holes can be easily punched in the metal stud with a hole cutter or knockout punch (*Figure 30*).

> **WARNING!**
>
> Cutting or punching metal studs can create sharp edges. Avoid contact that can result in cuts.

Boxes can be secured to the metal stud using self-tapping screws or one of the many types of box supports available. EMT conduit is supported by the metal studs using conduit straps or other approved methods. It is important that the conduit be properly supported to facilitate pulling the conductors through the tubing. Boxes are mounted on the metal studs so that the box will be flush with the finished walls. You must know what the finished wall thickness is going to be to properly secure the boxes to the metal studs. For example, if the finished wall will be $\frac{5}{8}$" (16 mm) drywall, then the box must be fastened so that it protrudes $\frac{5}{8}$" (16 mm) from the metal stud.

> **WARNING!**
>
> When using a screw gun or cordless drill to mount boxes to studs, keep the hand holding the box away from the gun/drill to avoid injury.

According to *NEC Section 300.4(B)(1)*, NM cable run through metal studs must be protected by listed bushings or listed grommets (*Figure 31*). This protects the cables from the friction of pulling during installation and from the weight of the cable and vibrations following the installation.

1.6.3 Wood Frame Environment

At one time, the use of rigid conduit in partitions and ceilings was a time-consuming operation.

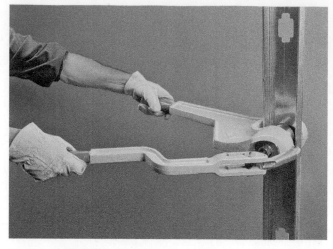

Figure 30 Metal stud punch.

Thinwall conduit makes an easier and quicker job, largely because of the types of fittings that are specially adapted to it.

Figure 32 shows two methods of running thinwall conduit in these locations: boring timbers and notching them. When boring, holes must be drilled large enough for the tubing to be inserted between the studs. The tubing is cut rather short, calling for multiple couplings. EMT can be bowed quite a bit while threading through holes in studs.

WARNING!	Always wear safety goggles when boring wood.

NEC Section 300.4 addresses the requirements to prevent physical damage to conductors and cabling in wood members. By keeping the edge of the drilled hole $1\frac{1}{4}$" (32 mm) from the closest edge of the stud, nails are not likely to penetrate the stud far enough to damage the cables. The building codes provide maximum requirements for bored or notched holes in studs.

NEC Section 300.4(A)(1) requires the use of a steel plate or bushing at least $\frac{1}{16}$" (1.6 mm) thick or a listed steel nail plate where wiring is installed through bored wooden members less than $1\frac{1}{4}$" (32 mm) from the nearest edge (*Figure 33*). Nail plates are also required to protect the conductors in all notched wooden members per *NEC Section 300.4(A)(2)*.

The exceptions in these sections permit IMC, RMC, PVC, and EMT to be installed through bored holes or laid in notches less than $1\frac{1}{4}$" (32

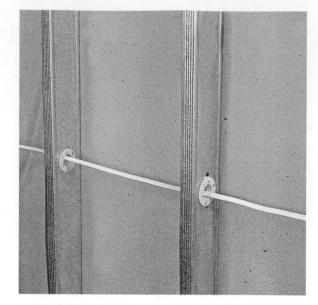

Figure 31 NM cable protected by grommets.

mm) from the nearest edge without a steel plate or bushing.

Because of its weakening effect upon the structure, notching should be resorted to only where absolutely necessary. Notches should be as narrow as possible and in no case deeper than $\frac{1}{16}$ the stock of a bearing timber. A bearing timber supports floor joists or other weight.

NOTE	Always check with the architect before notching or drilling.

Some wood I-beams are manufactured with perforated knockouts in their web, approximately 12" (300 mm) apart. Never notch or drill through the beam flange or cut other openings in the web without checking the manufacturer's specification sheet. Also, do not drill or notch other types of engineered lumber without first checking the specification sheets.

1.6.4 Metal Buildings

Many commercial and industrial buildings are prefabricated structures with steel structural supports, and roofing and siding made of light-gauge metal sheets (*Figure 34*). Conduit can be routed across the structural members that support the roof. *NEC Section 300.4(E)* states that a cable, raceway, or box in exposed or concealed locations under metal-corrugated sheet roof decking must be installed and supported so the nearest outside

Horizontal runs of EMT may be supported and secured by openings in framing members at intervals not greater than 10 feet when securely fastened within a distance of 3 feet at each of its termination points.

NEC Section 358.30(B)

EMT may be run through wood joists where the edges of the bored holes are less than 1¼" from the nearest edge of the stud, or where the studs are notched without the need for a steel plate.

NEC Section 300.4(A)(1), Exception No. 1

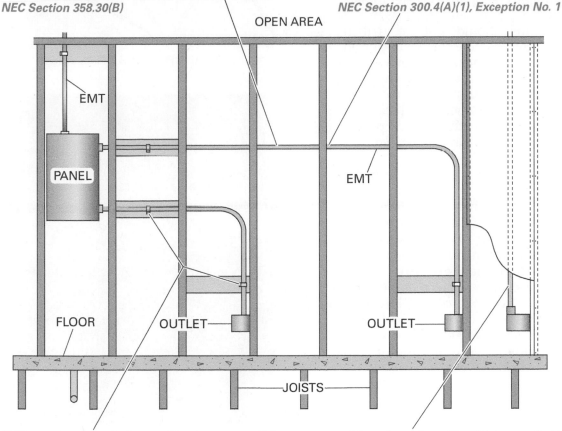

EMT must be securely fastened in place every 10 feet and within 3 feet of each outlet box, device box, cabinet, conduit body, or other termination.

NEC Section 358.30(A) and (B)

Unbroken lengths of EMT can be fastened at a distance of up to 5 feet from a termination point when structural members do not readily permit fastening within 3 feet.

NEC Section 358.30(A), Exception 1

Where fastening of EMT is impractical in finished buildings or prefinished walls, unbroken lengths of EMT may be fished.

NEC Section 358.30(A), Exception 2

Figure 32 Installing wire or conduit in a wood-frame building.

Figure 33 Steel nail plate.

Figure 34 Metal building.

surface of the cable or raceway is not less than $1\frac{1}{2}"$ (38 mm) from the nearest surface of the roof decking. The roof structure can consist of beams and purlins (*Figure 35*) or open-web steel joists (*Figure 36*).

Beams and purlins should not be drilled through; consequently, the conduit is supported from the metal beams by anchoring devices designed especially for that purpose. The supports attach to the beams or supports and have clamps to secure the conduit to the structure. All conduit runs should be plumb since they are exposed. Bends should be correct and have a neat and orderly appearance.

Rigid metal conduit is often required in metal buildings. If a large number of conduits are run along the same path, strut-type systems are used. These systems are sometimes referred to as Unistrut® systems (Unistrut® is a manufacturer of these systems). Another manufacturer of strut systems is B-Line systems. Both are very similar. These systems use a channel-type member that can support conduits from the ceiling by using threaded rod supports for the channel, as shown in *Figure 37*. Strut channel can also be secured to masonry walls to support vertical runs of conduit, wireways, and various types of boxes.

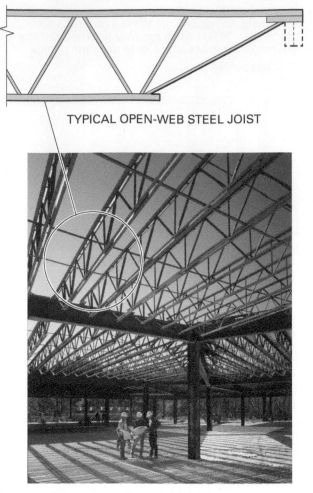

TYPICAL OPEN-WEB STEEL JOIST

Figure 36 Open-web steel joist roof supports.

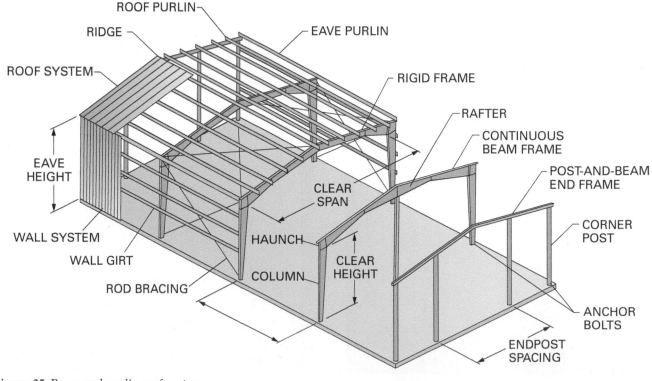

Figure 35 Beam and purlin roof system.

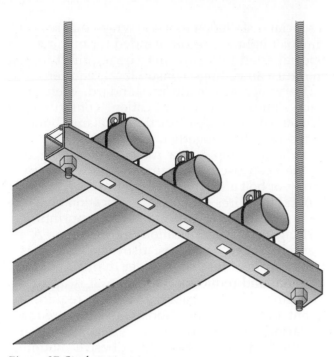

Figure 37 Steel strut system.

1.0.0 Section Review

1. Information on conduit fill for various conductors can be found in *NEC®* _____.

   a. *Informative Annex A*
   b. *Informative Annex B*
   c. *Informative Annex C*
   d. *Informative Annex D*

2. Metal raceways and other enclosures are bonded to _____.

   a. prevent leakage onto any electrical systems below
   b. ensure electrical continuity and safely conduct any fault current
   c. ensure electrical continuity and prevent leakage
   d. ensure mechanical continuity

3. A fitting that does not require support is a(n) _____.

   a. outlet box
   b. pull box
   c. junction box
   d. conduit body

4. If the conduit does *not* enter straight into the box, _____.

   a. it may result in a voltage drop
   b. the fittings will be too tight
   c. a kick is required
   d. cut the conduit short

5. Beam clamps are used with _____.

   a. straps
   b. standoff supports
   c. framing channels
   d. suspended hangers

6. When installing conduit in a masonry environment, it is best to use _____.

   a. 3' to 5' (900 mm to 1.5 m) lengths of conduit
   b. 6' (1.8 m) lengths of conduit
   c. 8' to 10' (2.4 m to 3 m) lengths of conduit
   d. 20' (6 m) lengths of conduit

2.0.0 FASTENERS AND ANCHORS FOR RACEWAY SYSTEMS

Objective

Select fasteners and anchors for the installation of raceway systems.

a. Select and install tie wraps.
b. Select and install screws.
c. Select and install hammer-driven pins and studs.
d. Identify the safety requirements for stud-type guns.
e. Select and install masonry anchors.
f. Select and install hollow-wall anchors.
g. Select and install epoxy anchoring systems.

Performance Tasks

2. Identify and select various types and sizes of raceways, fittings, and fasteners for a given application.
3. Demonstrate how to install a raceway system.
4. Terminate a selected raceway system.

Conduit and other types of raceways used to carry wiring and cables must be properly supported. This generally means attaching the raceway to the building structure. Depending on the type of construction, the raceways may have to be attached to wood, concrete, or metal. Each of these materials requires the use of fasteners designed for the specific use. Using the wrong fastener, or installing the right fastener incorrectly, can lead to a failure of the raceway support.

The project specifications and manufacturer's installation instructions may specify the type and size of fasteners to use and how to install them. In other instances, the electrician will be expected to select the right type of fastener for a given application. It is therefore important that every electrician be familiar with the different types of fasteners, their uses, and their limitations.

2.1.0 Tie Wraps

A tie wrap is a one-piece, self-locking cable tie, usually made of nylon, that is used to fasten a bundle of wires and cables together. Tie wraps can be quickly installed either manually or using a special installation tool. Tie wraps that resist ultraviolet light are recommended for outdoor use. Special rated tie wraps may be required for use in return air ceilings; consult local requirements.

Tie wraps are made in standard, cable strap and clamp, and identification configurations (*Figure 38*). All types function to clamp bundled wires or cables together. In addition, the cable strap and clamp has a molded mounting hole in the head used to secure the tie with a rivet, screw, or bolt after the tie wrap has been installed around the wires or cable. Identification tie wraps have a large flat area provided for imprinting or writing cable identification information. A releasable Velcro® wrap is also available. It is a nonpermanent tie that is wider than a traditional tie wrap and reduces cable strain. It is useful for bundling wires or cables that may require frequent additions or deletions. Cable ties are made in various lengths and colors. Tie wraps can also be attached to a variety of adhesive mounting bases made for that purpose. Always check the job specifications to ensure that you are using the correct cable wrap or tie.

> **CAUTION**
>
> Do not over-tighten any wrap or cable tie as it may result in electrical interference between conductors and/or damage to the cable.

2.2.0 Bolts and Screws

Bolts and screws are made in a variety of shapes and sizes for different fastening jobs. The finish or coating used on a bolt or screw determines whether it is for interior or exterior use, corrosion resistant, etc. Bolts and screws of all types have heads with different shapes and slots. Some have machine threads and are self-drilling. The size or diameter of the body or shank is given in gauge numbers ranging from No. 0 to No. 24, and in fractions of an inch for screws with diameters larger than $\frac{1}{4}$" (metric size M6). The higher the gauge number, the larger the diameter of the shank. Lengths are measured from the tip to the part of the head that is flush to the surface when driven in. When choosing a fastener for an application, you must consider the type and thickness of the materials to be fastened, the size of the fastener, the material it is made of, the shape of its head, and the type of driver. Because of the wide diversity in the types of fasteners and their applications, always follow the manufacturer's recommendations to select the right fastener for the job. To prevent damage to the fastener head or

Tie Wraps

Tie wraps are available in a wide variety of colors that can be used to color code different cable bundles.

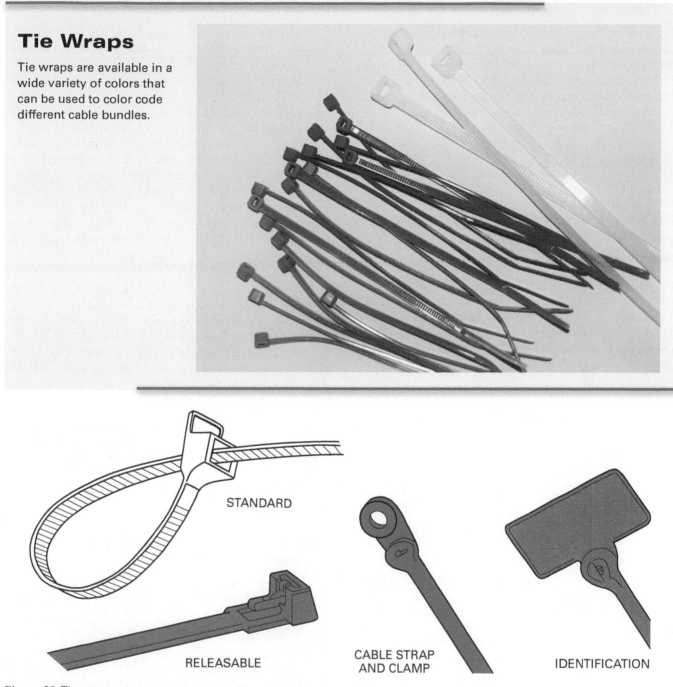

STANDARD

RELEASABLE

CABLE STRAP AND CLAMP

IDENTIFICATION

Figure 38 Tie wraps.

the material being fastened, always use a tool with the proper size and shape tip to fit the fastener.

Metric fasteners are sized differently than standard American (Imperial) fasteners. For example, a typical metric fastener size is M8 × 1. This means that the diameter of the threads is 8 mm and the thread pitch (space between the threads) is 1 mm. The length is indicated in centimeters. *Figure 39* compares the two types of fasteners.

Some of the more common types of fasteners are:

- Machine bolts
- Wood screws
- Lag screws
- Concrete/masonry screws
- Thread-forming and thread-cutting screws
- Drywall screws
- Drive screws

2.2.1 Machine Bolts

Machine bolts are used to assemble parts that do not require close tolerances. The tolerance is the amount of variation allowed from a standard. Machine bolts are made in various diameters and lengths. A machine bolt is tightened and released by turning the mating nut that is usually furnished along with the bolt.

2.2.2 Wood Screws

Wood screws are typically used to fasten boxes, panel enclosures, etc. to wood framing or structures where greater holding power is needed than can be provided by nails. They are also used to fasten equipment to wood in applications where it may occasionally need to be unfastened and removed. The shank size selected is normally determined by the size hole provided in the box, panel, etc. to be fastened. When determining the length of a wood screw to use, a good rule of thumb is to select screws long enough to allow about 2/3 of the screw length to enter the piece of wood that is being gripped.

2.2.3 Lag Screws and Shields

Lag screws (*Figure 40*) or lag bolts are heavy-duty wood screws with square- or hex-shaped heads that provide greater holding power. Lag screws are available in a wide variety of sizes and diameters. They are typically used to fasten heavy equipment to wood, but can also be used to fasten equipment to concrete when a lag shield is used.

A lag shield is a tube that is split lengthwise but remains joined at one end. It is placed in a predrilled hole in the concrete. When a lag screw is screwed into the lag shield, the shield expands in the hole, firmly securing the lag screw. In hard masonry, short lag shields may be used to minimize drilling time. In soft or weak masonry, longer lag shields should be used to achieve maximum holding strength.

Make sure to use the proper length lag screw to achieve proper expansion. The length of the lag screw used should be equal to the thickness of the component being fastened plus the length of the lag shield. Drill the hole in the masonry to a depth slightly longer than the shield being used (follow the manufacturer's instructions). If the head of a

Screws

In most applications, either threaded or nonthreaded fasteners such as nails could be used. However, threaded fasteners are sometimes preferred because they can usually be tightened and removed without damaging the surrounding material.

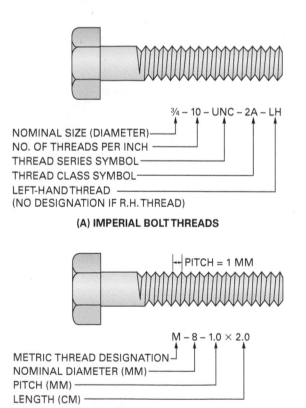

NOMINAL SIZE (DIAMETER)
NO. OF THREADS PER INCH
THREAD SERIES SYMBOL
THREAD CLASS SYMBOL
LEFT-HAND THREAD
(NO DESIGNATION IF R.H. THREAD)

¾ – 10 – UNC – 2A – LH

(A) IMPERIAL BOLT THREADS

PITCH = 1 MM

METRIC THREAD DESIGNATION
NOMINAL DIAMETER (MM)
PITCH (MM)
LENGTH (CM)

M – 8 – 1.0 × 2.0

(B) METRIC BOLT THREADS

Figure 39 Imperial and metric bolt sizes.

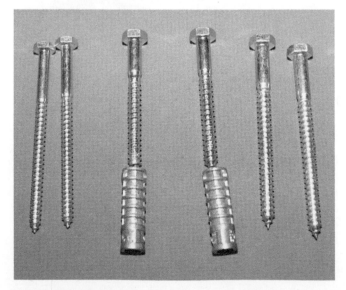

Figure 40 Lag screws and shields.

lag screw rests directly on wood when installed, a flat washer should be placed under the head to prevent the head from digging into the wood as the lag screw is tightened down. Be sure to take the thickness of any washers used into account when selecting the length of the screw.

2.2.4 Concrete/Masonry Screws

Concrete/masonry screws (*Figure 41*), commonly called self-threading anchors, are used to fasten a device or fixture to concrete, block, or brick. No anchor is needed. To provide a matched tolerance anchoring system, the screws are installed using specially designed carbide drill bits and installation tools made for use with the screws. These tools are typically used with a standard rotary drill hammer. The installation tool, along with an appropriate drive socket or bit, is used to drive the screws directly into predrilled holes that have a diameter and depth specified by the screw manufacturer. When being driven into the concrete, the widely spaced threads on the screws cut into the walls of the hole to provide a tight friction fit. Most types of concrete/masonry screws can be removed and reinstalled to allow for shimming and leveling of the fastened device.

> **WARNING!**
>
> Follow your company's rules for silica protection while drilling concrete.

2.2.5 Thread-Forming and Thread-Cutting Screws

Thread-forming screws (*Figure 42*), commonly called sheet metal screws, are made of hard metal. They form a thread as they are driven into the work. This thread-forming action eliminates the need to tap a hole before installing the screw. To achieve proper holding, it is important to make sure

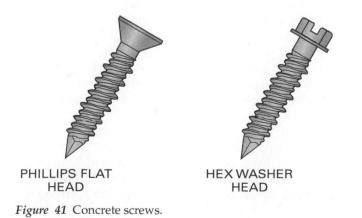

PHILLIPS FLAT
HEAD

HEX WASHER
HEAD

Figure 41 Concrete screws.

Installing Wood Screws

To maintain holding power, be careful not to drill your pilot hole too large. It's wise to drill a pilot hole deep enough to equal about two-thirds of the length of the threaded portion of the screw. Additionally, to lubricate screw threads, use soap, which makes the screw easier to drive.

to use the proper size bit when drilling pilot holes for thread-forming screws. The correct drill bit size used for a specific size screw is usually marked on the box containing the screws. Some types of thread-forming screws also drill their own holes, eliminating drilling, punching, and aligning parts. Thread-forming screws are primarily used to fasten light-gauge metal parts together. They are made in the same diameters and lengths as wood screws.

Hardened steel thread-cutting metal screws with blunt points and fine threads (*Figure 43*) are used to join heavy-gauge metals, metals of different gauges, and nonferrous metals. They are also used to fasten sheet metal to building structural members. These screws are made of hardened steel that is harder than the metal being tapped. They cut threads by removing and cutting a portion of the metal as they are driven into a pilot hole and through the material.

2.2.6 Drywall Screws

Drywall screws (*Figure 44*) are thin, self-drilling screws with bugle-shaped heads. Depending on the type of screw, it cuts through the wallboard and anchors itself into wood and/or metal studs,

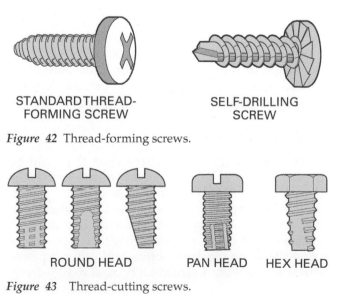

STANDARD THREAD-
FORMING SCREW

SELF-DRILLING
SCREW

Figure 42 Thread-forming screws.

ROUND HEAD   PAN HEAD   HEX HEAD

Figure 43 Thread-cutting screws.

COARSE THREAD

FINE THREAD

HIGH-AND-LOW THREAD

Figure 44 Drywall screws.

holding the wallboard tight to the stud. Coarse thread screws are normally used to fasten wallboard to wood studs. Fine thread and high-and-low thread types are generally used for fastening to metal studs. Some screws are made for use in either wood or metal. A Phillips or Robertson drive head allows the drywall screw to be countersunk without tearing the surface of the wallboard.

2.2.7 Drive Screws

Drive screws do not require that the hole be tapped. They are installed by hammering the screw into a drilled or punched hole of the proper size. Drive screws are mostly used to fasten parts that will not be exposed to much pressure. A typical use of drive screws is to attach permanent name plates on electric motors and other types of equipment. *Figure 45* shows a typical drive screw.

Figure 45 Drive screw.

2.3.0 Hammer-Driven Pins and Studs

Hammer-driven pins or threaded studs (*Figure 46*) can be used to fasten wood or steel to concrete or block without the need to predrill holes. The pin or threaded stud is inserted into a hammer-driven tool designed for its use. The pin or stud is inserted in the tool point end out with the washer seated in the recess. The pin or stud is then positioned against the base material where it is to be fastened and the drive rod of the tool tapped lightly until the striker pin contacts the pin or stud. Following this, the tool's drive rod is struck using heavy blows with about a two-pound engineer's hammer. The force of the hammer blows is transmitted through the tool directly to the head of the fastener, causing it to be driven into the concrete or block. For best results, the drive pin or stud should be embedded a minimum of $\frac{1}{2}$" (13 mm) in hard concrete to $1\frac{1}{4}$" (32 mm) in softer concrete block.

2.4.0 Safety Requirements for Stud-Type Guns

Stud-type guns powered by powder, gas, rocket fuel, or spring power (*Figure 47*) can be used to drive a wide variety of specially designed pin and threaded stud-type fasteners into masonry and steel. Powder-actuated tools look and fire like a gun and use the force of a detonated gunpowder load (typically .22, .25, or .27 caliber) to drive the fastener into the material. The depth to which the pin or stud is driven is controlled by the density of the base material in which the pin or stud is being installed and by the power level or strength of the cased powder load.

Powder loads and their cases are designed for use with specific types and/or models of powder-actuated tools and are not interchangeable. Typically, powder loads are made in 12 increasing power or load levels used to achieve the proper penetration. The different power levels are identified by a color-code system and load case types. Note that different manufacturers may use different color codes to identify load strength.

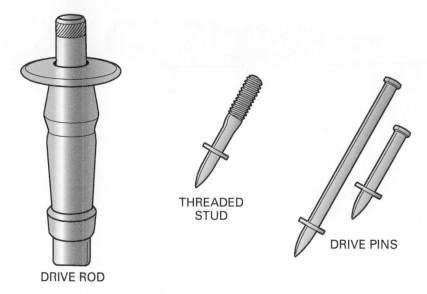

THREADED
STUD

DRIVE PINS

DRIVE ROD

Figure 46 Hammer-driven pins and installation tool.

Power level 1 is the lowest power level while 12 is the highest. Higher number power levels are used when driving into hard materials or when a deeper penetration is needed. Powder loads are available as single-shot units for use with single-shot tools. They are also made in multi-shot strips or disks for semiautomatic tools.

Think About It

Self-Drilling Screws

Can you name an electrical application for self-drilling screws?

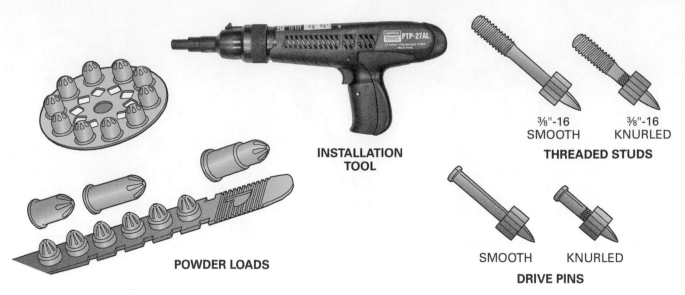

INSTALLATION TOOL

POWDER LOADS

⅜"-16 SMOOTH   ⅜"-16 KNURLED

THREADED STUDS

SMOOTH   KNURLED

DRIVE PINS

Figure 47   Powder-actuated installation tools and fasteners.

OSHA Standard 29 CFR 1926.302(e) governs the use of powder-actuated tools and states that only those individuals who have been trained in the operation of the particular powder-actuated tool in use be allowed to operate it. Authorized instructors available from the various powder-actuated tool manufacturers generally provide such training and licensing. Trained operators must take precautions to protect both themselves and others in the area when using a powder-actuated driver tool:

- Always use the tool in accordance with the published tool operation instructions.
- Instructions should be kept with the tool. Never attempt to override the safety features of the tool.
- Never place your hand or other body parts over the front muzzle end of the tool.
- Use only fasteners, powder loads, and tool parts specifically made for use with the tool. Use of other materials can cause improper and unsafe functioning of the tool.
- Operators and bystanders must wear eye and hearing protection along with hard hats. Other personal safety gear, as required, must also be used.
- Always post warning signs that state Powder-Actuated Tool in Use within 50' (15 m) of the area where tools are used.
- Before using a tool, make sure it is unloaded and perform a proper function test. Check the functioning of the unloaded tool as described in the published tool operation instructions.
- Do not guess before fastening into any base material; always perform a center punch test.

- Always make a test firing into a suitable base material with the lowest power level recommended for the tool being used. If this does not set the fastener, try the next higher power level. Continue this procedure until the proper fastener penetration is obtained.
- Always point the tool away from operators or bystanders.
- Never use the tool in an explosive or flammable area.
- Never leave a loaded tool unattended. Do not load the tool until you are prepared to complete the fastening. Should you decide not to make a fastening after the tool has been loaded, always remove the powder load first, then the fastener. Always unload the tool before cleaning or servicing, when changing parts, prior to work breaks, and when storing the tool.
- Always hold the tool perpendicular to the work surface and use the spall (chip or fragment) guard or stop spall whenever possible.
- Always follow the required spacing, edge distance, and base material thickness requirements.
- Never fire through an existing hole or into a weld area.
- In the event of a misfire, always hold the tool depressed against the work surface for at least 30 seconds. If the tool still does not fire, follow the published tool instructions. Never carelessly discard or throw unfired powder loads into a trash receptacle.
- Always store the powder loads and unloaded tool under lock and key.

Powder-Actuated Tools

A 22-year-old apprentice was killed when he was struck in the head by a nail fired from a high-velocity powder-actuated tool in an adjacent room. The tool operator was attempting to anchor plywood to a hollow wall and fired the gun, causing the nail to pass through the wall, where it traveled nearly 30 ft (9 m) before striking the victim. The tool operator had never received training in the proper use of the tool, and none of the employees in the area were wearing personal protective equipment.

The Bottom Line: Never use a powder-actuated tool to secure fasteners into easily penetrated materials; these tools are designed primarily for installing fasteners into masonry. The use of powder-actuated tools requires special training and certification. In addition, all personnel in the area must be aware that the tool is in use and must be wearing appropriate personal protective equipment.

2.5.0 Masonry Anchors

Mechanical anchors are devices used to give fasteners a firm grip in a variety of materials, where the fasteners by themselves would otherwise have a tendency to pull out. Anchors can be classified in many ways by different manufacturers. In this module, anchors have been divided into four broad categories:

- One-step anchors
- Bolt anchors
- Screw anchors
- Self-drilling anchors

2.5.1 One-Step Anchors

One-step anchors (*Figure 48*) are designed so that they can be installed through the mounting holes in the component to be fastened. This is because the anchor and the drilled hole into which it is installed have the same size diameter. They are available in various diameters and lengths. Common types of one-step anchors include the following:

- *Wedge anchors* – Wedge anchors are heavy-duty anchors supplied with nuts and washers. The drill bit size used to drill the hole is the same diameter as the anchor. The depth of the hole is not critical as long as the minimum length recommended by the manufacturer is drilled. After the hole is blown clean of dust and other material, the anchor is inserted into the hole and driven with a hammer far enough so that at least six threads are below the top surface of the component. Then, the component is fastened by tightening the anchor nut to expand the anchor and tighten it in the hole.
- *Stud bolt anchors* – Stud bolt anchors are heavy-duty threaded anchors. Because this type of anchor is made to bottom in its mounting hole,

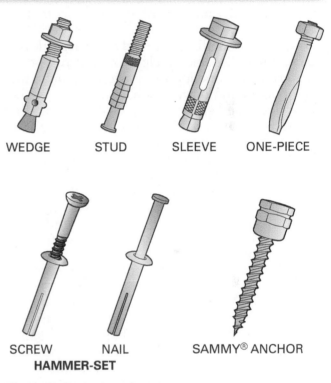

WEDGE   STUD   SLEEVE   ONE-PIECE

SCREW   NAIL   SAMMY® ANCHOR
HAMMER-SET

Figure 48 One-step anchors.

it is a good choice to use when jacking or leveling of the fastened component is needed. The depth of the hole drilled in the masonry must be as specified by the manufacturer in order to achieve proper expansion. After the hole is blown clean of dust and other material, the anchor is inserted in the hole with the expander plug end down. Following this, the anchor is driven into the hole with a hammer (or setting tool) to expand the anchor and tighten it in the hole. The anchor is fully set when it can no longer be driven into the hole. The component is fastened using the correct size and thread bolt for use with the anchor stud.

- *Sleeve anchors* – Sleeve anchors are multi-purpose anchors. The depth of the anchor hole is not critical as long as the minimum length recommended by the manufacturer is drilled. After the hole is blown clean of dust and other material, the anchor is inserted into the hole and tapped until flush with the component. Then, the anchor nut or screw is tightened to expand the anchor and tighten it in the hole.

- *One-piece anchors* – One-piece anchors are multi-purpose anchors. They work on the principle that as the anchor is driven into the hole, the spring force of the expansion mechanism is compressed and flexes to fit the size of the hole. Once set, it tries to regain its original shape. The depth of the hole drilled in the masonry must be at least deeper than the required embedment (follow the manufacturer's instructions). The proper depth is crucial. Overdrilling is as bad as underdrilling. After the hole is blown clean of dust and other material, the anchor is inserted through the component and driven with a hammer into the hole until the head is firmly seated against the component. It is important to make sure that the anchor is driven to the proper embedment depth. Note that manufacturers also make specially designed drivers and manual tools that are used instead of a hammer to drive one-piece anchors. These tools allow the anchors to be installed in confined spaces and help prevent damage to the component from stray hammer blows.

- *Hammer-set anchors* – Hammer-set anchors are made for use in concrete and masonry. There are two types: nail and screw. An advantage of the screw-type anchors is that they are removable. Both types have a diameter the same size as the anchoring hole. For both types, the anchor hole must be drilled to the diameter of the anchor and to a depth of at least $\frac{1}{4}$" (6 mm) deeper than that required for embedment (follow the manufacturer's instructions). After the hole is blown clean of dust and other material, the anchor is inserted into the hole through the mounting holes in the component to be fastened; then the screw or nail is driven into the anchor body to expand it. It is important to make sure that the head is seated firmly against the component and is at the proper embedment.

- *Threaded-rod anchors* – Threaded-rod anchors, such as the Sammy® anchor, are available for installation in concrete, steel, or wood. The anchor is designed to support a threaded rod, which is screwed into the head of the anchor after the anchor is installed. A special nut driver is available for installing the screws.

2.5.2 Bolt Anchors

Bolt anchors (*Figure 49*) are designed to be installed flush with the surface of the base material. They are used in conjunction with threaded machine bolts or screws. In some types, they can be used with threaded rod. Some commonly used types of bolt anchors include the following:

- *Drop-in anchors* – Drop-in anchors are typically used as heavy-duty anchors. There are two types of drop-in anchors. The first type is made for use in solid concrete and masonry, and has an internally threaded expansion anchor with a preassembled internal expander plug. The anchor hole must be drilled to the specific diameter and depth specified by the manufacturer. After the hole is blown clean of dust and other material, the anchor is inserted into the hole and tapped until it is flush with the surface. Following this, a setting tool supplied with the anchor is driven into the anchor to expand it. The component to be fastened is positioned in place and fastened by threading and tightening the correct size machine bolt or screw into the anchor.

  The second type, called a hollow-set drop-in anchor, is made for use in hollow concrete and masonry base materials. Hollow-set drop-in anchors have a slotted, tapered expansion sleeve and a serrated expansion cone. They come in various lengths compatible with the outer wall thickness of most hollow base materials. They can also be used in solid concrete and masonry. The anchor hole must be drilled to the specific diameter specified by the manufacturer. When installed in hollow base materials, the hole is drilled into the cell or void. After the hole is blown clean of dust and other material, the anchor is inserted into the hole and tapped until it is flush with the surface. Following this, the component to be fastened is positioned in place; then the proper size machine bolt or screw is threaded into the anchor and tightened to expand the anchor in the hole.

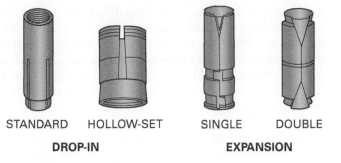

STANDARD　　HOLLOW-SET　　SINGLE　　DOUBLE

DROP-IN　　　　　　　　**EXPANSION**

Figure 49 Bolt anchors.

- *Single- and double-expansion anchors* – Single- and double-expansion anchors are both made for use in concrete and other masonry. The double-expansion anchor is used mainly when fastening into concrete or masonry of questionable strength. For both types, the anchor hole must be drilled to the specific diameter and depth specified by the manufacturer. After the hole is blown clean of dust and other material, the anchor is inserted into the hole, threaded cone end first. It is then tapped until it is flush with the surface. Following this, the component to be fastened is positioned in place; then the proper size machine bolt or screw is threaded into the anchor and tightened to expand the anchor in the hole.

2.5.3 Screw Anchors

Screw anchors are lighter-duty anchors made to be installed flush with the surface of the base material. They are used in conjunction with sheet metal, wood, or lag screws depending on the anchor type. Fiber and plastic anchors are common types of screw anchors (*Figure 50*). The lag shield anchor used with lag screws was described earlier in this module.

Fiber and plastic anchors are typically used in concrete and masonry. Plastic anchors are also commonly used in wallboard and similar base materials. The installation of all types is simple. The anchor hole must be drilled to the diameter specified by the manufacturer. The minimum depth of the hole must equal the anchor length. After the hole is blown clean of dust and other material, the anchor is inserted into the hole and tapped until it is flush with the surface. Following this, the component to be fastened is positioned in place; then the proper type and size screw is driven through the component mounting hole and into the anchor to expand the anchor in the hole.

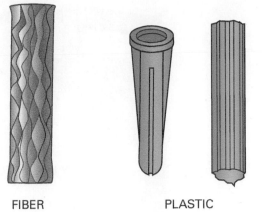

FIBER　　　　　　　　PLASTIC

(A) SCREW ANCHORS

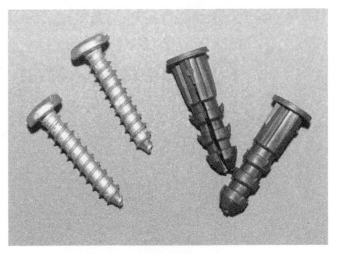

(B) SCREWS

Figure 50 Screw anchors and screws.

2.5.4 Self-Drilling Anchors

Some anchors made for use in masonry are self-drilling anchors. *Figure 51* is typical of those in common use. This fastener has a cutting sleeve that is first used as a drill bit and later becomes the expandable fastener itself. A rotary hammer is used to drill the hole in the concrete using the anchor sleeve as the drill bit. After the hole is drilled, the anchor is pulled out and the hole cleaned. This is followed by inserting the anchor's expander plug into the cutting end of the sleeve. The anchor sleeve and expander plug are driven back into the hole with the rotary hammer until they are flush with the surface of the concrete. As the fastener is hammered down, it hits the bottom, where the tapered expander causes the fastener to expand and lock into the hole. The anchor is then snapped off at the shear point with a quick lateral movement of the hammer. The component to be fastened can then be attached to the anchor using the proper size bolt.

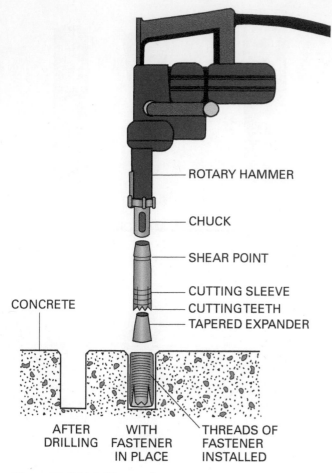

ROTARY HAMMER

CHUCK

SHEAR POINT

CUTTING SLEEVE
CUTTING TEETH
TAPERED EXPANDER

CONCRETE

AFTER
DRILLING

WITH
FASTENER
IN PLACE

THREADS OF
FASTENER
INSTALLED

Figure 51 Self-drilling anchor.

2.5.5 Drilling Anchor Holes in Hardened Concrete or Masonry

When selecting masonry anchors, regardless of the type, always take into consideration and follow the manufacturer's recommendations pertaining to hole diameter and depth, minimum embedment in concrete, maximum thickness of material to be fastened, and the pullout and shear load capacities.

When installing anchors and/or anchor bolts in hardened concrete, make sure the area where the equipment or component is to be fastened is smooth so that it will have solid footing. Uneven footing might cause the equipment to twist, warp, not tighten properly, or vibrate when in operation. Before starting, carefully inspect the rotary hammer or hammer drill and the drill bit(s) to ensure they are in good operating condition. Be sure to use the type of carbide-tipped masonry or percussion drill bits recommended by the drill/hammer or anchor manufacturer because these bits are made to take the higher impact of the masonry materials. Also, it is recommended that

the drill or hammer tool depth gauge be set to the depth of the hole needed. The trick to using masonry drill bits is not to force them into the material by pushing down hard on the drill. Use a little pressure and let the drill do the work. For large holes, start with a smaller bit, then change to a larger bit.

The methods for installing the different types of anchors in hardened concrete or masonry have been briefly described. Always install the selected anchors according to the manufacturer's directions. Here is an example of a typical procedure used to install many types of expansion anchors in hardened concrete or masonry (refer to *Figure 52* as you study the procedure):

> **WARNING!**
>
> Drilling in concrete generates noise, dust, and flying particles. Always wear safety goggles, ear protectors, and gloves. Make sure other workers in the area also wear protective equipment. Follow your company's silica rules to prevent exposure to concrete dust.

Step 1 Drill the anchor bolt hole the same size as the anchor bolt. The hole must be deep enough for six threads of the bolt to be below the surface of the concrete. Clean out the hole using a squeeze bulb.

Step 2 Drive the anchor bolt into the hole using a hammer. Protect the threads of the bolt with a nut that does not allow any threads to be exposed.

Step 3 Put a washer and nut on the bolt, and tighten the nut with a wrench until the anchor is secure in the concrete.

2.6.0 Hollow-Wall Anchors

Hollow-wall anchors are used in hollow materials such as concrete plank, block, structural steel, wallboard, and plaster. Some types can also be used in solid materials. Toggle bolts, sleeve-type wall anchors, wallboard anchors, and metal drive-in anchors are common anchors used when fastening to hollow materials.

When installing anchors in hollow walls or ceilings, regardless of the type, always follow the manufacturer's recommendations pertaining to use, hole diameter, wall thickness, grip range (thickness of the anchoring material), and the pullout and shear load capacities.

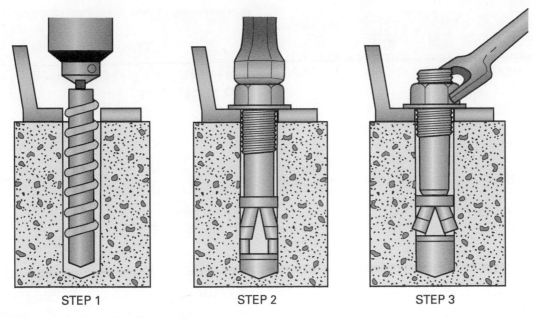

STEP 1 STEP 2 STEP 3

Figure 52 Installing an anchor bolt in hardened concrete.

2.6.1 Toggle Bolts

Toggle bolts (*Figure 53*) are used to fasten equipment, hangers, supports, and similar items into hollow surfaces such as walls and ceilings. They consist of a slotted bolt or screw and spring-loaded wings. When the bolt is inserted through the item to be fastened, then through a predrilled hole in the wall or ceiling, the wings spring apart and provide a firm hold on the inside of the hollow wall or ceiling as the bolt is tightened. Note that the hole drilled in the wall or ceiling should be just large enough for the compressed wing-head to pass through. Once the toggle bolt is installed, be careful not to completely unscrew the bolt because the wings will fall off, making the fastener useless. Screw-actuated plastic toggle bolts are also made. These are similar to metal toggle bolts, but they come with a pointed screw and do not require as large a hole. Unlike the metal version, the plastic wings remain in place if the screw is removed.

Toggle bolts are used to fasten a part to hollow block, wallboard, plaster, panel, or tile. The following general procedure can be used to install toggle bolts:

Step 1 Select the proper size drill bit or punch and toggle bolt for the job.

Step 2 Check the toggle bolt for damaged or dirty threads or a malfunctioning wing mechanism.

Step 3 Drill a hole completely through the surface to which the part is to be fastened.

Step 4 Insert the toggle bolt through the opening in the item to be fastened.

Step 5 Screw the toggle wing onto the end of the toggle bolt, ensuring that the flat side of the toggle wing is facing the bolt head.

Step 6 Fold the wings completely back and push them through the drilled hole until the wings spring open.

Step 7 Pull back on the item to be fastened in order to hold the wings firmly against the inside surface to which the item is being attached.

Step 8 Tighten the toggle bolt with a screwdriver until it is snug.

WARNING!

Follow all safety precautions when using an electric drill.

Safety

Be sure to wear safety goggles whenever you tackle any fastening project, regardless of how small the job may seem. Remember, you can never replace lost eyesight.

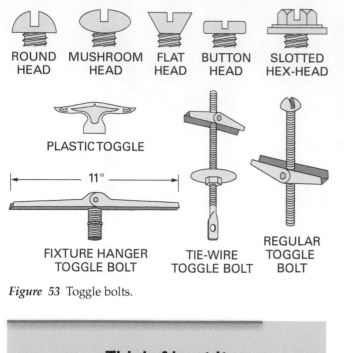

ROUND HEAD MUSHROOM HEAD FLAT HEAD BUTTON HEAD SLOTTED HEX-HEAD

PLASTIC TOGGLE

11"

FIXTURE HANGER TOGGLE BOLT TIE-WIRE TOGGLE BOLT REGULAR TOGGLE BOLT

Figure 53 Toggle bolts.

Think About It

Toggle Bolts

What will happen if you fasten a toggle bolt too tightly?

2.6.2 Sleeve-Type Wall Anchors

Sleeve-type wall anchors (*Figure 54*) are suitable for use in concrete, block, plywood, wallboard, hollow tile, and similar materials. The two types made are standard and drive. The standard type is commonly used in walls and ceilings and is installed by drilling a mounting hole to the required diameter. The anchor is inserted into the hole and tapped until the gripper prongs embed in the base material. Following this, the anchor's screw is tightened to draw the anchor tight against the inside of the wall or ceiling. Note that the drive-type anchor is hammered into the material without the need for drilling a mounting hole. After the anchor is installed, the anchor screw is removed, the component being fastened is positioned in place, then the screw is reinstalled through the mounting hole in the component and into the anchor. The screw is tightened into the anchor to secure the component.

2.6.3 Wallboard Anchors

Wallboard anchors (*Figure 54*) are self-drilling medium- and light-duty anchors used for fastening in wallboard. The anchor is driven into the wall with a Phillips head manual or cordless screwdriver until the head of the anchor is flush

Think About It

Sleeve-Type Drive Anchors

What happens when you remove the screw when a sleeve-type drive anchor is in place?

with the wall or ceiling surface. Following this, the component being fastened is positioned over the anchor, then secured with the proper size sheet metal screw driven into the anchor.

2.6.4 Metal Drive-In Anchors

Metal drive-in anchors (*Figure 54*) are used to fasten light to medium loads to wallboard. They have two pointed legs that stay together when the anchor is hammered into a wall and spread out against the inside of the wall when a sheet metal screw is driven in.

2.7.0 Epoxy Anchoring Systems

Epoxy resin compounds can be used to anchor threaded rods, dowels, and similar fasteners in solid concrete, hollow wall, and brick. For one manufacturer's product, a two-part epoxy is packaged in a two-chamber cartridge that keeps the resin and hardener ingredients separated until use. This cartridge is placed into a special tool similar to a caulking gun. When the gun handle is pumped, the epoxy resin and hardener components are mixed within the gun; then the epoxy is ejected from the gun nozzle.

To use the epoxy to install an anchor in solid concrete (*Figure 55*), a hole of the proper size is drilled in the concrete and cleaned using a nylon (not metal) brush. Following this, a small amount of epoxy is dispensed from the gun to make sure that the resin and hardener have mixed properly. This is indicated by the epoxy being of a uniform

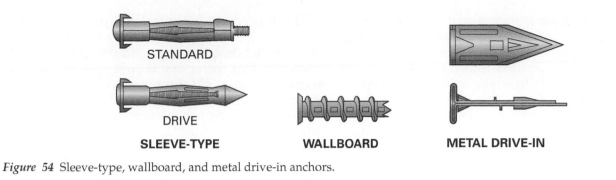

STANDARD

DRIVE

SLEEVE-TYPE

WALLBOARD

METAL DRIVE-IN

Figure 54 Sleeve-type, wallboard, and metal drive-in anchors.

Case History

Installation Requirements

In a college dormitory, battery-powered emergency lights were anchored to sheetrock hallway ceilings with sheetrock screws, with no additional support. These fixtures weigh 8-10 pounds each and might easily have fallen out of the ceiling, causing severe injury. When the situation was discovered, the contractor had to remove and replace dozens of fixtures.

The Bottom Line: Incorrect anchoring methods can be both costly and dangerous.

Think About It

Ceiling Installations

In the dormitory problem discussed earlier, which of the following fasteners could have been used to safely secure the emergency lights?

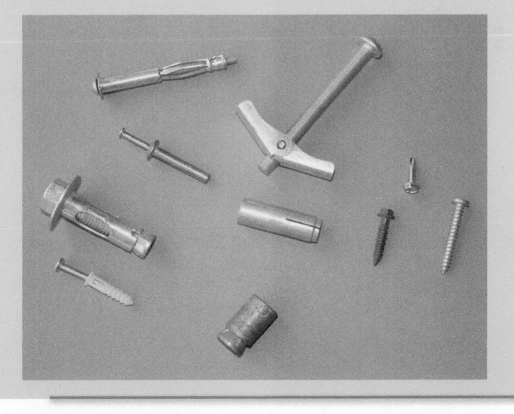

color. The gun nozzle is then placed into the hole, and the epoxy is injected into the hole until half the depth of the hole is filled. Following this, the selected fastener is pushed into the hole with a slow twisting motion to make sure that the epoxy fills all voids and crevices, then is set to the required plumb (or level) position. After the recommended cure time for the epoxy has elapsed, the fastener nut can be tightened to secure the component or fixture in place.

The procedure for installing a fastener in a hollow wall or brick using epoxy is basically the same as that described for solid concrete. The difference is that the epoxy is first injected into an anchor screen to fill the screen, then the anchor screen is installed into the drilled hole. Use of the anchor screen is necessary to hold the epoxy intact in the hole until the anchor is inserted into the epoxy.

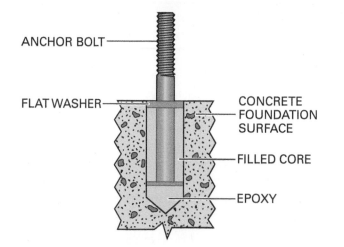

Figure 55 Fastener anchored in epoxy.

Use the Proper Tool for the Application

To avoid damaging fasteners, use the correct tool for the job. For example, don't use pliers to install bolts, and don't use a screwdriver that is too large or too small.

Using Epoxy

Once mixed, epoxy has a limited working time. Therefore, mix exactly what you need and work quickly. After the working time is up, epoxy requires a specific curing time. Always give epoxy its recommended curing time. Because epoxy is so strong and sets so quickly, you'll be tempted to stress the bond before it's fully cured.

2.0.0 Section Review

1. Tie wraps that will be used outdoors must be _____.

   a. white
   b. yellow
   c. releasable
   d. UV resistant

2. Which of the following is true regarding screw gauges and shank?

   a. The lower the gauge number, the larger the diameter of the shank.
   b. The lower the gauge number, the longer the length of the shank.
   c. The higher the gauge number, the larger the diameter of the shank.
   d. The higher the gauge number, the longer the length of the shank.

3. When installing hammer-driven pins and studs, _____.

   a. the predrilled hole should be slightly smaller than the pin or stud
   b. the predrilled hole should be the same size as the pin or stud
   c. the predrilled hole should be slightly larger than the pin or stud
   d. a predrilled hole is not required

4. When using a powder-actuated tool, _____.

   a. post warning signs within 50' (15 m) of use
   b. power level 12 is the lowest
   c. color codes are often used to indicate the manufacturer
   d. all loads are interchangeable among makes and models

5. A type of fastener that uses the anchor itself as a drill bit is a _____.

   a. wedge anchor
   b. stud anchor
   c. one-piece anchor
   d. self-drilling anchor

6. A type of fastener with spring-loaded wings is a _____.

   a. toggle bolt
   b. sleeve-type anchor
   c. drive-type anchor
   d. stud anchor

7. Epoxy anchoring systems are commonly used to anchor fasteners in _____.

   a. wood
   b. wallboard
   c. plaster
   d. brick

3.0.0 WIREWAYS AND OTHER SPECIALTY RACEWAYS

Objective

Select and install wireways and other specialty raceways.

a. Identify types of wireways and their components.
b. Install wireway supports.
c. Identify and install specialty raceways.

Trade Term

Trough: A long, narrow box used to house electrical connections that could be exposed to the environment.

A wireway is a sheet metal **trough** provided with a hinged or screw-on removable cover. Like other types of raceways, wireways are used for housing electric wires and cables. Wireways are available in various lengths to allow runs of different lengths without cutting the wireway ducts.

Metal wireways are covered in *NEC Article 376*. As listed in *NEC Section 376.22(A)*, the sum of the cross-sectional areas of all contained conductors and cables at any cross section of a wireway shall not exceed 20% of the interior cross-sectional area of the wireway. The derating factors in *NEC Table 310.15(C)(1)* shall be applied to wireways only where the number of current-carrying conductors exceeds 30, including neutral conductors classified as current-carrying under the provisions of *NEC Section 310.15(E)*. Conductors for signaling or controller conductors between a motor and its starter used only for starting duty shall not be considered current-carrying conductors.

It is also noted in *NEC Section 376.56(A)* that conductors, together with splices and taps, must not fill the wireway to more than 75% of its cross-sectional area. No conductor larger than that for which the wireway is designed shall be installed in any wireway. Be sure to check *NEC Article 378* for the requirements of nonmetallic wireways.

NEC Section 376.23(A) requires that the dimensions of *NEC Table 312.6(A)* be applied where insulated conductors are deflected in a wireway. *NEC Section 376.23(B)* requires that the provisions of *NEC Section 314.28(A)* apply where metal wireways are used as pull boxes.

An auxiliary gutter is a wireway that is intended to add to wiring space at switchboards, meters, and other distribution locations. Auxiliary gutters are dealt with in *NEC Article 366*. Even though the component parts of wireways and auxiliary gutters are identical, you should be familiar with the differences in their use. Auxiliary gutters are used as parts of complete assemblies of apparatus such as switchboards, distribution centers, and control equipment. However, an auxiliary gutter may only contain conductors or busbars, even though it looks like a surface metal raceway that may contain devices and equipment. Unlike auxiliary gutters, wireways represent a type of wiring because they are used to carry conductors between points located considerable distances apart.

The allowable ampacities for insulated conductors in wireways and gutters are given in *NEC Tables 310.16 and 310.18*. It should be noted that these tables are used for raceways in general. These *NEC*® tables and the notes are often used to determine if the correct materials are on hand for an installation. They are also used to determine if it is possible to add conductors in an existing wireway or gutter.

In many situations, it is necessary to make extensions from the wireways to wall receptacles and control devices. In these cases, *NEC Section 376.70* specifies that these extensions be made using any wiring method presented in *NEC Chapter 3* that includes a means for equipment grounding. Finally, as required in *NEC Section 376.120*, wireways must be marked in such a way that their manufacturer's name or trademark will be visible.

As you can see in *Figure 56*, a wide range of fittings is required for connecting wireways to one another and to fixtures such as switchboards, power panels, and conduit.

3.1.0 Types of Wireways and Their Components

Rectangular duct-type wireways come as either hinged-cover or screw-cover troughs. They are available in a variety of lengths to avoid cutting.

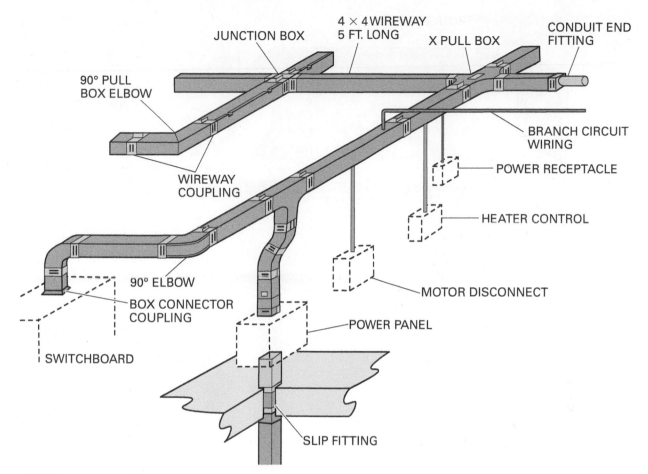

Figure 56 Wireway system layout.

Raintight troughs are used in environments where moisture is not permitted within the raceway. However, a raintight trough should not be confused with the raintight lay-in wireway, which has a hinged cover. *Figure 57* shows a raintight trough with a removable side cover.

Wireway troughs are exposed when first installed. Whenever possible, they are mounted on the ceilings or walls, although they may sometimes be suspended from the ceiling. Note that in *Figure 58*, the trough has knockouts similar to those found on junction boxes. After the wireway system has been installed, branch circuits are brought from the distribution panels using conduit. The conduit is joined to the wireway at the most convenient knockout possible.

Wireway components such as trough crosses, 90° internal elbows, and tee connectors serve the same function as fittings on other types of raceways. The fittings are attached to the duct using slip-on connectors. All attachments are made with nuts and bolts or screws. When assembling wireways, always place the head of the bolt on the inside and the nut on the outside so that the

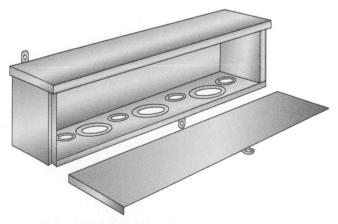

Figure 57 Raintight trough.

conductors will not be resting against a sharp edge. It is usually best to assemble sections of the wireway system on the floor, and then raise the sections into position. An exploded view of a section of wireway is shown in *Figure 59*. Both the wireway fittings and the duct come with screw-on, hinged, or snap-on covers to permit conductors to be laid in or pulled through.

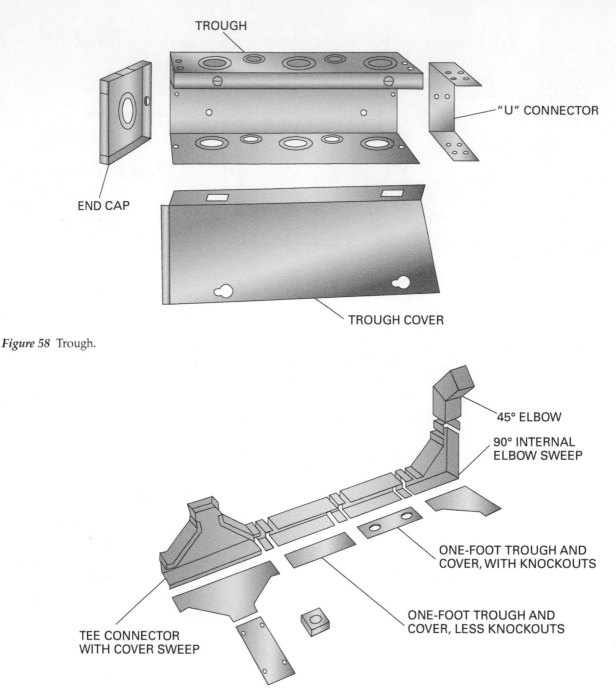

Figure 58 Trough.

Figure 59 Wireway sections.

The *NEC*® specifies that wireways may be used only for exposed work. Therefore, they cannot be used in underfloor installations. If they are used for outdoor work, they must be of an approved raintight construction. It is important to note that wireways must not be installed where they are subject to severe physical damage, corrosive vapors, or hazardous locations.

Wireway troughs must be installed so that they are supported at distances not exceeding 5' (1.5 m). When specially approved supports are used, the distance between supports must not exceed 10' (3 m).

3.1.1 Wireway Fittings

Many different types of fittings are available for wireways, especially for use in exposed, dry locations. The following sections explain fittings commonly used in the electrical craft.

Wireway or Trough?

A raintight lay-in wireway has a hinged cover, as shown here. A raintight trough simply has a removable cover.

3.1.2 Connectors

Connectors (*Figure 60*) are used to join wireway sections and fittings. Connectors are slipped inside the end of a wireway section and are held in place by small bolts and nuts. Alignment slots allow the connector to be moved until it is flush with the inside surface of the wireway. After the connector is in position, it can be bolted to the wireway. This helps to ensure a strong rigid connection. Connectors have a friction hinge that helps hold the wireway cover open when needed.

3.1.3 End Plates

End plates, or closing plates (*Figure 61*), are used to seal the ends of wireways. They are inserted into the end of the wireway and fastened by screws and bolts. End plates contain knockouts so that conduit or cable may be extended from the wireway.

3.1.4 Tees

Tee fittings (*Figure 62*) are used when a tee connection is needed in a wireway system. A tee connection is used where circuit conductors may branch in different directions. The tee fitting's covers and sides can be removed for access to splices and taps. Tee fittings are attached to other wireway sections using standard connectors.

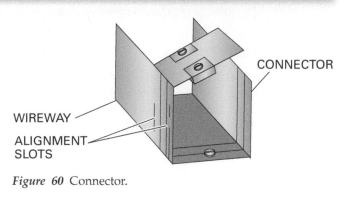

CONNECTOR
WIREWAY
ALIGNMENT SLOTS

Figure 60 Connector.

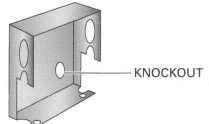

KNOCKOUT

Figure 61 End plate.

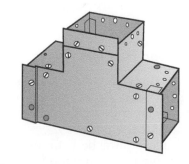

Figure 62 Tee.

3.1.5 Crosses

Crosses (*Figure 63*) have four openings and are attached to other wireway sections with standard connectors. The cover is held in place by screws and can be easily removed for laying in wires or for making connections.

3.1.6 Elbows

Elbows are used to make a bend in the wireway. They are available in angles of $22\frac{1}{2}°$, 45°, or 90°, and are either internal or external. They are attached to wireway sections with standard connectors. Covers and sides can be removed for wire installation. The inside corners of elbows are rounded to prevent damage to conductor insulation. An inside elbow is shown in *Figure 64*.

3.1.7 Telescopic Fittings

Telescopic or slip fittings are adjustable-length fittings that may be used between two lengths of wireway. Slip fittings are attached to standard lengths by setscrews and usually adjust from $\frac{1}{2}$" to $11\frac{1}{2}$" (12.7 mm to 292.1 mm). Slip fittings have a removable cover for installing wires and are similar in appearance to a nipple.

3.2.0 Wireway Supports

Horizontal wireway runs must be securely supported at each end and at intervals of no more than 5' (1.5 m) or for individual lengths greater than 5' (1.5 m) at each end or joint, unless listed for other support intervals. In no case shall the support distance be greater than 10' (3 m), in accordance with *NEC Section 376.30(A)*. If possible, wireways can be mounted directly to a surface. Otherwise, wireways are supported by hangers or brackets.

3.2.1 Suspended Hangers

In many cases, the wireway is supported from a ceiling, beam, or other structural member. In such installations, a suspended hanger (*Figure 65*) may be used to support the wireway.

The wireway is attached to or laid in the hanger. The hanger is suspended by a threaded rod. One end of the rod is attached to the hanger with hex nuts. The other end of the rod is attached to a beam clamp or anchor.

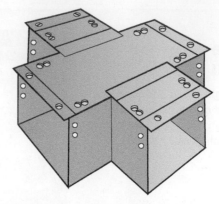

Figure 63 Cross.

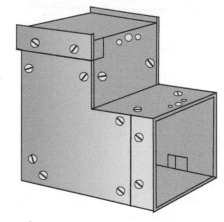

Figure 64 90° inside elbow.

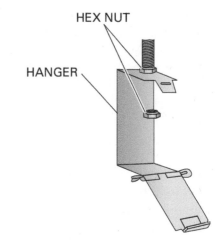

Figure 65 Suspended hanger.

3.2.2 Gusset Brackets

Another type of support used to mount wireways is a gusset bracket (*Figure 66*). This is an L-type bracket that is mounted to a wall. The wireway rests on the bracket and is attached by screws or bolts.

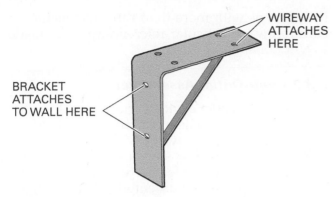

Figure 66 Gusset bracket.

3.2.3 Standard Hangers

Standard hangers (*Figure 67*) are made in two pieces. The two pieces are combined in different ways for different installation requirements. The wireway is attached to the hanger by bolts and nuts.

3.2.4 Wireway Hangers

When a larger wireway must be suspended, a wireway hanger may be used. A wireway hanger is made by suspending a piece of strut from a ceiling, beam, or other structural member. The strut is suspended by threaded rods attached to beam clamps or other ceiling anchors, as shown in *Figure 68*.

3.3.0 Specialty Raceways

Specialty raceways include enclosures such as surface metal and nonmetallic raceways, underfloor raceways, and underground ducts.

3.3.1 Surface Metal and Nonmetallic Raceways

Surface metal raceways consist of a wide variety of special raceways designed primarily to carry power and communications wiring to locations on the surface of ceilings or walls of building interiors.

Installation specifications of both surface metal raceways and surface nonmetallic raceways are listed in detail in *NEC Articles 386 and 388*, respectively. All these raceways must be installed in dry, interior locations. The number of conductors, their amperage, and the allowable cross-sectional area of the conductors, as well as regulations for combination raceways, are specified in *NEC Tables 310.16 and 310.18* and *NEC Articles 386 and 388*.

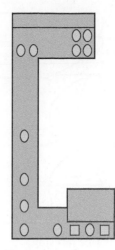

Figure 67 Standard hanger.

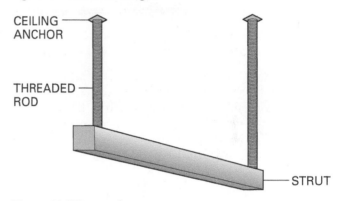

Figure 68 Wireway hanger.

One use of surface metal raceways is to protect conductors that run to non-accessible outlets.

Surface metal and nonmetallic raceways are divided into subgroups based on the specific purpose for which they are intended. There are three small surface raceways that are primarily used for extending power circuits from one point to another. In addition, there are six larger surface raceways that have a much wider range of applications. Typical cross sections of the first three smaller raceways are shown in *Figure 69*.

Additional surface metal raceway designs are referred to as pancake raceways, because their flat cross sections resemble pancakes. Their primary use is to extend power, lighting, telephone, or signal wire across a floor to locations away from the walls of a room without embedding them under the floor. Pancake raceways are shown in *Figure 70*.

There are also surface metal raceways available that house two or three different conductor raceways. These are referred to as twin-duct or triple-duct. These raceways permit different circuits, such as power and signal, to be placed within the same raceway.

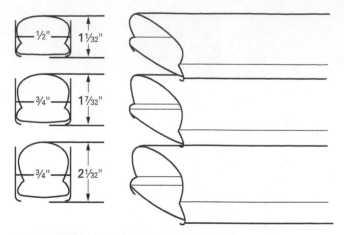

Figure 69 Smaller surface raceways.

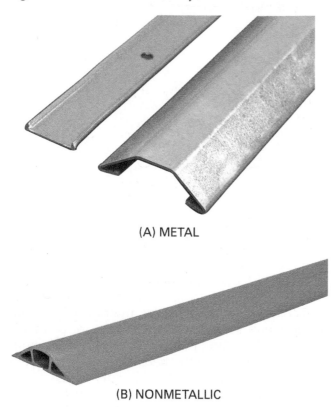

(A) METAL

(B) NONMETALLIC

Figure 70 Pancake raceways.

Nonmetallic raceways come in a variety of styles. The perimeter raceways shown in *Figure 71* are available in sizes ranging from ³⁄₄" (20 mm) to more than 7" (180 mm) wide. Many of these raceways contain barriers that allow them to carry both low voltage and power wiring.

The number and types of conductors permitted to be installed and the capacity of a particular surface raceway must be calculated and matched with *NEC®* requirements, as discussed previously. *NEC Tables 310.16 and 310.18* are used for surface raceways in the same manner in which they are used for wireways. For surface raceway installations with more than three conductors in each raceway, particular reference must be made to *NEC Table 310.15(C)(1)*.

3.3.2 Multi-Outlet Assemblies

Manufacturers offer a wide variety of multi-outlet surface raceways. Their function is to hold receptacles and other devices within the raceway. When surface raceways are used in this manner, the assembly is referred to as a multi-outlet assembly. Multi-outlet assemblies (*Figure 72*) are covered in *NEC Article 380*. Multi-outlet systems are either wired in the field or come pre-wired from the factory.

3.3.3 Pole Systems

In many situations, power and other electric circuits must be carried from overhead wiring systems to devices that are not located near existing wall outlets or control circuits. This type of wiring is typically used in open office spaces where cubicles are provided by temporary dividers. Poles are used to accomplish this. The poles usually come in lengths suitable for different ceiling heights. *Figure 73* shows a typical pole base.

3.3.4 Underfloor Systems

Underfloor raceway systems were developed to provide a practical means of bringing conductors for lighting, power, and signaling to cabinets and consoles. Underfloor raceways are available in 10' (3 m) lengths and widths of 4" and 8" (100 mm and 200 mm). The sections are made with inserts spaced every 24" (600 mm). The inserts can be removed for outlet installation. These are explained in *NEC Article 390*.

> **NOTE**
>
> Inserts must be installed so that they are flush with the finished grade of the floor.

Junction boxes are used to join sections of underfloor raceways. Conduit is also used with underfloor raceways by using a raceway-to-conduit connector (conduit adapter). A typical underfloor raceway duct with fittings is shown in *Figure 74*.

This wiring method makes it possible to place a desk or table in any location where it will always be over, or very near to, a duct line. The wiring method for lighting and power between

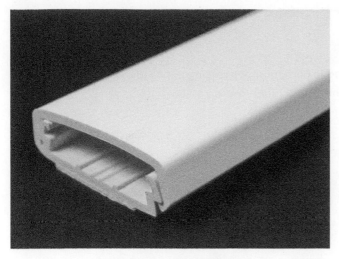

(A) SINGLE CHANNEL

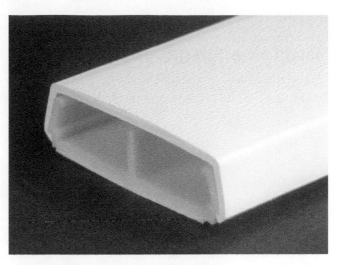

(B) DUAL CHANNEL

Figure 71 Examples of surface raceway.

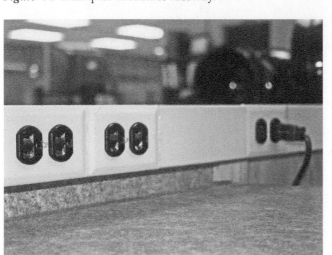

Figure 72 Multi-outlet assembly.

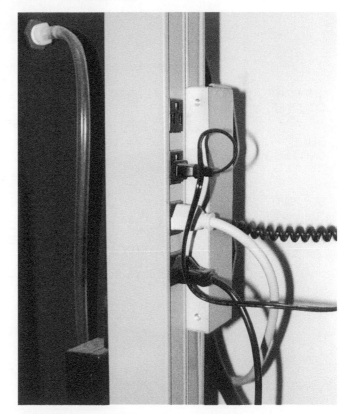

Figure 73 Power pole.

cabinets and the raceway junction boxes may be conduit, underfloor raceway, wall elbows, and cabinet connectors. *NEC Article 390* covers the installation of underfloor raceways.

3.3.5 Cellular Metal Floor Raceways

A cellular metal floor raceway is a type of floor construction designed for use in steel-frame buildings. In these buildings, the members supporting the floor between the beams consist of sheet steel rolled into shapes. These shapes are combined to form cells, or closed passageways, which extend across the building. The cells are of various shapes and sizes, depending upon the structural strength required. The cells of this type of floor construction form the raceways, as shown in *Figure 75*.

Connections to the cells are made using headers that extend across the cells. A header connects only to those cells to be used as raceways for conductors. A junction box or access fitting is necessary at each joint where a header connects to a cell. Two or three separate headers, connecting to different sets of cells, may be used for different systems. For example, light and power, signaling systems, and public telephones would each have a separate header. A special elbow fitting is used to extend the headers up to the distribution equipment on a wall or column. *NEC Article 374* covers the installation of cellular metal floor raceways.

Surface Raceways

Surface raceways with multiple channels are commonly used in computer networking applications to provide conductors for AC power to the computers, as well as telephone and other low-voltage wiring.

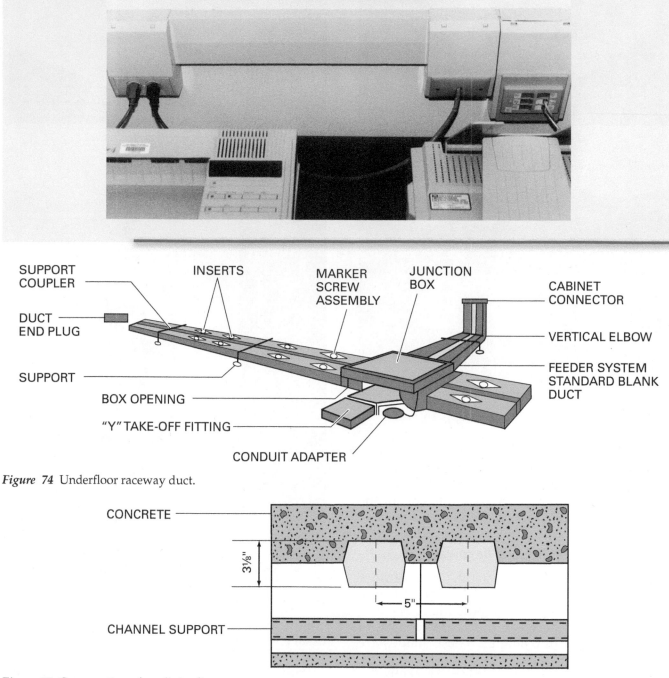

Figure 74 Underfloor raceway duct.

Figure 75 Cross section of a cellular floor.

3.3.6 Cellular Concrete Floor Raceways

The term *precast cellular concrete floor* refers to a type of floor used in steel-frame, concrete-frame, and wall-bearing construction. In this type of system, the floor members are precast with hollow voids that form smooth, round cells. The cells form raceways, which can be adapted, using fittings, for use as underfloor raceways. A precast cellular concrete floor is fire-resistant and requires no further fireproofing. The precast reinforced concrete floor members form the structural floor and are supported by beams or bearing walls. Connections to the cells are made with headers that are secured to the precast concrete floor. *NEC Article 372* covers the installation of cellular concrete floor raceways.

3.3.7 Duct Systems

In the common vocabulary of the electrical trade, a duct is a single enclosed raceway, or runway, through which conductors or cables can be led. Basically, ducting is a system of ducts. However, underground duct systems include manholes, transformer vaults, and risers.

There are several reasons for running power lines underground rather than overhead. In some situations, an overhead high-voltage line would be dangerous, or the space may not be adequate. For aesthetic reasons, architectural plans may require buried lines throughout a subdivision or a planned community. Tunnels may already exist, or be planned, for carrying steam or water lines. In any of these situations, underground installations are appropriate. Underground cables may be buried directly in the ground or run through tunnels or raceways, including conduit and recognized ducts.

In underground construction, a duct system provides a safe passageway for power lines, communication cables, or both. A duct consists of conduit or an approved duct system (such as HDPE) placed in a trench and covered with earth or concrete. The minimum depth at which the duct will be placed is determined using *NEC Table 300.5*. Encasing the duct in concrete or other materials provides mechanical strength and helps dissipate heat. *Figure 76* shows a duct bank in place and ready for backfill. In this case, it will be covered in concrete.

Manholes are set at intervals in an underground duct run. *Figure 77* shows a manhole with pull strings installed and tied off in preparation for the conductor installation. Manholes provide access through throats (sometimes called

Figure 76 Duct bank.

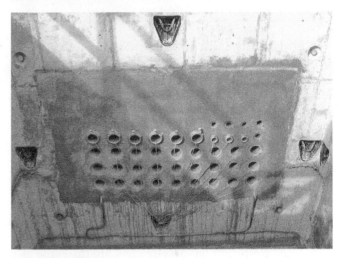

Figure 77 Manhole.

chimneys). At ground level, or street surface level, a manhole cover closes off the manhole area tightly. A duct line may consist of a single conduit or several, each carrying a cable length from one manhole to the next.

Manholes provide room for conductor installation and maintenance. Workers enter a manhole from above. In a two-way manhole, cables enter and leave in only two directions. There are also three-way and four-way manholes. Often manholes are located at the intersection of two streets so that they can be used for cables leaving in four directions. Manholes are usually constructed of brick or concrete. Their design must provide room for drainage and for workers to move around inside them. A similar opening known as a handhole is sometimes provided for splicing on lateral two-way duct lines.

Transformer vaults house power transformers, voltage regulators, network protectors, meters, and circuit breakers. A cable may end at a transformer vault. Other cables end at a customer's substation or terminate as risers that connect with overhead lines.

Underground duct lines can be made of fiber, vitrified tile, rigid metal or nonmetallic conduit, or poured concrete. The inside diameter of the ducting for a specific job is determined by the size of the cable that will be drawn into the duct. Inside diameters of 2" to 6" (MD 53 to MD 155) are available for most types of ducting.

WARNING!

Be careful when working with unfamiliar duct materials. In older installations, asbestos/cement duct may have been used. You must be certified to remove or disturb asbestos.

Rigid nonmetallic conduit may be made of PVC (polyvinyl chloride), PE (polyethylene), or styrene. Because this type of conduit is available in longer lengths, fewer couplings are needed than with other types of ducting. PVC is popular because it is easy to install, requires less labor than other types of conduit, and is low in cost.

Monolithic concrete duct is poured at the job site. Multiple duct lines can be formed using rubber tubing cores on spacers. The cores may be removed after the concrete has set. A die containing steel tubes, known as a boat, can also be used to form ducts. It is pulled slowly through the trench on a track as concrete is poured from the top. Poured concrete ducting made by either method is relatively expensive, but offers the advantage of creating a very clean duct interior with no residue that can decay. The rubber core method is especially useful for curving or turning part of a duct system.

One of the most popular duct types is the cable-in-duct. This type of duct comes from the manufacturer with cables already installed. The duct comes in a reel and can be laid in the trench with ease. The installed cables can be withdrawn in the future, if necessary. This type of duct, because of the form in which it comes, reduces the need for fittings and couplings. It is most frequently used for street lighting systems.

3.0.0 Section Review

1. Which of the following is true regarding wireways?

   a. Leave knockouts open in raintight troughs to allow for drainage.
   b. Raintight lay-in wireways have a hinged cover.
   c. Raintight lay-in wireways are the same as raintight troughs.
   d. Troughs do not require knockouts because the conductors are fully enclosed.

2. The maximum distance between wireway supports is _____.

   a. 5' (1.5 m)
   b. 10' (3 m)
   c. 15' (4.5 m)
   d. 20' (6 m)

3. Which of the following is true regarding surface metal raceways?

   a. They are typically used to carry service-entrance cable.
   b. They can be used indoors or outdoors.
   c. They are used in dry, interior locations only.
   d. They are typically installed on floors.

4.0.0 CABLE TRAYS

Objective

Select and install cable trays.
  a. Identify cable tray types and fittings.
  b. Install cable tray supports.

Trade Term

Cable trays: Rigid structures used to support electrical conductors.

Wire and cable installation in cable trays is defined by the *NEC®*. Read *NEC Article 392* to become familiar with the requirements and restrictions made by the *NEC®* for safe installation of wire and cable in a cable tray.

Metallic cable trays that support electrical conductors must be grounded as required by *NEC Article 250*. Where steel and aluminum cable tray systems are used as an equipment grounding conductor, all of the provisions of *NEC Section 392.60(B)* must be complied with.

> **WARNING!**
>
> Do not stand on, climb in, or walk on a cable tray.

Cable trays function as a support for conductors and tubing (*NEC Article 392*). A cable tray has the advantage of easy access to conductors, and thus lends itself to installations where the addition or removal of conductors is a common practice.

4.1.0 Cable Tray Types and Fittings

Cable trays are fabricated from aluminum, steel, and fiberglass. Cable trays are available in two basic forms: ladder and trough. Ladder tray, as the name implies, consists of two parallel channels connected by rungs. Trough consists of two parallel channels (side rails) having a corrugated, ventilated bottom, or a corrugated, solid bottom. There is also a special center-rail cable tray available for use in light-duty applications such as telephone and sound wiring. Cable trays are available in a variety of lengths, widths, and load depths to suit different applications.

Cable trays may be used in most electrical installations. Cable trays may be used in air handling ceiling space, but only to support the wiring methods permitted in such spaces by *NEC Section 300.22(C)(2)*. Also, cable trays may be used in Class 1, Division 2 locations according to *NEC Section 501.10(B)(1)(5)*. Cable trays may also be used above a suspended ceiling that is not used as an air handling space. Some manufacturers offer an aluminum cable tray that is coated with PVC for installation in caustic environments. A typical cable tray system with fittings is shown in *Figure 78*.

Cable tray fittings are part of the cable tray system and provide a means of changing the direction or dimension of the different trays. Some of the uses of horizontal and vertical tees, horizontal and vertical bends, horizontal crosses, reducers, barrier strips, covers, and box connectors are shown in *Figure 78*.

4.2.0 Installing Cable Tray Supports

Cable trays are usually supported in one of five ways: direct rod suspension, trapeze mounting, center hung, wall mounting, and pipe rack mounting.

4.2.1 Direct Rod Suspension

The direct rod suspension method of supporting cable tray uses threaded rods and hanger clamps. One end of the threaded rod is connected to the ceiling or other overhead structure. The other end

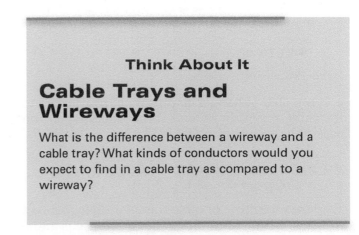

Think About It

Cable Trays and Wireways

What is the difference between a wireway and a cable tray? What kinds of conductors would you expect to find in a cable tray as compared to a wireway?

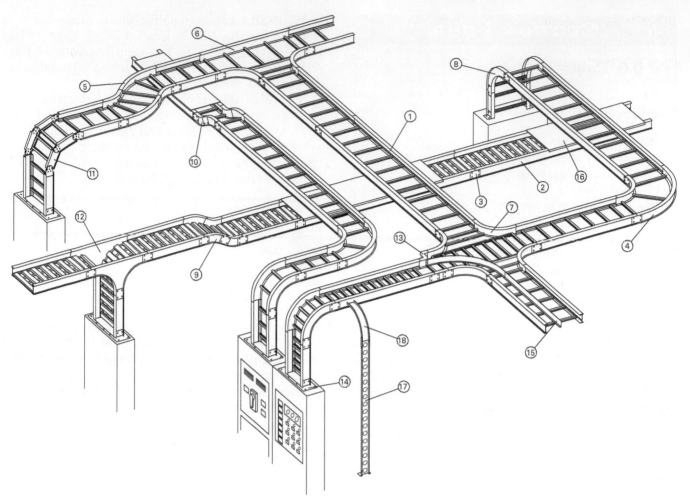

Legend

1.   LADDER CABLE TRAY
2.   VENTILATED TROUGH CABLE TRAY
3.   STRAIGHT SPLICE PLATE
4.   90° HORIZONTAL BEND, LADDER CABLE TRAY
5.   45° HORIZONTAL BEND, LADDER CABLE TRAY
6.   HORIZONTAL TEE, LADDER CABLE TRAY
7.   HORIZONTAL CROSS, LADDER CABLE TRAY
8.   90° VERTICAL OUTSIDE BEND, LADDER CABLE TRAY
9.   45° VERTICAL OUTSIDE BEND, VENTILATED CABLE TRAY

10. 30° VERTICAL INSIDE BEND, LADDER CABLE TRAY
11. VERTICAL BEND SEGMENT (VBS)
12. VERTICAL TEE DOWN, VENTILATED TROUGH CABLE TRAY
13. LEFT HAND REDUCER, LADDER CABLE TRAY
14. FRAME-TYPE BOX CONNECTOR
15. BARRIER STRIP STRAIGHT SECTION
16. SOLID FLANGED TRAY COVER
17. VENTILATED CHANNEL STRAIGHT SECTION
18. CHANNEL CABLE TRAY, 90° VERTICAL OUTSIDE BEND

Figure 78 Cable tray system.

is connected to hanger clamps that are attached to the cable tray side rails. A direct rod suspension assembly is shown in *Figure 79*.

4.2.2 Trapeze Mounting and Center Hung Support

Trapeze mounting of cable tray is similar to direct rod suspension mounting. The difference is in the method of attaching the cable tray to the threaded rods. A structural member, usually a steel chan-

nel or strut, is connected to the vertical supports to provide an appearance similar to a swing or trapeze. The cable tray is mounted to the structural member. Often, the underside of the channel or strut is used to support conduit. A trapeze mounting assembly is shown in *Figure 80*.

A method that is similar to trapeze mounting is a center hung tray support. In this case, only one rod is used and it is centered between the cable tray side rails.

4.2.3 Wall Mounting

Wall mounting is accomplished by supporting the cable tray with structural members attached to the wall (*Figure 81*). This method of support is often used in tunnels and other underground or sheltered installations where large numbers of conductors interconnect equipment that is separated by long distances.

4.2.4 Pipe Rack Mounting

Pipe racks are structural frames used to support piping that interconnects equipment in outdoor industrial facilities. Usually, some space on the rack is reserved for conduit and cable tray. Pipe rack mounting of cable tray is often used when power distribution and electrical wiring is routed over a large area.

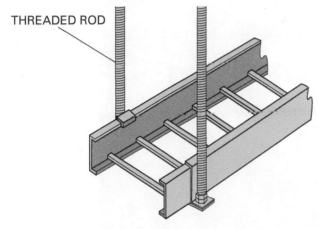

THREADED ROD

Figure 79 Direct rod suspension.

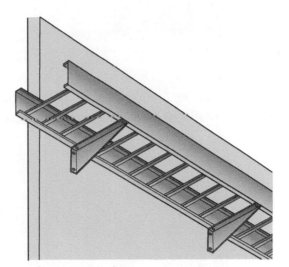

Figure 81 Wall mounting.

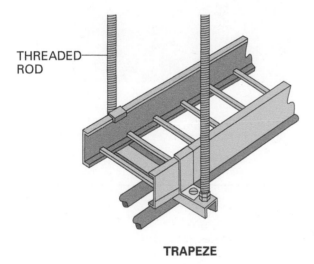

THREADED ROD

TRAPEZE

Figure 80 Trapeze mounting and center hung support.

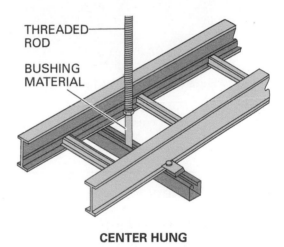

THREADED ROD

BUSHING MATERIAL

CENTER HUNG

Cable Tray Systems

Cable tray systems must be continuous and grounded. One of the advantages of using a cable tray system is that it makes it easy to expand or modify the wiring system following installation. Unlike conduit systems, wires can be added or changed by simply laying them into (or lifting them out of) the tray.

Figure Credit: Jim Mitchem

4.0.0 Section Review

1. Special light-duty center-rail cable tray is most often used for _____.

    a. industrial power requirements
    b. telephone or sound wiring
    c. outdoor power distribution
    d. large conductors

2. The type of cable tray support most likely to be used when power distribution and electrical wiring is routed over a large outdoor area is _____.

    a. trapeze mounting
    b. wall mounting
    c. pipe rack mounting
    d. direct rod suspension

5.0.0 HANDLING AND STORING RACEWAYS

Objective

Handle and store raceways.
a. Handle raceways.
b. Store raceways.

Proper and safe methods of storing conduit, wireways, raceways, and cable trays may sound like a simple task, but improper storage techniques can result in wasted time and damage to the raceways, as well as personal injury. Storing raceways correctly will help avoid costly damage, save time in identifying stored raceways, and reduce the chance of personal injury.

5.1.0 Handling Raceways

Raceway is made to strict specifications. It can be easily damaged by careless handling. From the time raceway is delivered to a job site until the installation is complete, use proper and safe handling techniques. These are a few basic guidelines for handling raceway that will help avoid damaging or contaminating it:

- Never drag raceway off a delivery truck or off other lengths of raceway.
- Never drag raceway on the ground or floor. Dragging raceway can cause damage to the ends.
- Keep the thread protection caps on when handling or transporting conduit raceway.
- Keep raceway away from any material that might contaminate it during handling.

- Flag the ends of long lengths of raceway when transporting it to the job site.
- Never drop or throw raceway when handling it.
- Never hit raceway against other objects when transporting it.
- Always use two people when carrying long pieces of raceway. Make sure that you both stay on the same side and that the load is balanced. Each person should be about one-quarter of the length of the raceway from the end. Lift and put down the raceway at the same time.

5.2.0 Storing Raceways

Pipe racks are commonly used for storing conduit. The racks provide support to prevent bending, sagging, distorting, scratching, or marring of conduit surfaces. Most racks have compartments where different types and sizes of conduit can be separated for ease of identification and selection. The storage compartments in racks are usually elevated to help avoid damage that might occur at floor level. Conduit that is stored at floor level is easily damaged by people and other materials or equipment in the area.

The ends of stored conduit should be sealed to help prevent contamination and damage. Conduit ends can be capped, taped, or plugged.

Always inspect raceway before storing it to make sure that it is clean and not damaged. It is discouraging to get raceway for a job and find that it is dirty or damaged. Also, make sure that the raceway is stored securely so that when someone comes to get it for a job, it will not fall in any way that could cause injury.

To prevent contamination and corrosion of stored raceway, it should be covered with a tarpaulin or other suitable covering. It should also be separated from noncompatible materials such as hazardous chemicals.

Wireways, surface metal raceways, and cable trays should always be stored off the ground on boards in an area where people will not step on it and equipment will not run over it. Stepping on or running over raceway bends the metal and makes it unusable.

Putting It All Together

Think about the effort that goes into the design of a large industrial installation. If you were to design a large complex, such as the one shown here, where would you start and why?

5.0.0 Section Review

1. When moving long pieces of raceway, it is best to _____.

    a. drag it
    b. use two people
    c. remove the caps
    d. use a forklift

2. When storing raceway, it is best to _____.

    a. leave it uncovered
    b. store it at ground level
    c. separate different types
    d. use it as a temporary walkway

1. The lightest duty and most widely used non-flexible metal conduit is _____.

    a. electrical metallic tubing
    b. rigid metal conduit
    c. aluminum conduit
    d. plastic-coated RMC

2. Which of the following is often referred to as thinwall conduit?

    a. IMC
    b. EMT
    c. RMC
    d. Galvanized rigid steel conduit

3. *NEC Article 358* covers _____.

    a. RMC
    b. IMC
    c. EMT
    d. ENT

4. In order to resist corrosion in wet environments, EMT is _____.

    a. galvanized
    b. rubber-coated
    c. enamel-coated
    d. dipped in aluminum

5. RMC is made of _____.

    a. cast iron
    b. steel or aluminum
    c. copper or aluminum
    d. PVC

6. A type of conduit that should be used where corrosion is a factor is _____.

    a. PVC
    b. black enamel steel conduit
    c. IMC
    d. RMC

7. Which of the following is true with regard to PVC?

    a. It can be used as an equipment grounding conductor.
    b. It increases the voltage drop of the conductors.
    c. It requires the use of expansion joints to avoid damage due to temperature changes.
    d. It must be threaded.

8. Type II PVC conduit is also known as _____.

    a. DB
    b. HDPE
    c. LFNC
    d. EB

9. A type of conduit used to connect machines that vibrate during operation is _____.

    a. aluminum
    b. black enamel steel conduit
    c. flexible metal conduit
    d. RMC

10. Which of the following is an acceptable combination of conduit bends between pull points?

    a. four 90-degree bends
    b. three 90-degree bends and three 45-degree bends
    c. six 45-degree bends and two 90-degree bends
    d. eight 30-degree bends and two 90-degree bends

11. The fitting used to protect conductors from the sharp edges of conduit where it enters a box is called a _____.

    a. bushing
    b. locknut
    c. coupling
    d. nipple

12. Ungrounded conductors entering a raceway must be protected by bushings when they are sized No. _____.
    a. 10 AWG and larger
    b. 8 AWG and larger
    c. 6 AWG and larger
    d. 4 AWG and larger

13. To avoid having to make an offset bend, use a fitting known as a(n) _____.
    a. nipple
    b. conduit body
    c. bushing
    d. hub

14. A Type LB conduit body has a cover on _____.
    a. the left
    b. the right
    c. the back
    d. both sides

15. A type of conduit body used to provide a pull point on a straight conduit run is _____.
    a. Type C
    b. Type L
    c. Type T
    d. Type X

16. A type of conduit body used at four intersecting conduits is _____.
    a. Type LR
    b. Type LL
    c. Type T
    d. Type X

17. A Type L conduit body that has a cover on both sides is a(n) _____.
    a. Type LL
    b. Type LRL
    c. LB
    d. LX

18. Where metal conduit is to be installed in a wet location, a clamp back strap can be used to maintain the minimum required distance from the wall surface, which is _____.
    a. $\frac{1}{4}$ inch
    b. $\frac{1}{2}$ inch
    c. $\frac{3}{4}$ inch
    d. 1 inch

19. The maximum depth of a box made for use in concrete construction is _____.
    a. 3 inches
    b. 4 inches
    c. 5 inches
    d. 6 inches

20. Which of the following is true when installing boxes in a metal stud environment?
    a. Boxes are always mounted flush with the studs.
    b. You must know the thickness of the finished surface before installing the boxes.
    c. PVC is the most common type of conduit used in metal stud environments.
    d. MC cable is not used in metal stud environments.

21. Which of the following regulations applies to drilling of wood joists and beams?
    a. An engineered beam cannot be drilled unless allowed by the manufacturer's specifications.
    b. They can only be drilled in the center third.
    c. The hole must be at least 1" from an edge.
    d. The hole diameter must not exceed one-half of the depth of the girder or joist.

22. Hammer-set anchors are designed for use in _____.
    a. wood
    b. metal studs
    c. concrete
    d. structural steel

23. *NEC Section 376.56(A)* limits wireway fill to no more than _____.

   a. 40% of the cross-sectional area of the wireway
   b. 60% of the cross-sectional area of the wireway
   c. 75% of the cross-sectional area of the wireway
   d. 90% of the cross-sectional area of the wireway

24. The cross-sectional areas of all conductors at any cross section of a wireway shall not exceed _____.

   a. 20% of the interior cross-section of the wireway
   b. 35% of the interior cross-section of the wireway
   c. 40% of the interior cross-section of the wireway
   d. 75% of the interior cross-section of the wireway

25. Raceways designed to extend conductors across a floor without embedding it in the floor are called _____.

   a. cellular raceways
   b. raceway ducts
   c. cellular ductways
   d. pancake raceways

Trade Terms Quiz

Fill in the blank with the correct term that you learned from your study of this module.

1. A(n) _____ area is one that can be reached for service or repair.

2. When something is in a(n) _____, it is not permanently closed in by the structure or finish of a building.

3. When materials meet a regulatory agency's requirements, the material is then said to be _____.

4. _____ is a regulatory agency that evaluates and approves electrical components and equipment.

5. A(n) _____ is used to make a continuous grounding path between equipment and ground.

6. _____ are rigid structures, either suspended or mounted, that are used to support electrical conductors.

7. Similar to pipe, _____ is a round raceway that houses conductors.

8. A(n) _____ is a bend made in a piece of conduit to alter its course.

9. _____ are enclosed channels that are used to house wires and cables.

10. _____ are steel troughs designed to carry electrical wire and cable.

11. A(n) _____ is the connection of two or more conductors.

12. An intermediate point on a main circuit where another wire is connected to supply electrical current to another circuit is called a(n) _____.

13. Electrical connectors that could be exposed to the environment are housed in a long, narrow box, or a _____.

Trade Terms

Accessible
Approved
Bonding wire
Cable trays
Conduit

Exposed location
Kick
Raceways
Splice
Tap

Trough
Underwriters Laboratories, Inc. (UL)
Wireways

1. Most electrical equipment that has a metal frame must be _____.

2. When installing EMT in a wet location, what type of fittings must be used?
   a. Setscrew
   b. Compression
   c. Raintight
   d. Steel

3. Conductors, along with splices and taps, must not fill a wireway to more than _____ of its cross-sectional area.
   a. 20%
   b. 30%
   c. 75%
   d. 90%

4. Which of the following covers the installation requirements for cable tray?
   a. *NEC Article 333*
   b. *NEC Article 338*
   c. *NEC Article 392*
   d. *NEC Article 394*

5. True or False? IMC conduit has the same internal diameter as RMC.

6. All of the following can be used as an equipment grounding conductor, *except* _____.
   a. rigid metal conduit
   b. rigid nonmetallic conduit
   c. electrical metallic tubing
   d. intermediate metal conduit

7. Flexible metal conduit can be connected to rigid metal conduit using a(n) _____ coupling.

8. Which of the following covers grounding provisions for metal boxes?
   a. *NEC Section 500.8(A)*
   b. *NEC Section 250.30(A)*
   c. *NEC Section 250.53(C)*
   d. *NEC Section 314.40(D)*

9. What type of rigid nonmetallic conduit may be installed where exposed to physical damage?
   a. Type DB Schedule 80
   b. Type DB Schedule 40
   c. Type EB
   d. Type 1

10. RMC is commonly used in _____ locations.

11. What type of conduit body is used to provide a junction point for three intersecting conduits?
    a. Type LL
    b. Type LR
    c. Type C
    d. Type T

12. When installing metal to conduit in a wet area, a(n) _____ air space must be provided between it and the supporting surface.

13. Conductors installed in PVC are subject to a(n) _____.
    a. increase in operating temperature
    b. decrease in operating temperature
    c. increase in ampacity
    d. reduced voltage drop

14. Flexible metal conduit must be supported within _____ of each end.

15. When installing a cable system near a metal corrugated sheet roofing deck, a spacing of _____ must be maintained from any point of the roof system.

Leonard "Skip" Layne

Rust Constructors Inc.

How did you choose a career in the electrical field?

I think the electrical field chose me. My father was a contractor for several years before closing shop and accepting a job as an electrical superintendent with the Rust Engineering Company. That happened when I was nine years old. After being moved around the country for the next several years and working as an apprentice on Dad's projects during my college summers, I couldn't think of anything that I would rather do.

Tell us about your apprenticeship experience.

I've never attended a formal apprenticeship school. There are probably several in our group who might say that they suspected this. My electrical education came from field work exposure and several electrical and engineering courses and seminars I've attended over the years.

I'm happy to say that I'm still learning and I've learned a great deal while working on the NCCER Electrical Committee and from my association with the other subject matter experts.

What positions have you held in the industry?

I started as a field apprentice on a tire plant in Madison, Tennessee, in 1959. I've held field positions as an apprentice, journeyman, field engineer, start-up manager, and superintendent. I spent a number of years estimating work, and I established the material control department for another major open-shop contractor several years ago. I managed the project controls group on a nuclear project for another open-shop contractor. I even spent a few years as vice president with an underground utility/treatment plant contractor.

What would you say is the primary factor in achieving success?

Keep learning. Work hard. I've had to work sixteen-hour days on the job site and in the office in order to meet the schedule and incorporate changes. Do what is asked of you and do it well.

What does your current job involve?

My job title says that I'm the Construction Engineering Manager for Rust, but the lack of a definitive job title means that I do whatever the company needs me to do at the time. I qualify the company's electrical licenses in seventeen states where we work.

Recently, Rust volunteered my services to the Gulf Coast Workforce Initiative, a business roundtable initiative to train 20,000 new construction workers for the Gulf Coast area devastated by hurricanes Katrina and Rita.

Do you have any advice for someone just entering the trade?

Get all of the classroom learning you can. Go through all four levels of the Electrical program while working in the field. Ask questions and try to get assigned to as many new and different tasks as you can. All of our larger ABC contractors have excellent supervisory training programs and you need to get into those after your craft training. Be adaptable and keep learning.

Al Hamilton
Willmar Electric Service

How did you become an electrician?

I was in college and met Ed, an electrician who told me about wiring buildings. I went to work for him part time at first and liked it which led to my becoming his apprentice. He was tough taskmaster who cared about his apprentices and he was an excellent communicator and teacher.

How did you get your training?

My training was 100% on the job. Everything was learned "hands on."

I was very fortunate to work for Ed and other experienced electricians who were true craftsmen.

What factor or factors have contributed to your success?

Work ethic, relationships, and reading. A good work ethic has allowed me to overcome mistakes and keep working to learn our craft. I was able to learn about electrical work, business, and life through relationships. I have always looked for successful people to listen and learn from. I discovered that many successful people are happy to share their ideas and that proved to be the key to personal growth. I was told long ago that if I could force myself to read just 10 pages a day that I could read a book a month because the average book is 300 pages. I have done this for many years and I read books on many subjects including the Bible, business, history, biographies, and money and, of course, the National Electrical Code. Through reading you can educate yourself and become a more knowledgeable and interesting person who others will look to for advice.

What does your current job entail?

I work for a leading electrical company, Willmar Electric Service Corp. and my responsibilities are business development and estimating. This includes maintaining relationships with our existing customers and finding new customers to work for. Our estimating team uses estimating software to bid jobs. We do competitive bidding to win jobs and we also estimate for design/build projects and projects that are negotiated with customers.

Any advice for apprentices just beginning their careers?

Work for an organization that shares your values and will recognize and reward your efforts. Don't ever give up! Who you become will depend largely on the people you meet and the books that you read. Find honest, ethical people who have demonstrated success and get to know them. If you have not been reading, start today—make yourself do it. Get involved in helping others by becoming a leader in your company, helping out at church, and passing on what you have learned.

Trade Terms Introduced in This Module

Accessible: Able to be reached, as for service or repair.

Approved: Meeting the requirements of an appropriate regulatory agency.

Bonding wire: A wire used to make a continuous grounding path between equipment and ground.

Cable trays: Rigid structures used to support electrical conductors.

Conduit: A round raceway, similar to pipe, that houses conductors.

Exposed location: Not permanently closed in by the structure or finish of a building; able to be installed or removed without damage to the structure.

Kick: A bend in a piece of conduit, usually less than 45°, made to change the direction of the conduit.

Raceways: Enclosed channels designed expressly for holding wires, cables, or busbars, with additional functions as permitted in the *NEC®*.

Splice: Connection of two or more conductors.

Tap: Intermediate point on a main circuit where another wire is connected to supply electrical current to another circuit.

Trough: A long, narrow box used to house electrical connections that could be exposed to the environment.

Underwriters Laboratories, Inc. (UL): An agency that evaluates and approves electrical components and equipment.

Wireways: Steel troughs designed to carry electrical wire and cable.

Additional Resources

This module presents thorough resources for task training. The following resource material is suggested for further study.

Benfield Conduit Bending Manual, Latest Edition. Overland Park, KS: EC&M Books.
Concrete Fastening Systems. **www.confast.com**.
National Electrical Code® Handbook, Latest Edition. Quincy, MA: National Fire Protection Association.

Figure Credits

Section Review Answer Key

Section 1.0.0

Answer	Section Reference	Objective
1. c	1.1.0	1a
2. b	1.2.0	1b
3. d	1.3.5	1c
4. c	1.4.0	1d
5. d	1.5.4	1e
6. a	1.6.1	1f

Section 2.0.0

Answer	Section Reference	Objective
1. d	2.1.0	2a
2. c	2.2.0	2b
3. d	2.3.0	2c
4. a	2.4.0	2d
5. d	2.5.4	2e
6. a	2.6.1	2f
7. d	2.7.0	2g

Section 3.0.0

Answer	Section Reference	Objective
1. b	3.1.0	3a
2. b	3.2.0	3b
3. c	3.3.1	3c

Section 4.0.0

Answer	Section Reference	Objective
1. b	4.1.0	4a
2. c	4.2.4	4b

Section 5.0.0

Answer	Section Reference	Objective
1. b	5.1.0	5a
2. c	5.2.0	5b

This page is intentionally left blank.

NCCER CURRICULA — USER UPDATE

NCCER makes every effort to keep its textbooks up-to-date and free of technical errors. We appreciate your help in this process. If you find an error, a typographical mistake, or an inaccuracy in NCCER's curricula, please fill out this form (or a photocopy), or complete the online form at **www.nccer.org/olf**. Be sure to include the exact module ID number, page number, a detailed description, and your recommended correction. Your input will be brought to the attention of the Authoring Team. Thank you for your assistance.

Instructors – If you have an idea for improving this textbook, or have found that additional materials were necessary to teach this module effectively, please let us know so that we may present your suggestions to the Authoring Team.

NCCER Product Development and Revision
13614 Progress Blvd., Alachua, FL 32615

Email: curriculum@nccer.org
Online: www.nccer.org/olf

❏ Trainee Guide ❏ Lesson Plans ❏ Exam ❏ PowerPoints Other _____

Craft / Level: _____ Copyright Date: _____

Module ID Number / Title: _____

Section Number(s): _____

Description: _____

_____

_____

_____

Recommended Correction: _____

_____

_____

_____

Your Name: _____

Address: _____

_____

Email: _____ Phone: _____

This page is intentionally left blank.

Conductors and Cables

OVERVIEW

As an electrician, you will be required to select the proper wire and/or cable for a job. You will also be required to pull this wire or cable through conduit runs in order to terminate it. This module discusses conductor types, cable markings, color codes, and ampacity derating. It also describes how to install conductors using fish tape and power conduit fishing systems.

Module 26109-20

Trainees with successful module completions may be eligible for credentialing through the NCCER Registry. To learn more, go to **www.nccer.org** or contact us at 1.888.622.3720. Our website, **www.nccer.org**, has information on the latest product releases and training.

   Your feedback is welcome. You may email your comments to **curriculum@nccer.org**, send general comments and inquiries to **info@nccer.org**, or fill in the User Update form at the back of this module.

This information is general in nature and intended for training purposes only. Actual performance of activities described in this manual requires compliance with all applicable operating, service, maintenance, and safety procedures under the direction of qualified personnel. References in this manual to patented or proprietary devices do not constitute a recommendation of their use.

Objectives

When you have completed this module, you will be able to do the following:

1. Classify conductors by wire size, insulation, and application.
   a. Identify wire sizes.
   b. Determine conductor ampacities.
   c. Identify conductor materials.
   d. Identify conductor insulation.
   e. Identify fixture wiring.
   f. Identify cable types and applications.
   g. Identify instrumentation control wiring.
2. Install conductors in a conduit system.
   a. Install conductors using fish tape.
   b. Install conductors using pulling equipment.

Performance Task

Under the supervision of the instructor, you should be able to do the following:

1. Install conductors in a raceway system.

Trade Terms

Ampacity
Capstan
Fish tape

Mouse
Wire grip

Industry Recognized Credentials

If you are training through an NCCER-accredited sponsor, you may be eligible for credentials from NCCER's Registry. The ID number for this module is 26109-20. Note that this module may have been used in other NCCER curricula and may apply to other level completions. Contact NCCER's Registry at 888.622.3720 or go to **www.nccer.org** for more information.

NOTE

NFPA 70®, *National Electrical Code*® and *NEC*® are registered trademarks of the National Fire Protection Association, Quincy, MA.

Contents

Figures and Tables

This page is intentionally left blank.

1.0.0 CONDUCTORS

Objective

Classify conductors by wire size, insulation, and application.

   a. Identify wire sizes.
   b. Determine conductor ampacities.
   c. Identify conductor materials.
   d. Identify conductor insulation.
   e. Identify fixture wiring.
   f. Identify cable types and applications.
   g. Identify instrumentation control wiring.

Trade Term

Ampacity: The maximum current in amperes a conductor can carry continuously under the conditions of use without exceeding its temperature rating.

The term *conductor* is used in two ways. It is used to describe the current-carrying portion of a wire or cable, and it is used to describe a wire or cable composed of the current-carrying portion and an outer covering (insulation). In this module, the term *conductor*, if not specified otherwise, is used to describe the wire assembly, which includes the insulation and the current-carrying portion of the wire.

Conductors are uniquely identified by size and insulation material. Size refers to the cross-sectional area of the current-carrying portion of the wire. The ampacity is affected by the conductor material and size, insulation, and installation location.

1.1.0 Wire Sizes

Wire sizes are expressed in gauge numbers. The standard system of wire sizes in the United States is the American Wire Gauge (AWG) system.

1.1.1 AWG System

The AWG system uses numbers to identify the different sizes of wire and cable (*Figure 1*). The larger the number, the smaller the cross-sectional area of the wire. The larger the cross-sectional area of the current-carrying portion of a conductor, the higher the amount of current the wire can conduct. The AWG numbers range from 50 to 1; then 0, 00, 000, and 0000 (one aught [1/0], two aught [2/0], three aught [3/0], and four aught [4/0]). Any wire larger than 0000 is identified by its area in thousands of circular mils (kcmil). Wire sizes smaller than No. 18 AWG are usually solid, but may be stranded in some cases. Wire sizes of No. 6 AWG or larger are stranded. For wire sizes larger than No. 16 AWG, the wire size is marked on the insulation (*Figure 2*).

NEC Chapter 9, Table 8 has descriptive information on wire sizes. Again, note that all wires smaller than No. 6 are available as solid or stranded. Wire sizes of No. 6 or larger are shown only as stranded.

1.1.2 Stranding

According to *NEC Chapter 9, Table 8*, wire sizes No. 18 to No. 2 have seven strands; wire sizes No. 1 to No. 4/0 have 19 strands; and wire sizes between 250 kcmil and 500 kcmil have 37 strands. The purpose of stranding is to increase the flexibility of the wire. Terminating solid wire sizes larger than No. 8 in pull boxes, disconnect switches, and panels would not only be very difficult, but might also result in damage to equipment and wire insulation.

Pulling solid wire in conduit around bends could pose a major problem and cause damage to equipment for wire sizes larger than No. 8. The reason for choosing 7, 19, and 37 strands for stranded conductors is that it is necessary to provide a flexible, almost round conductor. In order for a conductor to be flexible, individual strands must not be too large. *Figure 3* shows how these conductors are configured. Aluminum conductors are designed with compact (compressed) stranding (*Figure 4*).

Think About It

Wire Size

Why is wire size a critical factor in a wiring system? Other than load, what other factors may dictate wire size? What can happen when a wire isn't properly sized for the load?

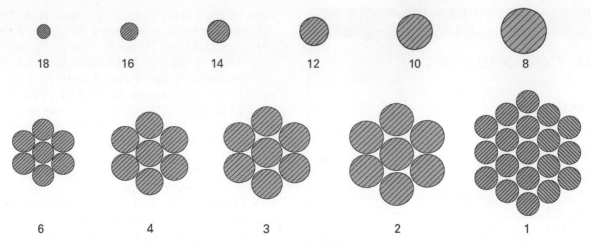

Figure 1 Comparison of wire sizes (enlarged) from No. 18 to No. 1 AWG.

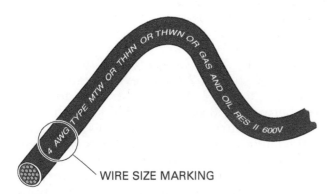

WIRE SIZE MARKING

Figure 2 Wire size marking.

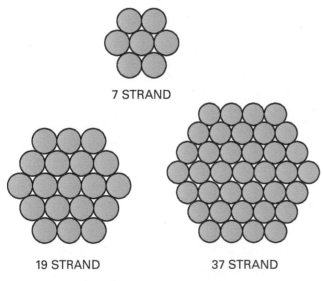

7 STRAND

19 STRAND   37 STRAND

Figure 3 Strand configurations.

1.1.3 Circular Mils

A circular mil is a circle that has a diameter of 1 mil. A mil is 0.001 inch. One kcmil is equal to 1,000 mils. When a wire size is 250 kcmil, the cross-sectional area of the current-carrying portion of the wire is the same as 250,000 circles having a diameter of 0.001 inch. This may seem to be a rather clumsy way of sizing wire at first; however, the alternative would be to size the wire as a function of its cross-sectional area expressed in square inches.

According to *NEC Chapter 9, Table 8*, the cross-sectional area of a 250 kcmil conductor is 0.260 square inch.

If a conductor is to be sized by cross-sectional area, it is much easier to express the wire size in circular mils (or thousands of circular mils) than in square inches.

1.2.0 Conductor Ampacities

Ampacity is the maximum current in amperes a conductor can carry continuously under the conditions of use without exceeding its temperature rating. The ampacity of conductors for given conditions are listed in *NEC Tables 310.16 through 310.21*. *NEC Table 310.16* covers conductors rated 2,000 volts and below where not more than three conductors are installed in a raceway or cable or are directly buried in the earth, based on an ambient temperature of 30°C (86°F).

Figure 4 Aluminum conductors.

NEC Table 310.17 covers both copper conductors and aluminum or copper-clad aluminum conductors 2,000 volts and below where conductors are used as single conductors in free air, based on an ambient temperature of 30°C (86°F).

NEC Tables 310.18 and 310.19 apply to conductors rated at 150°C to 250°C (302°F to 482°F), used either in raceway or cable or as single conductors in free air, based on an ambient temperature of 40°C (104°F).

Example:

Determine the ampacity of a No. 8 Cu (copper) THW conductor installed in a raceway in a 30°C (86°F) environment.

Solution:

50 amps [from *NEC Table 310.16*].

1.3.0 Conductor Materials

The most common conductor material is copper. Copper is used because of its excellent conductivity (low resistance), ease of use, and value. The value of a material as an ingredient for wire is determined by several factors, including conductivity, cost, availability, and workability.

1.3.1 Conductivity

Conductivity is a word that describes the ease (or difficulty) of travel of an electric current through a conductor. If a conductor has a low resistance, it has a high conductivity. Silver is one of the best conductors since it has very low resistance and high conductivity. Copper has high conductivity and a lower price than silver. Aluminum, another material with good conductivity, is also a good choice for conductor material. The conductivity of aluminum is approximately two-thirds that of copper.

1.3.2 Cost

Cost is always an issue that contributes to the selection of a material to be used for a given application. Often, a material that has low cost may be selected as a conductor material even though it has physical properties that are inferior to the more expensive material. Such is the case in the selection of copper over platinum. Here, the cost of platinum is very high, and very little thinking is required to determine that copper is a better choice. The choice between copper and aluminum is often more difficult to make.

1.3.3 Availability

The availability of some material is often a concern when selecting components for a job. As applied to wire, the mining industry often controls the availability of raw materials, which could produce shortages of some material. The availability of a substance such as copper or aluminum affects the price of the finished product (copper or aluminum wire).

1.3.4 Workability

It is a good idea to select a material that requires less expense for tools and is easier to work with. Aluminum conductors are lighter than copper conductors of the same size. They are also much more flexible than copper conductors and, in general, are easier to work with. However, terminating aluminum conductors often requires special tools and treatment of termination surfaces with an anti-oxidation material. Splicing and terminating aluminum conductors often requires a higher degree of training on the part of the electrician than do similar efforts with copper wire. This is partly due to the fact that aluminum expands and contracts with heat more than copper.

1.4.0 Conductor Insulation

The first attempt to insulate wire was made in the early 1800s during the development of the telegraph. This insulation was designed to provide physical protection rather than electrical protection. Electrical insulation was not an important issue because the telegraph operated at low-voltage DC. This early form of insulation was a substance composed of tarred hemp or cotton fiber and shellac and was used primarily for weatherproofing long-distance distribution lines to mines, industrial sites, and railroads.

Some early electrical distribution systems utilized the knob-and-tube technique of installing wire. The wire was often bare and was pulled between and wrapped around ceramic knobs that were affixed to the building structure. When it was necessary to pull wire through structural members, it was pulled through ceramic tubes. The structural member (usually wood) was drilled, the tube was pressed into the hole, and the wire was pulled through the hole in the tube. As dangerous as this may appear, older homes still exist that have knob-and-tube wiring that was installed in the early 1900s and is still operational. Knob-and-tube wiring was revised to use insulated conductors and was in use up to 1957 in some areas.

The grounded or neutral conductor in overhead services may be bare. Furthermore, the concentric grounded conductor in Type SE cable may be bare when used as a service-entrance cable. However, all current-carrying conductors (including the grounded conductor) must be insulated when used on the inside of buildings, or after the first overcurrent protection device.

NEC Table 310.4(A) presents application and construction data on the wide range of 600-volt insulated, individual conductors recognized by the *NEC*®, with the appropriate letter designation used to identify each type of insulated conductor.

1.4.1 Thermoplastic

Thermoplastic is a popular and effective insulation material. The following thermoplastics are widely used:

- *Polyvinyl chloride (PVC)* – The base material used for the manufacture of TW and THW insulation.
- *Polyethylene (PE)* – An excellent weatherproofing material used primarily for insulation of control and communications wiring. It is not used for high-voltage conductors (those exceeding 5,000 volts).
- *Cross-linked polyethylene (XLP)* – An improved PE with superior heat- and moisture-resistant qualities. Used for THHN, THWN, and THHW wiring as well as many high-voltage cables.
- *Nylon* – Primarily used as jacketing material. THHN building wire has an outer coating of nylon.
- *Teflon*® – A high-temperature insulation. Widely used for telephone wiring in a plenum (where other insulated conductors require conduit routing).

1.4.2 Thermoset

Many thermoplastic materials deform when heated. Thermoset materials maintain their form when heated. Thermoset insulations include RHH, RHW, XHH, XHHW, and SIS.

1.4.3 Letter Coding

Conductor insulation as applied to building wire is coded by letters. The letters generally, but not always, indicate the type of insulation or its environmental rating. The types of conductor insulation described in this module will be those indicated at the top of *NEC Table 310.16*. The various insulation designations are shown in *Table 1*.

> **NOTE**
>
> Any conductor used in a wet location (see definition under Location, Wet, in *NEC Article 100*) must be listed for use in wet locations. Any conduit run underground is assumed to be subject to water infiltration and is, therefore, in a wet location.

Table 1 Insulation Coding

Letter	Description
B	Braid
E	Ethylene or Entrance
F	Fluorinated or Feeder
H	Heat-Rated or Flame-Retardant
N	Nylon
P	Propylene
R	Rubber
S	Silicon or Synthetic
T	Thermoplastic
U	Underground
W	Weather-Rated
X	Cross-Linked Polyethylene
Z	Modified Ethylene Tetrafluoroethylene
TW	Weather-Rated Thermoplastic (60°C/140°F)
FEP	Fluorinated Ethylene Propylene
FEPB	Fluorinated Ethylene Propylene with Glass Braid
MI	Mineral Insulation
MTW	Moisture, Heat, and Oil-Resistant Thermoplastic
PFA	Perfluoroalkoxy
RHH	Flame-Retardant Heat-Rated Rubber
RHW	Weather-Rated, Heat-Rated Ruber (75°C/167°F)
SA	Silicon
SIS	Synthetic Heat-Resistant
TBS	Thermoplastic Braided Silicon
TFE	Extended Polytetrafluoroethylene
THHN	Heat-Resistant Thermoplastic
THHW	Moisture and Heat-Resistant Thermoplastic
THW	Moisture and Heat-Resistant Thermoplastic
THWN	Weather-Rated, Heat-Rated Theromplastic with Nylon Cover
UF	Underground Feeder
USE	Underground Service Entrance
XHH	Thermoset
XHHW	Heat-Rated, Flame-Retardant, Weather-Rated Thermoset
ZW	Weather-Rated Modified Ethylene Tetrafluoroethylene

Think About It

Conductor Insulation

What are the functions of conductor insulation? Under what conditions does the *NEC®* allow uninsulated conductors?

Terminating Aluminum Wire

Care must be taken to use listed connectors when terminating aluminum wire. All aluminum connections also require the use of anti-oxidizing compound. Some connectors are precoated with compound; others require the addition of it. Be sure to check the connectors before beginning the installation.

1.4.4 Color Coding

A color code is used to help identify wires by the color of the insulation. This makes it easier to install and properly connect the wires. A typical color code is as follows:

- *Two-conductor cable* – One white or gray wire (neutral), one black wire (hot), and a grounding wire (usually bare)
- *Three-conductor cable* – One white or gray, one black, one red, and a grounding wire
- *Four-conductor cable* – Same as three-conductor cable plus fourth wire (blue)
- *Five-conductor cable* – Same as four-conductor cable plus fifth wire (yellow)

The grounding conductor may be bare, green, or green with a yellow stripe. Power cable color codes are shown in *Figure 5*.

The *NEC®* does not require color coding of ungrounded conductors except where more than one nominal voltage system is present or on branch circuits supplied from direct-current systems [*NEC Section 210.5(C)(2)*]. The ungrounded conductors may be any color with the exception of white, gray, or green; however, it is a good practice to color code conductors as described here. In fact, many construction specifications require color coding. Furthermore, on a four-wire, delta-connected secondary where the midpoint of one phase winding is grounded to supply lighting and similar loads, the phase conductor having the higher voltage to ground must be identified by an outer finish that is orange in color, by tagging, or by other effective means. Such identification must be placed at each point where a connection is made if the grounded conductor is also present (*NEC Section 110.15*). In most cases, orange tape is used at all termination points when such a condition exists.

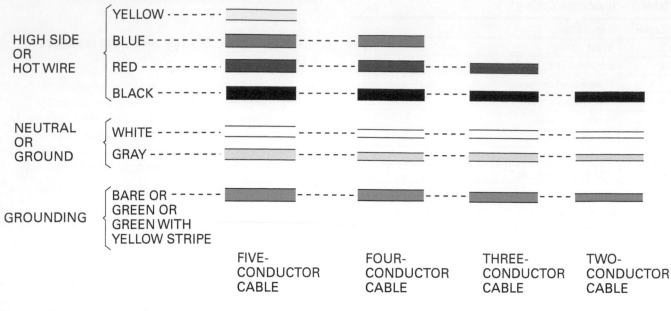

HIGH SIDE OR HOT WIRE
- YELLOW
- BLUE
- RED
- BLACK

NEUTRAL OR GROUND
- WHITE
- GRAY

GROUNDING
- BARE OR GREEN OR GREEN WITH YELLOW STRIPE

FIVE-CONDUCTOR CABLE
FOUR-CONDUCTOR CABLE
THREE-CONDUCTOR CABLE
TWO-CONDUCTOR CABLE

Figure 5 Typical power cable insulation color codes.

Insulation Types

Use *NEC Table 310.4(A)* to identify two types of insulation that are suitable for use in wet locations.

1.4.5 Wire Ratings

A critical factor in selecting conductors is the conductor's maximum operating temperature. Consider how and where a conductor will be used so that the conductor's limiting (maximum) temperature rating will not be exceeded. A conductor's operating temperature is determined by the ambient temperature, current flow in the conductor (including harmonic current), current flow in bundled conductors (which raises the ambient temperature), and how fast or slow heat is dissipated into the surrounding medium (which is affected by the conductor insulation).

Another significant factor to consider when selecting and installing conductors is where the conductor will be terminated. The temperature rating of the termination may limit the allowable ampacity of the conductor.

The amount of current a conductor can safely carry, and thus the maximum safe temperature the conductor can reach, is determined in general by conductor size (diameter in circular mils), the ambient (surrounding) temperature, the number of conductors in a bundle, and where the conductors are installed (raceways, conduit, ducts, underground, etc.).

Conductor selection is based largely on the temperature rating of the wire. This requirement is extremely important and is the basis of safe operation for insulated conductors. As shown in *NEC Table 310.4(A)*, conductors have various temperature ratings. Since *NEC Tables 310.16 and 310.17* are based on an assumed ambient temperature of 30°C (86°F), conductor ampacities are based on the ambient temperature plus the heat (I^2R) produced by the conductor while carrying current. Therefore, the type of insulation used on the conductor is the first consideration in determining the maximum permitted conductor ampacity.

For example, a No. 3/0 THW copper conductor for use in a raceway has an ampacity of 200A according to *NEC Table 310.16*. In a 30°C (86°F) ambient temperature, the conductor is subjected to this temperature when it carries no current. Since a THW-insulated conductor is rated at 75°C (167°F), this leaves a wide margin for increased temperature due to current flow. If the ambient

Think About It

Color Coding Ungrounded Conductors

Although the *NEC®* does not require the use of color-coded ungrounded current-carrying conductors, why might it be a good idea to use them anyway?

temperature exceeds 30°C (86°F), the conductor maximum load-current rating must be reduced proportionally per *NEC Table 310.15(B)(1)* so that the total temperature (ambient plus conductor temperature rise due to current flow) will not exceed the temperature rating of the conductor insulation. For the same reason, the allowable ampacity must be reduced when more than three conductors are contained in a raceway or cable. See *NEC Section 310.15(C)(1)*.

Using the ampacity tables – An important step in circuit design is the selection of the type of conductor to be used (TW, THW, THWN, RHH, THHN, XHHW, etc.). The various types of conductors are covered in *NEC Article 310*, and the ampacities of conductors are given in *NEC Tables 310.16 through 310.21* for the varying conditions of use (e.g., in a raceway, in open air, at normal or higher-than-normal ambient temperatures). Conductors must be used in accordance with the data in these tables and notes.

1.5.0 Fixture Wiring

Fixture wire is used for the interior wiring of fixtures and for wiring fixtures to a power source. Guidelines concerning fixture wire are given in *NEC Article 402*. The list of approved types of fixture wire is given in *NEC Table 402.3*. *Figure 6* shows one example of fixture wire. The wires are composed of insulated conductors with or without an outer jacket. The conductors range in size from No. 18 to No. 10 AWG.

The decision of which fixture wire to use depends primarily upon the operating temperature that is expected within the fixture. Therefore, it is the character of the insulation that will determine the wire selected. For instance, fixture wires insulated with perfluoroalkoxy (PFA) or extruded polytetrafluoroethylene (PTF) would be selected if the operating temperature of the fixture is expected to reach a maximum of 250°C (482°F). This is the highest operating temperature allowed for any fixture wire.

As indicated by *NEC Section 402.3*, fixture wires are suitable for service at 600 volts unless otherwise specified in *NEC Table 402.3*. The allowed ampacities of fixture wire are given in *NEC Table 402.5*.

Although the primary use for fixture wire is the internal wiring of fixtures, several of the wires listed in *NEC Table 402.3* may be used for wiring remote-control, signaling, or power-limited circuits in accordance with *NEC Section 725.49*. Fixture wires may never be used as substitutes for branch circuit conductors.

1.6.0 Cable Types and Applications

Cables are two or more insulated wires and may contain a grounding wire covered by an outer jacket or sheath. Cable is usually classified by the type of covering it has, either nonmetallic (plastic) or metallic, also called armored cable.

Cable may also be classified according to where it can be used (see *NEC Table 400.4*). Because water is such a good conductor of electricity, moisture on conductors can cause power loss or short circuits. For this reason, cables are classified for dry, damp, or wet locations. Cables can also be classified regarding exposure to sunlight and rough use.

All cables are marked to show important properties and uses. Cable markings show the wire size, number of conductors, cable type, and voltage rating. In addition, a marking may be included to signify approved service or applications. This information is printed on nonmetallic cable (*Figure 7*). On metallic cable, marking information is usually included on a tag. The *NEC®* requires that all cables and associated fittings be listed.

1.6.1 Nonmetallic-Sheathed Cable

Nonmetallic-sheathed cable (Type NM and Type NMC) is widely used for branch circuits and feeders in residential and commercial systems. See *Figure 8*. Both types are commonly called Romex®, even though the cable manufacturer only calls Type NM cable Romex®. Guidelines for the use of

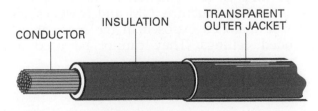

Figure 6 Fixture wire.

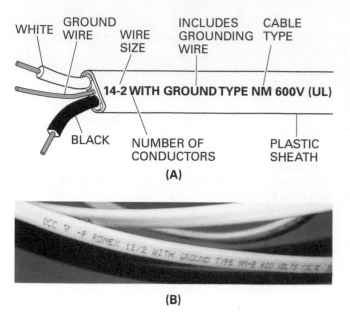

WHITE | **GROUND WIRE** | **WIRE SIZE** | **INCLUDES GROUNDING WIRE** | **CABLE TYPE**

14-2 WITH GROUND TYPE NM 600V (UL)

BLACK | **NUMBER OF CONDUCTORS** | **PLASTIC SHEATH**

(A)

(B)

Figure 7 Nonmetallic cable markings.

nonmetallic-sheathed cable are given in *NEC Article 334*. This cable consists of two or more insulated conductors and one bare conductor enclosed in a nonmetallic sheath. The conductors may be wrapped individually with paper, and the spaces between the conductors may be filled with jute, paper, or other material to protect the conductors and help the cable keep its shape. The sheath covering both Type NM cable and Type NMC cable is flame-retardant and moisture-resistant. The sheath covering Type NMC cable has the additional characteristics of being fungus- and corrosion-resistant.

NEC Article 334 lists the allowed and prohibited uses for Type NM cable and Type NMC cable. Both may be installed in either exposed or concealed work. The primary difference is that Type NM cable is suitable for dry locations only, while Type NMC is permitted for dry, moist, damp, or corrosive locations.

Types NM and NMC cables may be used in one- and two-family dwellings, and in certain multifamily dwellings, depending on the type of construction. See *NEC Section 334.10*.

In general, NM and NMC cable cannot be used in ducts or plenums because toxic gases from burning cable insulation would be spread throughout the structure. NM and NMC cables cannot be installed exposed in the space above suspended ceilings. They cannot be used as

service-entrance cable, embedded in concrete, or in hazardous locations. There are many other requirements for NM and NMC specified in *NEC Article 334*. Before installing this type of cable, be sure you read and understand the applicable sections of the *NEC®*.

1.6.2 Type UF Cable

Guidelines for the use of Type UF (underground feeder and branch circuit) cable are given in *NEC Article 340*. Type UF cable is very similar in appearance, construction, and use to Type NMC cable. The main difference between these two cables is that Type UF cable is suitable for direct burial, whereas Type NMC cable is not.

Some of the permitted uses of Type UF cable are: underground and direct burial; as a single-conductor cable; in wet, dry, or corrosive conditions; as a nonmetallic-sheathed cable; in solar photovoltaic systems; and in cable trays.

Typically, Type UF cable may not be used as service-entrance cable, in commercial garages, in theaters, in hoistways or elevators, or in hazardous locations. Type UF cable cannot generally be embedded in poured cement, concrete, or aggregate, exposed to sunlight (unless designed for that use), or used as an overhead cable. Refer to *NEC Article 340* for specifics on where and when to use Type UF cable.

1.6.3 Type MV Cable

Type MV (medium-voltage) cable is covered in *NEC Article 311*. It consists of one or more insulated conductors encased in an outer jacket. This cable is suitable for use with voltages ranging from 2,001 to 35,000 volts. It may be installed in wet and dry locations and may be buried directly in the earth. See *Figure 9*.

Cable Selection

There are two factors to be considered when determining the type of cable to be used for a specific application: the type of conductor insulation and the cable jacket. Both must be appropriate for the application.

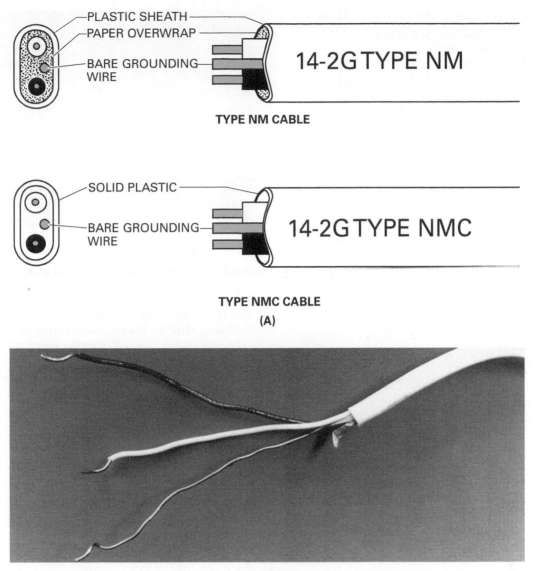

PLASTIC SHEATH
PAPER OVERWRAP
BARE GROUNDING WIRE

14-2G TYPE NM

TYPE NM CABLE

SOLID PLASTIC
BARE GROUNDING WIRE

14-2G TYPE NMC

TYPE NMC CABLE

(A)

(B)

Figure 8 Nonmetallic-sheathed cable.

1.6.4 Type MC Cable

Type MC (metal clad) cable consists of one or more insulated conductors encased in a metal tape or a metallic sheath. *NEC Article 330* covers Type MC cable. Further information can be found in *UL 1569, Standard for Metal Clad Cables*.

MC cable is used in a wide variety of applications, from small instrumentation cable up to medium voltage feeders. The conductors are coated with a thermoset or thermoplastic insulation. Type MC cable can also be a composite of electrical conductors and optical fiber conductors.

Typical markings on the cable include the maximum rated voltage, AWG size (or circular mil area), and insulation type. If the outer covering will not accept markings, the markings will be on

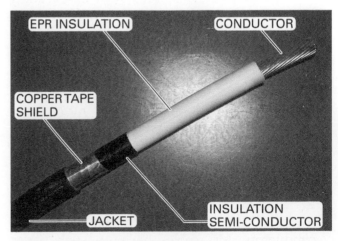

EPR INSULATION   CONDUCTOR

COPPER TAPE SHIELD

JACKET   INSULATION SEMI-CONDUCTOR

Figure 9 Type MV cable.

a tape inside the cable along the entire length of the cable. If on the outside, the markings typically have a 24" (61 cm) spacing.

The three types of MC cable are: interlocked metal tape, corrugated metal tube, and smooth metal tube. Cables with special uses will be marked accordingly. The outer covering may be a nonmetallic jacket over the metal sheath. One type of MC cable is shown in *Figure 10*.

Some of the typical uses for the three types of MC cable are for services, feeders, and branch circuits; for power, lighting, control, and signal circuits; indoors or outdoors; exposed or concealed; direct burial (if identified for that use); in any raceway; and other uses specified in *NEC Article 330*.

Type MC cable may not be used in corrosive or damaging conditions unless the metal cladding protects the conductors, or some other protective material is used. Uses typically not permitted are in areas where the cable is subject to physical damage, direct burial, in concrete, or where subject to caustic materials.

Both armored (AC) and MC cables provide advantages during installation. The flexible metal sheath protects the conductors and allows them to bend around corners without kinking or damage to the conductor. In addition, since the conductors are already protected by the sheathing, there is no need to pull conductors into a raceway, nor is there concern about conductor contact with pipes or other hard surfaces. Other advantages of metal clad cables are their relatively easy installation without the need for wire pullers, fish tapes, or lubricants.

There are some fundamental differences between Types AC and MC cables. The significant differences are:

- AC cable has a maximum of four conductors, plus a grounding conductor, and comes in sizes from 14 AWG to 1 AWG. Conversely, MC cable has no limitations on the number of conductors, and is sized from 18 AWG to 2,000 kcmil.
- AC cable has a bonding strip (16 AWG). This strip is in constant contact with the armor and, with the armor, forms an equipment ground. MC cable has no bonding strip. The MC cladding is not a ground, although it can supplement the ground.
- AC cable uses moisture-resistant and fire-retardant paper wraps on individual conductors. MC cables have no such paper wrap, but do incorporate a polyester tape used on the assembly.

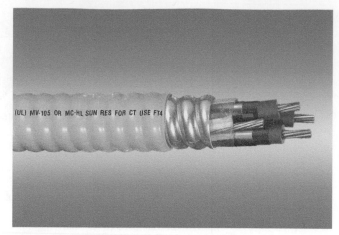

Figure 10 Type of MC cable.

1.6.5 High-Voltage Shielded Cable

Shielding of high-voltage cables protects the conductor assembly against surface discharge or burning due to corona discharge in ionized air, which can be destructive to the insulation and jacketing.

Electrostatic shielding of cables makes use of both nonmetallic and metallic materials (*Figure 11*).

1.6.6 Channel Wire Assemblies

Channel wire assemblies (Type FC) comprise an entire wiring system, which includes the cable, cable supports, splicers, circuit taps, fixture hangers, and fittings (*Figure 12*). Guidelines for the use of this system are given in *NEC Article 322*. Type FC cable is a flat cable assembly with three or four parallel No. 10 special stranded copper conductors. The assembly is installed in an approved U-channel surface metal raceway with one side open. Tap devices can be inserted anywhere along the run. Connections from the tap devices to the flat cable assembly are made by pin-type contacts when the tap devices are fastened in place. The pin-type contacts penetrate the insulation of the cable assembly and contact the multi-stranded conductors in a matched phase sequence. These taps can then be wired to lighting fixtures or power outlets (*Figure 13*).

As indicated in *NEC Section 322.10*, this wiring system is suitable for branch circuits that only supply small appliances and lights. This system is suitable for exposed wiring only and may not be concealed within the building structure. It is ideal for quick branch circuit wiring at field installations.

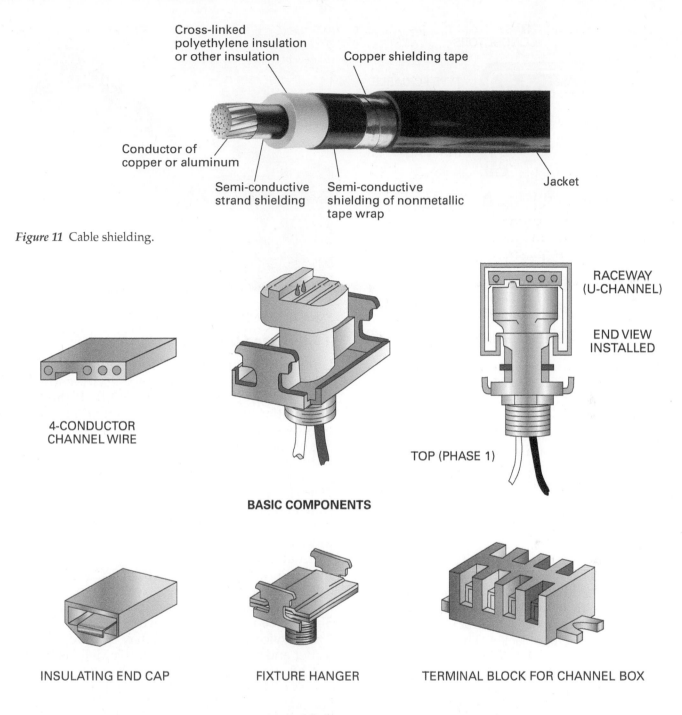

Cross-linked
polyethylene insulation
or other insulation

Copper shielding tape

Conductor of
copper or aluminum

Semi-conductive
strand shielding

Semi-conductive
shielding of nonmetallic
tape wrap

Jacket

Figure 11 Cable shielding.

4-CONDUCTOR
CHANNEL WIRE

RACEWAY
(U-CHANNEL)

END VIEW
INSTALLED

TOP (PHASE 1)

BASIC COMPONENTS

INSULATING END CAP

FIXTURE HANGER

TERMINAL BLOCK FOR CHANNEL BOX

ACCESSORIES

Figure 12 Channel wire components and accessories.

1.6.7 Flat Conductor Cable

Type FCC (flat conductor) cable comprises an entire branch wiring system similar in many respects to Type FC flat conductor assemblies. Guidelines for the use of this system are given in *NEC Article 324*. Type FCC cable consists of three to five flat conductors placed edge-to-edge, separated, and enclosed in a moisture-resistant and flame-retardant insulating assembly. Accesso-

ries include cable connectors, terminators, power source adapters, and receptacles.

This wiring system has been designed to supply floor outlets in office areas and other commercial and institutional interiors. It is meant to be run under carpets so that no floor drilling is required. This system is also suitable for wall mounting. As indicated in *NEC Section 324.42(B)*, telephone and other communications

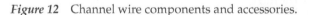

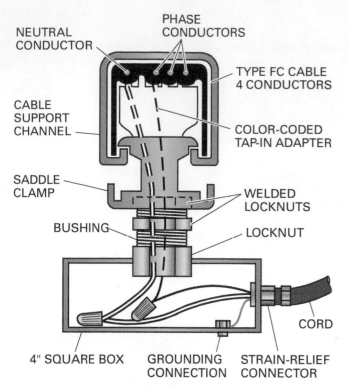

Figure 13 Type FC connection.

circuits may share the same enclosure as Type FCC flat cable. The main advantage of the system is its ease of installation. It is the ideal wiring system for use when remodeling or expanding existing office facilities.

1.6.8 Type TC Cable

Guidelines for the use of Type TC (power and control tray) cable are given in *NEC Article 336*. Type TC cable consists of two or more insulated conductors, with or without associated bare or fully insulated grounding conductors, and covered with a nonmetallic jacket. The cable is listed in conductor sizes No. 14 AWG to 1,000 kcmil copper or No. 12 AWG to 1,000 kcmil aluminum or copper-clad aluminum (*Figure 14*).

As the T in the letter designator indicates, this cable is tray cable. It can be used in cable trays and raceways. It may also be buried directly if the sheathing material is suitable for this use. Type TC cable is also good for use in sunlight when indicated by the cable markings.

Type TC-ER cable identified as JP (joist pull) may be used in one- and two-family dwellings and pulled through framing members.

Figure 14 Type TC cable.

1.6.9 SE and USE Cable

Guidelines for the use of Types SE (service-entrance) and USE (underground service-entrance) cable are given in *NEC Article 338*. If the type designation for the conductor is marked on the outside surface of the cable, the temperature rating of the cable corresponds to the rating of the individual conductor. When this marking does not appear, the temperature rating of the cable is 75°C (167°F). Type SE cable is for aboveground installation only.

For Type SE cable with ungrounded conductors sized 10 AWG and smaller where installed in thermal insulation, the ampacity shall be determined in accordance with the 60°C (140°F) conductor temperature rating.

When used as a service-entrance cable, Type SE must be installed as specified in *NEC Article 230*. Service-entrance cable may also be used as feeder and branch circuit cable. Guidelines for the use of service-entrance cable are given in *NEC Section 338.10*. *Figure 15* shows SE cable with a bare aluminum conductor.

Type USE cable is for underground installation including burial directly in the earth. Type USE cable in sizes No. 4/0 AWG and smaller with all conductors insulated is suitable for all of the underground uses for which Type UF cable is permitted by the *NEC®*.

Type USE cable may consist of either single conductors or a multi-conductor assembly provided with a moisture-resistant covering, but it is not required to have a flame-retardant covering.

Type MC Cable

Metal clad cable (Type MC) is a type of cable that is widely used in both commercial and industrial environments. It is available in many configurations, with or without an outer jacket. Some of the special applications of MC cable include homerun cables, super neutrals, direct burial, and fire alarm cable.

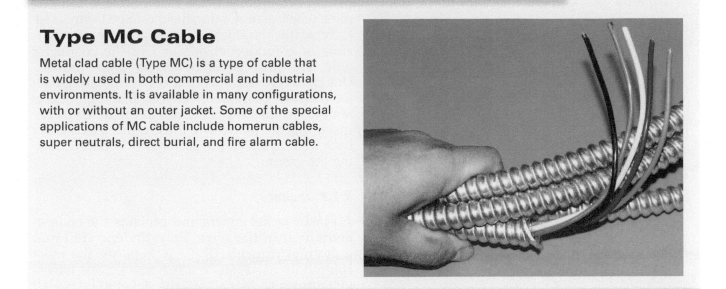

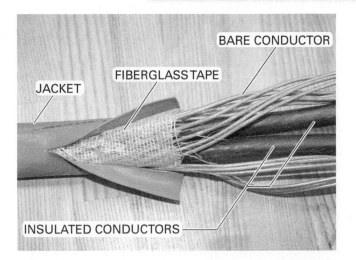

Figure 15 SE cable.

This type of cable may have a bare copper conductor cabled with the assembly. Furthermore, Type USE single, parallel, or cabled conductor assemblies recognized for underground use may have a bare copper concentric conductor applied. These constructions do not require an outer overall covering. Guidelines for the use of Type USE cable are specified in *NEC Article 338*.

When used as a service-entrance cable, Type USE cable must be installed as specified in *NEC Article 230*. Take the time to read *NEC Article 230* to ensure proper installation. Type USE service-entrance cable may also be used as feeder and branch circuit cable.

1.7.0 Instrumentation Control Wiring

Instrumentation control wiring links the field-sensing, controlling, printout, and operating devices that form an electronic instrumentation control system. The style and size of instrumentation control wiring must be matched to a specific job.

Instrumentation control wiring usually has two or more insulated conductor wires. These wires may also have a shield and a ground wire. An outer layer called the jacket protects the wiring (*Figure 16*). Instrumentation conductor wires come in pairs. The number of pairs in a multi-conductor cable depends on the size of the wire used. A multi-pair cable typically has 12, 24, or 36 pairs of conductors. More information on the application of instrumentation control wiring can be found in *NEC Articles 720, 725, and 727*.

1.7.1 Shields

Shields are provided on instrumentation control wiring to protect the electrical signals traveling through the conductors from electrical interference or noise. Shields are usually constructed of aluminum foil bonded to a plastic film (*Figure 17*). If the wiring is not properly shielded, electrical noise may cause erratic or erroneous control signals, false indications, and improper operation of control devices.

1.7.2 Shield Drain

A shield drain is a bare copper wire used in continuous contact with a specified grounding terminal. A shield drain allows connection of all the instruments within a loop to a common grounding

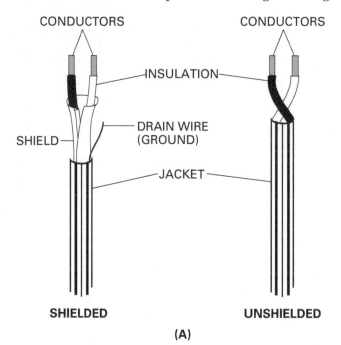

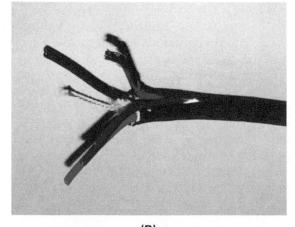

Figure 16 Instrumentation control cable.

point. Always refer to the loop diagram to determine whether or not the shield is to be terminated.

Typically, the shielding in instrumentation circuits is grounded at one end of the conductor only. The purpose of this is to drain induced charges to ground but not allow a circulating path for the flow of induced current. If the ground is not to be connected at the end of the wire you are installing, do not remove the ground wire. Fold it back and tape it to the cable. This is called floating the ground.

1.7.3 Jackets

A plastic jacket covers and protects the components within the wire. Polyethylene (PE) and polyvinyl chloride (PVC) jackets are the most commonly used (*Figure 18*). Some jackets have a nylon rip cord that allows the jacket to be peeled back without the use of a knife or cable cutter. This eliminates nicking of the conductor insulation when preparing for termination.

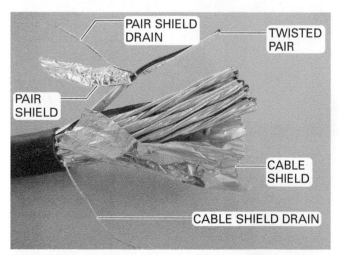

Figure 17 Multi-conductor instrumentation control cable with overall cable shield and individually shielded pairs.

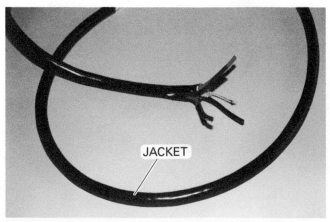

Figure 18 Wire jacket.

1.0.0 Section Review

1. According to *NEC Chapter 9, Table 8*, No. 10 AWG wiring has _____.

   a. 3 strands
   b. 7 strands
   c. 19 strands
   d. 37 strands

2. Using *NEC Table 310.16*, the ampacity of a No. 6 AWG, Type UF copper cable is _____.

   a. 40 amps
   b. 55 amps
   c. 65 amps
   d. 75 amps

3. If a conductor has a low resistance, it _____.

   a. will conduct poorly
   b. cannot be used in service-entrance cable
   c. has a high conductivity
   d. is suitable only for use in low-temperature applications

4. Telephone wiring in a plenum space is likely to use _____.

   a. Teflon® insulation
   b. PVC insulation
   c. PE insulation
   d. XLP insulation

5. Using *NEC Table 402.5*, the allowable ampacity of No. 14 AWG fixture wire is _____.

   a. 6 amps
   b. 8 amps
   c. 17 amps
   d. 23 amps

6. The temperature rating of service-entrance cable when the cable is not marked otherwise is _____.

   a. 25°C (77°F)
   b. 40°C (104°F)
   c. 55°C (131°F)
   d. 75°C (167°F)

7. Which of the following is the most common type of jacket?

   a. Polyethylene
   b. Rubber
   c. Fiberglass
   d. Silicone

SECTION TWO

2.0.0 INSTALLING CONDUCTORS IN CONDUIT SYSTEMS

Objective

Install conductors in a conduit system.

 a. Install conductors using fish tape.
 b. Install conductors using pulling equipment.

Performance Task

 1. Install conductors in a raceway system.

Trade Terms

Capstan: The turning drum of the cable puller on which the rope is wrapped and pulled.

Fish tape: A hand device used to pull a wire through a conduit run.

Mouse: A cylinder of foam rubber that fits inside the conduit and is then propelled by compressed air or vacuumed through the conduit run, pulling a line or tape.

Wire grip: A device used to link pulling rope to cable during a pull.

Conductors are installed in all types of conduit by pulling them through the conduit. This is done by using **fish tape**, pull lines, and pulling equipment.

2.1.0 Fish Tape

Fish tape can be made of flexible steel or nylon and is available in coils of 25' to 240' and in metric sizes between 5 m and 60 m. It should be kept on a reel to avoid twisting. Fish tape has a hook or loop on one end to attach to the conductors to be pulled (*Figure 19*). Broken or damaged fish tape should not be used. To prevent electrical shock, conductive or metallic fish tape should not be used near or in live circuits.

Fish tape is fed through the conduit from its reel. The tape usually enters at one outlet or junction box and is fed through to another outlet or junction box (*Figure 20*).

Sometimes fish tape can get hung up in very long conduit runs. These situations call for a rigid fishing tool known as a rodder. Rodders are available in various sizes and in lengths up to 1,000' (300 m). A typical rodder is shown in *Figure 21*.

2.1.1 Power Conduit Fishing Systems

String lines can be installed by using different types of power systems. The power system is similar to an industrial vacuum cleaner and pulls a string or rope attached to a piston-like plug (sometimes called a **mouse**) through the conduit (*Figure 22*). Once the string emerges at the opposite end, either the conductor or a pull rope is then attached and pulled through the conduit, either manually or with power tools.

The hose connection on these vacuum systems can also be reversed to push the mouse through the conduit. In other words, the system can either suck or blow the mouse through the conduit, depending on which method is best in a given situation. In either case, a fish tape is then attached to the string for retrieving through the conduit.

2.1.2 Connecting Wire to a String Line

Once the string is installed in the conduit run, a fish tape is connected to it and pulled back through the conduit. Conductors are then attached to the hooked end of the fish tape or else connected to a basket grip. In most cases, all required conductors are pulled at one time.

A **wire grip** is used to attach the cable to the pull tape. One type of wire grip used is a basket grip (sometimes called Chinese Fingers). A basket grip is a steel mesh basket that slips over the end of a large wire or cable (*Figure 23*). The fish tape hooks onto the end and the pull on the fish tape tightens the basket over the conductor.

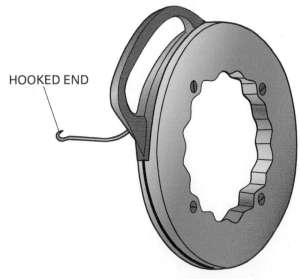

HOOKED END

Figure 19 Fish tape.

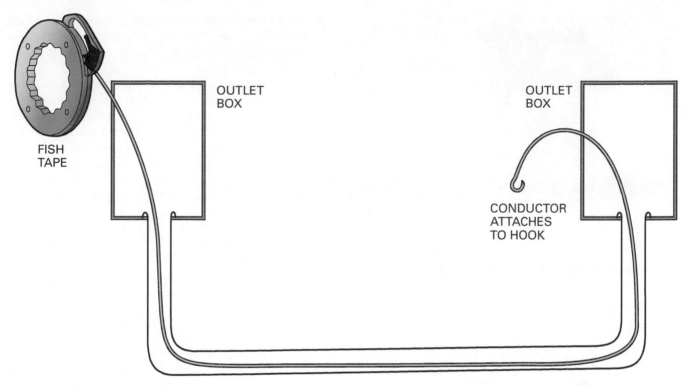

Figure 20 Fish tape installation.

Figure 21 Rodder.

2.1.3 Pull Lines

If a pull is going to be difficult because of bends in the conduit, the size of the conductors, or the length of the pull, a pull line should be used rather than a fish tape.

> **WARNING!**
>
> When using pull lines, exercise extreme caution and never stand in a direct line with the pulling rope. If the rope breaks, the line will whip back with great force. This can result in serious injury or death.

A pull line is usually made of nylon or some other synthetic fiber. It is made with a factory-spliced eye for easy connection to fish tape or conductors.

2.1.4 Safety Precautions

The following are several important safety precautions that will help to reduce the chance of being injured while pulling cable:

- To avoid electrical shock, never use fish tape near or in live circuits.
- Read and understand both the operating and safety instructions for the pull system before pulling cable.
- When moving reels of cable, use mechanical lifts for longer spools. For smaller spools, avoid back strain by using your legs to lift (rather than your back) and asking for help with heavy loads. Also, when manually pulling wire, spread your legs to maintain your balance and do not stretch.
- Be careful to avoid any pinch points in the capstans and sheaves.
- Select a rope that has a pulling load rating greater than the estimated forces required for the pull.

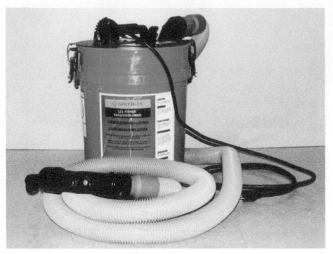

(A) VACUUM BLOWER UNIT

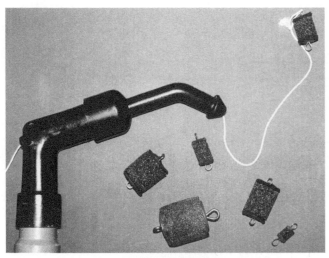

(B) FOAM PLUGS

Figure 22 Power fishing system.

Figure 23 Basket grip.

- Use only low-stretch rope such as multiplex and double-braided polyester for cable pulling. High-stretch ropes store energy much like a stretched rubber band. If there is a failure of the rope, pulling grip, conductors, or any other component in the pulling system, this potential energy will suddenly be unleashed. The whipping action of a rope can cause considerable damage, serious injury, or death.

- Inspect the rope thoroughly before use. Make sure there are no cuts or frays in the rope. Remember, the rope is only as strong as its weakest point.
- When designing the pull, keep the rope confined in conduit wherever possible. Should the rope break or any other part of the pulling system fail, releasing the stored energy in the rope, the confinement in the conduit will work against the whipping action of the rope by playing out much of this energy within the conduit.
- Do not stand in a direct line with the pulling rope. Some equipment is designed so that you may stand to one side for safety.
- Wrap up the pulling rope after use to prevent others from tripping over it.

2.2.0 Pulling Equipment

Many types of pulling equipment are available to help pull conductors through conduit. Pulling equipment can be operated both manually and electrically (*Figure 24*). A manually operated puller is used mainly for smaller pulling jobs where hand pulling is not possible or practical. It is also used in many locations where hand pulling would put an unnecessary strain on the conductors because of the angle of the pull involved. Wire pullers that attach to cordless drills are also available (*Figure 25*). They are used for small pulls in tight spaces, such as junction boxes.

Electrical or hydraulic power pullers are used where long runs, several bends, or large conductors are involved.

The main parts of a power puller are the electric motor, the chain or sprocket drive, the capstan, the sheave, and the pull line.

The pull line is routed over the sheave to ensure a straight pull. The pull line is wrapped around the capstan two or three times to provide a good grip on the capstan. The capstan is driven by the electric motor and does the actual pulling. The pull line is pulled by hand at the same speed at which the capstan is pulling. This eliminates the need for a large spool on the puller to wind the pull line. Ideally, the pull will be completed in a single smooth, steady operation and only stopped when complete, or if there is any doubt about the safety of continuing. Random stops during a pull can cause the conductor to get stuck and make pulling very difficult.

Attachments to power pullers, such as special application sheaves and extensions, are available for most pulling jobs. Follow the manufacturer's instructions for setup and operation of the puller.

2.2.1 Feeding Conductors into Conduit

After the fish tape or pull line is attached to the conductors, it must be pulled back through the conduit. As the fish tape is pulled, the attached conductors must be properly fed into the conduit. Usually, more than one conductor is fed into the conduit during a wire pull. It is important to keep the conductors straight and parallel, and free from kinks, bends, and crossovers. Conductors that are allowed to cross each other will form a bulge and make pulling difficult. This could also damage the conductors.

Spools and rolls of conductors must be set up so that they unwind easily, without kinks and bends.

When several conductors must be fed into the conduit at the same time, a reel cart is used (*Figure 26*). The reel cart will allow the spools to turn freely and help prevent the wires from tangling.

2.2.2 Conductor Lubrication

When conductors are fed into long runs of conduit or conduit with several bends, both the conduit and the wires are lubricated with a compound designed for wire lubrication.

Several types of formulated compounds designed for wire lubrication are available in either dry powder, paste, or gel form. These compounds must be noncorrosive to the insulation material of the conductor and to the conduit itself. The compounds are applied by hand to the conductors as they are fed into the conduit. Battery-operated pumps are also available to lubricate the conduit prior to installing the conductors.

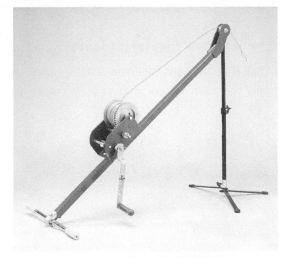

(A) MANUAL WIRE PULLER

(B) POWER PULLER

Figure 24 Pulling equipment.

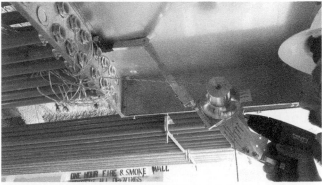

Figure 25 Cordless drill with puller attachment

Fish Tape Selection

Metal fish tape (A) generally comes in longer lengths and is the type used most often. Nylon fish tape (B) generally comes in shorter lengths and is more flexible than metal fish tape.

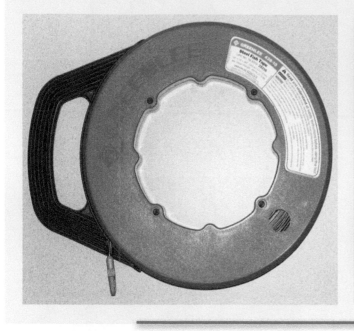

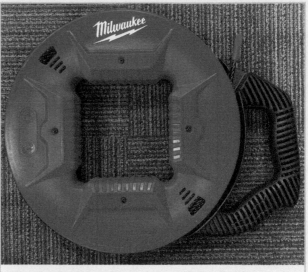

Figure Credit: Photo by Josiha Schuh

2.2.3 Conductor Termination

The amount of free conductor at each junction or outlet box must meet certain *NEC®* specifications. For example, there must be sufficient free conductor so that bends or terminations inside the box, cabinet, or enclosure may be made to a radius as specified in the *NEC®*. The *NEC®* specifies a minimum of 6" (150 mm) for connections made to wiring devices or for splices per *NEC Section 300.14*. Where conductors pass through junction or pull boxes, enough slack should be provided for splices at a later date.

When a box is used as a pull box, the conductors are not necessarily spliced. They may merely enter the pull box via one conduit run and exit via another conduit run. The purpose of a pull box, as the name suggests, is to facilitate pulling conductors on long runs. A junction box, however, is not only used to facilitate pulling conductors through the raceway system, but it also provides an enclosure for splices in the conductors.

Straightening a Bent Fish Tape

To straighten a bent fish tape, drive five 16-penny (16d) nails into a 2 × 4 about 1 inch (25 mm) apart in a straight line. Then wind the fish tape through the nails in a slalom fashion, and pull it through. This will straighten the tape.

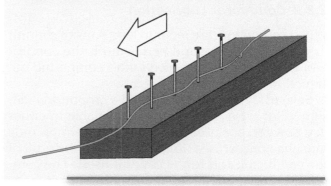

Figure 26 Reel cart.

Think About It

Putting It All Together

Think about the design of conductor installations. How does the location of pull points affect the ease of the pull?

2.0.0 Section Review

1. Which of the following is true when pulling conductors?
   a. Pull lines are typically made of natural cotton.
   b. Rodders can only be used in very short runs.
   c. In multiple-conductor installations, always pull conductors one at a time.
   d. If a pull is expected to be difficult, a pull line should be used rather than a fish tape.

2. Large conductors are likely to require the use of _____.
   a. electrically driven power pullers
   b. hand pulling for additional precision
   c. two or more power pullers
   d. multiple stops during the pulling operation

---

1. The maximum size solid wire that should be terminated in pull boxes and disconnects is _____.
   a. No. 6
   b. No. 8
   c. No. 10
   d. No. 12

2. Compact stranding is used _____.
   a. when installing larger conduit sizes
   b. when decreasing the ampacity of an existing service
   c. with aluminum conductors
   d. in corrosive environments

3. The ampacity ratings of conductors can be found in _____.
   a. *NEC Chapter 1*
   b. *NEC Articles 348 through 352*
   c. *NEC Tables 310.16 through 310.21*
   d. *NEC Chapter 9*

4. To determine the ampacity of copper conductors where conductors are used as a single conductor in free air at ambient temperatures of 30°C (86°F), use _____.
   a. *NEC Table 310.16*
   b. *NEC Table 310.17*
   c. *NEC Table 310.18*
   d. *NEC Table 310.19*

5. Polyethylene (PE) is used primarily for _____.
   a. the manufacture of TW and THW insulation
   b. insulation of control and communication wiring
   c. high-voltage cables
   d. high-temperature insulation

6. A type of thermoset insulation is _____.
   a. RHW
   b. THW
   c. THHW
   d. PVC

7. What letter must be included in the marking of a conductor if it is to be used in a wet, outdoor application?
   a. D
   b. O
   c. I
   d. W

8. What service conductor may be bare, green, or green with a yellow stripe?
   a. The grounding conductor of a multi-conductor cable
   b. The neutral conductor of a multi-conductor cable
   c. The ungrounded conductor of a multi-conductor cable
   d. The high leg of a four-wire, delta-connected secondary

9. What are the colors of insulation on the conductors for a three-conductor NM cable?
   a. One white or gray, one red, and one black
   b. Two white or gray and one red
   c. One white or gray, one red, one black, and a grounding conductor
   d. One green, one white or gray, one blue, and a grounding conductor

10. In some cases, fixture wires may be used for _____.
    a. 240V appliance wiring
    b. signaling wiring
    c. branch circuit wiring
    d. SE cable

11. The difference between AC cable and MC cable is _____.
    a. AC cable has a maximum of four conductors; MC has a maximum of six
    b. AC cable has a bonding strip; the MC ground is its metal cladding
    c. AC cable has fire-retardant paper wraps on individual conductors; MC cable does not
    d. AC cable is sized from 10 AWG to 4/0; MC is sized from 18 AWG to 2,000 kcmil

12. Type USE cable can be used for _____.
    a. aboveground installation only
    b. underground installation within a special PVC pipeline
    c. underground installation including direct burial
    d. indoor applications only

13. When used as a service-entrance cable, Type SE cable must be installed as specified in _____.
    a. *NEC Section 110.12*
    b. *NEC Section 210.5(C)*
    c. *NEC Article 230*
    d. *NEC Section 402.3*

14. To prevent unwanted ground loops, instrumentation cable shielding is _____.
    a. not grounded at both ends of the wire
    b. grounded at both ends of the wire
    c. floated on both ends
    d. always ungrounded

15. Ideally, a pull should be made in _____.
    a. one continuous operation
    b. two or more short segments
    c. four or more short segments
    d. six or more short segments

Trade Terms Quiz

Fill in the blank with the correct term that you learned from your study of this module.

1. A conductor's _____ is the current the conductor can carry continuously without exceeding its temperature rating.

2. On a cable puller, the _____ is the part on which the pulling rope is wrapped and pulled.

3. During a pull, cable may be attached to the pulling rope using a(n) _____.

4. A(n) _____ is a manually operated device that is used to pull a wire through conduit.

5. Composed of foam rubber, a(n) _____ fits inside a piece of conduit and is propelled by compressed air or vacuumed through the conduit run, pulling a line or tape.

Trade Terms

Ampacity
Capstan
Fish tape

Mouse
Wire grip

1. The designation for one thousand circular mils is _____.

2. _____ is the maximum current in amperes a conductor can carry continuously under the conditions of use without exceeding its temperature rating.

3. True or False? *NEC Table 310.16* covers conductors rated up to 2,000V where not more than three conductors are installed in a raceway or cable.

4. True or False? Type NMC is suitable for use as service-entrance cable.

5. Which of the following represents the largest wire size?

    a. 50
    b. 10
    c. 5
    d. 4/0

6. A(n) _____ is a circle that has a diameter of 1 mil.

7. True or False? All current-carrying conductors (including the grounded conductor) must be insulated when used on the inside of buildings.

8. True or False? Polyethylene is the base material used for the manufacture of TW and THW insulation.

9. Polyethylene is used primarily for insulation of _____.

    a. high-voltage conductors
    b. control and communications wiring
    c. THHN wiring
    d. XHHW wiring

10. What does the letter R stand for with regard to insulation coding? _____

11. When conductors are installed where the ambient temperature is above _____, the conductor ampacity must be reduced proportionally with the increase in temperature.

    a. 25°C (77°F)
    b. 30°C (86°F)
    c. 45°C (113°F)
    d. 60°C (140°F)

12. Type UF cable _____.

    a. is commonly called Romex®
    b. is suitable for direct burial
    c. is commonly referred to as BX®
    d. consists of five conductors

13. Type MV cable is suitable for _____.

    a. use in wet locations
    b. use in dry locations
    c. direct burial
    d. all of the above.

14. Some jackets have a(n) _____ that allows the jacket to be peeled back without the use of a knife or cable cutter.

15. A(n) _____ is a piston-like plug that is attached to a string or rope for pulling through conduit using a power fishing system.

Lawrence Joseph (L. J.) LeBlanc, Sr.

Senior Electrician/Instructor
Specialized Services Inc./Baton Rouge
Community College

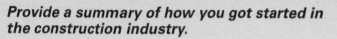

Provide a summary of how you got started in the construction industry.

While attending Louisiana State University majoring in Industrial Technology, I also attended Louisiana State Vo-tech taking a self-taught basic electrical course (with a proctor). I completed a four-year course in three years. I worked in the construction industry during vacations and the summer.

Who inspired you to enter the industry? Why?

My dad Leo LeBlanc, Sr., was an electrician educated at Cohn's Electrical School in Chicago, Illinois in 1941. Observing him gave me a thirst for the electrical industry.

What do you enjoy most about your job?

I love the challenge in fixing something that is not working. Troubleshooting problems in an ever-changing field of electrical work provides constant updating, education, and new challenges for me. My job as an electrical instructor is rewarding because I can see that it makes a difference in student's lives by affording them a career in a dynamic field.

Do you think training and education are important in construction? If so, why?

Education and training make a big difference! Watching something that works is one thing, but to know how it works enables one to fix it. A new code every three years makes it a must to stay abreast of changes. Not knowing the code can cost thousands of dollars.

How important are NCCER credentials to your career?

NCCER credentials are important because you can use them to enhance your resume. They are a part of your education and experience which helps to determine your pay scale. NCCER credentials prove who and what you are to anyone looking to be employed in the electrical industry.

How has training/construction impacted your life and your career?

In every career, education and experience are the core. How would you feel about your surgeon having an education in a meat market? My training and experience in construction have been life-giving and gratifying. My career has given me various opportunities to meet and work with people in the field and bring the experience to the classroom.

Would you suggest construction as a career to others? If so, why?

The answer is yes, because there is a world of opportunities waiting to be explored and learned. It is a wonderful way to make a good living. I believe the construction industry is a field where you determine your own fate. We can be taught amps, ohms, volts, etc. We can't teach integrity, morality, and work ethics. If a person brings the last three qualities and is willing to learn, their career can grow way beyond their expectations.

How do you define craftsmanship?

A person signs their name with their work. The job they do represents not only them, but the company they work for. See *NEC Section 110.12.*

Trade Terms Introduced in This Module

Ampacity: The maximum current in amperes a conductor can carry continuously under the conditions of use without exceeding its temperature rating.

Capstan: The turning drum of the cable puller on which the rope is wrapped and pulled.

Fish tape: A hand device used to pull a wire through a conduit run.

Mouse: A cylinder of foam rubber that fits inside the conduit and is then propelled by compressed air or vacuumed through the conduit run, pulling a line or tape.

Wire grip: A device used to link pulling rope to cable during a pull.

Additional Resources

This module presents thorough resources for task training. The following reference material is recommended for further study.

National Electrical Code® *Handbook, Latest Edition.* Quincy, MA: National Fire Protection Association.

Figure Credits

Greenlee / A Textron Company, Module Opener, Figures 21, 24
Tim Dean, Figures 4, 15
Jim Mitchem, Figure 8, 9
The Okonite Company, Figure 10
General Cable, Figure 11
Mike Powers, Figure 25, 26

Section Review Answer Key

SECTION 1.0.0

Answer	Section Reference	Objective
1. b	1.1.2; *NEC Chapter 9, Table 8*	1a
2. b	1.2.0; *NEC Table 310.16*	1b
3. c	1.3.1	1c
4. a	1.4.1	1d
5. c	1.5.0; *NEC Table 402.5*	1e
6. d	1.6.9	1f
7. a	1.7.3	1g

SECTION 2.0.0

Answer	Section Reference	Objective
1. d	2.1.3	2a
2. a	2.2.0	2b

This page is intentionally left blank.

NCCER CURRICULA — USER UPDATE

NCCER makes every effort to keep its textbooks up-to-date and free of technical errors. We appreciate your help in this process. If you find an error, a typographical mistake, or an inaccuracy in NCCER's curricula, please fill out this form (or a photocopy), or complete the online form at **www.nccer.org/olf**. Be sure to include the exact module ID number, page number, a detailed description, and your recommended correction. Your input will be brought to the attention of the Authoring Team. Thank you for your assistance.

Instructors – If you have an idea for improving this textbook, or have found that additional materials were necessary to teach this module effectively, please let us know so that we may present your suggestions to the Authoring Team.

NCCER Product Development and Revision

13614 Progress Blvd., Alachua, FL 32615

Email: curriculum@nccer.org
Online: www.nccer.org/olf

❑ Trainee Guide   ❑ Lesson Plans   ❑ Exam   ❑ PowerPoints   Other _____

Craft / Level: _____ Copyright Date: _____

Module ID Number / Title: _____

Section Number(s): _____

Description: _____

_____

_____

_____

_____

Recommended Correction: _____

_____

_____

_____

_____

Your Name: _____

Address: _____

_____

Email: _____ Phone: _____

This page is intentionally left blank.

Basic Electrical Construction Documents

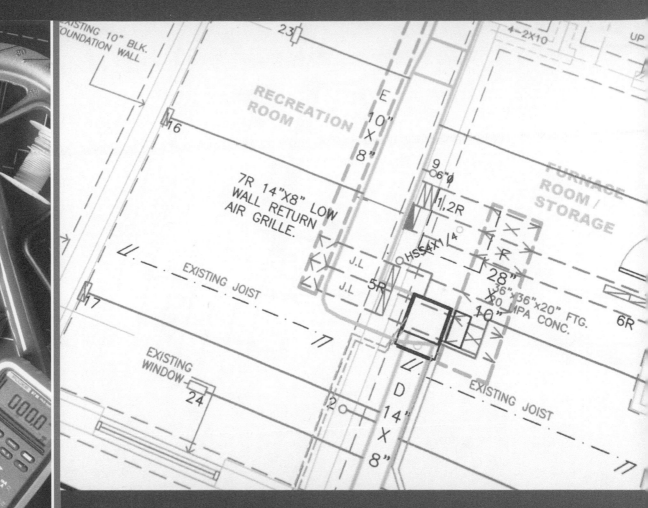

OVERVIEW

In all large construction projects and in many of the smaller ones, a professional is commissioned to prepare the construction documents—the complete working drawings and specifications for the project. The drawings include site plans, floor plans, detail drawings, lighting plans, power riser diagrams, and various schedules. The construction documents, along with the contractor's bid, are legal documents binding on the various parties in the construction project and must be maintained and kept correct. This module describes how to interpret electrical drawings, including the use of architect's and engineer's scales.

Module 26110-20

26110-20 V10.0

26110-20
BASIC ELECTRICAL CONSTRUCTION DOCUMENTS

Objectives

When you have completed this module, you will be able to do the following:

1. Interpret construction drawings.
   a. Locate the information found in drawing blocks.
   b. Identify the information found on different types of drawings.
   c. Interpret drafting lines.
2. Measure items on scale drawings.
   a. Use an architect's scale.
   b. Use an engineer's scale.
   c. Use a metric scale.
3. Apply the information on electrical drawings.
   a. Interpret electrical symbols.
   b. Check a set of residential electrical drawings.
   c. Locate information within a commercial plan set.
   d. Read schedules, block diagrams, and schematic diagrams.
4. Integrate specifications with electrical drawings.
   a. Select relevant information from written specifications.
   b. Compare two formats for written specifications.
   c. Identify document changes and the need for them.

Performance Tasks

Under the supervision of the instructor, you should be able to do the following:

1. Using an architect's scale, state the actual dimensions of a given drawing component.
2. Make a material takeoff of the luminaires specified in the provided drawing. The takeoff requires that all luminaires be counted, and where applicable, the total number of lamps for each luminaire type must be calculated. (Fill these in on the provided Performance Profile Task 2 Worksheet.)

Trade Terms

Architect's scale
Architectural drawings
As-built drawings
Block diagram
Blueprint
Change order
Contour lines
Detail drawing
Dimensions
Electrical drawing
Elevation view
Engineer's scale

Floor plan
One-line diagram
Plan
Power-riser diagram
Request for information
Scale
Schedule
Schematic diagram
Sectional view
Shop drawing
Site plan
Written specifications

Industry Recognized Credentials

If you are training through an NCCER-accredited sponsor, you may be eligible for credentials from NCCER's Registry. The ID number for this module is 26110-20. Note that this module may have been used in other NCCER curricula and may apply to other level completions. Contact NCCER's Registry at 888.622.3720 or go to **www.nccer.org** for more information.

Contents

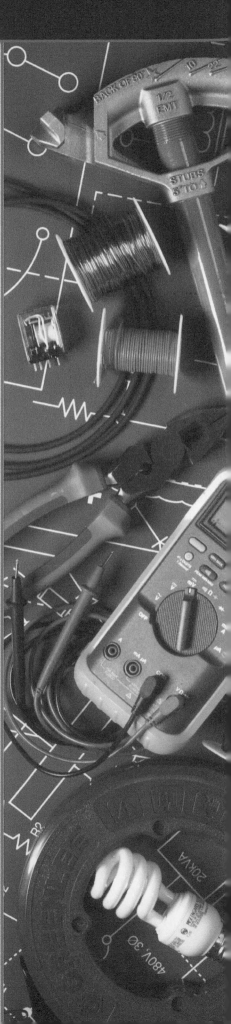

Figures and Tables

This page is intentionally left blank.

SECTION ONE

1.0.0 TYPES OF CONSTRUCTION DRAWINGS

Objective

Interpret construction drawings.
  a. Locate the information found in drawing blocks.
  b. Identify the information found on different types of drawings.
  c. Interpret drafting lines.

Trade Terms

Architectural drawings: Working drawings consisting of site plans, floor plans, elevations, sectional views, details, and other information necessary for the construction of a building.

Blueprint: An exact copy or reproduction of an original drawing.

Contour lines: Curving lines on a site plan, following a given elevation. The space between contour lines tells the slope of the property, such as steep when they are close together or fairly level when they are widely separated.

Detail drawing: An enlarged, detailed view taken from an area of a drawing and shown in a separate view.

Dimensions: Sizes or measurements printed on a drawing.

Electrical drawing: A means of conveying a large amount of exact, detailed information in an abbreviated form understood by electricians. Consists of blocks, lines, symbols, dimensions, and notations to accurately convey the designs to those who install the electrical system on a job.

Elevation view: An architectural drawing showing height and width, but not depth; usually showing the front, rear, or sides of a building or object.

Floor plan: A drawing of a building as if a horizontal cut were made through a building at about window level, and the top portion removed. The floor plan is what would appear if the remaining structure were viewed from above.

Plan: A drawing made as though the viewer were looking straight down on an object from above.

Scale: On a drawing, the size relationship between an object's actual size and the size it is drawn. Scale also refers to the measuring tool used to determine this relationship.

Schedule: A systematic method of presenting equipment lists on a drawing in tabular form.

Sectional view: A cutaway drawing that shows the inside of an object or building.

Site plan: A drawing showing the location of a building or buildings on the building site. Such drawings frequently show topographical lines, electrical and communication lines, water and sewer lines, sidewalks, driveways, and similar information.

In all large construction projects and in many of the smaller ones, an architect is commissioned to prepare complete working drawings and specifications for the project. The set of drawings usually includes a site plan, a floor plan, an elevation drawing, and a sectional view. Depending on the complexity of the installation, a detail drawing or multiple drawings may also be required.

For larger projects, the architect usually hires consulting engineers to prepare structural, electrical, and various types of mechanical drawings, including pipefitting, instrumentation, plumbing, and heating, ventilating, and air conditioning (HVAC) drawings.

1.1.0 Drawing Blocks

Although a strong effort has been made to standardize drawing practices in the building construction industry, a drawing or blueprint prepared by one architectural or engineering firm will rarely be identical to a drawing prepared by another firm. Similarities, however, will exist between most sets of drawings. With a little experience, you should have no trouble interpreting any set of drawings that might be encountered.

Every drawing has an area reserved for one or more blocks that tell the origin, identification, content, use, scale, and history of the drawing. This area is usually located in the drawing's bottom right corner but may be along the right-hand edge.

1.1.1 Title Blocks

Each drawing sheet has border lines framing the overall drawing and one or more title blocks, as shown in *Figure 1*. The type and size of title blocks varies with each firm preparing the drawings. In addition, some drawing sheets will also contain a revision block near the title block, and perhaps an approval block. This information is normally found on each drawing sheet, regardless of the type of project or the information contained on the sheet.

The architect's title block for a drawing is usually a box in the lower right-hand corner of the drawing sheet; the size of the block varies with the size of the drawing and with the information required (*Figure 2*).

In general, the title block of an **electrical drawing** should contain the following information:

- Name of the project
- Address of the project
- Name of the owner or client
- Name of the architectural firm
- Date of the drawing
- Scale(s)
- Initials of the drafter, checker, and designer, with dates under each
- Job number
- Drawing number
- General description of the drawing

Often, the consulting engineering firm will also be listed, which means that an additional title block will be applied to the drawing, usually next to the architect's title block. *Figure 3* shows completed architectural and engineering title blocks as they appear on an actual drawing.

Check the Title Block

Always refer to the title block for scale and be aware that it often changes from drawing to drawing within the same set. Also check for the current revision, and when you replace a drawing with a new revision, be sure to remove and file the older version.

1.1.2 Drawing Number Block

To avoid confusion, every job's drawing has a unique drawing number, generally located in a box within the architect's title block. This drawing number usually follows a standardized form, of a letter and a dash (or two letters) and one to three (sometimes more) numbers. The first letter gives the discipline or craft and the second (if used) gives additional subdivisions within the

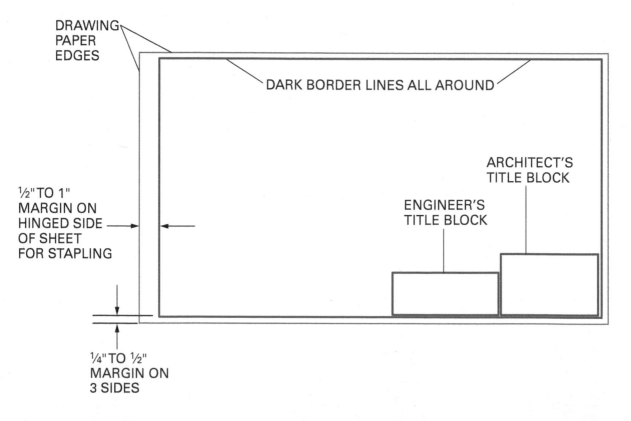

Figure 1 Typical drawing layout.

discipline. When used, the first number is usually the type of sheet or the floor of a multi-story building. The following numbers are in sequential order (often with numbers being skipped to allow for changes or additions later). Over 20 disciplines or crafts have specific letter codes. Some of these as well as some of the sheet types (if used) are given in *Table 1*. As an electrician, you will usually be working with drawings starting with an E, but you may often have to look at sheets for other crafts, and topics (such as architectural or the overall site layout) to ensure that your work is going where it needs to go and does not conflict with another craft's work.

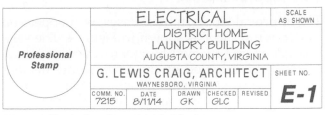

Figure 2 Typical architect's title block.

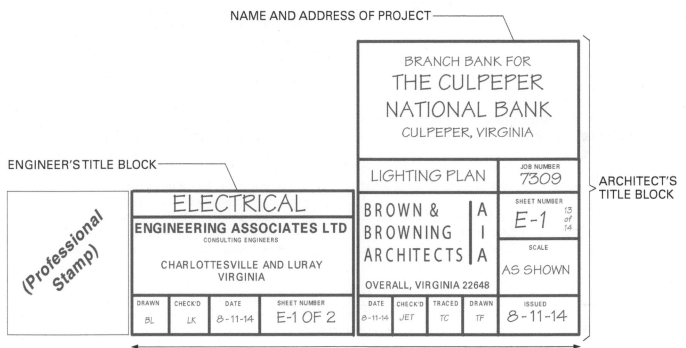

Figure 3 Title blocks.

Table 1 Commonly Used Drawing Identifications

First Letter (in usual order)	First Number
(blank) – Cover Sheet	0 – General usage
C – Civil	1 – Plans
S – Structural	2 – Elevations
A – Architectural	3 – Sections
I – Interiors	4 – Large Scale (but not details)
P – Plumbing	5 – Details
M – Mechanical	6 – Schedules and Diagrams
E – Electrical	
Z – Shop Drawings	

1.1.3 Approval Block

The approval block, in most cases, will appear on the drawing sheet as shown in *Figure 4*. The various types of approval blocks (drawn, checked, etc.) will be initialed by the appropriate personnel. This type of approval block is usually part of the title block and appears on each drawing sheet.

On some projects, authorized signatures are required before certain systems may be installed, or even before the project begins. An approval block such as the one shown in *Figure 5* indicates that all required personnel have checked the drawings for accuracy, and that the set meets with everyone's approval. Such an approval block usually appears on the front sheet of the blueprint set and may include:

- *Professional stamp* – Registered seal of approval by the licensed architect or consulting engineer.
- *Design supervisor* – Signature of the person who is overseeing the design.
- *Drawn (by)* – Signature or initials of the person who drafted the drawing and the date it was completed.
- *Checked (by)* – Signature or initials of the person who reviewed the drawing and the date of approval.
- *Approved* – Signature or initials of the architect/engineer and the date of the approval.
- *Owner's approval* – Signature of the project owner or the owner's representative along with the date signed.

COMM. NO. 7215	DATE 8/11/14	DRAWN GK	CHECKED GLC	REVISED

Figure 4 Typical approval block.

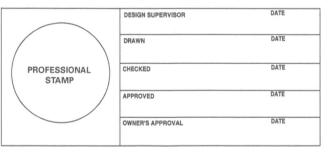

Figure 5 Alternate approval block.

Applying Your Skills

Once you learn how to interpret construction drawings, you can apply that knowledge to any type of construction, from simple residential applications to large industrial complexes.

1.1.4 Revision Block

Sometimes electrical drawings will have to be partially redrawn or modified during the construction of a project. It is extremely important that such modifications are noted and dated on the drawings to ensure that the workers have an up-to-date set of drawings to work from. In some situations, sufficient space is left near the title block for dates and descriptions of revisions, as shown in *Figure 6*. In other cases, a revision block is provided (again, near the title block), as shown in *Figure 7*. The area on the drawing where the revision has been made will often be outlined with a cloud shape.

> **NOTE**
>
> Architects, engineers, designers, and drafters have their own methods of showing revisions, so expect to find deviations from those shown here.

> **CAUTION**
>
> When a set of electrical working drawings has been revised, always make certain that the most up-to-date set is used for all future layout work. Either destroy the old, obsolete set of drawings or else clearly mark on the affected sheets, *Obsolete Drawing—Do Not Use*. Also, when working with a set of working drawings and written specifications for the first time, thoroughly check each page to see if any revisions or modifications have been made to the originals. Doing so can save much time and expense to all concerned with the project.

1.1.5 Legend Block

The legend block can be called by other names such as "Symbols Used" or "Symbols List" (as seen later in *Figure 32*). It can be placed in any available empty space on a drawing, and is in the form of a list, with the symbols or lines in a column on the left and their meaning in a matched column on the right. Typically, the legend block only explains symbols or lines used on the drawing on which it is placed. For a drawing only containing tables such as a panel schedule, there is no need for a legend block.

1.2.0 Drawing Types

A set of electrical drawings will include many types. Although we live in a three-dimensional world, almost all drawings display their information in only two dimensions. A **plan** is a drawing that uses the width and length **dimensions**

Professional Stamp	ELECTRICAL			
	DISTRICT HOME			
	LAUNDRY BUILDING			
	AUGUSTA COUNTY, VIRGINIA			
	G. LEWIS CRAIG, ARCHITECT			SHEET NO.
	WAYNESBORO, VIRGINIA			E-1
	COMM. NO. 7215	DATE 8/11/14	DRAWN GK	CHECKED GLC	REVISED TF

Figure 6 One method of showing revisions on working drawings.

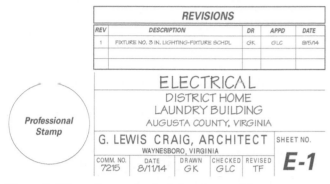

REVISIONS				
REV	DESCRIPTION	DR	APPD	DATE
1	FIXTURE NO. 3 IN. LIGHTING-FIXTURE SCHDL	GK	GLC	9/5/14

Professional Stamp	ELECTRICAL			
	DISTRICT HOME			
	LAUNDRY BUILDING			
	AUGUSTA COUNTY, VIRGINIA			
	G. LEWIS CRAIG, ARCHITECT			SHEET NO.
	WAYNESBORO, VIRGINIA			E-1
	COMM. NO. 7215	DATE 8/11/14	DRAWN GK	CHECKED GLC	REVISED TF

Figure 7 Alternative method of showing revisions on working drawings.

(but not height), depicting the objects in terms of their position on (or above) a specified surface. Depending on their intent and content, they are almost always site plans or floor plans.

In contrast to these, views are a type of drawing using height as one of its two dimensions and either a width or length as the other. These typically are elevations or sections. An additional type of drawing (commonly seen as a small detailed portion of a larger sheet) is a detail drawing. Depending on the drawing the detail is connected to, it can be a plan or a view.

Occasionally, a drawing set may include a projection of a three-dimensional object onto the drawing's two-dimensional surface, to show the relationships between the different parts of the object. Such a drawing is often called a *perspective drawing*.

1.2.1 Site Plans

A site plan (or plot plan) indicates the location of the building on the property, along with related important information including:

- The property boundaries
- The new contour lines (after grading)
- New and existing roadways
- All utility lines
- The drawing scale
- Other pertinent details

Descriptive notes may also be found on the site plan identifying the adjacent properties, the land surveyor, and the date of the survey. A legend or symbol list is also included so that anyone who must work with the site plan can readily read the information. See *Figure 8*.

It is usually the owner's responsibility to furnish the architect/engineer with property and topographic surveys, which are made by a certified land surveyor or civil engineer. These surveys show:

- All property lines
- Existing public utilities and their location on or near the property (e.g., electrical lines, sanitary sewer lines, gas lines, water-supply lines, storm sewers, manholes, telephone lines, etc.)

A land surveyor does the property survey using information obtained from a deed description of the property. A property survey shows only the property lines and their lengths, as if the property were perfectly flat.

The topographic survey shows both the property lines and the physical characteristics of the land by using contour lines, notes, and symbols. The physical characteristics may include:

- The direction of the land slope
- Whether the land is flat, hilly, wooded, swampy, high, or low, and other features of its physical nature

All of this information is necessary so that the architect can properly design a building to fit the property. The electrical engineer also needs this information to locate existing electrical utilities, route the new service to the building, and provide outdoor lighting and circuits.

1.2.2 Floor Plans

A floor plan shows the rooms, walls, doors, and windows for each floor or level. The floor plan of a building is drawn as if a horizontal cut were made through the building—at about window height—and then the top portion removed to reveal the bottom part (see *Figure 9*). If the cut is through the second-floor windows, the plan would be called the *second-floor plan* or *upper level*. See *Figure 10*.

A reflected ceiling plan is a mirror image of the floor plan. This plan shows the objects on the ceiling (such as architectural details, luminaires, etc.), as well as the architectural locations of walls, windows, and doors. A typical reflected ceiling plan is shown in *Figure 11*.

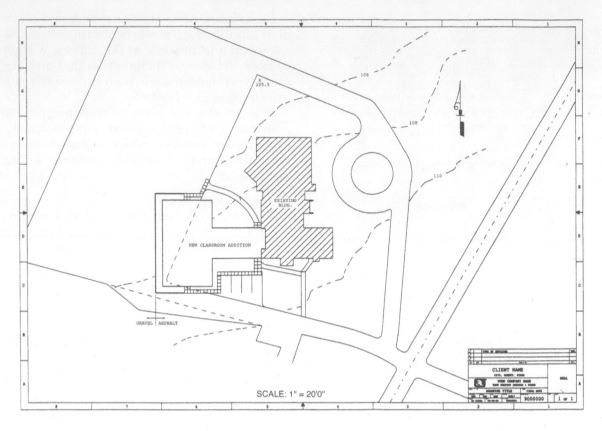

SCALE: 1" = 20'0"

Figure 8 Typical site plan.

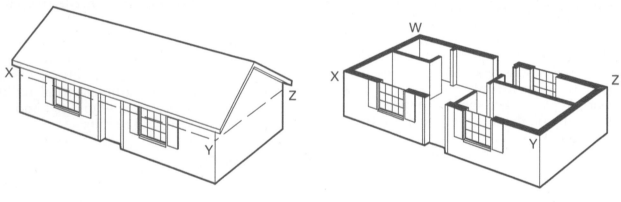

PERSPECTIVE VIEW SHOWING SECTION CUTS

TOP HALF OF SECTION REMOVED

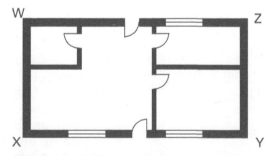

RESULTING FLOOR PLAN IS WHAT THE REMAINING
STRUCTURE LOOKS LIKE WHEN VIEWED FROM ABOVE

Figure 9 Principles of floor plan layout.

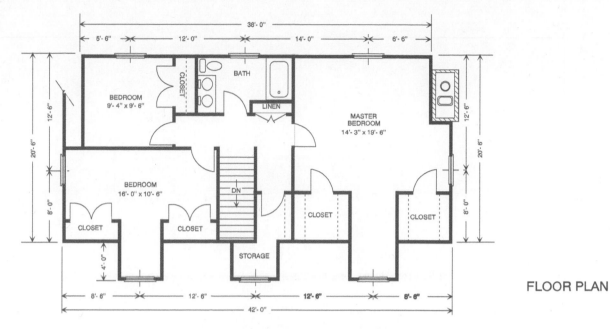

FLOOR PLAN

UPPER LEVEL

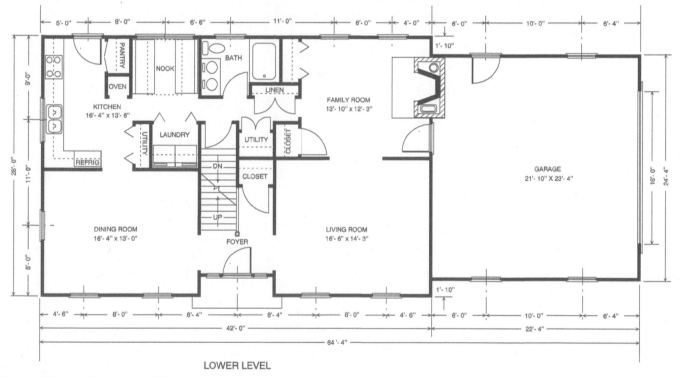

Figure 10 Floor plans of a building.

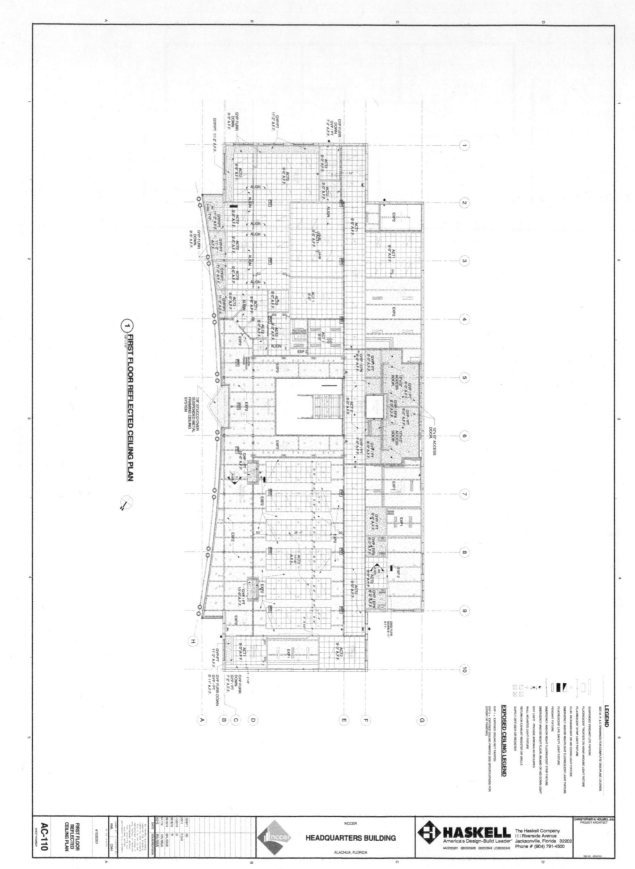

Figure 11 Reflected ceiling plan of a building.

Using a Drawing Set

Always treat a drawing set with care. It is best to keep two sets, one for the office and one for field use. Be sure to use the most current revision. After you use a sheet from a set of drawings, if the sheet is folded, refold the sheet with the title block facing up.

1.2.3 Elevation Views

An **elevation view** shows the exterior face of the building. Remember, an elevation is an outline of an object that shows heights and the length or width of a particular side, but not depth. Refer to *Figure 12* and *Figure 13*.

> **NOTE**
>
> Elevation views may also show the heights of windows, doors, and porches, the pitch of roofs, locations of outside lights, etc., because all of these measurements cannot be shown conveniently on floor plans.

1.2.4 Sectional Views

A section or sectional view (*Figure 14*) is a cut-away view that shows otherwise hidden parts of a structure. Sectional views are used to illustrate floor levels and details of footings, foundations, walls, floors, ceilings, and roof construction. A longitudinal section is taken lengthwise, while a cross section is usually taken straight across the width of an object. Sometimes, a section is taken along a zigzag line to show important parts of an object.

The place on the plan or elevation showing where the imaginary cut has been made is indicated by a section line, which is usually a dashed line. The special symbol used at the ends of the section line is made of a circle and an overlapping triangle. The triangle points in the direction you are looking in the sectional view. The circle identifies the section (such as Section A-A, or Section B-B) and the drawing sheet the section is located on (if not the on the current sheet). On the top of *Figure 14*, sectional view A-A is located on sheet A-5 and is looking toward the right. If the section is on the same sheet as the drawing, the symbol at the ends of the section lines is often abbreviated to a small arrow with an adjacent letter.

FRONT ELEVATION

REAR ELEVATION

Figure 12 Front and rear elevation views.

Wall sections are nearly always made vertically so that the cut edge is exposed from top to bottom. Wall sections show how each wall is constructed and usually indicate the material to be used. This information is important when determining wiring methods, such as box mounting depth.

1.2.5 Details

A detail is an enlargement of a portion of the main drawing. On the main drawing this portion is often bounded by a circle, or other closed line shape. Close to this boundary line is a special symbol that refers to the associated detail. You can see this on the left side of *Figure 14*, with detail B located on sheet A-5. For a detail there is no change in direction of view, so the special symbol does not have an arrow.

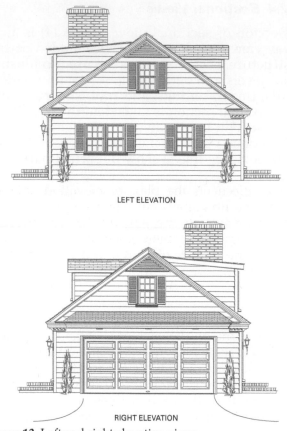

Figure 13 Left and right elevation views.

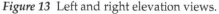

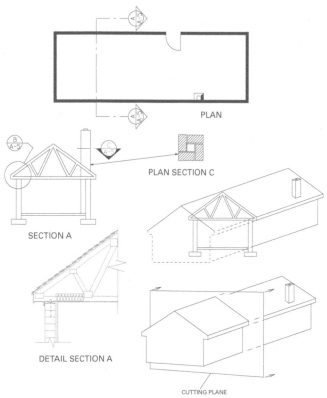

Figure 14 Sectional views.

Then and Now

For hundreds of years drawings were made by hand or on drafting tables with pencil or ink on paper or vellum. They were then copied onto specially treated paper that was developed in a machine with ammonia to create an exact copy with a blue background and white lines wherever the pencil or ink had been on the original. These copies were called *blueprints* and the term is still used today. Modern blueprints usually have black or colored lines against a white background and are generated using computer-aided design (CAD) programs. CAD programs allow three-dimensional modeling, use of layers for different crafts or purposes, and many other enhanced features that were impossible in the past.

Think About It
Using All of the Drawings

Look back over the information on floor plans, elevations, and sections. What kinds of information would an electrician get from each of these drawings? What could a sectional view show that a floor plan could not?

1.3.0 Drafting Lines

You will encounter many types of drafting lines. To specify the meaning of each type of line, contrasting lines are made by varying the width of the lines or breaking the lines in a uniform way.

Figure 15 shows common lines used on **architectural drawings**. Architects and engineers have endeavored to create a common standard for the past century, but their goal has yet to be reached. Therefore, you will find variations in lines and symbols from drawing to drawing, so always consult the legend or symbol list when referring to any drawing. Also, carefully inspect each drawing to ensure that line types are used consistently.

Interpreting Electrical Drawings

A good example of when an electrician must interpret the drawings is when wiring a log cabin. The drawings will show the receptacle and switch locations in branch circuits as usual, but the electrician must figure out how to route wires and install boxes where there is no hollow wall and sometimes exposed beams instead of a ceiling.

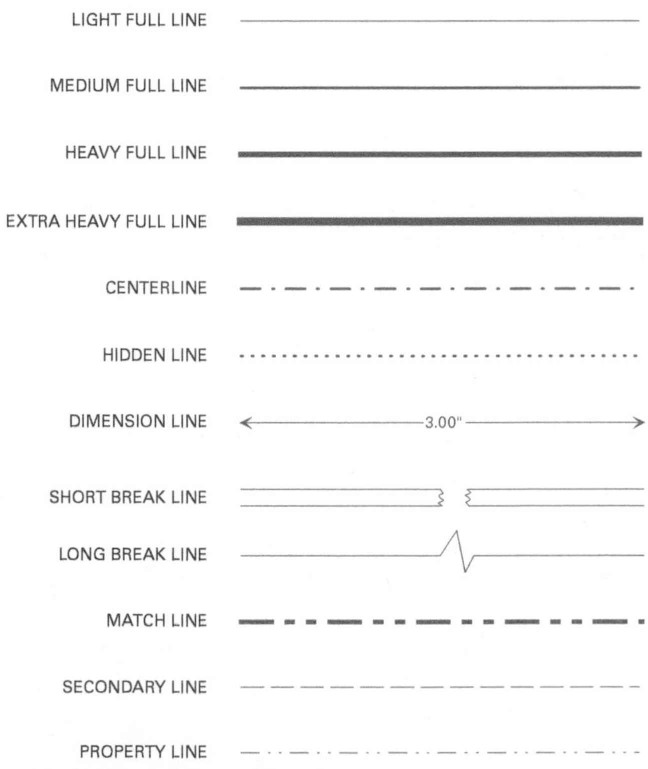

LIGHT FULL LINE	———————————
MEDIUM FULL LINE	———————————
HEAVY FULL LINE	———————————
EXTRA HEAVY FULL LINE	———————————
CENTERLINE	— · — · — · — · — · —
HIDDEN LINE	··················
DIMENSION LINE	←———— 3.00" ————→
SHORT BREAK LINE	≈
LONG BREAK LINE	⌁
MATCH LINE	▬ ▬ ▬ ▬ ▬ ▬
SECONDARY LINE	– – – – – –
PROPERTY LINE	— · · — · · — · · —

Figure 15 Typical drafting lines.

The drafting lines shown in *Figure 15* are used as follows:

- *Light full line* – Used for section lines, building background (outlines), and similar uses where the object to be drawn is secondary to the system being shown (e.g., HVAC or electrical).
- *Medium full line* – Frequently used for lettering on drawings. It is further used for some drawing symbols, circuit lines, etc.
- *Heavy full line* – Used for borders around title blocks, schedules, and for lettering drawing titles. Some types of symbols are frequently drawn with a heavy full line.
- *Extra heavy full line* – Used for border lines on architectural/engineering drawings. Sometimes drawings will have letters and numbers embedded in the border lines to enable quick location of a specific object on the drawing.
- *Centerline* – A centerline is a broken line made up of alternately spaced long and short dashes. It indicates the centers of objects such as holes, or pillars. Sometimes, the centerline indicates the dimensions of a finished floor.
- *Hidden line* – A hidden line consists of a series of short dashes that are closely and evenly spaced. It shows the edges of objects that are behind others in a particular view. The object outlined by hidden lines in one drawing is often fully pictured in another drawing.
- *Dimension line* – Thin lines used to show the extent and direction of dimensions. The actual dimension is usually placed in a break inside the dimension lines. Normal practice is to place the dimension lines outside the object's outline. However, it may sometimes be necessary to draw the dimensions inside the outline.
- *Break line (short or long)* – Shows where an object has been broken off to save space on the drawing or avoid visual interference with other drawing objects.
- *Match line* – Used to show the position of the cutting plane. Therefore, it is also called the *cutting plane line*. A match or cutting plane line is a heavy line with long dashes alternating with two short dashes. It is used on drawings of large structures to show where one sheet of the drawing stops and the next begins.
- *Secondary line* – Frequently used to outline pieces of equipment or to indicate reference points of a drawing that are secondary to the drawing's purpose.
- *Property line* – A light line made up of one long and two short dashes that are alternately spaced. It indicates land boundaries on the site plan.

Other uses of the lines just mentioned include the following:

- *Extension lines* – Lightweight lines that start about $\frac{1}{16}$" (1.6 mm) away from the edge of an object and extend out. A common use of extension lines is to create a boundary for dimension lines. Dimension lines meet extension lines with arrowheads, slashes, or dots. Extension lines pointing from a note or other reference to a particular feature on a drawing are called *leaders*. They usually end in either an arrowhead or a dot and may include an explanatory note at the end.
- *Section lines* – These are often referred to as *cross-hatch lines*. Drawn at a 45° angle, these lines show where an object has been cut away to reveal the inside.

- *Phantom lines* – Solid, light lines that show where an object will be installed. A future door opening or a future piece of equipment can be shown with phantom lines.

Check the Legend

Be sure to check the legend on every drawing set. Symbols and abbreviations often vary widely from one drawing set to another drawing set.

1.0.0 Section Review

1. A drawing contains an area circled in a cloud shape. This information is often detailed in the _____.
   a. legend block
   b. approval block
   c. revision block
   d. engineer's title block

2. Which of the following shows the face of a building?
   a. section view
   b. floor plan
   c. site plan
   d. elevation view

3. A drawing contains a match line. What does this mean?
   a. It indicates that the drawing continues on another sheet.
   b. It indicates that the structure is symmetrical.
   c. It connects together two points on the same drawing.
   d. It indicates reference points that are secondary to the drawing's purpose.

2.0.0 SCALE DRAWINGS

Objective

Measure items on scale drawings.
  a. Use an architect's scale.
  b. Use an engineer's scale.
  c. Use a metric scale.

Performance Task

  1. Using an architect's scale, state the actual dimensions of a given drawing component.

Trade Terms

Architect's scale: A special ruler with various measurement scales that can be used when drafting or making measurements on architectural drawings. Architect's scales measure distances in inches and fractions thereof.

Engineer's scale: A special ruler with various measurement scales that can be used when drafting or making measurements on architectural drawings. Engineer's scales measure distances in decimal units.

In most electrical drawings, the components are so large that it would be impossible to draw them to actual size. Consequently, drawings are made to some reduced scale; that is, all the distances are drawn smaller than the actual dimensions of the object itself, with all dimensions being reduced in the same proportion. For example, if a floor plan of a building is drawn to a scale of $\frac{1}{4}$" = 1'–0", each $\frac{1}{4}$" on the drawing would equal 1' on the building itself; if the scale is $\frac{1}{8}$" = 1'–0", each $\frac{1}{8}$" on the drawing equals 1' on the building, and so forth.

When architectural and engineering drawings are produced, the selected scale is very important. Where dimensions must be held to extreme accuracy, the scale drawings should be made as large as practical with dimension lines added. Where dimensions require only reasonable accuracy, the object may be drawn to a smaller scale (with dimension lines possibly omitted).

In dimensioning drawings, the dimensions written on the drawing are the actual dimensions of the building, not the distances that are measured on the drawing. To further illustrate

this point, look at the floor plan in *Figure 16*; it is drawn to a scale of $\frac{1}{2}$" = 1'–0". One of the walls is drawn to an actual length of $3\frac{1}{2}$" on the drawing paper, but since the scale is $\frac{1}{2}$" = 1'–0" and since $3\frac{1}{2}$" contains 7 halves of an inch ($7 \times \frac{1}{2}$" = $3\frac{1}{2}$"), the dimension shown on the drawing will therefore be 7'–0" on the actual building.

> **CAUTION**
>
> When drawings are printed or plotted, they are very often printed to a size that fits the paper and not to the size equal to the scale used when the drawing was made. Always double-check the scale against a marked dimension on the drawing to get a corrected drawing scale. This will avoid errors in layout and material ordering.

When a measurement is made on the drawing, it is made with the scaled ruler; when a measurement is made on the building, it is made with the standard foot rule. The most common scaled rulers used in electrical drawings are the **architect's scale** and the **engineer's scale**. Drawings may sometimes be encountered that use a metric scale; using this scale is similar to using the engineer's scale.

2.1.0 Architect's Scale

Architect's scales are usually made on flat scales (often called *rulers*), with one, two, or even four scales, on triangular scales with three faces and four scales on each face. The quality of architect's scales also varies from cheap plastic scales (costing a few dollars) to high-quality tools that are calibrated to precise standards. A flat ruler with a single 1" = 1'–0" scale is shown in *Figure 17*. Note that on the one-inch scale in *Figure 17*, the longer marks to the right of the zero (with a numeral beneath) represent feet. Therefore, the distance between the zero and the numeral 1 equals one foot. To the left of the zero, the shortest lines are half-inches and the longer lines are inches, with the 3-, 6-, and 9-inch divisions numbered.

After some practice, you will be able to tell the exact measurement at a glance. For example, the measurement A in *Figure 17* represents 5" because it is the fifth inch mark to the left of the zero. All scales also have their scale marked at the extreme end of the scale; in this figure, one inch on the scale equals one foot.

The triangular scale is frequently found in drafting and estimating departments or engineering and electrical contracting firms, while the flat scales are more convenient to carry on the job site. For a triangular scale, each edge has two

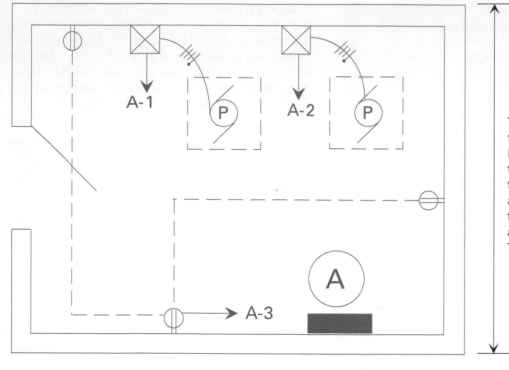

A-1 A-2

The distance between the arrowheads to the left measures 3½" on the drawing, but since the drawing is made to a scale of ½" = 1'–0", this measurement actually represents 7'–0".

A-3

PUMP HOUSE FLOOR PLAN
½" = 1'–0"

Figure 16 Typical floor plan showing drawing scale.

scales that start on opposite ends but share the markings in between. The scales on each edge are as follows:

- Common foot rule, and $\frac{1}{16}$" = 1'–0"
- $\frac{3}{32}$" = 1'–0", and $\frac{3}{16}$" = 1'–0"
- $\frac{1}{8}$" = 1'–0", and $\frac{1}{4}$" = 1'–0"
- $\frac{3}{8}$" = 1'–0", and $\frac{3}{4}$" = 1'–0"
- $\frac{1}{2}$" = 1'–0", and 1" = 1'–0"
- $1\frac{1}{2}$" = 1'–0", and 3" = 1'–0"

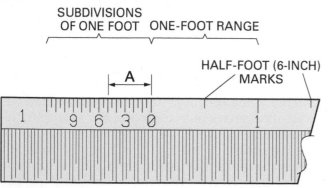

Figure 17 One-inch architect's scale.

Two separate scales on one face may seem confusing at first, but after some experience, reading these scales becomes second nature. For example, on the one-inch face, the one-inch scale is read from left to right, starting from its zero mark (on the left). The half-inch scale is read from right to left, again starting from its zero mark (on the right).

Figure 19 shows all the scales found on the triangular architect's scale.

The partial floor plan shown in *Figure 20* is drawn to a scale of $\frac{1}{8}$" = 1'–0". The dimension in question is found by placing the $\frac{1}{8}$" architect's scale on the drawing and reading the figures. Two sets of numbers along the scale can be confusing at first—just remember to use the numbers that count up from the zero at the end that has the same scale marking as the one on the drawing. In this figure, you are using the numbers closest to the edge of the scale, with each line representing one foot; and not the numbers farther from the

Figure 18 Typical triangular architect's scale.

edge, with the longer lines representing one foot. The dimension in question is 25'–3".

Every drawing should have the scale to which it is drawn plainly marked on it as part of the drawing title. However, it is not uncommon to have several different drawings or details on one blueprint sheet—all with different scales. Therefore, always check the scale of each different view found on a drawing sheet.

2.2.0 Engineer's Scale

Site plans are drawn to a scale using the civil engineer's scale rather than the architect's scale. The engineer's scale is used mainly with site plans to determine distances between property lines, manholes, duct runs, direct-burial cable runs, and the like. The engineer's scale is used in basically the same manner as the architect's scale, with the principal difference being that the graduations on the engineer's scale are decimal units, rather than feet and inches as on the architect's scale.

The engineer's scale is used by placing it on the drawing, aligned in the direction of the required measurement. Then, by looking down at the scale, the dimension is read.

Civil engineer's scales commonly show the following graduations:

- 1" = 10 units
- 1" = 20 units
- 1" = 30 units
- 1" = 40 units
- 1" = 50 units
- 1" = 60 units

On small lots, a scale of 1" = 10' or 1" = 20' is used. For a 1:10 scale, this means that 1" (the actual measurement on the drawing) is equal to 10' on the land itself, or 2.4" equals 24' on the land. On larger drawings, where a large area must be covered, the scale could be 1" = 100' or 1" = 1,000', or any other integral power of 10. On drawings with the scale in multiples of 10, the engineer's

Architect's Scale

Measurements are usually made on architectural drawings using an architect's scale rather than a standard ruler. Architect's scales, like the ones on the left, are divided into feet and inches and usually consist of several scales on one rule. Architect's scales also come in other forms such as tapes or with wheels, like the one shown on the right.

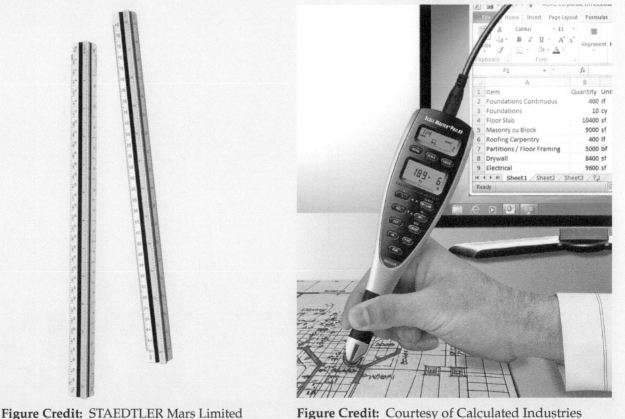

Figure Credit: STAEDTLER Mars Limited

Figure Credit: Courtesy of Calculated Industries

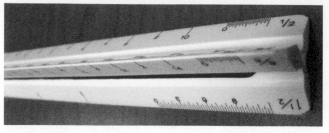
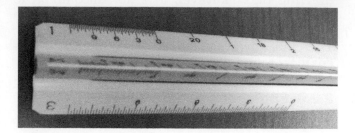

Figure 19 Various scales on a triangular architect's scale.

scale marked 10 is used. If the scale is 1" = 200', the engineer's scale marked 20 is used, and so on.

Although site plans appear reduced in scale, depending on the size of the object and the size of the drawing sheet to be used, the actual dimensions must be shown on the drawings at all times. When you are reading the drawing plans to scale, think of each dimension in its full size and not in the reduced scale it happens to be on the drawing (*Figure 21*).

2.3.0 Metric Scale

Metric scales are also calibrated in units of 10 (*Figure 22*), but they are calibrated in metric dimensions instead of English dimensions of inches and feet. The two common length measurements used in the metric scale on architectural drawings are the meter and the millimeter, the millimeter being $1/1,000$ of a meter. On drawings drawn to scales between 1:1 and 1:100, the millimeter is typically used. On drawings drawn to scales between 1:200

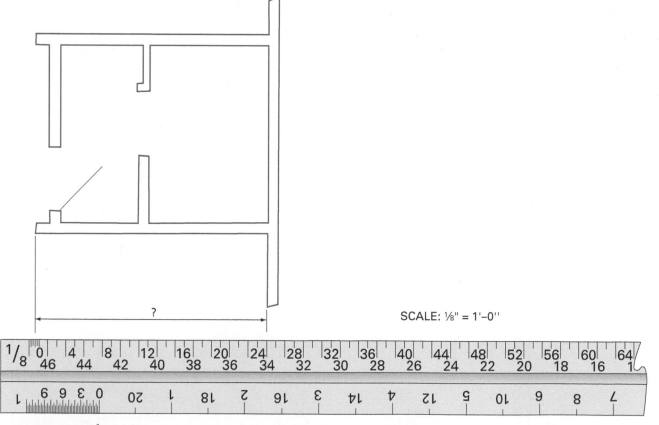

SCALE: ⅛" = 1'–0"

Figure 20 Using the ⅛" architect's scale to determine the dimensions on a drawing.

and 1:2,000, the meter is generally used. Many contracting firms that deal in international trade have adopted a dual-dimension system with measurements in both metric and English units of distance. Drawings prepared for government projects may also require metric dimensions. A metric conversion chart is provided in the *Appendix*.

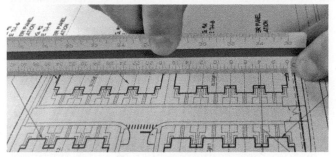

Figure 21 Practical use of the engineer's scale.

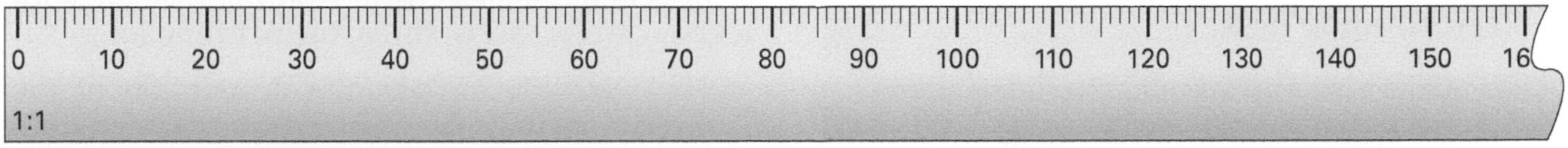

Figure 22 Typical metric scale.

2.0.0 Section Review

1. If a floor plan of a building is drawn to a scale of ⅛" = 1'–0", a distance of 1⅜" on the drawing would equal _____.

   a. 1'–3" on the building itself
   b. 8'–0" on the building itself
   c. 11'–0" on the building itself
   d. 13'–8" on the building itself

2. Using a 1:20 scale on an engineer's scale, a measurement of 2" on a drawing would equal _____.

   a. 5'
   b. 10'
   c. 20'
   d. 40'

3. A metric drawing scale of 1:1,000 typically depicts units in _____.

   a. millimeters
   b. centimeters
   c. decimeters
   d. meters

3.0.0 Electrical Drawings

Objective

Apply the information on electrical drawings.

  a. Interpret electrical symbols.
  b. Check a set of residential electrical drawings.
  c. Locate information within a commercial plan set
  d. Read schedules, block diagrams, and schematic diagrams.

Performance Task

  2. Make a material takeoff of the luminaires specified in the provided drawing. The takeoff requires that all luminaires be counted, and where applicable, the total number of lamps for each luminaire type must be calculated. (Fill these in on the provided Performance Profile Task 2 Worksheet.)

Trade Terms

Block diagram: A single-line diagram used to show electrical equipment and related connections. See power-riser diagram.

One-line diagram: A drawing that shows, by means of lines and symbols, the path of an electrical circuit or system of circuits along with the various circuit components. Also called a *single-line diagram*.

Power-riser diagram: A single-line block diagram used to indicate the electric service equipment, service conductors and feeders, and subpanels. Notes are used on power-riser diagrams to identify the equipment; indicate the size of conduit; show the number, size, and type of conductors; and list related materials. A panelboard schedule is usually included with power-riser diagrams to indicate the exact components (panel type and size), along with fuses, circuit breakers, etc., contained in each panelboard.

Schematic diagram: A detailed diagram showing complicated circuits, such as control circuits. A schematic diagram has no scale but shows the components and their electrical connections.

Shop drawing: A drawing that is usually developed by manufacturers, fabricators, or contractors to show specific dimensions and other pertinent information concerning a particular piece of equipment and its installation methods.

An electrical drawing shows in a clear, concise manner exactly what is required of the electricians. The amount of data shown on such drawings should be sufficient, but not overdone. This means that a complete set of electrical drawings could consist of only one $8\frac{1}{2}$" × 11" sheet, or it could consist of several dozen 24" × 36" (or larger) sheets, depending on the size and complexity of a given project. A set of working drawings for an industrial installation may contain dozens of drawing sheets detailing the electrical system for lighting and power, along with equipment, motor controls, and wiring diagrams. A schematic diagram and equipment schedule may contain a host of other pertinent data.

In general, the electrical working drawings for a given project serve three distinct functions:

- They provide an accurate description of the project so that electrical contractors can estimate materials and labor to calculate a total cost of the project for bidding purposes.
- They provide workers on the project with information as to how and where the electrical system is to be installed.
- They provide a map of the electrical system once the job is completed to aid in maintenance and troubleshooting for years to come.

Electrical drawings from consulting engineering firms will vary in quality from sketchy, incomplete drawings to neat, precise drawings that are easy to understand. Few, however, will cover every detail of the electrical system. Therefore, a good knowledge of installation practices must go hand-in-hand with interpreting electrical working drawings.

Sometimes electrical contractors will have electrical drafters prepare a special supplemental drawing for use by the contractors' employees, often called a shop drawing or *vendor drawing*. On certain projects, these supplemental drawings can save supervision time in the field once the project has begun.

3.1.0 Electrical Symbols and Lines

The electrician must be able to correctly read and understand electrical working drawings. This skill requires a thorough knowledge of electrical symbols and their applications.

It would be much simpler if all architects, engineers, electrical designers, and drafters used the same symbols; however, this is not the case. Although standardization is getting closer to a reality, existing symbols are still modified, and new symbols are being created for new products and projects.

An electrical symbol is a figure or mark that represents a component used in the electrical system. The electrical symbols described in the following paragraphs are found on actual electrical working drawings throughout the United States and Canada. Many are ones recommended by the American National Standards Institute (ANSI) and the Consulting Engineers Council (CEC), while others are not. Understanding how these symbols were devised will help you to interpret unknown electrical symbols in the future.

Three simple shapes are the starting point for most symbols you will see—a circle for a small box, an S for a switch, and a rectangle for a large box or enclosure. *Figure 23* shows a short list of electrical symbols that are currently recommended by ANSI. Most begin with the simple shape of a circle, an S, or a rectangle, but their meaning and purpose is made more precise by a slight variation. A good procedure to follow in learning symbols is to first learn the basic form and then apply the variations to obtain the different meanings.

The receptacle symbols in *Figure 24* each start with a circle, but the addition of one or more lines crossing the circle and a few abbreviations give individual meanings.

A light fixture (luminaire) symbol also starts with a circle, which represents a ceiling-mounted luminaire. The addition of a short line indicates wall mounting and a rectangle or long line indicates fluorescent or other strip-type luminaires. The type of luminaire is usually identified by a numeral placed inside a triangle or other shape placed near the luminaire to be identified. A complete description of the luminaires identified by the symbols should be found in the luminaire schedule on the drawings, or in the written specifications and fixture submittals. This description should include the manufacturer, catalog number, number and type of lamps, voltage, finish, mounting, and any other information needed for the electrician to properly install the luminaire.

Switches used to control luminaires or switched receptacles usually begin with the letter S, followed by numbers or letters to define the exact type of switch. For example, S_3 indicates a three-way switch; S_4 identifies a four-way switch; and S_p indicates a single-pole switch with a pilot light. A curving line or a subscript letter can be used to match a switch to the luminaire(s) it controls.

A rectangle or box is used for main distribution centers, panelboards, transformers, safety switches, and other large electrical components. A detailed description of the service equipment is usually given in the panelboard schedule or in the written specifications. However, on small projects, the service equipment is sometimes indicated only by a symbol and notes on the drawings. Other symbols are simplified pictographs, or a box combined with a pair of lines (representing an external handle) and abbreviations, as shown in *Figure 25*.

Communication systems, such as telephone, security, nurse call, and many more, also use many of these common shapes, but with additional shapes or letters to indicate their exact meaning. You will also see many more symbols in fields such as industrial control or outdoor line work.

Lines on the drawings are used to represent circuits and their related components. Most circuits concealed in the ceiling or wall are indicated by a solid line; a broken line is used for circuits concealed in the floor or ceiling below; and exposed raceways are indicated by short dashes or by the letter E placed at intervals along the same plane with the circuit line. The number of conductors in a conduit or raceway system may be indicated by short slashes on the drawing, or in the panelboard schedule. To avoid clutter, home runs are frequently shown with a short line pointing toward the panelboard, and arrowheads at the end of the line; the panel and circuit numbers may be marked nearby.

The various symbols are placed on the drawings as close as possible to their intended location. Where extreme accuracy is required in locating outlets and equipment, exact locations are given on larger-scale detail drawings or in the specifications. All individual symbols used on the drawings should be included in the symbol list or legend. You should become familiar with these symbols.

The electrical symbols shown in *Figure 26* were selected by a consulting engineering firm for use on a small industrial electrical installation. Five groups of electrical symbols, recommended by CEC, are in *Figure 27A* and *Figure 27B*. Additional groups of these symbols are in *Figure 27C, Figure 27D, Figure 27E, Figure 27F,* and *Figure 27G*. As you progress in your learning and experience, you will repeatedly encounter and use electrical symbols.

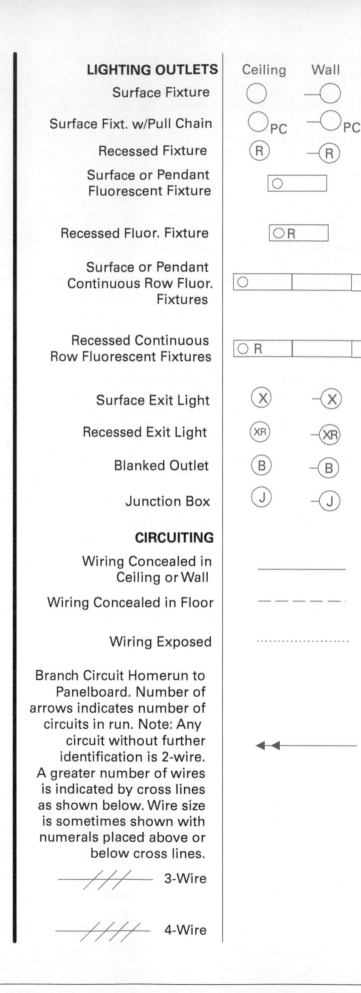

SWITCH OUTLETS

Single-Pole Switch	S
Double-Pole Switch	S$_2$
Three-Way Switch	S$_3$
Four-Way Switch	S$_4$
Key-Operated Switch	S$_K$
Switch w/Pilot	S$_P$
Low-Voltage Switch	S$_L$
Switch & Single Receptacle	⊖$_S$
Switch & Duplex Receptacle	⊖$_S$
Door Switch	S$_D$
Momentary Contact Switch	S$_{MC}$

RECEPTACLE OUTLETS

Single Receptacle

Duplex Receptacle

Triplex Receptacle

Split-Wired Duplex Recep.

Single Special-Purpose Recep.

Duplex Special-Purpose Recep.

Range Receptacle

Special Purpose Connection or Provision for Connection. Subscript letters indicate Function (DW - Dishwasher; CD - Clothes Dryer, etc.)

Clock Receptacle w/Hanger

Fan Receptacle w/Hanger

Single Floor Receptacle

Note: A numeral or letter within the symbol or as a sub-script keyed to the list of symbols indicates type of receptacle or usage.

LIGHTING OUTLETS Ceiling Wall

Surface Fixture

Surface Fixt. w/Pull Chain

Recessed Fixture

Surface or Pendant Fluorescent Fixture

Recessed Fluor. Fixture

Surface or Pendant Continuous Row Fluor. Fixtures

Recessed Continuous Row Fluorescent Fixtures

Surface Exit Light

Recessed Exit Light

Blanked Outlet

Junction Box

CIRCUITING

Wiring Concealed in Ceiling or Wall

Wiring Concealed in Floor

Wiring Exposed

Branch Circuit Homerun to Panelboard. Number of arrows indicates number of circuits in run. Note: Any circuit without further identification is 2-wire. A greater number of wires is indicated by cross lines as shown below. Wire size is sometimes shown with numerals placed above or below cross lines.

3-Wire

4-Wire

Figure 23 ANSI electrical symbols.

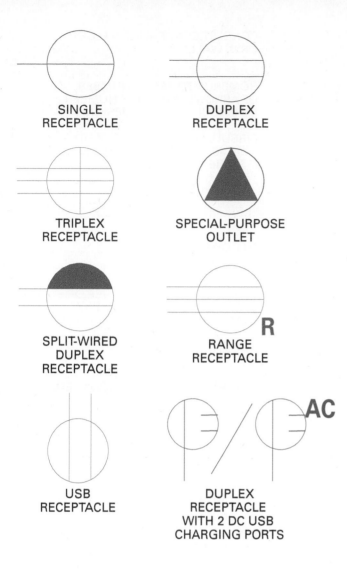

SINGLE RECEPTACLE

DUPLEX RECEPTACLE

TRIPLEX RECEPTACLE

SPECIAL-PURPOSE OUTLET

SPLIT-WIRED DUPLEX RECEPTACLE

RANGE RECEPTACLE

USB RECEPTACLE

DUPLEX RECEPTACLE WITH 2 DC USB CHARGING PORTS

Figure 24 Various receptacle symbols used on electrical drawings.

3.2.0 Residential Electrical Drawings

In the past, with the exception of very large residences, the electrical systems were laid out by the architect in the form of a sketched arrangement or laid out by the electrician on the job as the work progressed. However, many technical developments in residential electrical use—such as electric heat with sophisticated control wiring, increased use of electrical appliances, various electronic alarm systems, new lighting techniques, and the need for energy conservation techniques—have greatly expanded the demand and extended the complexity of today's residential electrical systems. Homes with very complex installations may be provided with complete electrical working drawings and specifications, similar to those frequently provided for commercial and industrial projects. Still, these are more the exception than the rule. Most residential projects will have a very basic set of drawings.

The most practical way to learn how to read electrical construction documents is to analyze an existing set of drawings. A set of residential electrical drawings may have one to three sheets, which makes them simpler to read and use than commercial or industrial drawings. However, the drawing details such as lines, symbols, and branch circuiting apply to all.

3.2.1 Branch Circuits

The required number of branch circuits for lighting and receptacles depends on the building use or *occupancy*. *NEC Article 220, Part II* provides information on calculating lighting loads for various occupancies. For residential

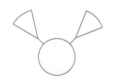

DOUBLE FLOODLIGHT FIXTURE

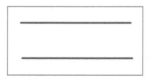

INFRARED ELECTRIC HEATER WITH TWO QUARTZ LAMPS

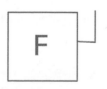

FUSIBLE SAFETY SWITCH

NON-FUSIBLE SAFETY SWITCH

DOUBLE-THROW SAFETY SWITCH

Figure 25 Pictographic symbols used on electrical drawings.

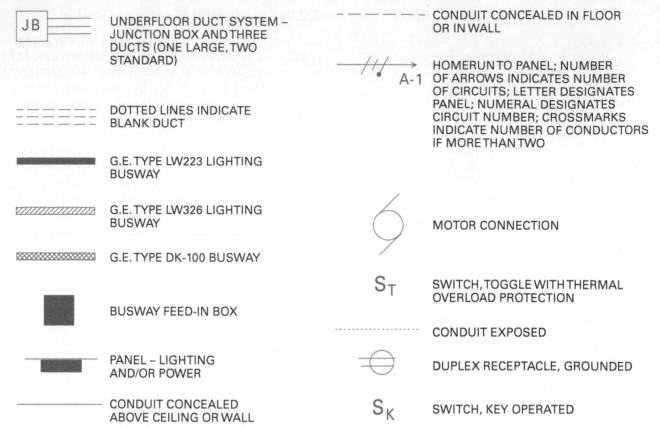

JB — UNDERFLOOR DUCT SYSTEM – JUNCTION BOX AND THREE DUCTS (ONE LARGE, TWO STANDARD)	CONDUIT CONCEALED IN FLOOR OR IN WALL
DOTTED LINES INDICATE BLANK DUCT	A-1 HOMERUN TO PANEL; NUMBER OF ARROWS INDICATES NUMBER OF CIRCUITS; LETTER DESIGNATES PANEL; NUMERAL DESIGNATES CIRCUIT NUMBER; CROSSMARKS INDICATE NUMBER OF CONDUCTORS IF MORE THAN TWO
G.E. TYPE LW223 LIGHTING BUSWAY	
G.E. TYPE LW326 LIGHTING BUSWAY	MOTOR CONNECTION
G.E. TYPE DK-100 BUSWAY	
BUSWAY FEED-IN BOX	S$_T$ SWITCH, TOGGLE WITH THERMAL OVERLOAD PROTECTION
	CONDUIT EXPOSED
PANEL – LIGHTING AND/OR POWER	DUPLEX RECEPTACLE, GROUNDED
CONDUIT CONCEALED ABOVE CEILING OR WALL	S$_K$ SWITCH, KEY OPERATED

Figure 26 Electrical symbols used by one consulting engineering firm.

occupancies, *NEC Section 220.14(J)* specifies 3 volt-amperes per square foot, which includes both the lighting and general use receptacles. Since these branch circuits are considered continuous loads, *NEC Section 210.19(A)(1)* states that the minimum conductor size shall have an ampacity of not less than 125% of the continuous load. For the maximum load on the circuit, it is the reciprocal of 125% ($^1/_{1.25}$ = 0.80) or 80%. For a 15A branch circuit, the rated capacity it can carry (in volt-amperes) is as follows:

$$15A \times 120V = 1{,}800VA$$

But since it can only be loaded to 80% of its rated capacity, the maximum initial connected load should be no more than 1,440VA, as shown by the calculation below:

$$1{,}800VA \times 0.80 = 1{,}440VA$$

Each point at which electrical equipment is connected to the branch circuit is commonly called an *outlet*. There are many classifications of outlets: lighting, receptacle, motor, appliance, and so forth. This section will deal with the power and lighting outlets normally found in residential electrical wiring systems.

Branch circuits are shown on electrical drawings by means of a single line drawn from the panelboard to the outlet (or by home run arrowheads indicating that the circuit goes to the panelboard), and from outlet to outlet where there is more than one outlet on the circuit.

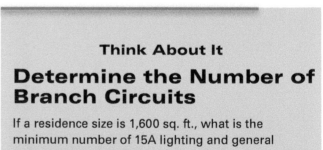

Think About It

Determine the Number of Branch Circuits

If a residence size is 1,600 sq. ft., what is the minimum number of 15A lighting and general receptacle branch circuits?

1,600 ft^2 x 3 watt/ft^2 x 1 circuit/1,440 watts = 3.3 circuits

Round the answer to the next whole number—4 circuits are required.

SWITCH OUTLETS		RECEPTACLE OUTLETS	
Single-Pole Switch	S	Where weatherproof, explosionproof, or other specific types of devices are to be required, use the upper-case subscript letters to specify. For example, weatherproof single or duplex receptacles would have the upper-case WP subscript: letters noted alongside the symbol. All outlets must be grounded.	
Double-Pole Switch	S_2		
Three-Way Switch	S_3		
Four-Way Switch	S_4	Single Receptacle Outlet	
Key-Operated Switch	S_K	Duplex Receptacle Outlet	
Switch and Fusestat Holder	$S_F H$	Triplex Receptacle Outlet	
Switch and Pilot Lamp	S_P	Quadruplex Receptacle Outlet	
Fan Switch	S_F	Duplex Receptacle Outlet Split Wired	
Switch for Low-Voltage Switching System	S_L	Triplex Receptacle Outlet Split Wired	
Master Switch for Low-Voltage Switching System	S_{LM}	250-Volt Receptacle/Single Phase Use subscript letter to indicate function (DW – Dishwasher, RA – Range) or numerals (with explanation in symbols schedule).	
Switch and Single Receptacle	S		
Switch and Duplex Receptacle	S	250-Volt Receptacle/Three Phase	
Door Switch	S_D	Clock Receptacle	Ⓒ
Time Switch	S_T	Fan Receptacle	Ⓕ
Momentary Contact Switch	S_{MC}	Floor Single Receptacle Outlet	
Ceiling Pull Switch	Ⓢ	Floor Duplex Receptacle Outlet	
"Hand-Off-Auto" Control Switch	HOA	Floor Special-Purpose Outlet	⊿ *
Multi-Speed Control Switch	M	Floor Telephone Outlet – Public	◀
Pushbutton	▪	Floor Telephone Outlet – Private	◁

* Use numeral keyed explanation of symbol usage.

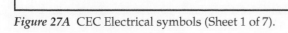

Figure 27A CEC Electrical symbols (Sheet 1 of 7).

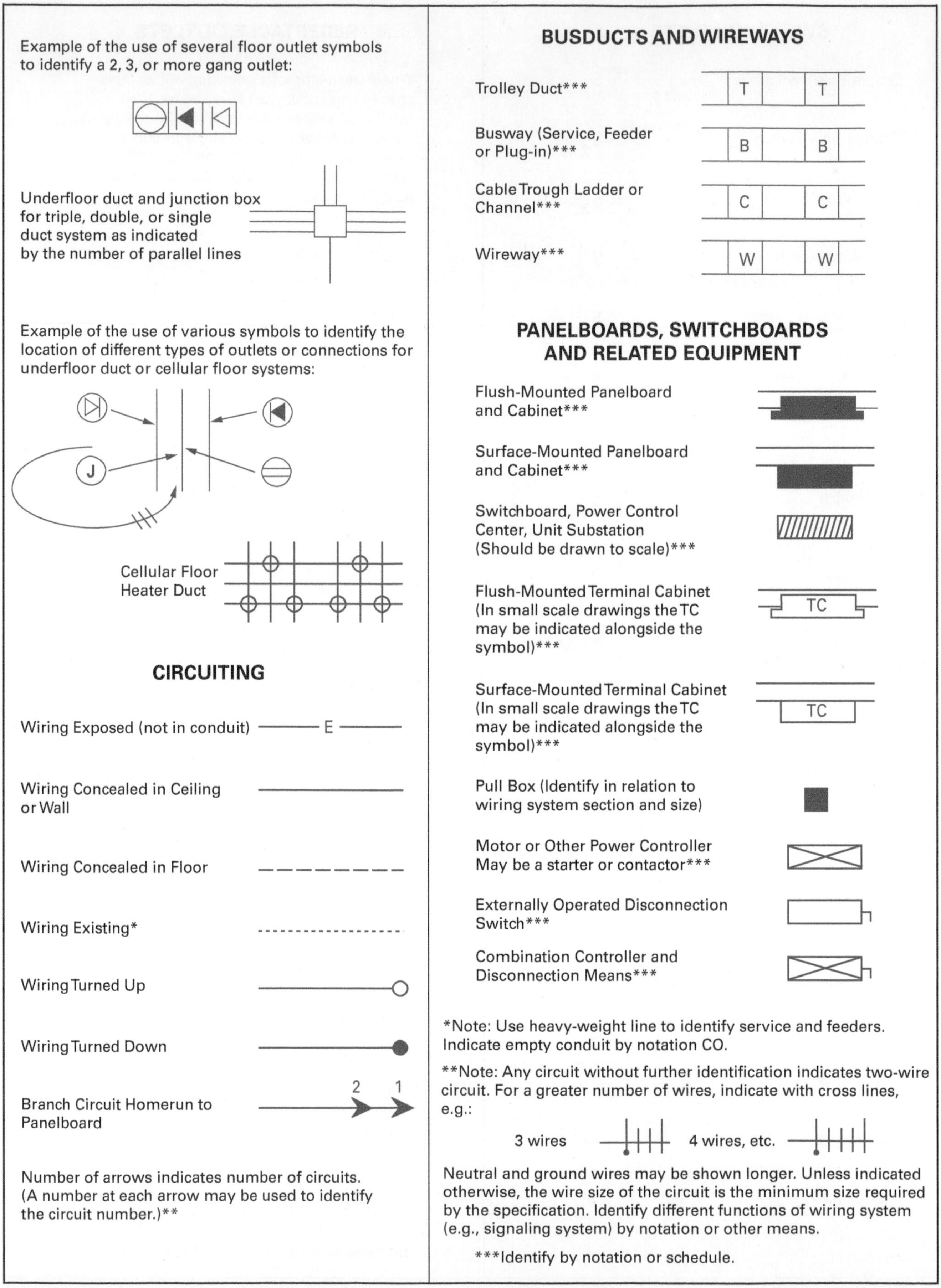

Example of the use of several floor outlet symbols to identify a 2, 3, or more gang outlet:

Underfloor duct and junction box for triple, double, or single duct system as indicated by the number of parallel lines

Example of the use of various symbols to identify the location of different types of outlets or connections for underfloor duct or cellular floor systems:

Cellular Floor Heater Duct

CIRCUITING

Wiring Exposed (not in conduit) —— E ——

Wiring Concealed in Ceiling or Wall

Wiring Concealed in Floor

Wiring Existing*

Wiring Turned Up

Wiring Turned Down

Branch Circuit Homerun to Panelboard

2 1

Number of arrows indicates number of circuits. (A number at each arrow may be used to identify the circuit number.)**

BUSDUCTS AND WIREWAYS

Trolley Duct***	T	T
Busway (Service, Feeder or Plug-in)***	B	B
Cable Trough Ladder or Channel***	C	C
Wireway***	W	W

PANELBOARDS, SWITCHBOARDS AND RELATED EQUIPMENT

Flush-Mounted Panelboard and Cabinet***

Surface-Mounted Panelboard and Cabinet***

Switchboard, Power Control Center, Unit Substation (Should be drawn to scale)***

Flush-Mounted Terminal Cabinet (In small scale drawings the TC may be indicated alongside the symbol)***

TC

Surface-Mounted Terminal Cabinet (In small scale drawings the TC may be indicated alongside the symbol)***

TC

Pull Box (Identify in relation to wiring system section and size)

Motor or Other Power Controller May be a starter or contactor***

Externally Operated Disconnection Switch***

Combination Controller and Disconnection Means***

*Note: Use heavy-weight line to identify service and feeders. Indicate empty conduit by notation CO.

**Note: Any circuit without further identification indicates two-wire circuit. For a greater number of wires, indicate with cross lines, e.g.:

3 wires   4 wires, etc.

Neutral and ground wires may be shown longer. Unless indicated otherwise, the wire size of the circuit is the minimum size required by the specification. Identify different functions of wiring system (e.g., signaling system) by notation or other means.

***Identify by notation or schedule.

Figure 27B CEC Electrical symbols (Sheet 2 of 7).

POWER EQUIPMENT

Electric Motor (HP as indicated)	1/4
Power Transformer	
Pothead (cable termination)	
Circuit Element e.g., circuit breaker	CB
Circuit Breaker	
Fusible Element	
Single-Throw Knife Switch	
Double-Throw Knife Switch	
Ground	
Battery	
Contactor	C
Photoelectric Cell	PE
Voltage Cycles, Phase	EX: 480/60/3
Relay	R
Equipment Connection (as noted)	

REMOTE CONTROL STATIONS FOR MOTORS OR OTHER EQUIPMENT

Pushbutton Station	PB
Float Switch - Mechanical	F
Limit Switch - Mechanical	L
Pneumatic Switch - Mechanical	P
Electric Eye - Beam Source	
Electric Eye - Relay	
Temperature Control Relay Connection (3 denotes quantity)	R 3
Solenoid Control Valve Connection	S
Pressure Switch Connection	P
Aquastat Connection	A
Vacuum Switch Connection	V
Gas Solenoid Valve Connection	G
Flow Switch Connection	F
Timer Connection	T
Limit Switch Connection	L

LIGHTING OUTLETS

	CEILING	WALL
Incandescent Fixture (surface or pendant)		
Incandescent Fixture with Pull Chain (surface or pendant)	PC	PC

Figure 27C CEC Electrical symbols (Sheet 3 of 7).

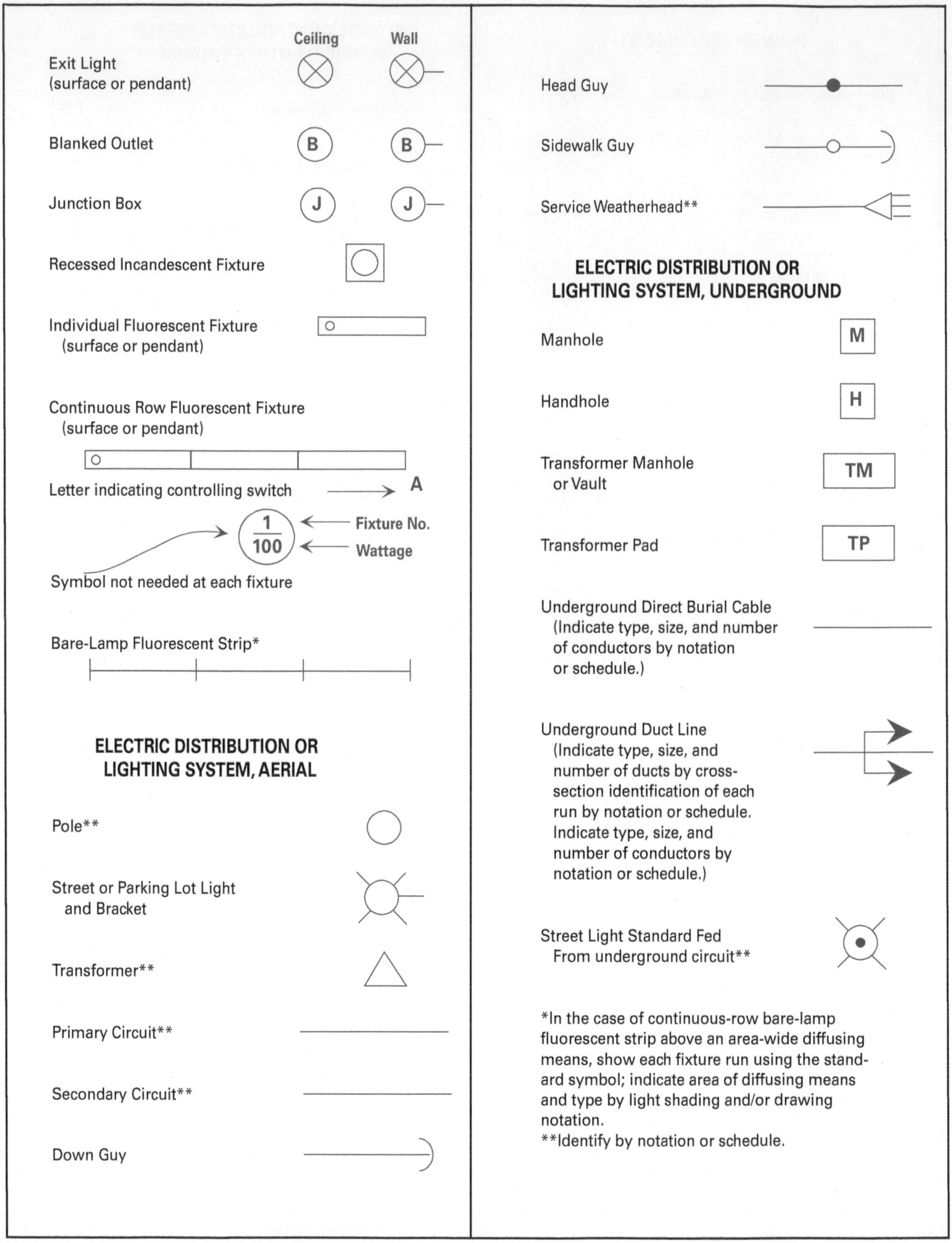

	Ceiling	Wall
Exit Light (surface or pendant)	⊗	⊗—
Blanked Outlet	Ⓑ	Ⓑ—
Junction Box	Ⓙ	Ⓙ—
Recessed Incandescent Fixture	▢ (○)	
Individual Fluorescent Fixture (surface or pendant)	▭	
Continuous Row Fluorescent Fixture (surface or pendant)	▭▭▭	

Letter indicating controlling switch →→ **A**

Fixture No. ←
Wattage ←

$\dfrac{1}{100}$

Symbol not needed at each fixture

Bare-Lamp Fluorescent Strip*

ELECTRIC DISTRIBUTION OR LIGHTING SYSTEM, AERIAL

Pole**	○
Street or Parking Lot Light and Bracket	⊕
Transformer**	△
Primary Circuit**	——
Secondary Circuit**	——
Down Guy	——⟩

Head Guy	——●——
Sidewalk Guy	——○—⟨
Service Weatherhead**	——◁=

ELECTRIC DISTRIBUTION OR LIGHTING SYSTEM, UNDERGROUND

Manhole	M
Handhole	H
Transformer Manhole or Vault	TM
Transformer Pad	TP
Underground Direct Burial Cable (Indicate type, size, and number of conductors by notation or schedule.)	——
Underground Duct Line (Indicate type, size, and number of ducts by cross-section identification of each run by notation or schedule. Indicate type, size, and number of conductors by notation or schedule.)	⇉→
Street Light Standard Fed From underground circuit**	⊛

*In the case of continuous-row bare-lamp fluorescent strip above an area-wide diffusing means, show each fixture run using the standard symbol; indicate area of diffusing means and type by light shading and/or drawing notation.
**Identify by notation or schedule.

Figure 27D CEC Electrical symbols (Sheet 4 of 7).

SIGNALING SYSTEM OUTLETS

INSTITUTIONAL, COMMERCIAL, AND INDUSTRIAL OCCUPANCIES

I NURSE CALL SYSTEM DEVICES (Any Type)

Basic Symbol

(Examples of individual item identification. Not a part of standard.)

Nurses' Annunciator
(Add a number after it as
—① 24 to indicate number
of lamps)   —①

Call Station, Single Cord,
Pilot Light   —②

Call Station, Double Cord,
Microphone Speaker   —③

Corridor Dome Light
1 Lamp   —④

Transformer   —⑤

Any other item on same
system use numbers as
required.   —⑥

II PAGING SYSTEM DEVICES

Basic Symbol

(Examples of individual item identification. Not a part of standard.)

Keyboard   1

Flush Annunciator   2

2-Face Annunciator   3

Any other item on same
system use numbers as
required.   4

III FIRE ALARM SYSTEM DEVICES (Any Type) Including Smoke and Sprinkler Alarm Devices

Basic Symbol

(Examples of individual item identification. Not a part of standard.)

Control Panel   —[1]

Station   —[2]

10" Gong   —[3]

Pre-Signal Chime   —[4]

Any other item on same
system use numbers as
required.   —[5]

IV STAFF REGISTER SYSTEM DEVICES (Any Type)

Basic Symbol

(Examples of individual item identification. Not a part of standard.)

Phone Operators' Register   ◇1

Entrance Register - Flush   ◇2

Staff Room Register   ◇3

Transformer   ◇4

Any other item on same system
use number as required.   ◇5

V ELECTRIC CLOCK SYSTEM DEVICES (Any Type)

Basic Symbol

(Examples of individual item identification. Not a part of standard.)

Figure 27E CEC Electrical symbols (Sheet 5 of 7).

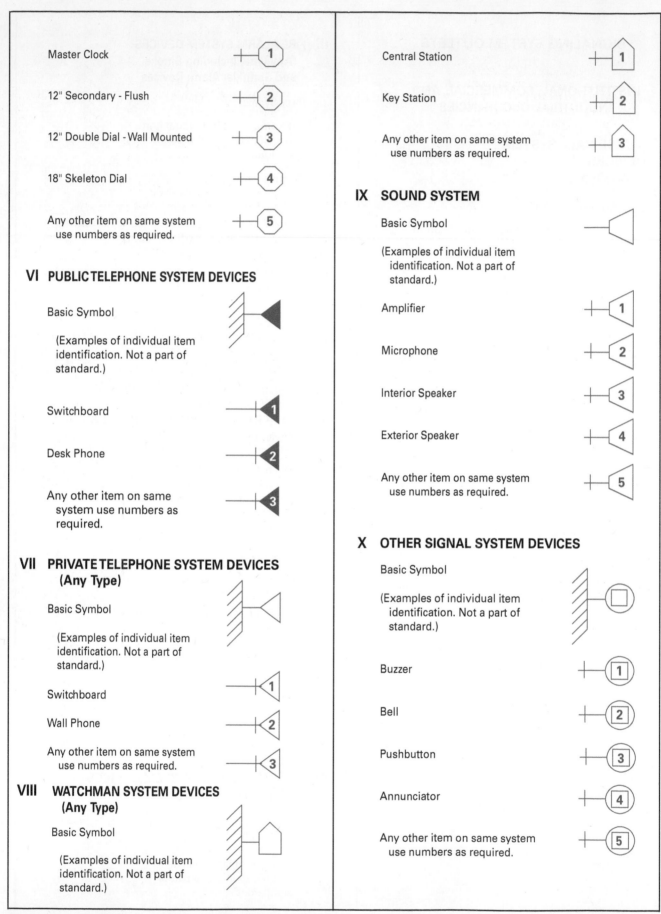

Master Clock ———1

12" Secondary - Flush ——2

12" Double Dial - Wall Mounted ——3

18" Skeleton Dial ——4

Any other item on same system
use numbers as required. ——5

VI PUBLIC TELEPHONE SYSTEM DEVICES

Basic Symbol

(Examples of individual item
identification. Not a part of
standard.)

Switchboard ——1

Desk Phone ——2

Any other item on same
system use numbers as
required. ——3

VII PRIVATE TELEPHONE SYSTEM DEVICES
(Any Type)

Basic Symbol

(Examples of individual item
identification. Not a part of
standard.)

Switchboard ——1

Wall Phone ——2

Any other item on same system
use numbers as required. ——3

VIII WATCHMAN SYSTEM DEVICES
(Any Type)

Basic Symbol

(Examples of individual item
identification. Not a part of
standard.)

Central Station ——1

Key Station ——2

Any other item on same system
use numbers as required. ——3

IX SOUND SYSTEM

Basic Symbol

(Examples of individual item
identification. Not a part of
standard.)

Amplifier ——1

Microphone ——2

Interior Speaker ——3

Exterior Speaker ——4

Any other item on same system
use numbers as required. ——5

X OTHER SIGNAL SYSTEM DEVICES

Basic Symbol

(Examples of individual item
identification. Not a part of
standard.)

Buzzer ——1

Bell ——2

Pushbutton ——3

Annunciator ——4

Any other item on same system
use numbers as required. ——5

Figure 27F CEC Electrical symbols (Sheet 6 of 7).

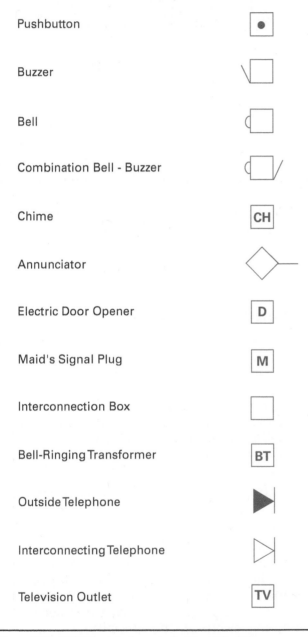

RESIDENTIAL OCCUPANCIES

Signaling system symbols identify standardized residential-type signal system items on residential drawings where a descriptive symbol list is not included on the drawing. When other signal system items are to be identified, use the basic symbols below for such items together with a descriptive symbol list.

Pushbutton

Buzzer

Bell

Combination Bell - Buzzer

Chime

Annunciator

Electric Door Opener

Maid's Signal Plug

Interconnection Box

Bell-Ringing Transformer

Outside Telephone

Interconnecting Telephone

Television Outlet

Figure 27G CEC Electrical symbols (Sheet 7 of 7).

As mentioned earlier, the lines indicating branch circuits can be solid to show that the conductors are to be run concealed in the ceiling or wall; dashed to show that the conductors are to be run in the floor or ceiling below; or dotted to show that the wiring is to be run exposed. *Figure 28* shows examples of these three types of branch circuit lines.

In *Figure 28*, No. 12 indicates the wire size. The slash marks shown through the circuit lines in *Figure 28* indicate the number and use of current-carrying conductors in each circuit. A short line is the hot wire, a longer line is the neutral, and a line ending with a dot is the ground. A single circuit can be shown with or without slash marks. Wiring runs with more than one circuit are always indicated on electrical working drawings by slash marks for each conductor, by a note, or similar means.

NEC Section 210.52(A) states the *minimum* requirements for the location of receptacles in dwelling units. It specifies that in each kitchen, family room, and dining room, receptacle outlets shall be installed so that no point along the floor line in any wall space is more than 6' (1.8 m), measured horizontally, from an outlet in that space, including any wall space 2' (600 mm) or more in width and the wall space occupied by fixed panels in walls, but excluding sliding panels. This means that the outlets will be no more than 12' (3.7 m) apart. When spaced in this manner, a 6' (1.8 m) fixture cord will reach a receptacle at any point along the wall line. Receptacle outlets in floors shall not be counted as part of the required number of receptacle outlets unless located within 18" (450 mm) of the wall. Remember the *NEC*® is a minimum standard, so with the ever-increasing use of electrical devices, a closer spacing of receptacle outlets is wise.

NEC Section 210.52(A)(2) defines wall space as a wall that is unbroken along the floor line by doorways, fireplaces, or similar openings and fixed cabinets that do not have countertops or similar work surfaces. Each wall space that is 2' (600 mm) or more in width must be treated individually and separately from other wall spaces within the room. The purpose of this section is to eliminate the hazards caused by using extension cords and running fixture cords across doorways, fireplaces, and similar openings.

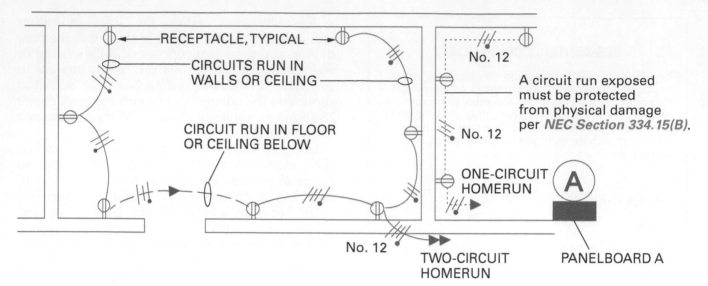

Figure 28 Types of branch circuit lines shown on electrical working drawings.

3.2.2 Residential Receptacle and Lighting Branch Circuits

Figure 29 shows the receptacle outlets for a sample residence. In laying out these receptacle outlets, the unbroken floor line of the wall is measured (also around corners) and the receptacle outlets are then spaced equal distances apart. Unlike residential receptacle circuits with no limit on the number of general use receptacles per branch circuit, commercial or industrial occupancies require 180VA per receptacle mounting yoke (the yoke is the part that attaches the receptacle to the box).

Figure 30 shows the lighting arrangement for the same sample residence. All luminaires are shown in their approximate physical locations as they should be installed. If incandescent or light-emitting diode (LED) luminaires are used in a closet, they must meet the requirements of *NEC Section 410.16(A)(1)* and be completely enclosed.

Reading Notes

The notes are crucial elements of the drawing set. Receptacles, for example, are hard to position precisely based on a scaled drawing alone, and yet the designer may call for exact locations. For example, the designer may want receptacles exactly 6" (152 mm) above the kitchen counter backsplash and centered on the sink.

Think About It
Can You Find the Error?

Check your drawings for errors! In *Figure 29*, a receptacle outlet is missing in Bedroom 2. Where should it be placed? Hint—look behind the...

3.3.0 Reading a Commercial Plan Set

A plan set contains many sections generally in order starting with civil at the front and electrical near the end. This is by priority—site layout first, then architectural plans that fit the owner's needs to the site, then structural plans to build what the architect drew, then the various craft contractors who will work within the structure (mechanical, plumbing, electrical, and specialty).

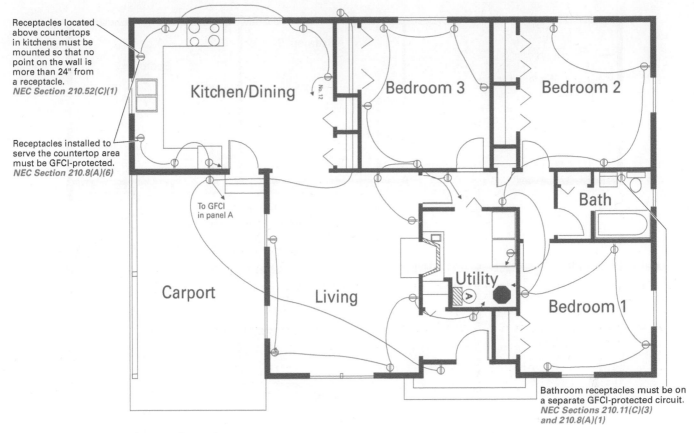

Receptacles located above countertops in kitchens must be mounted so that no point on the wall is more than 24" from a receptacle.
NEC Section 210.52(C)(1)

Receptacles installed to serve the countertop area must be GFCI-protected.
NEC Section 210.8(A)(6)

To GFCI in panel A

Kitchen/Dining

No. 12

Bedroom 3

Bedroom 2

Bath

Carport

Living

Utility

Bedroom 1

Bathroom receptacles must be on a separate GFCI-protected circuit.
NEC Sections 210.11(C)(3) and 210.8(A)(1)

Figure 29 Floor plan of a sample residence.

Engineers or electrical designers are responsible for the complete layout of electrical systems for most projects. Electrical drafters then transform the engineer's designs into working drawings, using either manual drafting instruments or computer-aided design (CAD) systems. The following is a brief outline of what usually takes place in preparation of the electrical design and the working drawings:

- The engineer meets with the architect and owner to discuss the electrical needs of the building or project and to discuss various recommendations made by all parties.
- After that, an outline of the architect's floor plan is laid out.
- The engineer then calculates the required power and lighting outlets for the project; these are later transferred to the working drawings.
- All communications and alarm systems are located on the floor plan, along with lighting and power panelboards.
- Circuit calculations are made to determine wire size and overcurrent protection.

- The main electric service and related components are determined and shown on the drawings.
- Equipment schedules with needed details are made and placed on the drawings.
- Wiring diagrams are made to show the workers how various electrical components are to be connected.
- A legend or electrical symbol list is drafted and shown on the drawings to identify all symbols used to indicate electrical outlets or equipment.
- Various large-scale electrical details are included, if necessary, to show exactly what is required of the electricians.
- Written specifications are made to give a description of the materials and installation methods.

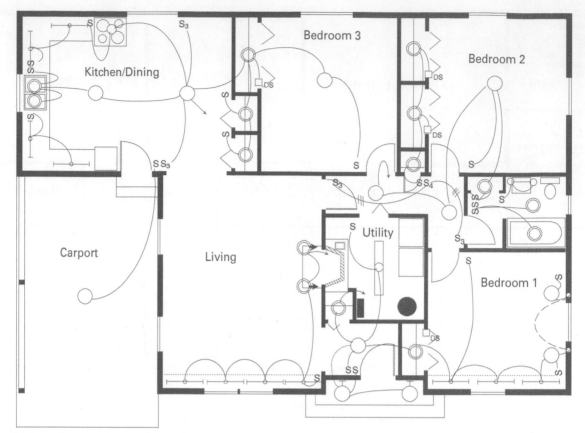

Figure 30 Lighting layout of a sample residence.

3.3.1 Electrical Site Plans

The site plan is the beginning point for new electrical construction work. Electrical site work is sometimes shown on the architect's plot plan. However, when site work involves many crafts and several utilities (e.g., gas, telephone, electric, television, water, and sewage), it can become confusing if all details are shown on one drawing sheet. In cases like these, it is best to have a separate site plan devoted entirely to the electrical work, as shown in *Figure 31*. This project is a small office/warehouse building for Virginia Electric, Inc. The electrical drawings consist of four large drawing sheets, along with a set of written specifications.

The electrical site or plot plan shown in *Figure 31* has the conventional architect's and engineer's title blocks in the lower right-hand corner of the drawing. These blocks identify the project and project owners, the architect, and the engineer. They also show how this drawing sheet relates to the entire set of drawings. Note the engineer's

professional stamp of approval to the left of the engineer's title block. Similar blocks appear on all four of the electrical drawing sheets.

When examining a set of electrical drawings for the first time, always look at the area around the title block. This is where most revision blocks or revision notes are placed. If revisions have been made to the drawings, make certain that you have a clear understanding of what has taken place before proceeding with the work.

Refer again to the drawing in *Figure 31* and note the north arrow in the upper left corner. A north arrow shows the direction of true north to help you orient the drawing to the site. Look directly down from the north arrow to the bottom of the page and notice the drawing title, *Plot Utilities*. Directly beneath the drawing title you can see that the drawing scale of 1" = 30' is shown. This is an engineer's scaling in which 1" on the drawing represents 30' on the actual job site. This scale holds true for all drawings on the page unless otherwise noted.

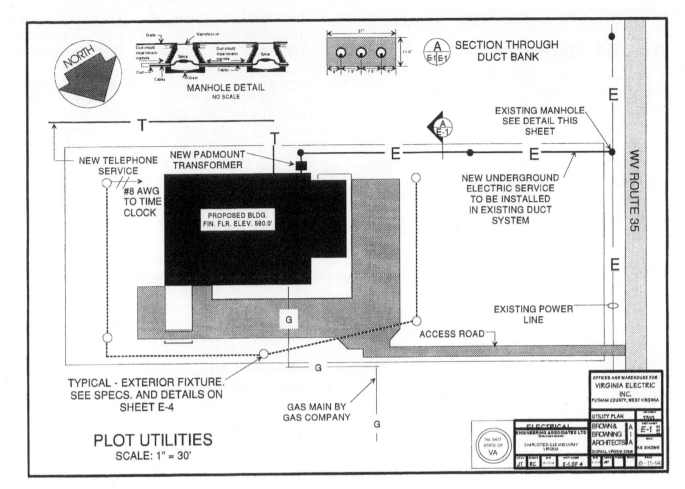

Figure 31 Typical electrical site plan.

An outline of the proposed building is indicated on the drawing along with a callout, *Proposed Bldg. Fin. Flr. Elev. 590.0*. This means that the finished floor level of the building is to be 590' above mean sea level, which on this site will be about 2' above the finished grade around the building. This information helps the electrician locate conduit sleeves and stub-ups to the correct height before the finished concrete floor is poured.

The shaded area represents asphalt paving for the access road, drives, and parking lot. Note that the access road leads into a highway, which is designated Route 35. This information further helps workers to orient the drawing to the building site.

Existing manholes are indicated by a solid circle, while an open circle is used to show the position of the five new pole-mounted luminaires that are to be installed around the new building. Existing power lines are shown with a light solid line with the letter E (for electrical) placed at intervals along the line. The new underground electric service is shown in the same way, except the lines are somewhat wider and darker on the drawing. Note that this new high-voltage cable terminates into a pad-mounted transformer near the proposed building. New telephone lines are similar, except the letter T is used to identify the telephone lines.

The direct-burial underground cable supplying the exterior luminaires is shown with dashed lines on the drawing, connecting the luminaire symbols (open circles). A home run for this circuit is also shown to a time clock.

The manhole detail shown to the right of the north arrow may seem to serve very little purpose on this drawing since the manholes have already been installed. However, the dimensions and details of their construction will help the electrical contractor or supervisor to better plan the pulling of the high-voltage cable. The same is true of the duct bank cross section shown at the top of the drawing. The electrical contractor knows that three empty ducts are available if it is discovered that one of them is damaged when the work begins.

Although the electrical work will not involve working with gas, the main gas line is shown on the electrical drawing to let the electrical workers know its approximate location while they are installing the direct-burial conductors for the exterior luminaires.

3.3.2 Power Plans

Sometimes, the physical locations of all wiring and outlets are shown on one drawing; that is, outlets for lighting, power, signal and communications, special electrical systems, and related equipment are shown on the same plan. However, this often clutters the drawing until it is unreadable, so most projects will have a separate drawing for power and another for lighting, etc. Riser diagrams and details may be shown on yet another drawing sheet or, if room permits, they may be shown on the lighting or power floor plan sheets.

The electrical power plan (*Figure 32*) shows the complete floor plan of the office/warehouse building with all interior partitions drawn to scale.

Look closely at the title blocks in the lower right corner of the drawing sheet. These blocks list both the architectural and engineering firms, along with information to identify the project and this drawing sheet (E-2). Note that the floor plan is drawn to a scale of $1/8" = 1'-0"$ and is titled *Floor Plan "B"—Power*. There are no revisions shown on this drawing sheet. Two important items to note on the power plan are the key plan and the electrical symbols list.

A small key plan appears on this drawing sheet immediately above the engineer's title block. The purpose of this key plan (enlarged in *Figure 33*) is to identify that part of the project to which this sheet applies. In this case, the project involves two attached parts: Part A and Part B. Since the outline of Part B is cross-hatched in the key plan, this is the building to which this drawing applies. Note that this key plan is not drawn to scale; only its approximate shape and orientation are shown.

Although Part A is also shown on this key plan, a note below the key plan title states that there is no electrical work required in Part A.

On some larger installations, the overall project may involve several buildings requiring appropriate key plans on each drawing to help the workers orient the drawings to the appropriate building. In some cases, separate drawing sheets may be used for each room or area in an industrial project—again requiring key plans on each drawing sheet to identify applicable drawings for each room.

A symbol list also appears on the electrical power plan (immediately above the architect's title block) to identify the various symbols used for both power and lighting on this project. In most cases, the only symbols listed are those that apply to the particular project. In other cases, however, a standard list of symbols is used for all the project's drawings with the following note:

"These are standard symbols and may not all appear on the project drawings; however, wherever the symbol on the project drawings occurs, the item shall be provided and installed."

Only electrical symbols that are actually used for the office/warehouse drawings are shown in the list on the example electrical power plan. A close-up look at these symbols appears in *Figure 34*.

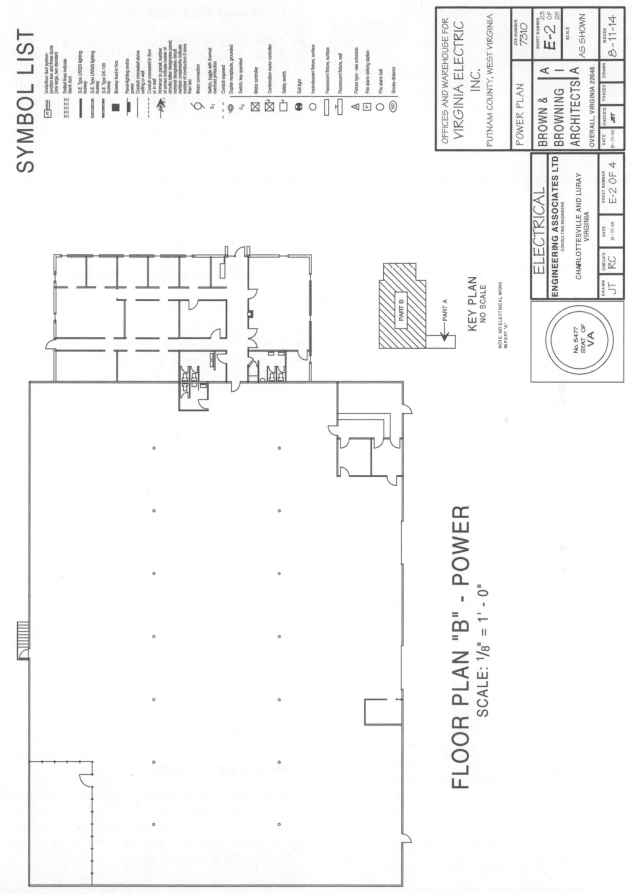

Figure 32 Electrical power plan.

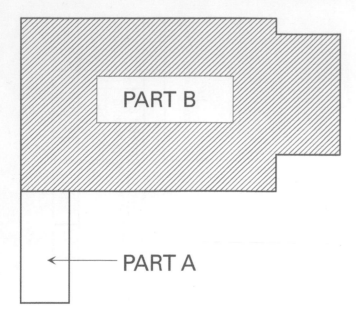

PART B

PART A

KEY PLAN
NO SCALE

NOTE: NO ELECTRICAL WORK
IN PART "A"

Figure 33 Key plan appearing on electrical power plan.

> **NOTE**
>
> When devices are to be located at heights specified above the finished floor (AFF), be sure to find out the actual thickness of the flooring to be installed. Some materials, such as ceramic tile, can add significantly to the height of the finished floor.

Symbols

Never assume that you know the meaning of any electrical symbol. Although great efforts have been made in recent years to standardize drawing symbols, architects, consulting engineers, and electrical drafters still modify existing symbols or devise new ones to meet their own needs. Always consult the symbol list or legend on electrical working drawings to verify your interpretation of the symbols used.

3.3.3 *Power Plan Details*

A somewhat enlarged view of the electrical floor plan drawing is shown in *Figure 35*. Due to the size and complexity of the drawing, it is still difficult to see very much detail. This illustration is meant to show the overall layout of the floor plan and how the symbols and notes are arranged.

In general, this plan shows the service equipment (in plan view), receptacles, underfloor duct system, motor connections, motor controllers, electric heat, busways, and similar details. The electric panels and other service equipment are drawn close to scale. The locations of other electrical outlets and similar components are only approximated on the drawings because they have to be exaggerated in size to show up on the prints. To illustrate, a common duplex receptacle is only about 3" (76 mm) wide. If such a receptacle were to be located on the floor plan of this building (drawn to a scale of $\frac{1}{8}$" = 1'–0"), even a small dot on the drawing would be too large to draw the receptacle exactly to scale. When such receptacles are scaled on the drawings to determine the proper location, a measurement is usually taken to the center of the symbol to determine the distance between outlets. Junction boxes, switches, and other electrical connections shown on the floor plan will be exaggerated in a similar manner.

The office/warehouse project utilizes three types of busways: two types of lighting busways and one power busway. Only the power busway is shown on the floor plan; the lighting busways would appear on the lighting plan.

Figure 35 shows two runs of busways: one running the length of the building on the south end (top wall on drawing), and one running the length of the north wall. The symbol list in *Figure 34* shows this busway as General Electric Type DK-100. These busways are fed from the main distribution panel (circuits MDP-1 and MDP-2) through GE No. DHIBBC41 tap boxes.

> **Think About It**
> ## Power Plans
>
> Study *Figure 35*. Where does the power enter, and how is it distributed and controlled? What is meant by each of the symbols and lines? Is every electrical connection marked or are some left to the discretion of the electrician?

JB ⊟	Underfloor duct system – junction box and three ducts (one large, two standard)
☰☰☰	Dotted lines indicate blank duct
▬▬▬	G.E. Type LW223 lighting busway
▭▱▭	G.E. Type LW326 lighting busway
▨▨▨	G.E. Type DK-100 busway
■	Busway feed-in box
▬	Panel-lighting and/or power
───	Conduit concealed above ceiling or wall
─ ─ ─	Conduit concealed in floor or in wall
→///▶ A-1	Homerun to panel; number of arrows indicate number of circuits; letter designates panel; numeral designates circuit number; crossmarks indicate number of conductors if more than two
⎔	Motor connection
S_T	Switch, toggle with thermal overload protection
┄┄┄	Conduit exposed
⊖	Duplex receptacle, grounded
S_K	Switch, key operated
⊠	Motor controller
⊠┤	Combination motor controller
□┤	Safety switch
⊗	Exit light
○	Incandescent fixture, surface
▭	Fluorescent fixture, surface
⊤▭	Fluorescent fixture, wall
Ⓐ	Fixture type – see schedule
F	Fire alarm striking station
○	Fire alarm bell
SD	Smoke detector

Figure 34 Electrical symbols list for the example building.

NEC Section 368.2 defines a busway as a metal enclosure containing factory-mounted, bare or insulated conductors, which are usually copper or aluminum bars, rods, or tubes.

The relationship of the busway and its hangers to the building construction should be checked prior to commencing the installation so that any problems due to space conflicts, inadequate or inappropriate supporting structure, openings through walls, etc., are worked out in advance so as not to incur lost time.

For example, the drawings and specifications may call for the busway to be suspended from brackets clamped or welded to steel columns. However, the spacing of the columns may be such that additional supplementary hanger rods suspended from the ceiling or roof structure may be necessary for the adequate support of the busway. To offer more assistance to workers on the office/warehouse project, the engineer may also provide an additional drawing that shows how the busway is to be mounted.

There are also several notes appearing at various places on the floor plan. These notes offer additional information to clarify certain aspects of the drawing. For example, only one electric heater is to be installed by the electrical contractor; this heater is in the building's vestibule. Rather than have a separate symbol in the symbol list for this one heater, a note is used to identify it on the drawing. Other notes on this drawing describe how certain parts of the system are to be installed. In the office area of the drawing (rooms 112, 113, and 114), you will see the following note: *CONDUIT UP AND STUBBED OUT ABOVE CEILING.* This empty conduit is for telephone/communications cables that will be installed later by the telephone company.

Other details include the general arrangement of the underfloor duct system, junction boxes and feeder conduit for the underfloor duct system, plan

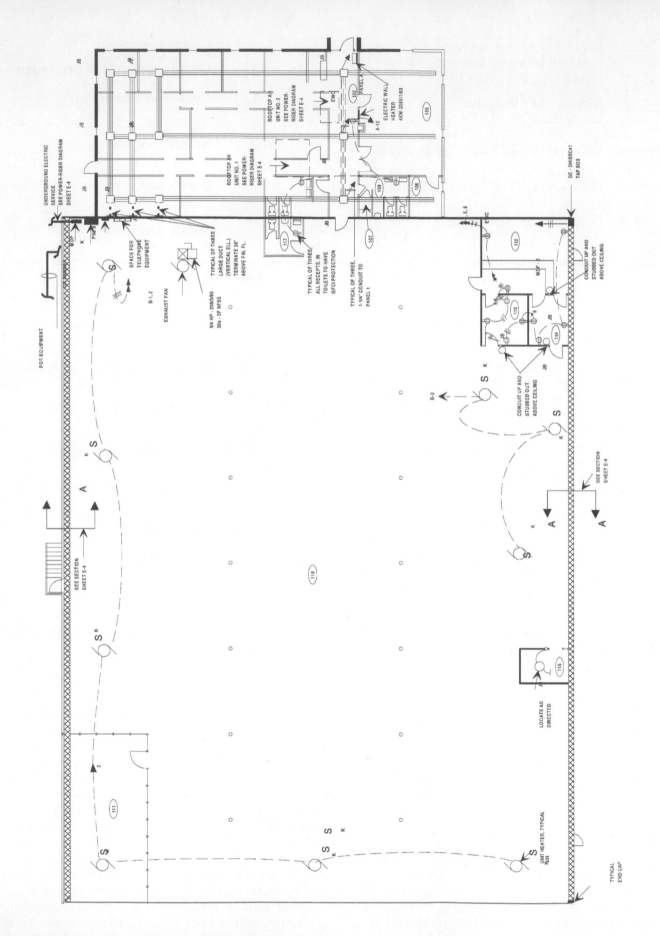

Figure 35 Floor plan for an office/warehouse building.

views of the service and telephone equipment, and duplex receptacle outlets. A note on the drawing requires all receptacles in the toilets to be provided with ground fault circuit interrupter (GFCI) protection.

A section of the drawing in *Figure 35* is shown in an enlarged view with additional details in *Figure 36*. The letters EWC next to the receptacle in the vestibule designate this receptacle for use with an electric water cooler. Notice the numbers placed inside an oval symbol in each room. These numbered ovals represent the room name or type and correspond to a room schedule in the architectural drawings. For example, on the room schedule for the floor plan in *Figure 36* (this room schedule is on a drawing sheet not provided), room number 112 is designated as the lobby, room number 113 is designated as office No. 1, and so on. On some drawings, these room symbols are omitted and the room names are written out on the drawings.

3.3.4 Identifying Luminaires on a Lighting Floor Plan

The next few figures are from a textbook publisher's office. For this building, a partial view of a lighting floor plan is shown in *Figure 37*. This plan is drawn to a scale of $1/8" = 1'-0"$. These symbols are drawn to indicate the physical shape and approximate size of the fixture. Different engineering professionals may use different ways to identify fixtures, such as seen in this drawing, or a code inside a triangle or other shape.

For this plan set, a partial luminaire symbol schedule is given in *Table 2*. The fixtures identified with an E at their end are for emergency lighting and are not separately listed on this schedule. The schedule gives the details needed for selecting and installing each fixture. On some projects, this schedule or details for it may be found only in the written specifications.

3.3.5 Correlating Information Between Plans

Very often the electrician will have to work in spaces congested by other mechanical or piping systems. This is particularly true near electrical and mechanical equipment rooms. The next four drawings are the different systems in a single area of the same publisher's building. *Figure 38* is the power; *Figure 39* is the lighting; *Figure 40* is the communications; and *Figure 41* is the mechanical (HVAC).

Don't Just Check the Electrical Plan

Always review all of the drawings in a drawing set, not just the electrical plan. Several drawings in the set will have information of relevance to the electrician. For example, you should review:

- Site plans for utility lines and elevation information
- Mechanical drawings for routing, clearances, and HVAC equipment and controls
- Architectural drawings for the type of construction (block, wood, metal stud, etc.), fire ratings, and special details.
- Finish drawings (e.g., reflected ceiling plans) for locations of luminaires, fans, and other devices
- Room finish schedules for ceiling heights and floor and wall finishing details
- Plumbing drawings for pumps, water service, and sprinklers

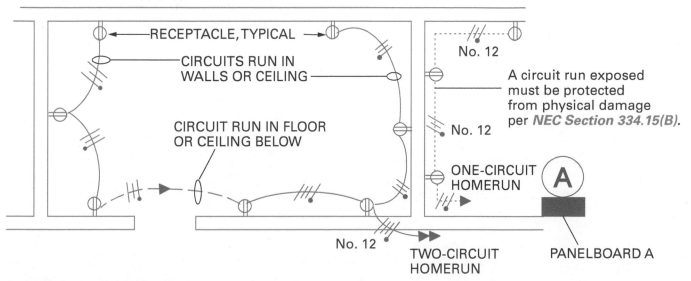

Figure 36 Partial floor plan for office/warehouse building.

Table 2 Partial Luminaire Symbol Schedule

Type	Mfg	Cat. #	Description	# lamps	Watts	Desc.	Input Watts	Mount
A2	Metalux	AC	2x4 recessed	2	17	T8	48	Grid
A4	Metalux	RDI	2x4 indirect recessed	3	32	T8	114	Grid
B1	Metalux	GR8	2x2 recessed	2	17	T8	48	Grid
B2	Metalux	GR8	2x4 recessed	2	32	T8	76	Grid
C1	Cooper	PD6	6" downlight	1	36	CFL-TRT	30	Grid
C2	Cooper	PD6	6" downlight	1	32	CFL-TRT	36	Grid
C3	Cooper	PD6	6" downlight, fresnel	1	18	CFL-TRT	22	Grid
C6	Spectrum	SPC0812CF	6" pendant	1	26	CFL-TRT	36	Pendant
D1	Spectrum	WN	2x4 wraparound	2	32	T8	76	Chain
D3	Spectrum	STPD3LED	3 in 72" black track	3	15	LED, incl.	45	Ceiling
G2	Metalux	SSL	staggered strip	1	32	T8	38	Surface
X1	T&B	LXN	exit, direction shown	1		LED, incl.		Box
X2	T&B	ELXN400	exit, direction shown	1		LED, incl.		Box

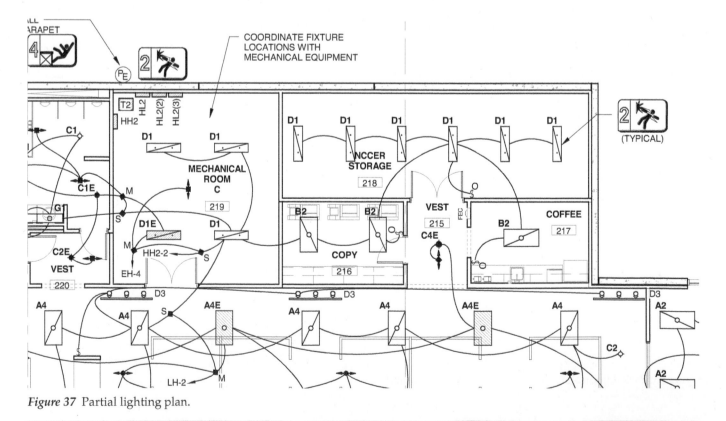

Figure 37 Partial lighting plan.

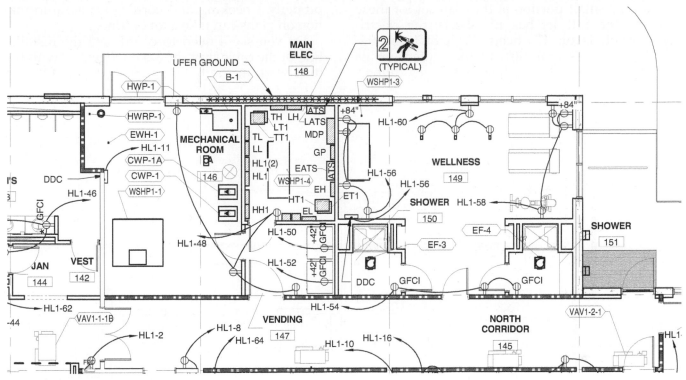

Figure 38 Partial power detail.

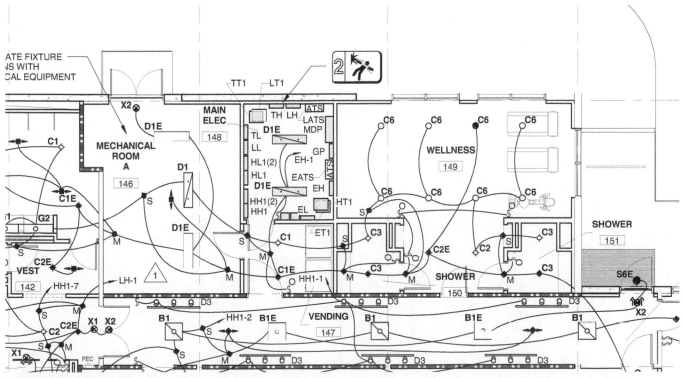

Figure 39 Partial lighting detail.

Figure 38, Figure 39, Figure 40, and Figure 41 are a very small portion of the plan set for these offices—the full set has 107 sheets; the electrical portion is on 17 sheets. Each of these figures shows less than 10% of the content on its full sheet. Other sheets show details, elevations, and schedules necessary for completing the work properly. Checking and correlating the information on plans can take a lot of time.

Do you see a need to coordinate the installation of the cable tray shown on *Figure 40* with the duct work in the hallway on *Figure 41*? On two

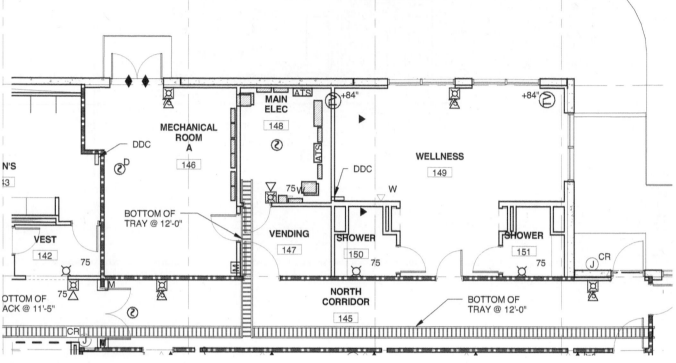

Figure 40 Partial communication detail.

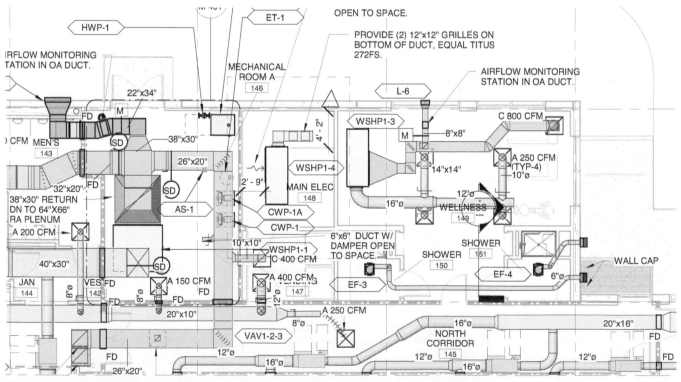

Figure 41 Partial mechanical detail.

drawings, you will find the air handler marked WSHP1-4 in the main electric room (room number 148), but the lighting drawing (*Figure 39*) shows a fixture in the same location but doesn't even show the air handler. This emphasizes the importance of examining the drawings thoroughly before completing the installation.

3.4.0 Schedules, Block Diagrams, and Schematic Diagrams

A schedule presents specific types of related information on a drawing in the form of a table. Electrical diagrams are drawings that show different electrical components and their connections together. They are seldom, if ever, drawn to scale.

3.4.1 Drawing Schedules

When properly organized and thoroughly understood, schedules are powerful timesaving devices for those who prepare the drawings and the workers on the job. For example, a luminaire (lighting fixture) schedule lists the luminaires and identifies each luminaire type on the drawing by number. It shows the manufacturer and catalog number of each luminaire, along with the number, size, and type of lamp required.

Panelboard schedules usually indicate the panel ID, type of mounting (either flush or surface), panel mains (ampere and voltage rating), phase (single- or three-phase), and number of wires. A four-wire panel, for example, is for a three-phase panel with a solid neutral. For each branch circuit, the schedule lists the type of overcurrent protection (number of poles, trip rating, and catalog number), wire size, and load being fed. The panelboard schedule also frequently gives the calculated load for each circuit and the total load on the panelboard feeders. See *Figure 42* for a small typical panelboard schedule.

Other schedules that are frequently found on electrical working drawings include: connected load schedule, electric heat schedule, and kitchen equipment schedule. Depending on the type of project, these schedules may also include lists of equipment such as motors, motor controllers, and similar items. The information in these schedules provides the details for wiring identified pieces of equipment shown on the drawings.

The electrical drawings for the publishing company's office building include the following:

- Luminaire schedule
- Transformer schedule
- Feeder schedule
- Mechanical equipment connections schedule
- 15 panelboard schedules

At times, all the information found in schedules will be duplicated in the written specifications; but combing through page after page of written specifications can be time consuming. Workers do not always have access to the specifications while on the job, whereas they usually do have access to the working drawings. Therefore, the schedule is an excellent way to provide essential information in a clear, compact, and accurate manner. This allows the workers to carry out their assignments in the least amount of time.

PANELBOARD SCHEDULE

| PANEL No. | CABINET TYPE | PANEL MAINS | | | BRANCHES | | | | | ITEMS FED OR REMARKS |
		AMPS	VOLTS	PHASE	1P	2P	3P	PROT.	FRAME	
MDP	SURFACE	600A	120/208	3φ,4-W	-	-	1	225A	25,000	PANEL "A"
					-	-	1	100A	18,000	PANEL "B"
					-	-	1	100A		POWER BUSWAY
					-	-	1	60A		LIGHTING BUSWAY
					-	-	1	70A		ROOFTOP UNIT #1
					-	-	1	70A	↓	SPARE
					-	-	1	600A	42,000	MAIN CIRCUIT BRKR

Figure 42 Typical panelboard schedule.

3.4.2 Block Diagrams

A single-line block diagram, also called a one-line diagram, is typically used to show the arrangement of electric service equipment. A power-riser diagram (*Figure 43*) is a common example. Power-riser diagrams show all pieces of electrical equipment as well as the connecting lines used to indicate service-entrance conductors and feeders. Notes are used to identify the equipment, indicate the size of conduit necessary for each feeder, and show the number, size, and type of conductors in each conduit.

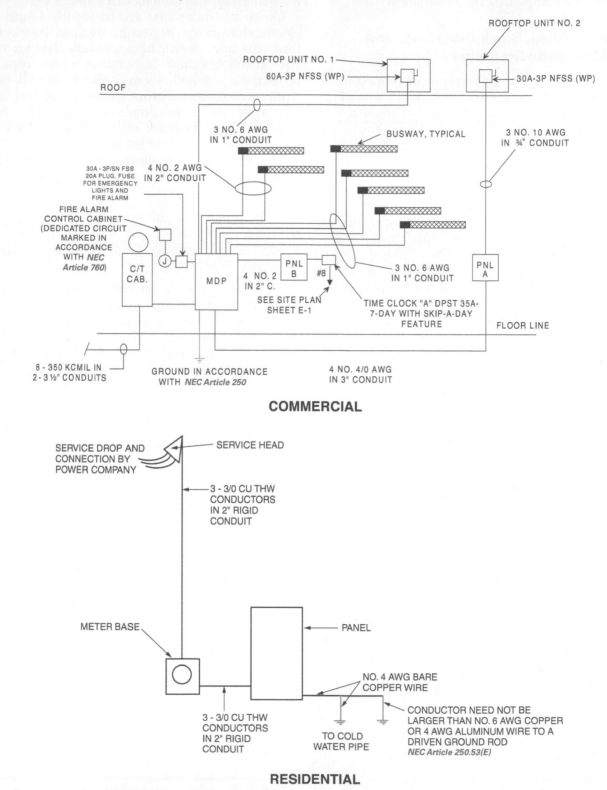

Figure 43 Commercial and residential power-riser diagrams.

3.4.3 Schematic Diagrams

Complete schematic wiring diagrams are normally used only in complicated electrical systems, such as control circuits.

For complex pieces of equipment and control systems, two different types of schematic diagrams may be used. One type is similar in form to a one-line block diagram; the components are within blocks and the wiring is abbreviated to a single line between these blocks or group of blocks. When drawn like this, the individual wires are uniquely identified where they branch from the single line "assembly" onto the components within their blocks. This type of diagram is very helpful for locating components when troubleshooting or for installing the wiring between components in separate areas of a building or sections of equipment.

The second type of schematic diagram typically appears like a ladder and is called a *ladder diagram*, with vertical lines for power flow and horizontal "rungs" for each separate control function. When drawn like this, the wires are shown individually and the components are often split into separate parts that may appear on separate rungs or even separate drawing sheets. Dashed or dotted lines and break lines are used to show the physical relationship of components.

Figure 44 shows a complete schematic wiring diagram for a three-phase, AC magnetic non-reversing motor starter.

Note that this diagram shows the various devices in symbol form and indicates the actual connections of all wires between the devices. The motor power flow starts at the disconnect (on terminals typically marked L1, L2, and L3) then through the contactor and the thermal overload sensor (using terminals typically marked T1, T2, and T3) to the motor. The control wiring takes power from L1 ahead of the contactor through both normally closed stop pushbuttons (wired in series) to both normally open start pushbuttons (wired in parallel). Pressing either start button puts control power through the contactor coil and normally closed overload contact to L2. When the contactor closes, its auxiliary contact (wired in parallel with the start pushbuttons) seals in the operation of the contactor. The power is applied to the motor until either stop pushbutton is pressed or an overload occurs.

Any number of additional pushbutton stations may be added to this control circuit. When adding pushbutton stations, the stop buttons are always connected in series and the start buttons are always connected in parallel.

Schematic wiring diagrams have only been touched upon in this module; there are many other details that you will need to know to perform your work in a proficient manner. Later modules cover wiring diagrams in more detail.

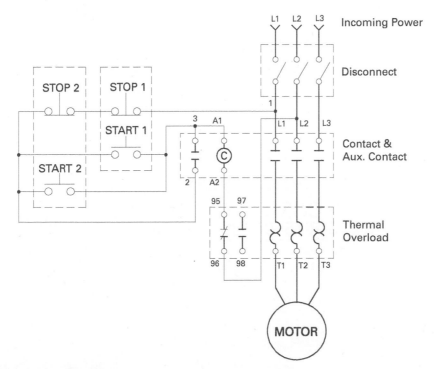

Figure 44 Motor starter wiring diagram.

3.4.4 Drawing Details

A detail drawing is a drawing of a separate item or portion of an electrical system, giving a complete and exact description of its use and all the details needed to show the electrician exactly what is required for its installation. Detail drawings are indicated using sectional cuts with alphanumeric designations (e.g., A3) to identify their location. *Figure 45* is a door security wiring detail drawing for the publishing company's office building.

A set of electrical drawings will sometimes require large-scale drawings of certain areas that are not indicated with sufficient clarity on the small-scale drawings. For example, the site plan may show wiring the electrician must install prior to the concrete contractor's pouring the concrete base for the exterior pole-mounted luminaires.

Understanding Contact Symbols

When a drawing shows normally open (N.O.) or normally closed (N.C.) contacts, the word *normally* refers to the condition of the contacts in their de-energized or shelf state.

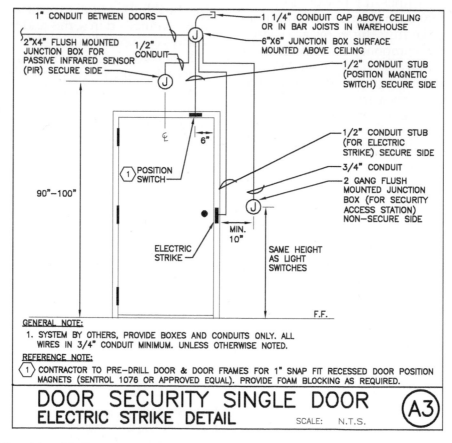

Figure 45 Door security wiring details.

3.0.0 Section Review

1. The letter designation SL on an electrical drawing indicates a _____.

    a. limit switch
    b. low-voltage switch
    c. load switch
    d. light switch

2. The load on a branch circuit should not exceed ____% of its rated capacity.

    a. 100
    b. 90
    c. 80
    d. 70

3. The abbreviation AFF means _____.

    a. above finished floor
    b. as found formerly
    c. after final finish
    d. attach front first

4. A tabular format for giving information on a drawing is called a _____.

    a. symbol list
    b. submittal
    c. schematic diagram
    d. schedule

4.0.0 SPECIFICATIONS AND CHANGE ORDERS

Objective

Integrate specifications with electrical drawings.

a. Select relevant information from written specifications.
b. Compare two formats for written specifications.
c. Identify document changes and the need for them.

Trade Terms

As-built drawings: A marked-up set of drawings, also called *red-lines*, made at the completion of the project, showing all changes made during the construction project.

Change order: A formal document from the project manager or owner, specifying one or more changes to the drawings, specifications, or project scope. It will also state any changes in project cost.

Request for information: A formal document from the contractor, seeking clarification when errors, omissions, and discrepancies are found in or between the construction documents.

Written specifications: A written description of what is required by the owner, architect, and engineer in the way of materials and workmanship. Together with working drawings, the specifications form the basis of the contract requirements for construction.

The written specifications for a building or project are the written descriptions of work and duties required of the owner, architect, and consulting engineer. Together with the working drawings, these specifications form the basis of the contract requirements for the construction of the building or project. A small project may have only a handful of drawings and few (if any) written specifications, but a large or complex project may have well over a hundred drawings and hundreds of pages of specifications. The specifications, combined with the working drawings, govern practically all the important decisions that are made during the construction span of every project.

Written specifications supplement the related working drawings in that they contain details not shown on the drawings. They can have a major impact on the electrician's work and job costs, such as a requirement for a certain type of conduit to be used. This type of information is used to purchase the various items of hardware needed to accomplish the installation in accordance with the contractual requirements.

4.1.0 Content of Specifications

Writing accurate and complete specifications for building construction is a serious responsibility for those who design the buildings. Compiling and writing these specifications is not a simple task, even for those who have had considerable experience in preparing such documents. A set of written specifications for a single project will draw its information from thousands of products, parts, and components, as well as their installation requirements. No one can memorize all of these necessary items. In addition, one must rely upon reference materials such as manufacturer's data and instructions, catalogs, and checklists. This is pointed out in *NEC Section 110.3.*

A high-quality specification will:

- Define and clarify the scope of the job, including work to be done by each craft.
- Describe the specific types and characteristics of the components that are to be used on the job and the requirements for their installation.
- Identify required components by the manufacturer's model and part numbers.

For example, suppose the drawings show light switches by the doors in an apartment, and a note specifies that this apartment is handicap accessible. The specifications reference the Americans with Disabilities Act reach range of an adult in a wheelchair (15"–48") and then specify that on this job, switches will be mounted at a maximum height of 44" and receptacles at a minimum height of 18". Typical mounting heights for switches and receptacles are outside of these specified ranges, so use of the specification contents will save errors and rework.

4.2.0 Format of Specifications

For convenience, speed, and ease of use, specifications are formatted into a series of sections dealing with the construction requirements for the various crafts and areas of work on a construction project. People who use the specifications must be able to find all the information they need without spending too much time looking for it.

Over the last 150 years, construction materials and methods have expanded greatly. The written specifications for construction work have similarly expanded from a few pages to documents that can easily exceed 500 pages or more. After World War II, specifications were standardized and separated into 16 divisions, as follows:

- Division 1 — General Requirements
- Division 2 — Site Construction
- Division 3 — Concrete
- Division 4 — Masonry
- Division 5 — Metals
- Division 6 — Wood and Plastics
- Division 7 — Thermal and Moisture Protection
- Division 8 — Doors and Windows
- Division 9 — Finishes
- Division 10 — Specialties
- Division 11 — Equipment
- Division 12 — Furnishings
- Division 13 — Special Construction
- Division 14 — Conveying Systems
- Division 15 — Mechanical
- Division 16 — Electrical

This list expanded after 2004 into the most common specification writing format used in North America, the MasterFormat®, with five major groupings and 49 divisions. For each division, the first two digits are the division number, the next two digits are subsections of the division, and the two remaining digits are the sub-subsections. For example, the number 262716 represents division 26 (electrical), subsection 27 (low-voltage distribution equipment), and sub-subsection 16 (electrical cabinets and enclosures). If needed, an additional level of detail is marked by a decimal and a number added to the end of the six-digit number. The 2016 version of MasterFormat® is shown in *Figure 46*.

Under the new standard, the Facility Services Subgroup contains the divisions that are most important to the electrician. These include the following divisions:

- Division 25 – Integrated Automation
- Division 26 – Electrical
- Division 27 – Communications
- Division 28 – Electronic Safety and Security

Figure 47A and *Figure 47B* is a breakdown of the electrical division into its subsections, with an example of the sub-subsection numbers for subsection 262700.

4.3.0 Changes

Those who use construction drawings and specifications must always be alert to discrepancies between the working drawings and the written specifications. These are some situations where discrepancies may occur:

- Architects or engineers use standard or prototype specifications and attempt to apply them without any modification to specific working drawings.
- Previously prepared standard drawings are changed or amended by reference in the specifications only and the drawings themselves are not changed.
- Items are duplicated in both the drawings and specifications, but an item is subsequently amended in one and overlooked in the other contract document.

In such instances, the person in charge of the project has the responsibility to ascertain whether the drawings or the specifications take precedence. Such questions must be resolved by a **request for information**, preferably before the work begins, to avoid added costs to the owner, architect/engineer, or contractor.

Often, a change in specified materials or methods can save cost or improve the finished job. At other times, a specified product or component is no longer available or has been replaced by one which better fits the requirements of the job.

In all of these cases, a formal request for information or a less formal conference between the parties will result in a **change order** that officially authorizes a change in the work. This change order may be accompanied by one or more revisions to the drawings, properly identified in the revisions block and as a cloud or balloon figure on the drawings, or by a change to the written specifications.

When the electrical work is completed, one task remains—completing the **as-built drawings**. Traditionally, these changes have been marked with a red pencil or pen and are therefore called *red-lines*. Typically, the foreman will mark the corrections on the site drawings and then convey them back to the draftsman or designer to make the final set. The best way to do this is to mark up two sets of drawings, keeping one onsite for reference until the formal as-built drawings are provided.

MasterFormat Groups, Subgroups, and Divisions

PROCUREMENT AND CONTRACTING REQUIREMENTS GROUP

Division 00 – Procurement and Contracting
                    Requirements
     Introductory Information
     Procurement Requirements
     Contracting Requirements

SPECIFICATIONS GROUP

GENERAL REQUIREMENTS

Division 01 – General Requirements

FACILITY CONSTRUCTION SUBGROUP

Division 02 – Existing Conditions
Division 03 – Concrete
Division 04 – Masonry
Division 05 – Metals
Division 06 – Wood, Plastics, and Composites
Division 07 – Thermal and Moisture Protection
Division 08 – Openings
Division 09 – Finishes
Division 10 – Specialties
Division 11 – Equipment
Division 12 – Furnishings
Division 13 – Special Construction
Division 14 – Conveying Equipment
Division 15 – Reserved for Future Expansion
Division 16 – Reserved for Future Expansion
Division 17 – Reserved for Future Expansion
Division 18 – Reserved for Future Expansion
Division 19 – Reserved for Future Expansion

FACILITY SERVICES SUBGROUP

Division 20 – Reserved for Future Expansion
Division 21 – Fire Suppression

Division 22 – Plumbing
Division 23 – Heatin, Ventilating, and
                    Air-Conditioning (HVAC)

Division 24 – Reserved for Future Expansion
Division 25 – Integrated Automation
Division 26 – Electrical
Division 27 – Communications
Division 28 – Electronic Safety and Security
Division 29 – Reserved for Future Expansion

SITE AND INFRASTRUCTURE SUBGROUP

Division 30 – Reserved for Future Expansion
Division 31 – Earthwork
Division 32 – Exterior Improvements
Division 33 – Utilities
Division 34 – Transportation
Division 35 – Waterway and Marine Construction
Division 36 – Reserved for Future Expansion
Division 37 – Reserved for Future Expansion
Division 38 – Reserved for Future Expansion
Division 39 – Reserved for Future Expansion

PROCESS EQUIPMENT SUBGROUP

Division 40 – Process Interconnections
Division 41 – Material Processing and Handling
                    Equipment
Division 42 – Process Heating, Cooling, and
                    Drying Equipment
Division 43 – Process Gas and Liquid Handling,
                    Purification, and Storage Equipment
Division 44 – Pollution and Waste Control
                    Equipment
Division 45 – Industry-Specific Manufacturing
                    Equipment
Division 46 – Water and Wastewater Equipment
Division 47 – Reserved for Future Expansion
Division 48 – Electrical Power Generation
Division 49 – Reserved for Future Expansion

Figure 46 2016 MasterFormat®.

DIVISION 26 – ELECTRICAL

26 00 00   Electrical

may be used as division level section title.

See: 02 41 19 for selective demolition of existing electrical systems.
03 30 00 for cast-in-place concrete equipment bases.
07 84 00 for firestopping.
07 92 00 for joint sealants.
08 31 00 for access doors and panels.
09 91 00 for field painting.
31 23 33 for trenching and backfilling.

26 01 00   Operation and Maintenance of Electrical Systems

Includes: maintenance, repair, rehabilitation, replacement, restoration, preservation, etc. of electrical systems. medium voltage: 2400 V to 69 kV. low voltage: 600 V and less.

Notes: Definitions medium voltage: 2400 V to 69 kV. low voltage: 600 V and less.

Level 4 Numbering Recommendation: following numbering is recommended for the creation of Level 4 titles:
.51-.59 for maintenance.
.61-.69 for repair.
.71-.79 for rehabilitation.
.81-.89 for replacement.
.91-.99 for restoration.

26 01 10   Operation and Maintenance of Medium-Voltage Electrical Distribution
26 01 20   Operation and Maintenance of Low-Voltage Electrical Distribution
26 01 26   Maintenance Testing of Electrical Systems
26 01 30   Operation and Maintenance of Facility Electrical Power Generating and Storing Equipment
26 01 40   Operation and Maintenance of Electrical and Cathodic Protection Systems
26 01 50   Operation and Maintenance of Lighting
　　　　　26 01 50.51   Luminaire Relamping
　　　　　26 01 50.81   Luminaire Replacement

26 05 00   Common Work Results for Electrical

Includes: subjects common to multiple titles in Division 26. raceway and boxes includes conduit, tubing, surface raceways, and electrical boxes. medium voltage: 2400 V to 69 kV. low voltage: 600 V and less. control voltage: 50 V

311

Figure 47A Breakdown of the electrical division (1 of 2).

and less.

Alternate Terms/Abbreviations: EMT: electrical metallic tubing.

Notes: Definitions medium voltage: 2400 V to 69 kV. low voltage: 600 V and less. control voltage: 50 V and less.

See 01 80 00 for performance requirements of subjects common to multiple titles.
05 35 00 for raceway decking assemblies.
05 45 16 for electrical metal supports.
13 48 00 for sound, vibration, and seismic control.
25 05 13 for conductors and cables for integrated automation.
25 05 26 for grounding and bonding for integrated automation.
25 05 28 for pathways for integrated automation.
25 05 48 for vibration and seismic control for integrated automation.
25 05 53 for identification for integrated automation.
27 05 28 for pathways for communications systems.
27 05 46 for utility poles for communications systems.
27 05 48 for vibration and seismic controls for communications.
27 05 53 for identification for communications.
28 05 13 for conductors and cables for electronic safety and security.
28 05 26 for grounding and bonding for electronic safety and security.
28 05 28 for pathways for electronic safety and security.
28 05 48 for vibration and seismic controls for electronic safety and security.
28 05 53 for identification for electronic safety and security.
33 71 16 for electrical utility poles.
33 71 19 for electrical utility underground ducts and manholes.

Number	Title
26 05 13	Medium-Voltage Cables
26 05 13.13	Medium-Voltage Open Conductors
26 05 13.16	Medium-Voltage, Single- and Multi-Conductor Cables
26 05 19	Low-Voltage Electrical Power Conductors and Cables
26 05 19.13	Undercarpet Electrical Power Cables
26 05 19.23	Manufactured Wiring Assemblies
26 05 23	Control-Voltage Electrical Power Cables
26 05 26	Grounding and Bonding for Electrical Systems
26 05 29	Hangers and Supports for Electrical Systems
26 05 33	Raceway and Boxes for Electrical Systems
26 05 33.13	Conduit for Electrical Systems
26 05 33.16	Boxes for Electrical Systems
26 05 33.23	Surface raceways for Electrical Systems

312

Figure 47B Breakdown of the electrical division (2 of 2).

Trending

For many years, construction documents were created on computers but transmitted to the job site in printed or plotted formats. The widespread adoption of smart phones and portable and very powerful computers is rapidly making paper copies of construction documents obsolete. For many new jobs, everything from bids through change orders are being done with electronic documents. This allows for a quicker response to questions; quick access to all documents; and the cost-effective implementation of design-build contracts (the job is bid and contracts awarded before drawings are completed, with the subcontractors having direct input into the drawing content).

Computer Assisted Drafting (CAD) software will continue to be used, but it is also being integrated with Building Information Modeling (BIM) software. BIM is an intelligent 3D model-based process that gives architecture, engineering, and construction professionals the insight and tools to more efficiently plan, design, construct, and manage buildings and infrastructure. Within BIM, every component in a building can be modeled in 3D with additional information regarding its design life, maintenance needs and history, energy efficiency, space and access requirements, materials used, and many more data points. As with CAD software, BIM modeling software is constantly being expanded and adapted to new uses and applications. BIM modeling has become a requirement for new infrastructure and major construction projects in a growing number of countries around the world.

4.0.0 Section Review

1. A well-written specification document will contain _____.

   a. elevation drawings
   b. model and part numbers
   c. *NEC®* references for installation
   d. personal protective equipment (PPE) requirements

2. Specifications for medium-voltage electrical distribution equipment are currently located in MasterFormat® division _____.

   a. 261000
   b. 262000
   c. 263000
   d. 264000

3. A conflict between the drawings and the specifications is resolved by _____.

   a. the equipment supplier
   b. the electrical contractor
   c. the as-built drawings
   d. the change order

1. A dashed line showing the location of an imaginary cut on a drawing is called a _____.

   a. section line
   b. match line
   c. hidden line
   d. break line

2. A section line on a drawing shows _____.

   a. the north orientation
   b. the location of the section on the plan
   c. where to locate receptacles in that section
   d. the section scale

3. In a drawing with dimension lines, the values shown inside the lines are _____.

   a. for general reference only
   b. reduced to scale
   c. always shown in decimal values
   d. the actual dimensions

4. On an architect's triangular scale, the ³⁄₈" = 1'–0"and ¾" = 1'–0" scales will start _____.

   a. at the same end of the same edge
   b. on the same edge as a scale marked 1:38
   c. on opposite ends of the same edge
   d. on opposite ends of different edges

5. A _____ is a special drawing prepared by a vendor for use by their employees.

   a. shop drawing
   b. detail drawing
   c. perspective drawing
   d. plan view

6. The symbol shown in Figure RQ01 represents a _____.

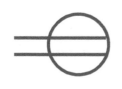

Figure RQ01

   a. single receptacle outlet
   b. duplex receptacle outlet
   c. duplex receptacle outlet split wired
   d. triplex receptacle outlet

7. The symbol shown in Figure RQ02 represents a _____.

Figure RQ02

   a. temperature control relay
   b. transformer
   c. incandescent luminaire (surface or pendant)
   d. incandescent luminaire with pull chain (surface or pendant)

8. The symbol shown in Figure RQ03 represents a _____.

Figure RQ03

   a. single receptacle outlet
   b. duplex receptacle outlet
   c. receptacle outlet split wired
   d. triplex receptacle outlet

9. The symbol shown in Figure RQ04 represents a _____.

Figure RQ04

   a. single receptacle outlet
   b. duplex receptacle outlet
   c. duplex receptacle outlet split wired
   d. quadruplex receptacle outlet

10. The symbol shown in Figure RQ05 represents a _____.

CEILING   WALL

Figure RQ05

   a. temperature control relay
   b. transformer
   c. incandescent surface-mounted luminaire
   d. incandescent surface-mounted luminaire with pull chain

11. What electrical symbol is shown in Figure RQ06?

Figure RQ06

   a. Panelboard
   b. Underfloor bus junction box
   c. Motor controller
   d. Fusible safety switch

12. Dashed lines used to represent a branch circuit on a drawing mean that the wiring is to be _____.
   a. concealed in the ceiling or wall
   b. concealed in the floor
   c. exposed
   d. turned up

13. An electrical drafting line with an arrowhead represents _____.
   a. wiring concealed in the floor
   b. wiring turned down
   c. wiring concealed in a ceiling or wall
   d. a branch circuit homerun

14. To meet general recommendations, a residential branch circuit rated for 2,400VA should have a connected load of no more than _____.
   a. 1,680VA
   b. 1,920VA
   c. 2,040VA
   d. 2,160VA

15. A commercial floor power plan will show _____.
   a. luminaire locations
   b. fire alarm details
   c. service size calculations
   d. electrical panels

16. Power-riser diagrams are used to show the _____.
   a. connections between pieces of electrical equipment
   b. branch circuit layout for power
   c. branch circuit layout for lighting
   d. panelboard schedule

17. The symbols T1, T2, and T3 in a typical motor starter schematic represent _____.
   a. voltage supply lines
   b. auxiliary contacts
   c. motor terminals
   d. line contacts

18. The current MasterFormat® section covering communications systems is under _____.
   a. Division 16
   b. Division 27
   c. Division 37
   d. Division 48

19. The current MasterFormat® section covering integrated automation is _____.
   a. Division 12
   b. Division 18
   c. Division 25
   d. Division 28

20. A change order is made because of _____.
   a. a request for information
   b. a construction error
   c. an as-built drawing
   d. a late material delivery

Trade Terms Quiz

Fill in the blank with the correct term that you learned from your study of this module.

1. _____ typically include the following information: a site plan, floor plans, elevations of all exterior faces of the building, and large-scale detail drawings.

2. A _____ is an exact copy or reproduction of an original drawing.

3. A simple, single-line diagram used to show electrical equipment and related connections is a _____.

4. A _____ shows the path of an electrical circuit or system of circuits, along with the circuit components.

5. To convey a substantial amount of detailed information to installation electricians, an engineer will use an _____.

6. Shown in a separate view, a _____ view is an enlarged, detailed view taken from an area of a drawing.

7. A cutaway drawing that shows the inside of an object or building is a _____.

8. The sizes or measurements that are printed on a drawing are called _____.

9. The relationship between an object's size in a drawing and the object's actual size is the _____.

10. The height of the front, rear, or sides of a building is shown in an _____.

11. A building's location on the site is shown in a _____.

12. A drawing that has a top-down view of a building is a _____.

13. A drawing that has a top-down view of a single object is a _____.

14. A _____ is a single-line block diagram used to indicate the electric service equipment, service conductors and feeders, and subpanels.

15. Owners, architects, and engineers use _____ to specify material and workmanship requirements.

16. A _____ is a systematic way of presenting equipment lists on a drawing in tabular form.

17. Complicated circuits, such as control circuits, are shown in a _____.

18. Usually developed by manufacturers, fabricators, or contractors, a _____ shows specific dimensions and other information about a piece of equipment and its installation methods.

19. A special ruler with increments in inches is the _____.

20. A special ruler with increments in decimal units is the _____.

21. To resolve a conflict between a drawing and specifications you would use a _____.

22. _____ are final drawings that include all work including all changes.

23. _____ show the slope of land on a site plan.

24. A formal modification of the construction documents, scope of work or contract price will be given in a _____.

Trade Terms

Architect's scale
Architectural drawings
As-built drawings
Block diagram
Blueprint
Change order

Contour lines
Detail drawing
Dimensions
Electrical drawing
Elevation view
Engineer's scale
Floor plan

One-line diagram
Plan
Power-riser diagram
Request for information
Scale
Schedule

Schematic diagram
Sectional view
Shop drawing
Site plan
Written specifications

Supplemental Exercises

1. A(n) _____ indicates the location of the building on the property.

2. The _____ show the walls and partitions for each floor or level.

3. What are the three main functions of electrical drawings?
   _____
   _____

4. The title block of an electrical drawing should contain the following ten items:
   _____
   _____

5. Match the following names to their corresponding electrical drafting lines.

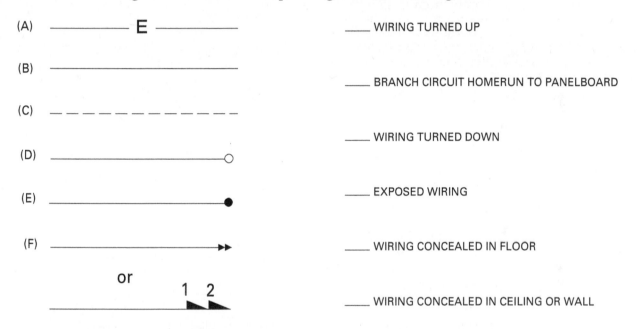

   (A)   ———————— E ————————   ____ WIRING TURNED UP

   (B)   ————————————————   ____ BRANCH CIRCUIT HOMERUN TO PANELBOARD

   (C)   — — — — — — — — — —   ____ WIRING TURNED DOWN

   (D)   ————————————————○   ____ EXPOSED WIRING

   (E)   ————————————————●   ____ WIRING CONCEALED IN FLOOR

   (F)   ———————————————▶▶   ____ WIRING CONCEALED IN CEILING OR WALL

           or
                    1   2
           _____▰▰

6. What does the abbreviation NF stand for in reference to safety switches? _____

7. On a floor plan with a scale of $\frac{1}{2}$" = 1'–0", what would be the equivalent distance if you measured $3\frac{3}{4}$" on the drawing? _____

8. The purpose of a(n) _____ is to identify that part of the project to which the sheet applies.

9. Single-line block diagrams are also known as _____.

10. Division _____ of the current CSI specification covers electrical work.

Wayne Stratton
Associated Builders and Contractors

How did you choose a career in the electrical field?

Three events in my childhood created the desire to learn the electrical craft. At age six, the farmhouse we lived in was destroyed by fire. The cause was electrical. As a young teen, a local electrician had incorrectly wired a heating element and electrocuted several pigs. In 1973, my father hired this electrician to install a motor starter on a grain conveyor. He could not figure it out. I wanted to learn how to do this type of work and do it safely.

Tell us about your apprenticeship experience.

My education is from a technical school. I have attended several manufacturers' training sessions. I had to gain the hands-on experience after learning the craft. My observation of the apprenticeship programs is this: you get hands-on experience while you learn.

What positions have you held in the industry?

I worked as a plant industrial electrician responsible for motor control, DC motors, co-generation, and medium voltage distribution. Later, I began working for an electrical contractor who wanted to expand his business into the industrial field. I worked as a PLC technician designing and installing control systems. In 1987, I began teaching apprenticeship classes.

What would you say is the primary factor in achieving success?

The desire to learn all that I can learn, the ability to think outside the box, and the opportunities to gain a variety of experiences. All this helps me continue to learn and share with trainees.

What does your current job involve?

I teach electrical apprenticeship levels one through four at two different locations in Iowa. My other responsibilities involve task training for electrical licensing, fire alarm, and code updates.

Do you have any advice for someone just entering the trade?

Continue to learn. Completing an apprenticeship program or acquiring an electrician's license is not the end of learning. With code changes every 3 years, there is always more to learn. If you don't understand something, ask! Observe and learn from experienced individuals.

John Mueller
Master Electrician

How did you choose a career in the electrical field?

I enjoy and am good at working with my hands, building things, and problem solving. Within the industry, the work I do most can be highly technical, such as programming controllers (thinking logically), doing infrared scans (problem solving, learning, identifying technical causes and explanations for temperature imbalances), and layout of work (fitting into available space, avoiding conflicts with other work, maintaining future access, and code compliance).

Who inspired you to enter the industry?

Watching and helping my dad when I was under ten, while he rewired a 1905 home and installed a new 200-amp service in the 1950s. When I left the healthcare field, I needed a job.

What positions have you held in the industry?

Master electrician, senior service electrician, controls engineer, and electrical contractor.

How important is education and training in construction?

Vital. If you don't learn from mistakes (both your own and those made by others), you will repeat them. If you are not open to learning as you work, you will not be able to work effectively or efficiently, and you will often have to tear out your work to do it right. Knowing the codes, ways to use materials, and safety practices will allow you to take instructions and a pile of stuff and transform it into a good-looking and usable finished job.

Do you have any advice for someone just entering the trade?

You will never know it all. Everyone can teach you something, even if it is how not to do a job. Listen, listen, listen. Ask questions. Document what you have done so others can follow and build on it, and so you can come back to it and know why you did it a certain way. Be honest. Be responsible. Be safe—for yourself and for others; safe work practices should be habits and not just check-boxes on a form.

How do you define craftsmanship?

Doing things so the completed job will work, be safe, meet the user's needs, meet or exceed the minimum in applicable codes and standards, and look good.

METRIC CONVERSION CHART

METRIC CONVERSION CHART

INCHES Fractional	Decimal	METRIC mm	INCHES Fractional	Decimal	METRIC mm	INCHES Fractional	Decimal	METRIC mm
.	0.0039	0.1000	.	0.5512	14.0000	.	1.8898	48.0000
.	0.0079	0.2000	9/16	0.5625	14.2875	.	1.9291	49.0000
.	0.0118	0.3000	.	0.5709	14.5000	.	1.9685	50.0000
1/64	0.0156	0.3969	37/64	0.5781	14.6844	2	2.0000	50.8000
.	0.0157	0.4000	.	0.5906	15.0000	.	2.0079	51.0000
.	0.0197	0.5000	19/32	0.5938	15.0813	.	2.0472	52.0000
.	0.0236	0.6000	39/64	0.6094	15.4781	.	2.0866	53.0000
.	0.0276	0.7000	.	0.6102	15.5000	.	2.1260	54.0000
1/32	0.0313	0.7938	5/8	0.6250	15.8750	.	2.1654	55.0000
.	0.0315	0.8000	.	0.6299	16.0000	.	2.2047	56.0000
.	0.0354	0.9000	41/64	0.6406	16.2719	.	2.2441	57.0000
.	0.0394	1.0000	.	0.6496	16.5000	2 1/4	2.2500	57.1500
.	0.0433	1.1000	21/32	0.6563	16.6688	.	2.2835	58.0000
3/64	0.0469	1.1906	.	0.6693	17.0000	.	2.3228	59.0000
.	0.0472	1.2000	43/64	0.6719	17.0656	.	2.3622	60.0000
.	0.0512	1.3000	11/16	0.6875	17.4625	.	2.4016	61.0000
.	0.0551	1.4000	.	0.6890	17.5000	.	2.4409	62.0000
.	0.0591	1.5000	45/64	0.7031	17.8594	.	2.4803	63.0000
1/16	0.0625	1.5875	.	0.7087	18.0000	2 1/2	2.5000	63.5000
.	0.0630	1.6000	23/32	0.7188	18.2563	.	2.5197	64.0000
.	0.0669	1.7000	.	0.7283	18.5000	.	2.5591	65.0000
.	0.0709	1.8000	47/64	0.7344	18.6531	.	2.5984	66.0000
.	0.0748	1.9000	.	0.7480	19.0000	.	2.6378	67.0000
5/64	0.0781	1.9844	3/4	0.7500	19.0500	.	2.6772	68.0000
.	0.0787	2.0000	49/64	0.7656	19.4469	.	2.7165	69.0000
.	0.0827	2.1000	.	0.7677	19.5000	2 3/4	2.7500	69.8500
.	0.0866	2.2000	25/32	0.7813	19.8438	.	2.7559	70.0000
.	0.0906	2.3000	.	0.7874	20.0000	.	2.7953	71.0000
3/32	0.0938	2.3813	51/64	0.7969	20.2406	.	2.8346	72.0000
.	0.0945	2.4000	.	0.8071	20.5000	.	2.8740	73.0000
.	0.0984	2.5000	13/16	0.8125	20.6375	.	2.9134	74.0000
7/64	0.1094	2.7781	.	0.8268	21.0000	.	2.9528	75.0000
.	0.1181	3.0000	53/64	0.8281	21.0344	.	2.9921	76.0000
1/8	0.1250	3.1750	27/32	0.8438	21.4313	3	3.0000	76.2000
.	0.1378	3.5000	.	0.8465	21.5000	.	3.0315	77.0000
9/64	0.1406	3.5719	55/64	0.8594	21.8281	.	3.0709	78.0000
5/32	0.1563	3.9688	.	0.8661	22.0000	.	3.1102	79.0000
.	0.1575	4.0000	7/8	0.8750	22.2250	.	3.1496	80.0000
11/64	0.1719	4.3656	.	.8858	22.5000	.	3.1890	81.0000
.	0.1772	4.5000	57/64	.89063	22.6219	.	3.2283	82.0000
3/16	0.1875	4.7625	.	.9055	23.0000	.	3.2677	83.0000
.	0.1969	5.0000	29/32	.90625	23.0188	.	3.3071	84.0000
13/64	0.2031	5.1594	59/64	.92188	23.4156	.	3.3465	85.0000
.	0.2165	5.5000	.	.9252	23.5000	.	3.3858	86.0000
7/32	0.2188	5.5563	15/16	.93750	23.8125	.	3.4252	87.0000
15/64	0.2344	5.9531	.	.9449	24.0000	.	3.4646	88.0000
.	0.2362	6.0000	61/64	.95313	24.2094	3 1/2	3.5000	88.9000
1/4	0.2500	6.3500	.	.9646	24.5000	.	3.5039	89.0000
.	0.2559	6.5000	31/32	.96875	24.6063	.	3.5433	90.0000
17/64	0.2656	6.7469	.	.9843	25.0000	.	3.5827	91.0000
.	0.2756	7.0000	63/64	.98438	25.0031	.	3.6220	92.0000
9/32	0.2813	7.1438	1	1.000	25.40	.	3.6614	93.0000
.	0.2953	7.5000	.	1.0039	25.5000	.	3.7008	94.0000
19/64	0.2969	7.5406	.	1.0236	26.0000	.	3.7402	95.0000
5/16	0.3125	7.9375	.	1.0433	26.5000	.	3.7795	96.0000
.	0.3150	8.0000	.	1.0630	27.0000	.	3.8189	97.0000
21/64	0.3281	8.3344	.	1.0827	27.5000	.	3.8583	98.0000
.	0.3346	8.5000	.	1.1024	28.0000	.	3.8976	99.0000
11/32	0.3438	8.7313	.	1.1220	28.5000	.	3.9370	100.0000
.	0.3543	9.0000	.	1.1417	29.0000	4	4.0000	101.6000
23/64	0.3594	9.1281	.	1.1614	29.5000	.	4.3307	110.0000
.	0.3740	9.5000	.	1.1811	30.0000	4 1/2	4.5000	114.3000
3/8	0.3750	9.5250	.	1.2205	31.0000	.	4.7244	120.0000
25/64	0.3906	9.9219	1 1/4	1.2500	31.7500	5	5.0000	127.0000
.	0.3937	10.0000	.	1.2598	32.0000	.	5.1181	130.0000
13/32	0.4063	10.3188	.	1.2992	33.0000	.	5.5118	140.0000
.	0.4134	10.5000	.	1.3386	34.0000	.	5.9055	150.0000
27/64	0.4219	10.7156	.	1.3780	35.0000	6	6.0000	152.4000
.	0.4331	11.0000	.	1.4173	36.0000	.	6.2992	160.0000
7/16	0.4375	11.1125	.	1.4567	37.0000	.	6.6929	170.0000
.	0.4528	11.5000	.	1.4961	38.0000	.	7.0866	180.0000
29/64	0.4531	11.5094	1 1/2	1.5000	38.1000	.	7.4803	190.0000
15/32	0.4688	11.9063	.	1.5354	39.0000	.	7.8740	200.0000
.	0.4724	12.0000	.	1.5748	40.0000	8	8.0000	203.2000
31/64	0.4844	12.3031	.	1.6142	41.0000	.	9.8425	250.0000
.	0.4921	12.5000	.	1.6535	42.0000	10	10.0000	254.0000
1/2	0.5000	12.7000	.	1.6929	43.0000	20	20.0000	508.0000
.	0.5118	13.0000	.	1.7323	44.0000	30	30.0000	762.0000
33/64	0.5156	13.0969	1 3/4	1.7500	44.4500	40	40.0000	1016.000
17/32	0.5313	13.4938	.	1.7717	45.0000	60	60.0000	1524.000
.	0.5315	13.5000	.	1.8110	46.0000	80	80.0000	2032.000
35/64	0.5469	13.8906	.	1.8504	47.0000	100	100.0000	2540.000

TO CONVERT TO MILLIMETERS, MULTIPLY INCHES × 25.4
TO CONVERT TO INCHES, MULTIPLY MILLIMETERS × 0.03937*
*FOR SLIGHTLY GREATER ACCURACY WHEN CONVERTING TO INCHES, DIVIDE MILLIMETERS BY 25.4

Architect's scale: A special ruler with various measurement scales that can be used when drafting or making measurements on architectural drawings. Architect's scales measure distances in inches.

Architectural drawings: Working drawings consisting of site plans, floor plans, elevations, sectional views, details, and other information necessary for the construction of a building.

As-built drawings: A marked-up set of drawings, also called *red-lines*, is made at the completion of the project, showing all changes made during the construction project.

Block diagram: A single-line diagram used to show electrical equipment and related connections. See power-riser diagram.

Blueprint: An exact copy or reproduction of an original drawing.

Change order: A formal document from the project manager or owner, specifying one or more changes to the drawings, specifications, or project scope. It will also state any changes in project cost.

Contour lines: Curving lines on a site plan, following a given elevation. The space between contour lines tells the slope of the property, such as steep when they are close together or fairly level when they are widely separated.

Detail drawing: An enlarged, detailed view taken from an area of a drawing and shown in a separate view.

Dimensions: Sizes or measurements printed on a drawing.

Electrical drawing: A means of conveying a large amount of exact, detailed information in an abbreviated language. Consists of lines, symbols, dimensions, and notations to accurately convey an engineer's designs to electricians who install the electrical system on a job.

Elevation view: An architectural drawing showing height, but not depth; usually the front, rear, and sides of a building or object.

Engineer's scale: A special ruler with various measurement scales that can be used when drafting or making measurements on architectural drawings. Engineer's scales measure distances in decimal units.

Floor plan: A drawing of a building as if a horizontal cut were made through a building at about window level, and the top portion removed. The floor plan is what would appear if the remaining structure were viewed from above.

One-line diagram: A drawing that shows, by means of lines and symbols, the path of an electrical circuit or system of circuits along with the various circuit components. Also called a *single-line diagram*.

Plan: A drawing made as though the viewer were looking straight down (from above) on an object.

Power-riser diagram: A single-line block diagram used to indicate the electric service equipment, service conductors and feeders, and subpanels. Notes are used on power-riser diagrams to identify the equipment; indicate the size of conduit; show the number, size, and type of conductors; and list related materials. A panelboard schedule is usually included with power-riser diagrams to indicate the exact components (panel type and size), along with fuses, circuit breakers, etc., contained in each panelboard.

Request for information: A formal document from the contractor, seeking clarification when errors, omissions, and discrepancies are found in or between the construction documents.

Scale: On a drawing, the size relationship between an object's actual size and the size it is drawn. Scale also refers to the measuring tool used to determine this relationship.

Schedule: A systematic method of presenting equipment lists on a drawing in tabular form.

Schematic diagram: A detailed diagram showing complicated circuits, such as control circuits.

Sectional view: A cutaway drawing that shows the inside of an object or building.

Shop drawing: A drawing that is usually developed by manufacturers, fabricators, or contractors to show specific dimensions and other pertinent information concerning a particular piece of equipment and its installation methods.

Site plan: A drawing showing the location of a building or buildings on the building site. Such drawings frequently show topographical lines, electrical and communication lines, water and sewer lines, sidewalks, driveways, and similar information.

Written specifications: A written description of what is required by the owner, architect, and engineer in the way of materials and workmanship. Together with working drawings, the specifications form the basis of the contract requirements for construction.

Additional Resources

This module presents thorough resources for task training. The following resource material is suggested for further study.

Electronics Fundamentals: Circuits, Devices, and Applications, Thomas L. Floyd. New York: Prentice Hall.
National Electrical Code® Handbook, Latest Edition. Quincy, MA: National Fire Protection Association.
Principles of Electric Circuits, Thomas L. Floyd. New York: Prentice Hall.

Figure Credits

iStock@Branislav, Module Opener

Mike Powers, Figures 19, 21

©2018 The Construction Specifications Institute, Inc. (CSI). *Master Format*® excerpt and trademarks used under permission from CSI, Figures 46, 47A, 47B.

Section Review Answer Key

Section 1.0.0

Answer	Section Reference	Objective
1. c	1.1.4	1a
2. d	1.2.3	1b
3. a	1.3.0	1c

Section 2.0.0

Answer	Section Reference	Objective
1. c	2.1.0	2a
2. d	2.2.0	2b
3. d	2.3.0	2c

Section 3.0.0

Answer	Section Reference	Objective
1. b	3.1.0; Figure 23	3a
2. c	3.2.1	3b
3. a	3.3.2	3c
4. d	3.4.0	3d

Section 4.0.0

Answer	Section Reference	Objective
1. b	4.1.0	4a
2. a	4.2.0; Figure 48	4b
3. d	4.3.0	4c

NCCER CURRICULA — USER UPDATE

NCCER makes every effort to keep its textbooks up-to-date and free of technical errors. We appreciate your help in this process. If you find an error, a typographical mistake, or an inaccuracy in NCCER's curricula, please fill out this form (or a photocopy), or complete the online form at **www.nccer.org/olf**. Be sure to include the exact module ID number, page number, a detailed description, and your recommended correction. Your input will be brought to the attention of the Authoring Team. Thank you for your assistance.

Instructors – If you have an idea for improving this textbook, or have found that additional materials were necessary to teach this module effectively, please let us know so that we may present your suggestions to the Authoring Team.

NCCER Product Development and Revision
13614 Progress Blvd., Alachua, FL 32615

Email: curriculum@nccer.org
Online: www.nccer.org/olf

❏ Trainee Guide   ❏ Lesson Plans   ❏ Exam   ❏ PowerPoints   Other _____

Craft / Level: _____ Copyright Date: _____

Module ID Number / Title: _____

Section Number(s): _____

Description: _____

_____

_____

_____

_____

Recommended Correction: _____

_____

_____

_____

_____

Your Name: _____

Address: _____

_____

Email: _____ Phone: _____

This page is intentionally left blank.

Residential Wiring

OVERVIEW

When planning any electrical system, there are certain general steps to be followed, regardless of the type of construction. Residential electrical systems are essential for many everyday necessities, including heating and air conditioning, lighting, and household appliances. This module discusses basic load calculations and *NEC*® requirements for residential electrical systems. It also describes how to lay out branch circuits, install wiring, size outlet boxes, and install wiring devices.

Module 26111-20

Trainees with successful module completions may be eligible for credentialing through the NCCER Registry. To learn more, go to **www.nccer.org** or contact us at 1.888.622.3720. Our website, **www.nccer.org**, has information on the latest product releases and training.

Your feedback is welcome. You may email your comments to **curriculum@nccer.org**, send general comments and inquiries to **info@nccer.org**, or fill in the User Update form at the back of this module.

This information is general in nature and intended for training purposes only. Actual performance of activities described in this manual requires compliance with all applicable operating, service, maintenance, and safety procedures under the direction of qualified personnel. References in this manual to patented or proprietary devices do not constitute a recommendation of their use.

26111-20 V10.0

Objectives

When you have completed this module, you will be able to do the following:

1. Size the electric service for a dwelling.
   a. Calculate the electric service load.
   b. Apply demand factors.
   c. Calculate appliance loads.
   d. Size the load center.
2. Identify the grounding requirements for a residential electrical system.
   a. Size grounding electrodes.
   b. Size the main bonding jumper.
   c. Install the equipment grounding system.
3. Install service-entrance equipment.
   a. Identify the service drop location.
   b. Select the panelboard location.
4. Identify wiring methods for various types of residences.
   a. Select and install cable systems.
   b. Select and install raceways.
5. Lay out branch circuits and size outlet boxes.
   a. Complete the branch circuit layout for power.
   b. Complete the branch circuit layout for lighting.
   c. Install outlet boxes.
6. Select and install various wiring devices.
   a. Select and install receptacles.
   b. Select and install switches.
   c. Install devices near residential swimming pools, spas, and hot tubs.

Performance Tasks

Under the supervision of the instructor, you should be able to do the following:

1. For a residential dwelling of a given size and equipped with a given list of major appliances, demonstrate or explain how to:

   - Compute lighting, small appliance, and laundry loads.
   - Compute the loads for large appliances.
   - Determine the number of branch circuits required.
   - Size and select the service-entrance conductors, panelboard, and protective devices.

2. Using an unlabeled diagram of a panelboard (Performance Profile Task 2 Worksheet), label the lettered components.
3. Select the proper type and size of outlet box needed for a given set of wiring conditions.

Trade Terms

Appliances
Bonding bushing
Bonding jumper
Branch circuit
Feeder
Load center
Metal-clad (Type MC) cable
Nonmetallic-sheathed (Type NM and NMC) cable
Romex®

Roughing in
Service drop
Service entrance
Service-entrance conductors
Service-entrance equipment
Service lateral
Switch
Switch leg

Industry Recognized Credentials

If you are training through an NCCER-accredited sponsor, you may be eligible for credentials from NCCER's Registry. The ID number for this module is 26111-20. Note that this module may have been used in other NCCER curricula and may apply to other level completions. Contact NCCER's Registry at 888.622.3720 or go to **www.nccer.org** for more information.

> **NOTE**
>
> NFPA 70®, *National Electrical Code®* and *NEC®* are registered trademarks of the National Fire Protection Association, Quincy, MA.

Contents

Contents (continued)

Figures

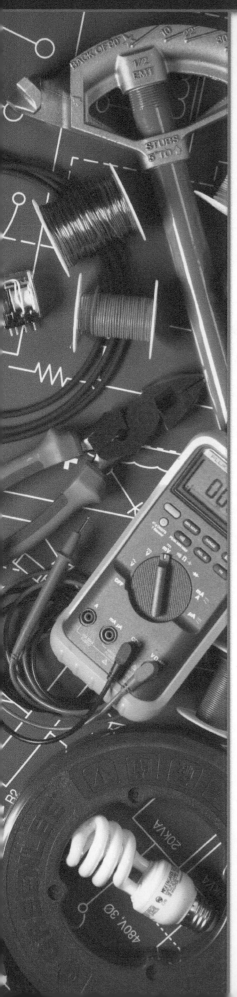

Figures (continued)

SECTION ONE

1.0.0 SIZING THE ELECTRICAL SERVICE

Objective

Size the electric service for a dwelling.

a. Calculate the electric service load.
b. Apply demand factors.
c. Calculate appliance loads.
d. Size the load center.

Performance Task

1. For a residential dwelling of a given size, and equipped with a given list of major appliances, demonstrate or explain how to:
   - Compute lighting, small appliance, and laundry loads.
   - Compute the loads for large appliances.
   - Determine the number of branch circuits required.
   - Size and select the service-entrance conductors, panelboard, and protective devices.

Trade Terms

Appliances: Equipment designed for a particular purpose (for example, using electricity to produce heat, light, or mechanical motion). Appliances are usually self-contained, are generally available for applications other than industrial use, and are normally produced in standard sizes or types.

Branch circuit: The portion of a wiring system extending beyond the final overcurrent device protecting a circuit.

Feeder: Any circuit conductor between the service equipment, the source of a separately derived system, or other power supply source and the final branch circuit overcurrent device.

Load center: A type of panelboard that is normally located at the service entrance of a residential installation. It sometimes contains the main disconnect.

Service entrance: The point where power is supplied to a building (including the equipment used for this purpose). The service entrance includes the service main switch or panelboard, metering devices, overcurrent protective devices, and conductors/raceways for connecting to the power company's conductors.

Service-entrance conductors: The conductors between the point of termination of the overhead service drop or underground service lateral and the main disconnecting device in the building.

Switch: A mechanical device used for turning an electrical circuit On and Off.

The use of electricity in houses began shortly after the opening of the California Electric Light Company in 1879 and Thomas Edison's Pearl Street Station in New York City in 1882. These two companies were the first to enter the business of producing and selling electric service to the public. In 1886, the Westinghouse Electric Company secured patents that resulted in the development and introduction of alternating current; this paved the way for rapid acceleration in the use of electricity.

The primary use of early home electrical systems was to provide interior lighting, but today's uses of electricity include:

- Heating and air conditioning
- Electrical appliances
- Interior and exterior lighting
- Communications systems
- Alarm systems

When planning any electrical system, there are certain general steps to be followed, regardless of the type of construction. In planning a residential electrical system, the electrician must take certain factors into consideration. These include:

- Wiring method
- Overhead or underground electrical service
- Type of building construction
- Type of service entrance and equipment
- Grade of wiring devices and lighting fixtures
- Selection of lighting fixtures
- Type of heating and cooling system
- Control wiring for the heating and cooling system
- Signal and alarm systems
- Presence of alternative electrical systems, if any

The experienced electrician readily recognizes, within certain limits, the type of system that will be required. However, always check the local code requirements when selecting a wiring method. The *NEC*® provides minimum requirements for the practical safeguarding of persons and property from hazards arising from the use of electricity. These minimum requirements are not necessarily efficient, convenient, or adequate for good service or future expansion of electrical use. Some local building codes require electrical installations that surpass the requirements of the *NEC*®. For example, *NEC Section 230.51(A)* requires that service cable be secured by means of cable straps placed every 30" (750 mm) and within 12" (300 mm) of every service head, gooseneck, or connection to a raceway or enclosure. The electrical inspection department in one area requires these cable straps to be placed at a minimum distance of 18" (450 mm).

If more than one wiring method may be practical, a decision as to which type to use should be made prior to beginning the installation.

> **NOTE**
>
> See the *Appendix* for other codes and electrical standards that apply to residential electrical installations. Always refer to the latest editions of codes in effect in your area.

In a residential occupancy, a 120/240V, single-phase service entrance will be provided by the utility company. The service and **feeder** will be three-wire, each **branch circuit** will be either two- or three-wire, and each safety **switch**, the service equipment, and panelboards will be three-wire with a solid neutral. On each project, the electrician must consult with the local utility to determine the point of attachment for overhead connections and the location of the metering equipment.

It may be difficult to decide at times which comes first, the layout of the outlets or the sizing of the electric service. In many cases, the service (main disconnect, panelboard, service conductors, etc.) can be sized using the *NEC*® before the outlets are actually located. In other cases, the outlets will have to be laid out first. However, in either case, the service entrance and panelboard locations will have to be determined before the circuits can be installed—so the electrician will know in which direction (and to which points) the circuit homeruns will terminate. In this module, a typical residence will be used as a model to size the electric service according to the latest edition of the *NEC*®.

1.1.0 Calculating the Electric Service Load

The first step in calculating a service is to review the floor plan. A floor plan is a drawing that shows the length and width of a building and the rooms that it contains. A separate plan is made for each floor.

Figure 1 shows how a floor plan is developed. An imaginary cut is made through the building as shown in the view on the left. The top half of this cut is removed (top right), and the resulting floor plan (bottom) is what the remaining structure looks like when viewed directly from above.

Figure 2 shows the floor plan for a small residence. This building is constructed on a concrete slab with no basement or crawl space. There is an unfinished attic above the living area and an open carport just outside the kitchen entrance. Appliances include a 12 kilovolt-ampere (kVA) electric range, a 4.5kVA water heater, a ½hp 120V disposal, and a 1.5kVA dishwasher.

There is also a washer/dryer (rated at 5.5kVA) in the utility room. A gas furnace with a ⅓hp 120V blower supplies the heating. In this module, the electrical requirements of this example building will be computed.

1.1.1 General Lighting Loads

General lighting loads are calculated on the basis of *NEC Section 220.14(J)*. For residential occupancies, 3 volt-amperes (watts) per square foot of living space (33VA per square meter) is the figure to use. This includes non-appliance duplex receptacles into which lamps, televisions, etc., may be connected. Therefore, the area of the building must be calculated first. If the building is under construction, the dimensions can be determined by scaling the working drawings used by the builder. If the residence is an existing building with no drawings, actual measurements will have to be made on the site.

Using the floor plan of the residence in *Figure 2* as a guide, an architect's scale is used to measure the longest width of the building (using outside dimensions). It is determined to be 33 feet. The longest length of the building is 48 feet. These two measurements multiplied together give 33 x 48 = 1,584 square feet of living area. However, there is an open carport on the lower left of the drawing. This carport area will have to be calculated and then deducted from 1,584 to give the true amount of living space. This open area (carport) is 12 feet wide by 19.5 feet long: 12 x 19.5 = 234 square feet. Subtract the carport area from 1,584 square feet: 1,584 − 234 = 1,350 square feet of living area.

A standard calculation worksheet for a single-family dwelling is shown in *Figure 3*. This form contains numbered blank spaces to be filled in while making the service calculation.

The total area of our sample dwelling (1,350 square feet) is entered in the appropriate space (Box 1) on the form and multiplied by 3 volt-amperes (VA) for a total general lighting load of 4,050VA (Box 2).

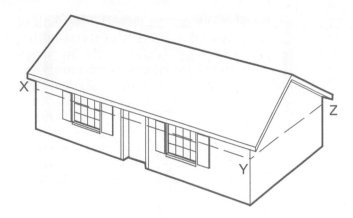

PERSPECTIVE VIEW SHOWING SECTION CUTS

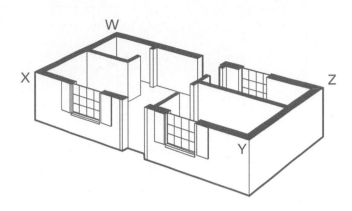

TOP HALF OF SECTION REMOVED

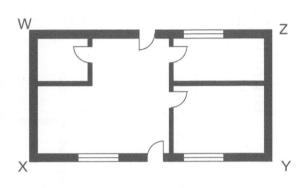

RESULTING FLOOR PLAN IS WHAT THE REMAINING
STRUCTURE LOOKS LIKE WHEN VIEWED FROM ABOVE

Figure 1 Principles of floor plan layout.

When using the floor area to determine lighting loads for buildings, *NEC Section 220.11* requires the floor area for each floor to be computed from the outside dimensions. When calculating lighting loads for residences, the computed floor area must not include open porches, carports, garages, or unused or unfinished spaces that are not adaptable to future use.

1.1.2 Small Appliance Loads

NEC Section 210.11(C)(1) requires at least two 120V, 20A small appliance branch circuits to be installed for the small appliance loads in each kitchen area of a dwelling. Kitchen areas include the dining area, breakfast nook, pantry, and similar areas where small appliances will be used. *NEC Section 220.52(A)* gives further requirements for residential small appliance circuits; that is, the load for those circuits is to be computed at 1,500VA each. Because our example dwelling has only one kitchen area, the number 2 is entered in Box 3 for the number of required kitchen small appliance branch circuits. Multiply the number of these circuits by 1,500 and enter the result in Box 4.

1.1.3 Laundry Circuit

NEC Section 210.11(C)(2) requires an additional 20A branch circuit to be provided for the exclusive use of the laundry area (Box 5). This circuit must not have any other outlets connected except for the laundry receptacle(s) and is calculated at 1,500VA per *NEC Section 220.52(B)*. Therefore, enter 1,500VA in Box 6 on the form.

So far, there is enough information to complete the first portion of the service calculation form:

- General lighting: **4,050VA (Box 2)**
- Small appliance load: **3,000VA (Box 4)**
- Laundry load: **1,500VA (Box 6)**
- Total general lighting and appliance loads: **8,550VA (Box 7)**

1.1.4 Lighting Demand Factors

All residential electrical outlets are never used at one time. There may be a rare instance when all the lighting may be on for a short time every night, but even so, all the small appliances and receptacles throughout the house will never be used simultaneously. Knowing this, *NEC Section*

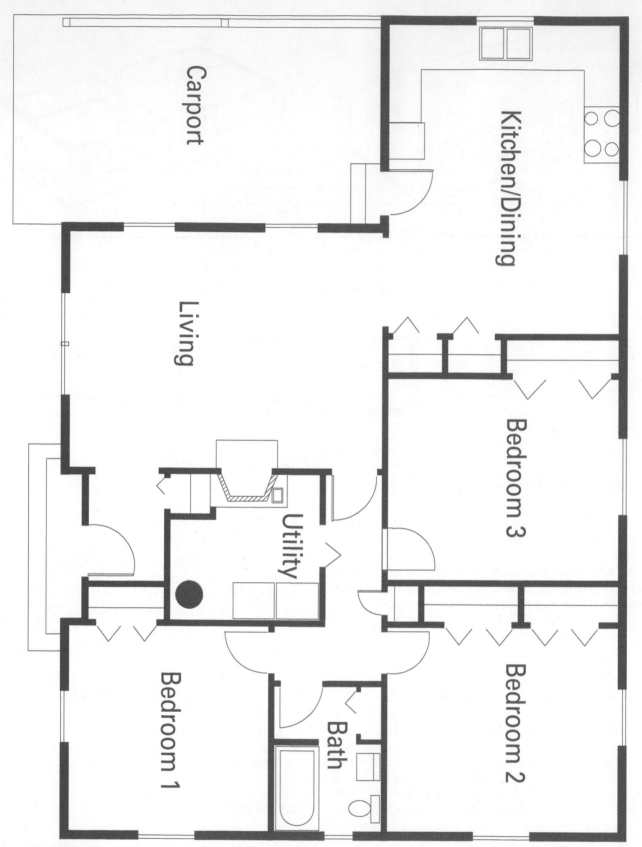

Figure 2 Floor plan of a typical residence.

General Lighting Load							Phase	Neutral
Square footage of the dwelling	[1]	1350	× 3VA =	[2]	4050	*NEC Section 220.14(J)*		
Kitchen small appliance circuits	[3]	2	× 1500 =	[4]	3000	*NEC Section 220.52(A)*		
Laundry branch circuit	[5]	1	× 1500 =	[6]	1500	*NEC Section 220.52(B)*		
Subtotal of gen. lighting loads				[7]	8550			
Subtract 1st 3000VA per *NEC Table 220.42*				[8]	3000	× 100% =	[9] 3000	
Remaining VA times 35% per *NEC Table 220.42*				[10]	5550	× 35% =	[11] 1943	
Total demand for general lighting loads =				[12]	4943		[13]	

Fixed Appliance Loads (Nameplate or *NEC®* FLA of motors) per *NEC Section 220.14(C)*		
Hot water tank, 4.5kVA, 240V	[14]	4500
Dishwasher 1.5kVA, 120V	[15]	1500
Disposal ½ hp, 120V per *NEC Table 430.248* = 9.8A	[16]	1176
Blower ⅓ hp, 120V per *NEC Table 430.248* = 7.2A	[17]	864
	[18]	
	[19]	
Subtotal of fixed appliances	[20]	8040

NEC Section 220.53 — If 3 or less fixed appliances take @ 100% =	[21]		[22]	
If 4 or more fixed appliances take @ 75% =	[23]	6030	[24]	

Other Loads per *NEC Section 220.14*				Phase	Neutral
Electric range per *NEC Section 220.55* [neutral @ 70% per *NEC Section 220.61(B)(1)*]			[25]	8000	[26]
Electric dryer per *NEC Section 220.54* [neutral @ 70% per *NEC Section 220.61(B)(1)*]			[27]	5500	[28]
Electric heat per *NEC Section 220.51*					
Air conditioning *NEC Section 220.82(C)*	omit smaller load per *NEC Section 220.60*		[29]		[30]
Largest Motor = 1176	× 25% (per *NEC Section 430.24*) =		[31]	294	[32]
	Total VA Demand =		[33]	24767	[34]
	(VA divided by 240V) **Amps** =		[35]	**103**	[36]
	Service OCD and minimum size grounding electrode conductor		[37]	125	[38]
	AWG per *NEC Table 310.12*, *NEC Section 220.61*, and *NEC Table 310.16* for neutral		[39]		[40]

Figure 3 Calculation worksheet for residential requirements.

220.42 allows a diversity or demand factor to be used when computing the general lighting load for services. Our calculation continues as follows:

- The first 3,000VA is rated at 100%: **3,000VA (Box 8)**
- The remaining 5,550VA (Box 10) may be rated at 35% (the allowable demand factor); therefore, 5,550 x 0.35 = **1,943VA (Box 11)**
- Net general lighting and small appliance load (rounded off): **4,943VA (Box 12)**

1.1.5 Fixed Appliances

NEC Section 220.53 permits the loads for four or more fixed appliances in a single-family dwelling only to be computed at 75% as long as they are not electric heating, air conditioning, electric cooking, or electric clothes dryer loads. To compute the load of the fixed appliances in this dwelling, list all the fixed appliances that meet *NEC Section 220.53*. Enter the nameplate rating of the appliance or VA for motors by using *NEC Table 430.248* to find the FLA of each motor. *NEC Section 220.5(A)* tells us to use 120V (not 115V) for calculation purposes. The fixed appliances would be as follows:

- Hot water tank: **4,500VA (Box 14)**
- Dishwasher: **1,500VA (Box 15)**
- ½hp 120V disposal (9.8A x 120V): **1,176VA (Box 16)**
- ⅓hp gas furnace blower (7.2A x 120V): **864VA (Box 17)**
- Add the loads for the fixed appliances: **8,040VA (Box 20)**
- Because there are four or more fixed appliances, multiply the total in Box 20 by 75%: **6,030VA (Box 23)**

1.1.6 Other Loads

The remaining loads of the dwelling are now computed in the Other Loads section in *Figure 3*. *NEC Section 220.14(B)* allows electric dryers to be computed as permitted in *NEC Table 220.54* and electric cooking appliances to be computed per *NEC Table 220.55*. For a single range rated over 8.75kVA, but not over 12kVA, Column C of *NEC Table 220.55* permits a demand of 8kVA for the range in this dwelling. Enter 8,000VA in Box 25.

The electric dryer must be computed at 5,000VA or the nameplate, whichever is greater, according to *NEC Section 220.54*. Up to four electric dryers must be taken at 100%. Enter 5,500VA in Box 27.

If this dwelling had electric space heating and/or air conditioning, it would be computed in this section using the larger of the two loads. Because they are typical noncoincidental loads, *NEC Section 220.60* permits the smaller of those loads to be omitted.

The final step in this calculation is to add in 25% of the largest motor in the dwelling. This dwelling unit has two motors: the disposal at 9.8A and the blower at 7.2A. (See *NEC Section 430.17*.) In this case, the larger motor is the disposal; therefore, we must add 25% of the rating to meet the requirements of *NEC Section 430.24*. Enter 294VA (1,176 × 25%) in Box 31. Adding together the individual loads as computed, we have a minimum demand of 24,767VA (Box 33) for the phase conductors.

1.1.7 Required Service Size

The conventional electric service for residential use is 120/240V, three-wire, single-phase. Services are sized in amperes, and when the volt-amperes are known on single-phase services, amperes may be found by dividing the total volt-amperes by the highest voltage. For example:

24,767VA ÷ 240V = 103A (Box 35)

The **service-entrance conductors** have now been calculated and must be rated at a minimum of 110A, which is a standard rating for overcurrent protection. However, this is not a typical trade size; therefore, we will use the more common rating of 125A as the size of our service.

If the demand for our dwelling unit had resulted in a load of less than 100A, *NEC Section 230.79(C)* would have required a minimum rating of 100A for this single-family dwelling service disconnect. *NEC Section 230.42(B)* would have required the ampacity of the service conductors to be equal to the rating of the 100A disconnect as well.

1.2.0 Applying Demand Factors

NEC Article 220, Part III provides the rules regarding the application of demand factors to certain types of loads. Recall that a demand factor is the maximum amount of volt-ampere load expected at any given time compared to the total connected load of the circuit. The maximum demand of a feeder circuit is equal to the connected load times the demand factor. The loads to which demand factors apply can be found in the *NEC®* as follows:

- Receptacle loads: *NEC Section 220.14(I)*
- Lighting loads: *NEC Table 220.42*
- Dryer loads: *NEC Table 220.54*
- Range loads: *NEC Table 220.55*

In addition to those demand factors listed in *NEC Article 220, Part III*, alternative (optional) methods for computing loads can be found in *NEC Article 220, Part IV*. They include the following:

- Dwelling unit loads: *NEC Section 220.82*
- Existing dwelling unit loads: *NEC Section 220.83*
- Multi-family dwelling unit loads: *NEC Section 220.84*

1.3.0 Calculating Appliance Loads

NEC Section 210.11(C) provides the number of branch circuits required for small appliances and laundry loads. Demand factors for dryers and ranges are found in *NEC Tables 220.54 and 220.55*.

1.3.1 Small Appliance Loads

The small appliance branch circuits required by *NEC Section 210.11(C)(1)* for small appliances supplied by 15A or 20A receptacles on 20A branch circuits for each kitchen area served are calculated at 1,500VA. If a dwelling has more than one kitchen area, the *NEC®* will require two small appliance branch circuits computed at 1,500VA for each kitchen area served. Where a dwelling with only one kitchen area has more than the required two small appliance branch circuits installed to serve a single kitchen area, only the first two required circuits need be computed. Additional circuits for countertops or refrigeration provide a separation of load, not additional loads. If a dwelling has two kitchen areas, then the total small appliance branch circuits required would be four at 1,500VA each. These loads are permitted to be included with the general lighting load and subjected to the demand factors of *NEC Table 220.42*.

1.3.2 Laundry Circuit Load

A 1,500VA feeder load is added to load calculations for each two-wire laundry branch circuit installed in a home. The branch circuit is required by *NEC Section 210.11(C)(2)*. This load may also be added to the general lighting load and subjected to the same demand factors provided in *NEC Section 220.42*.

1.3.3 Dryer Load

The dryer load for each electric clothes dryer is 5,000VA or the actual nameplate value of the dryer, whichever is larger. Demand factors listed in *NEC Table 220.54* may be applied for more than one dryer in the same dwelling. If two or more single-phase dryers are supplied by a three-phase, four-wire feeder, the total load is computed by using twice the maximum number connected between any two phases.

1.3.4 Range Load

Range loads and other cooking appliances are covered under *NEC Section 220.55*. The feeder demand loads for household electric ranges, wall-mounted ovens, countertop cooking units, and other similar household appliances individually rated over $1\frac{3}{4}$kW are permitted to be computed in accordance with *NEC Table 220.55*. If two or more single-phase ranges are supplied by a three-phase, four-wire feeder, the total load is computed by using twice the maximum number connected between any two phases.

1.3.5 Demand Factors for Electric Ranges

Ranges can be computed in various ways that depend on which part of *NEC Article 220* you are using and the occupancy type for the ranges involved. Note the demand factors permitted for the following occupancy types:

- Dwelling units per Part III: *NEC Section 220.55*
- Dwelling units per Part IV: *NEC Section 220.82*
- Additions to existing dwellings per Part IV: *NEC Section 220.83*
- Multi-family dwellings per Part III: *NEC Section 220.55*
- Multi-family dwellings per Part IV: *NEC Section 220.84*

Think About It

Demand Factors

Examine *NEC Table 220.55*. Why does the demand factor decrease as the number of appliances increases? Why does the demand factor decrease more for larger ranges than it does for smaller ones?

1.3.6 Demand Factors for Neutral Conductors

The neutral conductor of electrical systems generally carries only the maximum current imbalance of the phase conductors. For example, in a single-phase feeder circuit with one phase conductor carrying 50A and the other carrying 40A, the neutral conductor would carry 10A. Because the neutral in many cases will never be required to carry as much current as the phase conductors, the *NEC®* allows us to apply a demand factor. (See *NEC Section 220.61*.) Note that in certain circumstances such as electrical discharge lighting, data processing equipment, and other similar equipment, a demand factor cannot be applied to the neutral conductors because these types of equipment produce harmonic currents that increase the heating effect in the neutral conductor.

The next step is to size the neutral conductor. The neutral conductor in a three-wire, single-phase service carries only the imbalanced load between the two ungrounded (hot) wires or legs. Because there are several 240V loads in the above calculations, these 240V loads will be balanced and therefore reduce the load on the service neutral conductor. Consequently, in most cases, the service neutral does not have to be as large as the ungrounded (hot) conductors.

Think About It

Balanced Phase Conductors

The word *phase* is used in these modules to refer to a hot wire rather than a neutral one. Some electricians call these legs rather than phases. Why must the phase conductors be balanced?

In the previous example, the water heater does not have to be included in the neutral conductor calculation, because it is strictly 240V with no 120V loads. This takes the total number of fixed appliances on the neutral conductor down to three appliances. Therefore, each of the fixed appliance loads on the neutral must be computed at 100% (dishwasher at 1,500VA, plus disposal at 1,176VA, plus the blower at 864VA). The neutral loads of the electric range and clothes dryer are permitted by *NEC Section 220.61(B)* to be computed at 70% of the demand for the phase conductors because these appliances have both 120V and 240V loads. In this case, the largest motor is the same for the neutral conductors as it is for the phase conductors; therefore, it is computed in the same manner. Using this information, the neutral conductor may be sized accordingly:

- Net general lighting and small appliance load: **4,943VA (Box 13)**
- Fixed appliance loads: **3,540VA (Box 22)**
- Electric range (8,000VA × 0.70): **5,600VA (Box 26)**
- Clothes dryer (5,500VA × 0.70): **3,850VA (Box 28)**
- Largest motor: **294VA (Box 32)**
- Total: **18,227VA (Box 34)**

To find the total phase-to-phase amperes, divide the total volt-amperes by the voltage between phases:

$$18,227VA \div 240V = 75.9A \text{ or } 76A$$

The service-entrance conductors have now been calculated and are rated at 125A with a neutral conductor rated for at least 76A. Refer to *Figure 4* for a completed calculation form for the example residence.

In *NEC Section 310.12*, special consideration is given to 120/240V, single-phase residential services and feeders. Conductors are sized per *NEC Table 310.12*. The neutral conductor is sized per *NEC Section 310.12* or *NEC Table 310.16* using the appropriate column for the markings on the service equipment per *NEC Section 110.14(C)*. Assuming our service panel is marked as suitable for use with 75°C-rated conductors, the minimum size of the neutral would be a No. 4 AWG copper or No. 2 AWG aluminum. Per *NEC Table 250.66*, the grounding electrode conductor is No. 8 AWG.

When sizing the grounded conductor for services, the provisions stated in *NEC Sections 215.2(A)(2), 220.61, and 230.42* must be met, along with other applicable sections.

1.4.0 Sizing the Load Center

Each ungrounded conductor in all circuits must be provided with overcurrent protection in the form of either fuses or circuit breakers. If more than six such devices are used, a means of disconnecting the entire service must be provided using either a main disconnect switch or a main circuit breaker.

To calculate the number of fuse holders or circuit breakers required in the sample residence, look at the general lighting load first. The total general lighting load of 4,050VA can be divided by 120V to find the amperage:

$$4,050VA \div 120V = 33.75A$$

Either 15A or 20A circuits may be used for the lighting load. Two 20A circuits (2 × 20) equal 40A, so two 20A circuits would be adequate for the lighting. However, two 15A circuits total only 30A and 33.75A are needed. Therefore, if 15A circuits are used, three will be required for the total lighting load. In this example, three 15A circuits will be used.

In addition to the lighting circuits, the sample residence will require a minimum of two 20A circuits for the small appliance load and one 20A circuit for the laundry. So far, the following branch circuits can be counted:

- General lighting load: **Three 15A circuits**
- Small appliance load: **Two 20A circuits**
- Laundry load: **One 20A circuit**
- Total: **Six branch circuits**

Usually, the load center and panelboard are provided with an even number of circuit breaker spaces or fuse holders (for example, four, six, eight, or ten). But before the panelboard can be selected, space must be provided for the remaining loads. Each 240V load will require two spaces. In some existing installations, you might find a two-pole fuse block containing two cartridge fuses being used to feed a residential electric range. Each

---

Think About It

Common Loads

Which of the following devices uses the most power?

- Giant-screen television
- Typical hair dryer
- Curling iron
- Crockpot

General Lighting Load

						Phase	Neutral
Square footage of the dwelling	[1] 1350	× 3VA =	[2] 4050	*NEC Section 220.14(J)*			
Kitchen small appliance circuits	[3] 2	× 1500 =	[4] 3000	*NEC Section 220.52(A)*			
Laundry branch circuit	[5] 1	× 1500 =	[6] 1500	*NEC Section 220.52(B)*			
Subtotal of gen. lighting loads			[7] 8550				
Subtract 1st 3000VA per *NEC Table 220.42*			[8] 3000	× 100% =	[9] 3000		
Remaining VA times 35% per *NEC Table 220.42*			[10] 5550	× 35% =	[11] 1943		
Total demand for general lighting loads =						[12] 4943	[13] 4943

Fixed Appliance Loads (Nameplate or *NEC®* FLA of motors) per *NEC Section 220.14*

		Phase	Neutral
Hot water tank, 4.5kVA, 240V	[14]	4500	
Dishwasher 1.5kVA, 120V	[15]	1500	
Disposal ½ hp, 120V per *NEC Table 430.248* = 9.8A	[16]	1176	
Blower ⅓ hp, 120V per *NEC Table 430.248* = 7.2A	[17]	864	
	[18]		
	[19]		
Subtotal of fixed appliances	[20]	8040	
NEC Section 220.53 — If 3 or less fixed appliances take @ 100% =	[21]		[22] 3540
If 4 or more fixed appliances take @ 75% =	[23]	6030	[24]

Other Loads per *NEC Section 220.14*

		Phase		Neutral
Electric range per *NEC Section 220.55* [neutral @ 70% per *NEC Section 220.61(B)(1)*]		[25] 8000	[26]	5600
Electric dryer per *NEC Section 220.5* [neutral @ 70% per *NEC Section 220.61(B)(1)*]		[27] 5500	[28]	3850
Electric heat per *NEC Section 220.51*	omit smaller load per *NEC Section 220.60*	[29]	[30]	
Air conditioning *NEC Section 220.82(C)*				
Largest Motor = 1176	× 25% (per *NEC Section 430.24*) =	[31] 294	[32]	294
Total VA Demand =		[33] 24767	[34]	18227
(VA divided by 240V) **Amps** =		[35] **103**	[36]	**76**
Service OCD and minimum size grounding electrode conductor		[37] 125	[38]	8 AWG
AWG per *NEC Table 310.12*, *NEC Section 220.61*, and *NEC Table 310.16* for neutral		[39] 2 AWG	[40]	4 AWG

Figure 4 Completed calculation form.

120V load will require one space each. Thus, the remaining number of circuits for this example is as follows:

- Hot water heater: **One two-pole breaker**
- Dishwasher: **One single-pole breaker**
- Disposal: **One single-pole breaker**
- Blower: **One single-pole breaker**
- Electric range: **One two-pole breaker**
- Electric dryer: **One two-pole breaker**

These additional appliances will therefore require an additional nine spaces in the load center or panelboard. *NEC Section 210.11(C)(3)* requires that at least one separate 20A branch circuit be provided for the bathroom receptacles. While this circuit requires extra space within a load center, it does not add to the demand on the service for a dwelling unit. Adding the nine spaces for the other loads in the dwelling, plus one for a bathroom circuit, to the six required for the general lighting and small appliance loads requires at least a 16-space load center to handle the circuits.

1.4.1 Ground Fault Circuit Interrupters

Under certain conditions, the amount of current it takes to open an overcurrent protective device can be critical. Remember that a shock of less than 1A can be fatal. The overcurrent protection installed on services, feeders, and branch circuits protects only the conductors and equipment.

Because of this fact, the *NEC®* requires ground fault circuit interrupter (GFCI) protection for receptacle outlets and/or equipment in many locations and occupancies. The *NEC®* defines a GFCI as "a device intended for the protection of personnel that functions to de-energize a circuit or portion thereof within an established period of time when a ground-fault current exceeds the values established for a Class A device." Class A GFCIs trip when the current to ground has a value in the range of 4mA to 6mA.

For dwelling units, the majority of requirements to provide protection for 15A or 20A, 125V-rated receptacles can be found in *NEC Section 210.8(A)*. Additional requirements for

GFCI protection in dwelling units can be found in other *NEC*® articles such as *NEC Article 590* for temporary construction sites; *NEC Article 620* for special equipment such as elevators; or in *NEC Article 680* for special equipment such as swimming pools, hot tubs, and hydromassage tubs. These articles may also expand the requirements for GFCI protection to include circuits rated at more than 20A or operating at 240V.

According to *NEC Section 210.8(A)*, all 125V through 250V receptacles in a dwelling that require GFCI protection must be readily accessible and include the following:

• Bathrooms
• Receptacles within garages and accessory buildings, such as storage sheds or workshops, or similar uses with a floor at or below grade level
• Outdoor receptacles
• Crawl spaces at or below grade level
• Basements
• Receptacles serving the countertops in kitchens
• Receptacles within 6' (1.8 m) of the outside edge of wet bar sinks, utility, or laundry sinks
• Boathouses
• Bathtubs or shower stalls where receptacles are installed within 6' (1.8 m) of outside edge
• Laundry areas

One way to provide this GFCI protection is through the use of a GFCI circuit breaker. GFCI circuit breakers require the same mounting space as standard single-pole circuit breakers and provide the same branch circuit wiring protection as standard circuit breakers. They also provide Class A ground fault protection.

Listed GFCI circuit breakers are available in single- and two-pole construction; 15A, 20A, 25A, and 30A, 50A, and 60A ratings; and have a 10,000A interrupting capacity. Single-pole units are rated at 120VAC; two-pole units are rated at 120/240VAC.

GFCI breakers can be used not only in load centers and panelboards, but they are also available factory-installed in meter pedestals and power outlet panels for recreational vehicle (RV) parks and construction sites.

The GFCI sensor continuously monitors the current balance in the ungrounded or energized (hot) load conductor and the neutral load conductor. If the current in the neutral load wire becomes less than the current in the hot load wire, then a ground fault exists, because a portion of the current is returning to the source by some means other than the neutral load wire. When a current imbalance occurs, the sensor, which is a differential current transformer, sends a signal to the solid-state circuit, which activates the ground trip solenoid mechanism and breaks the hot load connection (*Figure 5*). A current imbalance as low as four milliamps (4mA) will cause the circuit breaker to interrupt the circuit. This is indicated by the trip indicator on the front of the device.

The two-pole GFCI breaker (*Figure 6*) continuously monitors the current balance between the two hot conductors and the neutral conductor. As long as the sum of these three currents is zero, the device will not trip; that is, if the A load wire is carrying 10A of current, the neutral is carrying 5A, and the B load wire is carrying 5A, then the sensor is balanced and will not produce a signal. A current imbalance from a ground fault condition as low as 4mA will cause the sensor to produce a signal of sufficient magnitude to trip the device.

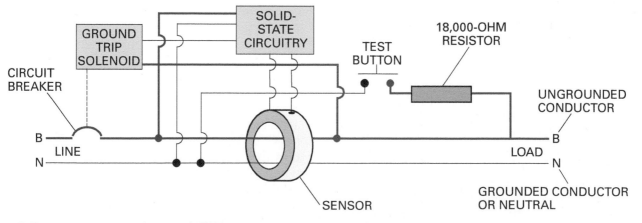

Figure 5 Operating circuitry of a typical GFCI.

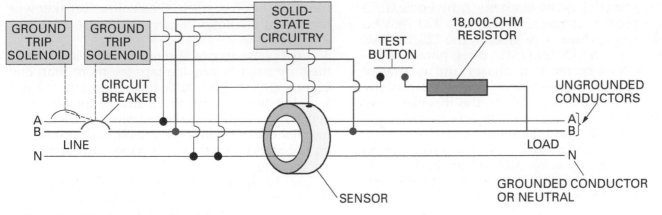

Figure 6 Operating characteristics of a two-pole GFCI.

Single-pole GFCI circuit breakers – The single-pole GFCI breaker has two load lugs and a white wire pigtail in addition to the line side plug-on or bolt-on connector. The line side hot connection is made by installing the GFCI breaker in the panel just as any other circuit breaker is installed. The white wire pigtail is attached to the panel neutral (S/N) assembly. Both the neutral and hot wires of the branch circuit being protected are terminated in the GFCI breaker. These two load lugs are clearly marked Load Power and Load Neutral in the breaker case. Also in the case is the identifying marking for the pigtail, Panel Neutral.

> **NOTE**
>
> Single-pole GFCI circuit breakers cannot be used on multiwire circuits.

Care should be exercised when installing GFCI breakers in existing panels. Be sure that the neutral wire for the branch circuit corresponds with the hot wire of the same circuit. Always remember that unless the current in the neutral wire is equal to that in the hot wire (within 4mA), the GFCI breaker senses this as being a possible ground fault (*Figure 7*).

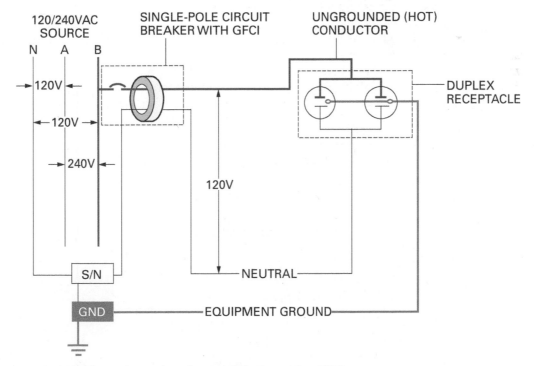

Figure 7 Operating characteristics of a single-pole circuit breaker with a GFCI.

Two-pole GFCI circuit breakers – A two-pole GFCI circuit breaker can be installed on a 120/240VAC single-phase, three-wire system; the 120/240VAC portion of a 120/240VAC three-phase, four-wire system; or the two phases and neutral of a 120/208VAC three-phase, four-wire system. Regardless of the application, the installation of the breaker is the same—connections are made to two hot buses and the panel neutral assembly. When installed on these systems, protection is provided for two-wire 240VAC or 208VAC circuits, three-wire 120/240VAC or 120/208VAC circuits, and 120VAC multiwire circuits.

The circuit in *Figure 8* illustrates the problems that are encountered when a common load neutral is used for two single-pole GFCI breakers. Either or both breakers will trip when a load is applied at the #2 duplex receptacle. The neutral current from the #2 duplex receptacle flows through breaker #1; this increase in neutral current through breaker #1 causes an imbalance in its sensor, thus causing it to produce a fault signal. At the same time, there is no neutral current flowing through breaker #2; therefore, it also senses a current imbalance. If a load is applied at the #1 duplex receptacle and there is no load at

the #2 duplex receptacle, then neither breaker will trip because neither breaker will sense a current imbalance.

Junction boxes can also present problems when they are used to provide taps for more than one branch circuit. Even though the circuits are not wired using a common neutral, sometimes all neutral conductors are connected together. Thus, parallel neutral paths are established, producing an imbalance in each GFCI breaker sensor, causing them to trip.

The two-pole GFCI breaker eliminates the problems encountered when trying to use two single-pole GFCI breakers with a common neutral. Because both hot currents and the neutral current pass through the same sensor, no imbalance occurs between the three currents, and the breaker will not trip.

Direct-wired GFCI receptacles – Direct-wired GFCI receptacles provide Class A ground fault protection on 120VAC circuits. They are available in both 15A and 20A arrangements. The 15A unit has an NEMA 5-15R receptacle configuration for use with 15A plugs only. The 20A device has an NEMA 5-20R receptacle configuration for use with 15A or 20A plugs. Both 15A and 20A units

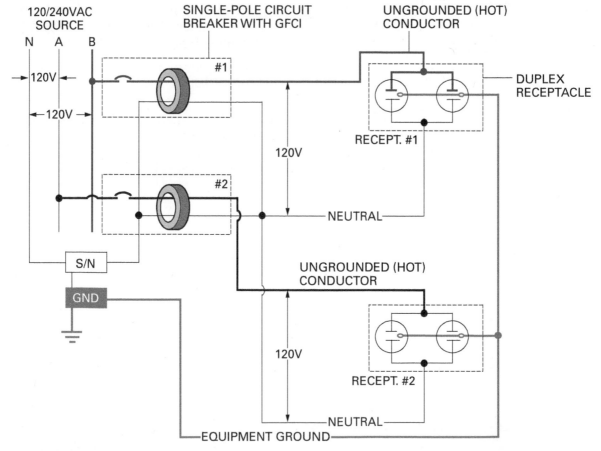

Figure 8 Circuit depicting the common load neutral.

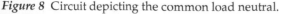

GFCI and AFCI Circuit Breakers

GFCI breakers protect against ground faults, while AFCI breakers protect against arc faults.

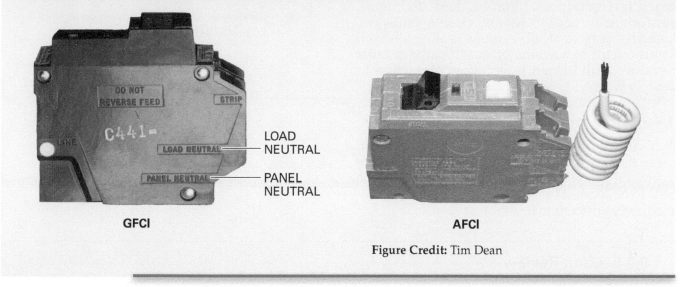

Figure Credit: Tim Dean

have a 120VAC, 20A circuit rating. This is to comply with *NEC Table 210.24*, which requires that 15A circuits use 15A receptacles but permits the use of either 15A or 20A receptacles on 20A circuits. Therefore, GFCI receptacle units that contain a 15A receptacle may be used on 20A circuits.

These receptacles have line terminals for the hot, neutral, and ground wires. In addition, they have load terminals that can be used to provide ground fault protection for other receptacles electrically downstream on the same branch circuit (*Figure 9*). All terminals will accept No. 14 to No. 10 AWG copper wire.

GFCI receptacles have a two-pole tripping mechanism that breaks both the hot and the neutral load connections.

When tripped, the Reset button pops out. The unit is reset by pushing the button back in.

GFCI receptacles have the additional benefit of noise suppression. Noise suppression minimizes false tripping due to spurious line voltages or radio frequency (RF) signals between 10 and 500 megahertz (MHz).

GFCI receptacles can be mounted without adapters in wall outlet boxes that are at least $1\frac{1}{2}$" (38 mm) deep.

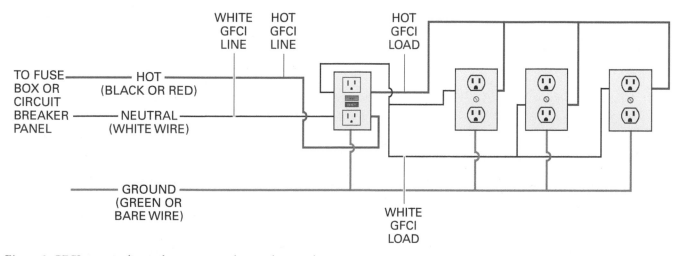

Figure 9 GFCI receptacle used to protect other outlets on the same circuit.

1.4.2 Arc Fault Circuit Interrupters

All branch circuits that supply the lighting and general-purpose receptacles or devices in dwelling unit kitchens, family rooms, dining rooms, living rooms, parlors, libraries, dens, bedrooms, sunrooms, recreation rooms, closets, hallways, laundry areas, or similar rooms or areas, as well as all circuit extensions or modifications, must have arc fault circuit interrupter protection to comply with *NEC Section 210.12(A)*.

1.4.3 Surge Protective Devices

NEC Section 230.67(A) requires all dwelling unit services to be provided with a surge protective device (SPD). *NEC Section 230.67(B)* requires the SPD to be located in the service equipment or immediately adjacent to it.

Think About It

GFCIs

Explain the difference(s) in the operation of single-pole and double-pole GFCIs.

1.0.0 Section Review

1. A small residence has a floor area of 900 square feet. What is its general lighting load?
   a. 900VA
   b. 1,800VA
   c. 2,700VA
   d. 3,600VA

2. Per *NEC Table 220.54*, a household with two dryers has a demand factor of _____.
   a. 100%
   b. 85%
   c. 75%
   d. 65%

3. If a dwelling has two kitchen areas, then the total small appliance branch circuits required would be _____.
   a. one at 1,500VA each
   b. two at 1,500VA each
   c. three at 1,500VA each
   d. four at 1,500VA each

4. A lighting load of 2,700VA at 120V has an amperage of _____.
   a. 20A
   b. 22.5A
   c. 30A
   d. 36.5A

2.0.0 GROUNDING

Objective

Identify the grounding requirements for a residential electrical system.

a. Size grounding electrodes.
b. Size the main bonding jumper.
c. Install the equipment grounding system.

Trade Terms

Bonding bushing: A special conduit bushing equipped with a conductor terminal to accept a bonding jumper. It also has a screw or other sharp device to bite into the enclosure wall to bond the conduit to the enclosure without a jumper when there are no concentric knockouts left in the wall of the enclosure.

Bonding jumper: A bare or green insulated conductor used to ensure the required electrical conductivity between metal parts required to be electrically connected. Bonding jumpers are frequently used from a bonding bushing to the service-equipment enclosure to provide a path around concentric knockouts in an enclosure wall, and they may also be used to bond one raceway to another.

Service drop: The overhead conductors, through which electrical service is supplied, between the last power company pole and the point of their connection to the service facilities located at the building.

Service-entrance equipment: Equipment that provides overcurrent protection to the feeder and service conductors, a means of disconnecting the feeders from energized service conductors, and a means of measuring the energy used.

Service lateral: The underground conductors through which service is supplied between the power company's distribution facilities and the first point of their connection to the building or area service facilities located at the building.

*N*EC *Section 250.4(A)* provides the general requirements for grounding and bonding of grounded electrical systems. In order to ensure systems are properly grounded and bonded, the prescriptive requirements of *NEC Article 250* must be followed.

The grounding system is a major part of the electrical system. Its purpose is to protect people and equipment against the various electrical faults that can occur. It is sometimes possible for higher-than-normal voltages to appear at certain points in an electrical system or in the electrical equipment connected to the system. Proper grounding ensures that the electrical charges that cause these higher voltages are channeled to the earth or ground and that an effective ground fault path is provided throughout the system so that overcurrent devices will open before people are endangered or equipment is damaged.

The word *ground* refers to ground potential or earth ground. If a conductor is connected to the earth or some conducting body that serves in place of the earth, such as a driven ground rod (electrode), the conductor is said to be grounded. The neutral conductor in a three- or four-wire service, for example, is intentionally grounded, and therefore becomes a grounded conductor. This is the path back to the source of supply for all ground faults in an electrical system. This conductor is intended to carry the imbalanced loads of an installation and also to provide a low-impedance path back to the source so that enough current will flow in the system to open the overcurrent devices. A wire that is used to connect this neutral conductor to a grounding electrode or electrodes is referred to as a grounding electrode conductor (GEC). Note the difference in the two meanings: one is grounded, while the other provides a means for grounding.

There are two general classifications of protective grounding:

- System grounding
- Equipment grounding

The system ground relates to the **service-entrance equipment** and its interrelated and bonded components; that is, the system and circuit conductors are grounded to limit voltages due to lightning, line surges, or unintentional contact with higher voltage and to stabilize the voltage to ground during normal operation per *NEC Sections 250.4(A)(1) and (2)*.

The noncurrent-carrying conductive parts of materials enclosing electrical conductors or equipment, or forming a part of such equipment, and electrically conductive materials that are likely to become energized are all connected together to the supply source in a manner that establishes an effective ground fault path per *NEC Sections 250.4(A)(3) and (4)*.

NEC Section 250.4(A)(5) defines the requirements for an effective ground path. It requires that electrical equipment and wiring and other electrically conductive materials likely to become energized shall be installed in a manner that creates a permanent, low-impedance circuit capable of safely carrying the maximum ground fault current likely to be imposed on it from any point on the wiring system where a ground fault may occur to the electrical supply source. The earth shall not be used as the sole equipment grounding conductor or effective ground fault current path.

To better understand a complete grounding system, a conventional residential system will be examined, beginning at the power company's high-voltage lines and transformer, as shown in *Figure 10*. The pole-mounted transformer is fed with a two-wire, single-phase 7,200V system, which is transformed and stepped down to a three-wire, 120/240V, single-phase electric service suitable for residential use. Note that the voltage between line A and line B is 240V. However, by connecting a third (neutral) wire on the secondary winding of the transformer—between the other two—the 240V is split in half, providing 120V between either line A or line B and the neutral conductor. Consequently, 240V is available for household appliances such as ranges, hot water heaters, and clothes dryers, while 120V is available for lights and small appliances.

Referring again to *Figure 10*, conductors A and B are ungrounded conductors, while the neutral is a grounded conductor. If only 240V loads were connected, the neutral (grounded conductor) would carry no current. In this instance, the neutral would be used to carry any ground fault currents from the load side of the service back to the utility instead of depending on the earth as the path back to the source. However, because 120V loads are present, the neutral will carry the imbalanced load and become a current-carrying conductor.

For example, if line A carries 60A and line B carries 50A, the neutral would carry only 10A (60A – 50A = 10A). This is why the *NEC®* allows the neutral conductor in an electric service to be smaller than the ungrounded conductors. However, it must be sufficient to carry fault currents back to the source. *NEC Section 250.24(C)(1)* requires that it must not be less than the required conductor size using *NEC Table 250.102(C)(1)* for service conductors up to 1,100 kcmil and not less than 12.5% of the area of the service-entrance conductors (or equivalent) larger than 1,100 kcmil. The typical pole-mounted **service drop** conductors are normally routed by a messenger cable from a point on the pole to a point on the building being served, terminating at the point where service-entrance conductors exit a weatherhead. Service-entrance conductors are then typically routed through metering equipment into the service disconnecting means.

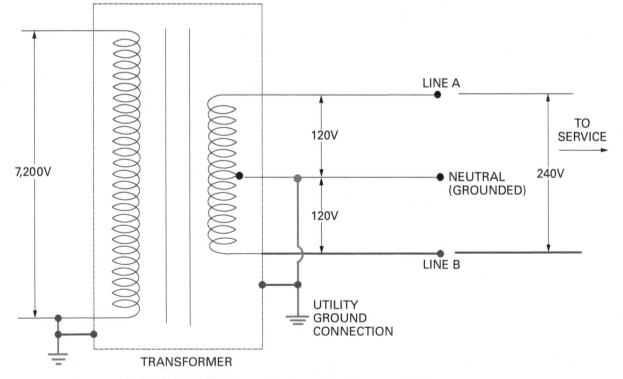

Figure 10 Wiring diagram of a 7,200V to 120/240V, single-phase transformer connection.

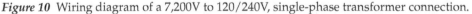

This is the point where most services are grounded. Refer to *Figure 11*. *NEC Section 250.24(A)(1)* requires that the grounding electrode for the structure connection to the neutral (grounded conductor) be at any accessible point from the load end of the service drop or service lateral up to and including the terminal or bus to which the neutral (grounded service conductor) is connected to the service disconnecting means.

> **NOTE**
>
> Effectively grounded means intentionally connected to earth through one or more ground connection(s) of sufficiently low impedance and having sufficient current-carrying capacity to prevent the buildup of voltages that may result in a hazard to people or connected equipment.

Grounding

Systematic grounding wasn't required by the *NEC®* until the mid-1950s; even then, electricians commonly grounded an outlet by wrapping an uninsulated wire around a cold-water pipe and taping it. Three-hole receptacles with grounding terminals became common in the 1960s, but in many older houses, you cannot assume that receptacles are grounded, even when you see a three-hole receptacle. Sometimes, new receptacles have simply been screwed onto old boxes where there is no equipment grounding conductor.

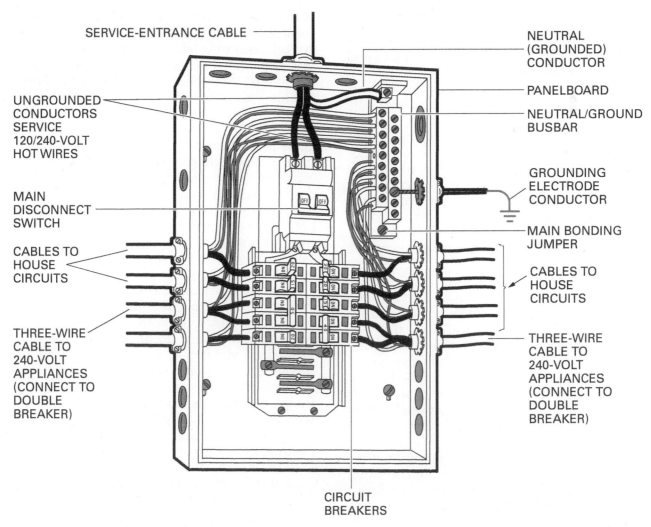

Figure 11 Interior view of panelboard showing connections.

2.1.0 Grounding Electrodes

Requirements for connecting an electric service to the grounding electrode system of a building or structure are provided in *NEC Article 250, Part III*. *NEC Section 250.50* requires, in general, that all of the electrodes described in *NEC Section 250.52(A)* be used (if present), and they must be bonded together to form the grounding electrode system. The electrodes listed in *NEC Section 250.52(A)* are as follows:

> **NOTE**
> Some of the *NEC®* metric conversions for grounding electrodes are much more precise than those listed in other areas of the *NEC®*. Always read and follow the applicable code sections when selecting and installing any electrical equipment or devices.

- Metal underground water pipe in direct contact with the earth for 10' (3.0 m) or more and electrically continuous (or made electrically continuous by bonding around insulating joints or insulating pipe) to the points of connection of the grounding electrode conductor and the bonding conductors. Interior metal water piping located more than 5' (1.52 m) from the point of entrance to the building shall not be used as part of the grounding electrode system or as a conductor to interconnect electrodes that are part of the grounding electrode system per *NEC Section 250.68(C)(1)*.
- Metal in-ground support structure that complies with *NEC Section 250.52(A)(2)*.
- An electrode encased by at least 2" (50 mm) of concrete may be used if it is located within and near the bottom of a concrete foundation or footing that is in direct contact with the earth. The electrode must be at least 20' (6.0 m) long and must be made of electrically conductive coated steel reinforcing bars or rods of not less than $\frac{1}{2}$" (13 mm) in diameter, or consisting of at least 20' (6.0 m) of bare copper conductor not smaller than No. 4 AWG wire size. See *NEC Section 250.52(A)(3)*.
- A ground ring encircling the building or structure, in direct contact with the earth, consisting of at least 20' (6.0 m) of bare copper conductor not smaller than No. 2 AWG. See *NEC Section 250.52(A)(4)*.
  Per *NEC Section 250.52(A)(5)*, rod and pipe electrodes shall not be less than 8' (2.44 m) in length and consist of either:Pipe or conduit not smaller than trade size $\frac{3}{4}$" (MD 21) and, where of iron or steel, shall have the outer surface galvanized or otherwise metal-coated for corrosion protection.

Rods of stainless steel and copper-zinc coated steel shall be at least $\frac{5}{8}$" (15.87 mm) in diameter, unless listed.

- Plate electrodes shall expose less than 2 square feet (0.186 square meter) of surface to exterior soil. Plates made of iron or steel shall be at least $\frac{1}{4}$" (6.4 mm) thick. Nonferrous metal plates shall be at least 0.06" (1.5 mm) thick. See *NEC Section 250.52(A)(7)*.
- Other local metal underground systems or structures such as piping systems and underground tanks. See *NEC Section 250.52(A)(8)*.

Often in residential construction, the only grounding electrode that is available is the metal underground water piping system. *NEC Section 250.53(D)(2)* requires that whenever water piping is used as an electrode, it must be supplemented. Any of the electrodes listed above can be used to supplement the water pipe electrode. *Figure 12* shows a typical residential electric service and the available grounding electrodes for this structure using a ground rod to supplement the water pipe electrode.

This house also has a metal underground gas piping system, but this may not be used as an electrode per *NEC Section 250.52(B)*. In some cases, a water pipe electrode, building steel, and a concrete-encased electrode are not available to be used as a part of the grounding electrode system. For example, a building may be fed by plastic water piping, be constructed of wood, and an electrician may not be present at the site when the foundation for the structure is poured. When that happens, *NEC Section 250.50* requires that rod, pipe, plate, or other local metal underground structures be used.

Some local jurisdictions do not recognize water piping as an electrode due to the rise in the use of nonmetallic piping for both new and replacement water systems. They do not want to rely on the maintained viability of an existing metallic water service and therefore require the use of other electrodes such as the concrete-encased or rod electrodes. This means electricians must be involved with the construction prior to the foundation being poured in order to utilize concrete-encased electrodes.

In most cases, the supplemental electrode used for a water pipe electrode will consist of either a driven rod or pipe electrode, the specifications for which are shown in *Figure 13*.

> **WARNING!**
> A metal underground gas piping system must never be used as a grounding electrode.

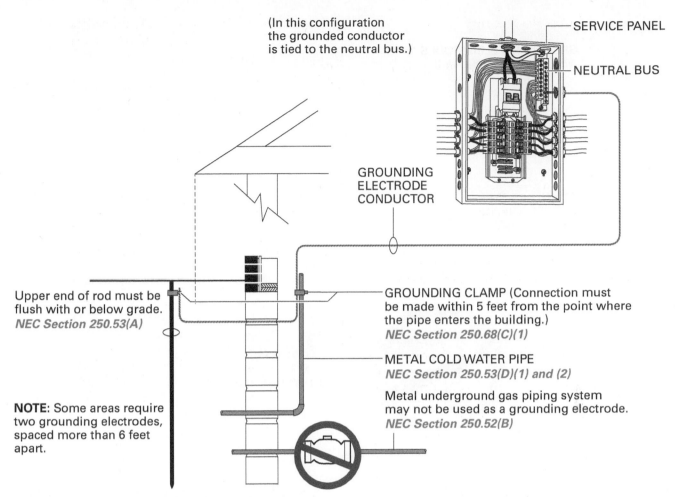

(In this configuration the grounded conductor is tied to the neutral bus.)

SERVICE PANEL

NEUTRAL BUS

GROUNDING ELECTRODE CONDUCTOR

Upper end of rod must be flush with or below grade. *NEC Section 250.53(A)*

GROUNDING CLAMP (Connection must be made within 5 feet from the point where the pipe enters the building.) *NEC Section 250.68(C)(1)*

METAL COLD WATER PIPE *NEC Section 250.53(D)(1) and (2)*

Metal underground gas piping system may not be used as a grounding electrode. *NEC Section 250.52(B)*

NOTE: Some areas require two grounding electrodes, spaced more than 6 feet apart.

Figure 12 Components of a residential grounding system.

2.1.1 Grounding Electrode Installations

NEC Section 250.53(A)(1) requires that rod, pipe, and plate electrodes, where practical, be buried below the permanent moisture level and that they are free from any nonconductive coatings, such as paint or enamel.

NEC Section 250.53(A)(4) permits a rod or pipe electrode to be driven at a 45-degree angle if rock bottom is encountered and prevents the rod or pipe from being driven vertically for at least 8' (2.44 m). Where driving a rod or pipe electrode at a 45-degree angle will not work, it is permitted to lay a rod or pipe horizontally in a trench that is at least 30" (750 mm) deep. For rod or pipe electrodes longer than 8' (2.44 m), it is permitted to have the upper end above ground level if a suitable means of protection is provided for the grounding electrode conductor attachment; otherwise, the upper end must be flush with the earth surface.

NEC Section 250.53(A)(2) requires a supplemental electrode when a single rod, pipe, or plate electrode of a type specified in *NEC Sections 250.52(A)(2) through 250.52(A)(8)* is used. If a single rod or plate grounding electrode that has a resistance of 25 ohms or less is used, the supplemental electrode shall not be required. Always check with the local inspection authority, including the local utility, for rules that surpass the requirements of the *NEC®*. Some general rules are as follows:

- Where multiple rod, pipe, or plate electrodes are installed to meet the requirements of this section, they shall not be less than 6' (1.83 m) apart per *NEC Section 250.53(A)(3)*.
- Plate electrodes must be buried at least 30" (750 mm) below the surface of the earth, according to *NEC Section 250.53(A)(5)*.
- Where two or more electrodes are effectively bonded together, they are treated as a single electrode system.

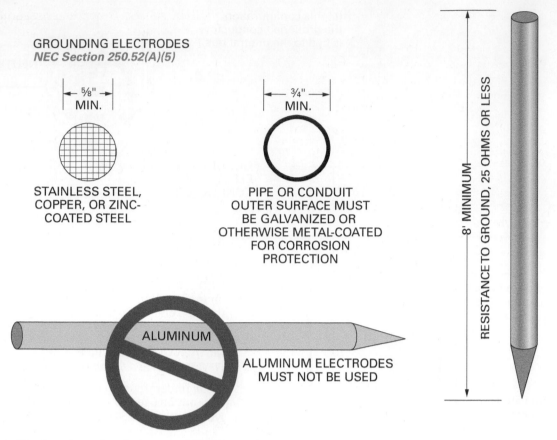

GROUNDING ELECTRODES
NEC Section 250.52(A)(5)

⅝"
MIN.

STAINLESS STEEL,
COPPER, OR ZINC-
COATED STEEL

¾"
MIN.

PIPE OR CONDUIT
OUTER SURFACE MUST
BE GALVANIZED OR
OTHERWISE METAL-COATED
FOR CORROSION
PROTECTION

ALUMINUM

ALUMINUM ELECTRODES
MUST NOT BE USED

8' MINIMUM

RESISTANCE TO GROUND, 25 OHMS OR LESS

Figure 13 Specifications for rod and pipe grounding electrodes.

2.1.2 Grounding Electrode Conductors (GECs)

The grounding electrode conductor (GEC) connecting the neutral (grounded conductor of the service) at the panelboard neutral bus to the grounding electrodes must meet the requirements of *NEC Section 250.62*. This requires that it be made of copper, aluminum, or copper-clad aluminum. The material selected shall be protected against corrosion. The GEC may be either solid or stranded and insulated, covered or bare. Note that the GEC is not an equipment-grounding conductor, and thus is not required to be identified by the use of the color green or green with yellow stripes, if insulated.

2.1.3 Installation of GECs

NEC Section 250.64 provides the installation requirements for GECs and does not permit bare aluminum or copper-clad aluminum grounding conductors to be used where in direct contact with concrete, where subject to corrosive conditions, or where used outside within 18" (450 mm) of the earth at the termination point. Other *NEC®* requirements include the following:

- A GEC or its enclosure is required to be securely fastened to the surface on which it is carried.
- A No. 6 AWG or larger copper or aluminum GEC is required to be protected if it will be exposed to severe physical damage.
- A No. 6 AWG or larger GEC that is free from exposure to physical damage is permitted to be run along the surface of the building without metal covering or protection if it is securely fastened to the building. Otherwise, it must be installed in rigid metallic conduit (RMC), intermediate metal conduit (IMC), Schedule 80 polyvinyl chloride (PVC), Type XW reinforced thermosetting resin conduit (RTRC-XW), electrical metallic tubing (EMT), or cable armor.
- GECs smaller than No. 6 AWG must be protected by RMC, IMC, Schedule 80 PVC, RTRC-XW, EMT, or cable armor.
- The GEC shall be installed in one continuous length without a splice or joint, unless it meets the requirements of *NEC Sections 250.64(C)(1) through (C)(4)*.
- Where a service consists of more than a single enclosure, as permitted in *NEC Section 230.71(B)*, grounding electrode connections

shall be made in accordance with *NEC Sections 250.64(D)(1), (D)(2), or (D)(3).*

- Ferrous metal enclosures for the GEC are required to be electrically continuous from the point of attachment to metal cabinets or metallic equipment enclosures to the GEC. They must also be securely fastened to the ground clamp or fitting.
- Ferrous metal enclosures for the GEC that are not physically continuous from a metal cabinet or metallic equipment enclosure to the grounding electrode must be made electrically continuous by bonding each to the enclosed GEC. Bonding methods must comply with *NEC Section 250.92(B)* for installations at service equipment locations and with *NEC Sections 250.92(B)(2) through (B)(4)* for other-than-service equipment locations.
- GECs may be run to any convenient grounding electrode available in the grounding electrode system, or to one or more grounding electrode(s) individually. The GEC shall be sized for the largest grounding electrode conductor required among all the electrodes connected together.

2.1.4 Methods of Connecting GECs

NEC Section 250.70 requires the GEC to be connected to electrodes using exothermic welding, listed pressure connectors, listed clamps, listed lugs, or other listed means. Connections that depend on solder must never be used. To prevent corrosion, the ground clamp must be listed for the material of the grounding electrode and the GEC.

Where used on a pipe, ground rod, or other buried electrodes, the fitting must be listed for direct soil burial or concrete encasement. More than one conductor is not permitted to be connected to the grounding electrode using a single clamp or fitting unless the clamp or fitting is specifically listed for the connection of more than one conductor.

For the connection to an electrode, you must use one of the following:

- A listed, bolted clamp of cast bronze or brass, or plain or malleable iron
- A pipe fitting, pipe plug, or other approved device that is screwed into a pipe or pipe fitting
- For indoor communication purposes only, a listed sheet metal strap-type ground clamp with a rigid metal base that sits on the electrode with a strap that will not stretch during or after installation
- An equally substantial approved means

The connection of a GEC or a **bonding jumper** to a grounding electrode must be accessible unless that connection is to the concrete-encased or buried grounding electrodes permitted in *NEC Section 250.68*. Where it is necessary to ensure the grounding path for metal piping used as a grounding electrode, effective bonding shall be provided around insulated joints and around any equipment likely to be disconnected for repairs or replacement. Bonding conductors shall be of sufficient length to permit removal of such equipment while retaining the integrity of the bond.

> **NOTE**
> The UL listing states that "strap-type ground clamps are not suitable for attachment of the grounding electrode conductor of an interior wiring system to a grounding electrode."

Think About It
Grounding Conductors

This residential application has a grounding electrode conductor connected to a rod that travels through the floor and into the ground at least 8' (2.44 m). In addition, it is also connected to a metal cold-water pipe (not shown). Which type of grounding system is provided at your home? Does it meet *NEC®* requirements?

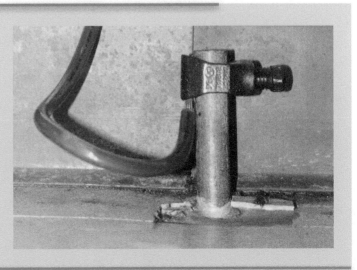

As required by *NEC Section 250.96(A)*, coatings on metal piping systems must be removed to ensure that a permanent and effective grounding path is provided. Grounding electrode conductors and bonding jumpers shall be permitted to be connected at the following locations and used to extend the connections to an electrode:

- Interior metal piping located not more than 5' (1.52 m) from the point of entrance of a building per *NEC Section 250.68(C)*
- Structural frame of a building directly connected to the GEC

For the example house, the point of connection to the water piping would be required to be accessible after any wall coverings are installed. Any nonconductive coatings on the water piping would also have been scraped off or removed prior to installing the clamp on the water pipe.

2.1.5 Sizing GECs

Grounding electrode conductors must be sized per *NEC Section 250.66*, which uses the area of the largest service-entrance conductor (or equivalent area for paralleled conductors). Except as noted below, *NEC Table 250.66* will provide the minimum size GEC and any bonding jumpers used to interconnect grounding electrodes used.

- Where connected to rod, pipe, or plate electrodes, that portion of the GEC that is the sole connection to the grounding electrode shall not be required to be larger than No. 6 AWG copper or No. 4 AWG aluminum.
- Where connected to a concrete-encased electrode, that portion of the GEC that is the sole connection to the grounding electrode shall not be required to be larger than No. 4 AWG copper wire.
- Where the GEC is connected to a ground ring, that portion of the conductor that is the sole connection to the grounding electrode shall not be required to be larger than the conductor used for the ground ring.
- Where multiple sets of service-entrance conductors are used as permitted in *NEC Section 230.40, Exception No. 2*, the equivalent size of the largest service-entrance conductor is required to be determined by the largest sum of the areas of the corresponding conductors of each set.
- Where there are no service-entrance conductors, the GEC size is required to be determined by the equivalent size of the largest service-entrance conductor required for the load to be served.

For our sample dwelling unit, the size of the service-entrance conductors is No. 2 AWG. Using *NEC Table 250.66*, we can determine that the size of the conductor coming from the service panel to the water pipe (the GEC) must be at least No. 8 AWG copper. This No. 8 AWG may continue on without a splice to the ground rod, or a separate No. 6 AWG could be installed for the ground rod(s). This conductor would have to be connected to the service panel as an individual run or with a separate connector to any portion of the No. 8 AWG. It may not be connected directly to the water piping.

2.1.6 Air Terminals (Lightning Protection)

Air terminal conductors and driven pipes, rods, or plate electrodes used for grounding air terminals are not permitted to be used in lieu of the grounding electrodes covered in *NEC Section 250.50* for grounding wiring systems and equipment. However, *NEC Section 250.106* requires that they be bonded to the wiring and equipment grounding electrode system for the structure.

2.2.0 Main Bonding Jumper

NEC Section 250.24(B) requires that an unspliced main bonding jumper (MBJ) shall be used to connect the equipment grounding conductor(s) and the service disconnect enclosure to the grounded conductor (neutral) of the system within the enclosure of each service disconnect.

The MBJ must be of copper or other corrosion-resistant material. An MBJ may be in the form of a wire, bus, screw, or similar suitable conductor.

Where an MBJ is in the form of a screw, it is required to be identified with a green finish so that the head of the screw is visible for inspection. An MBJ must be attached using exothermic welding, a listed pressure connector, listed clamp, or other listed means.

The MBJ cannot be smaller than the sizes given in *NEC Table 250.102(C)(1)*. See *NEC Table 250.102(C)(1), Note 1* for service conductors that exceed 1,100 kcmil.

What Does a Lightning Rod Do?

An interesting fact about grounding is that a lightning rod (air terminal) isn't meant to bring a bolt of lightning to ground. To do this, its conductors would have to be several feet (1.0 m or more) in diameter. The purpose of the rod is to dissipate the negative static charge that would cause the positive lightning charge to strike the house.

The MBJ is the means by which any ground fault in the branch circuits and feeders of the electrical system travels back to the source of supply at the utility. A ground fault will travel along the equipment conductors of the circuits back to the service disconnecting means. Where metallic raceways are used as equipment grounding conductors, there will be no connection to the grounded conductor at the service. Without the MBJ, the path back to the source would be through the grounding electrode system and the earth. This does not provide a low-impedance path, and thus will not allow enough current to flow in the circuit to let the overcurrent devices open.

For example, suppose a phase conductor makes contact with the metallic housing of a 120V, 15A appliance that was wired using EMT. Further suppose that the total combined resistance of the EMT being used as the equipment grounding conductor connected to the appliance and the resistance of the metal water piping and our ground rod in the sample house is 20Ω [less than the 25Ω permitted in *NEC Section 250.53(A)(2), Exception*]. The amount of current that could flow back to the utility source would be 120V ÷ 20 = 6A. The smallest overcurrent device in our electrical system is 15A and would not trip. With the MBJ installed, the path back to the utility source is through the MBJ to the grounded conductor of the service. This resistance will be much less than 1Ω, and thus would allow enough current to flow to open up the overcurrent devices within the system. The MBJ provides the path back to the source for faults that occur within the service disconnect means.

2.2.1 Bonding at the Service

Electrical continuity is required at the service per *NEC Section 250.92(A)*, which states that all of the following must be bonded:

- The service raceways, auxiliary gutters, or service cable armor or sheaths, except for underground metallic sheaths of continuously underground cables as noted in *NEC Section 250.84*
- All service enclosures containing service-entrance conductors, including meter fittings, boxes, or the like interposed in the service raceway or armor
- Any metallic raceway or armor enclosing a grounding electrode conductor as specified in *NEC Section 250.64(E)*

Bonding shall apply at each end and to all intervening raceways, boxes, and enclosures between the service equipment and the grounding electrode.

The items that typically require bonding include the mast and weatherhead, the meter enclosure, the armor of the SE cable (if it has armor), and the service disconnect.

2.2.2 Methods of Bonding at the Service

The electrical continuity of the service equipment, raceways, and enclosures will be ensured per *NEC Section 250.92(B)* through the use of the following methods:

- Bonding equipment to the grounded service conductor in a manner provided in *NEC Section 250.8*
- Connections utilizing threaded couplings or threaded bosses on enclosures where made up wrenchtight
- Threadless couplings and connectors where made up tight for metal raceways and metal-clad cables
- Other approved devices, such as a bonding-type locknut or **bonding bushing**

Bonding jumpers must be used around concentric or eccentric knockouts that are punched or otherwise formed so as to impair the electrical connection to ground. Standard locknuts or bushings shall not be the sole means for bonding.

2.2.3 Bonding and Grounding Requirements for Other Systems

An accessible means external to the service equipment enclosure is required for connecting intersystem bonding and grounding conductors and connections for the communications, radio and television (TV), community antenna television (CATV), and network-powered broadband communication system. The intersystem bonding termination must meet the following requirements per *NEC Section 250.94(A)*:

- Be accessible for connection and inspection
- Consist of a set of terminals with the capacity for connection of not less than three intersystem bonding conductors
- Not interfere with access to service, building, or structure disconnecting means or metering equipment
- At the service equipment, be securely mounted and electrically connected to the service equipment enclosure, meter enclosure, or exposed nonflexible metallic surface raceway, or mounted at one of these enclosures and connected to the enclosure or the GEC with a minimum No. 6 AWG copper conductor
- At the disconnecting means for a building or structure, be securely mounted and electrically connected to the metal enclosure or GEC with a minimum No. 6 AWG copper conductor
- Be listed as grounding and bonding equipment

2.2.4 Bonding of Water Piping Systems

Metallic water piping systems in or on a structure must be bonded as required by *NEC Section 250.104(A)*. The metallic water piping system(s) must be bonded by means of a bonding jumper sized in accordance with *NEC Table 250.102(C)(1)* and connected to one of the following:

- The service-entrance enclosures
- The grounded (neutral) conductor at the service
- The grounding electrode conductor where of sufficient size
- The grounding electrode(s) used

The points of attachment of the bonding jumper(s) shall be accessible. It shall be installed in accordance with *NEC Section 250.64(A), (B), and (E)*. Note that while this conductor is sized in the same manner as if the water piping system is a grounding electrode, the point of attachment to the water piping is permitted to be at any convenient point on the water piping system and not just within the first 5' (1.52 m) of where the water enters the building.

NEC Section 250.104(A)(2) states that in multifamily dwelling units (or other multiple occupancy buildings) where the metal water piping system(s) installed in or attached to a building or structure for the individual occupancies is metallically isolated from all other occupancies by use of nonmetallic water piping, the metal water piping system(s) for each occupancy shall be permitted to be bonded to the equipment grounding terminal of the panelboard or switchboard enclosure (other than service equipment) supplying that occupancy. The bonding jumper shall be sized in accordance with *NEC Section 250.102(D)*.

2.2.5 Bonding of Other Piping Systems

NEC Section 250.104(B) requires that other piping systems, where installed in or attached to a building or structure, including gas piping, that may become energized shall be bonded to one of the following:

- Equipment grounding conductor for the circuit that is likely to energize the piping system
- The service equipment enclosure
- The grounded conductor at the service
- The grounding electrode conductor where of sufficient size
- One or more grounding electrodes used

The bonding jumper(s) shall be sized in accordance with *NEC Table 250.122* using the rating of the circuit that may energize the piping system(s). The equipment grounding conductor for the circuit that may energize the piping shall be permitted to serve as the bonding means. The points of attachment of the bonding jumper(s) shall be accessible.

> **NOTE**
> Bonding all piping and metal air ducts within the premises will provide additional safety.

2.3.0 Installing the Equipment Grounding System

NEC Article 250, Part IV generally requires that all metallic enclosures, raceways, and cable armor be grounded. The exceptions in *NEC Sections 250.80 and 250.86* allow metal enclosures or short sections of raceways that are used to provide support or physical protection to be ungrounded under specific conditions.

NEC Article 250, Part VI covers equipment grounding and equipment grounding conductors. This section generally requires that the exposed noncurrent-carrying metal parts of fixed equipment likely to become energized be grounded under the following conditions per *NEC Section 250.110*:

- Where within 8' (2.5 m) vertically or 5' (1.5 m) horizontally of ground or grounded metal objects and subject to contact by occupants or others
- Where located in wet or damp locations
- Where in electrical contact with metal
- Where in hazardous (classified) locations as covered by *NEC Articles 500 through 517*
- Where supplied by a metal-clad, metal-sheathed, or metal raceway, or other wiring method that provides an equipment ground
- Where equipment operates with any terminal at over 150V to ground

Specific equipment that is required to be grounded regardless of the voltage is listed in *NEC Section 250.112* and includes equipment such as motors, motor controllers, and lighting fixtures (luminaires). Types of cord- and plug-connected equipment in dwelling units that are required to be grounded are found in *NEC Section 250.114* and include equipment such as refrigerators, freezers, air conditioners, information technology equipment (computers), clothes washers, clothes dryers, and dishwashing machines.

The types of equipment grounding conductors that are acceptable to be used are found in *NEC Section 250.118*. Note that among the list of wiring methods approved for use as equipment grounding conductors, both flexible metal conduit (FMC) and liquidtight flexible metal conduit (LFMC) are permitted to be used. Listed FMC is permitted to be used as an equipment grounding conductor only when the following conditions are met:

- The conduit is terminated in fittings listed for grounding.
- The circuit conductors contained in the conduit are protected by overcurrent devices rated at 20A or less.
- The size of the conduit does not exceed 1¼" (MD 35).
- The combined length of FMC, FMT, and LFMC in the same effective ground-fault current path does not exceed 6' (1.8 m).
- If the conduit is used where flexibility is necessary, a wire-type equipment grounding conductor shall be installed.

Type LFMC is also used in dwelling units and has slightly different requirements when used as an equipment grounding conductor:

- The conduit is terminated in fittings listed for grounding.
- For trade sizes ⅜" through ½" (MD 12 to MD 16), the circuit conductors contained in the conduit are protected by overcurrent devices rated at 20A or less.
- For trade sizes ¾" through 1¼" (MD 21 to MD 35), the circuit conductors contained in the conduit are protected by overcurrent devices rated at 60A or less and there is no FMC, FMT, or LFMC in trade sizes ⅜" through ½" (MD 12 to MD 16) in the effective ground-fault current path.
- The combined length of FMC, FMT, and LFMC in the same effective ground-fault current path does not exceed 6' (1.8 m).
- If the conduit is used where flexibility is necessary, a wire-type equipment grounding conductor shall be installed.

Where external bonding jumpers are used to provide the continuity of the fault current path, *NEC Section 250.102(E)(2)* limits the length to not more than 6' (1.8 m), except at outside pole locations for the purposes of bonding or grounding the isolated sections of metal raceways or elbows installed in exposed risers at those pole locations. When installing an equipment grounding conductor in a raceway, *NEC Table 250.122* is used to determine the size of the equipment grounding conductor. It is permitted to install one equipment grounding conductor in a raceway that has several circuits. In that case, the size of the equipment grounding conductor is based on the rating of the largest overcurrent device protecting the circuits contained in the raceway.

NEC Section 250.148 requires that where circuit conductors are spliced within a box or terminated on equipment within or supported by a box, all wire-type equipment grounding conductors associated with those circuit conductors shall be spliced or joined within the box or to the box with devices suitable for that use. *Figure 14* shows several types of fittings that are suitable for this purpose.

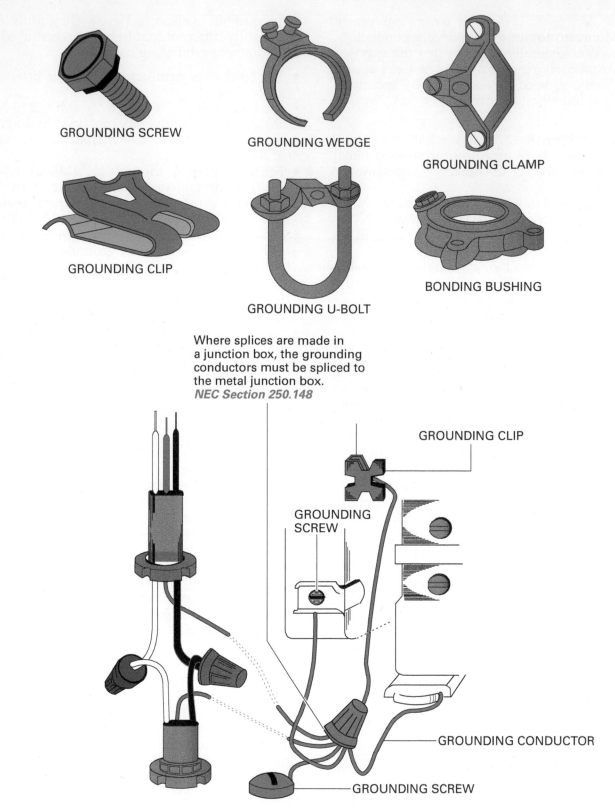

GROUNDING SCREW

GROUNDING WEDGE

GROUNDING CLAMP

GROUNDING CLIP

GROUNDING U-BOLT

BONDING BUSHING

Where splices are made in a junction box, the grounding conductors must be spliced to the metal junction box. *NEC Section 250.148*

GROUNDING CLIP

GROUNDING SCREW

GROUNDING CONDUCTOR

GROUNDING SCREW

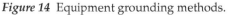

Figure 14 Equipment grounding methods.

2.0.0 Section Review

1. When used as a grounding electrode, a ground ring must be at least _____.

   a. 5' (1.5 m) long
   b. 10' (3.0 m) long
   c. 15' (4.5 m) long
   d. 20' (6.0 m) long

2. When a screw is used as a main bonding jumper, it must be colored _____.

   a. white
   b. green
   c. blue
   d. black

3. Information on equipment grounding for computers is found in _____ .

   a. *NEC Section 250.112*
   b. *NEC Section 250.114*
   c. *NEC Section 250.116*
   d. *NEC Section 250.118*

3.0.0 INSTALLING SERVICE-ENTRANCE EQUIPMENT

Objective

Install service-entrance equipment.
  a. Identify the service drop location.
  b. Select the panelboard location.

Performance Task

2. Using an unlabeled diagram of a panelboard (Performance Profile Task 2 Worksheet), label the lettered components.

Trade Term

Roughing in: The first stage of an electrical installation, when the raceway, cable, wires, boxes, and other equipment are installed. This is the electrical work that must be done before any finishing work can be done.

In practical applications, the electric service is normally one of the last components of an electrical system to be installed. However, it is one of the first considerations when laying out a residential electrical system. For instance:

- The electrician must know in which direction and to which location to route the circuit homeruns while roughing in the electrical wiring.
- Provisions must be made for sleeves through footings and foundations in cases where underground systems (service laterals) are used.
- The local power company must be notified as to the approximate size of service required so they may plan the best way to furnish a service drop to the property.

3.1.0 Service Drop Locations

The location of the service drop, electric meter, and load center should be considered first. It is always wise to consult the local power company to obtain their requirements; where you want the service drop and where they want it may not coincide. A brief meeting with the power company about the location of the service drop can prevent problems later on.

The service drop must be routed so that the service drop conductors have a clearance of not less than 3' (900 mm) horizontally and below windows that open, doors, porches, balconies, ladders, stairs, fire escapes, or similar locations per *NEC Section 230.9(A)*. Where overhead service conductors pass over rooftops, driveways, yards, and so forth, they must have vertical clearances as specified in *NEC Section 230.24*.

A plot plan (also called a *site plan*) is often available for new construction. The plot plan shows the entire property, with the building or buildings drawn in their proper location on the plot of land. It also shows sidewalks, driveways, streets, and existing utilities—both overhead and underground.

A plot plan of the sample residence is shown in *Figure 15*. In reviewing this drawing, you can see that the closest power pole is located across a public street from the house. By consulting with the local power company, it is learned that the service will be brought to the house from this pole by triplex cable, which will connect to the residence at a point on its left (west) end. The steel uninsulated conductor of triplex cable acts as both the grounded conductor (neutral) and as a support for the insulated (ungrounded) conductors. It is also suitable for overhead use.

When service-entrance cable is used, it will run directly from the point of attachment and service head to the meter base. However, because the carport in our sample residence is located on the west side of the building, a service mast (*Figure 16*) will have to be installed. Where raceway-type service masts are used, all raceway fittings must be identified for use with service masts per *NEC Section 230.28*.

The *NEC*® requires a clearance of not less than 8' (2.5 m) over rooftops, unless the roof has a slope of 4" in 12" (100 mm in 300 mm) or greater, in which case the clearance may be reduced to 3' (900 mm). Where the service drop conductors pass over only the overhang (eaves) of a roof, the clearance may be reduced to 18" (450 mm) as long as no more than 6' (1.8 m) of the conductors travel over no more than 4' (1.2 m) of the overhang (eave). This minimum height requirement extends beyond the roof for a distance of not less than 3' (900 mm) in all directions, except the final portion of the span where the service drop conductors attach to the sides of a building.

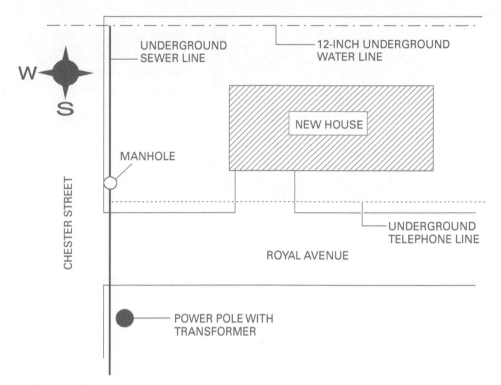

Figure 15 Plot plan of the sample residence.

3.1.1 Vertical Clearances of Service Drop

NEC Section 230.24(B) specifies the distances by which service drop conductors must clear the ground. These distances vary according to the surrounding conditions.

In general, service drop conductors must be at least 10' (3.0 m) above the ground or other accessible surfaces at all times. More distance is required under most conditions. For example, if the service conductors pass over residential property and driveways or commercial property that is not subject to truck traffic, the conductors must be at least 15' (4.5 m) above the ground. However, this distance may be reduced to 12' (3.7 m) when the voltage is limited to 300V to ground.

In other areas, such as public streets, alleys, roads, parking areas subject to truck traffic, driveways on other-than-residential property, the minimum vertical distance is 18' (5.5 m). The conditions of the sample residence are shown in *Figure 17.*

3.1.2 Service Drop (Overhead Service Conductor) Clearances for Building Openings

Service conductors that are installed as open conductors or multiconductor cable without an overall outer jacket must have a clearance of not less than 3' (900 mm) from windows that are designed to be opened, doors, porches, balconies, ladders, stairs, fire escapes, or similar locations (*NEC Section 230.9*). However, conductors run above the top level of a window are permitted to be less than 3' (900 mm) from the window opening.

The 3' (900 mm) of clearance is not applicable to raceways or cable assemblies that have an overall outer jacket approved for use as a service conductor. The intention of this requirement is to protect the conductors from physical damage and/ or physical contact with unprotected personnel when evacuating a structure through the window opening. The exception allows service conductors, including drip loops and service drop conductors, to be located just above the window openings because they would not interfere with ladders leaning against the structure to the right, left, or below the window opening when used to evacuate people from the building.

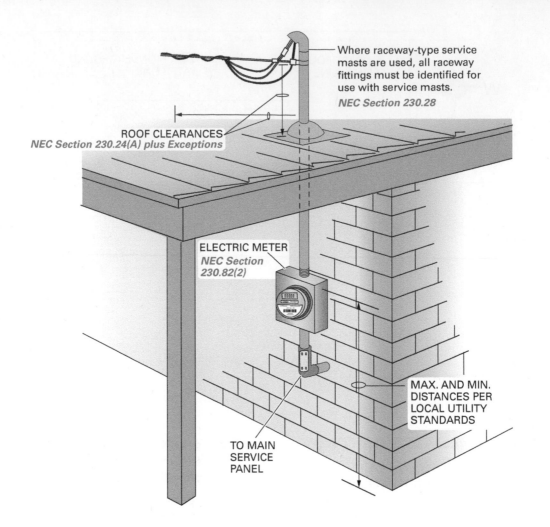

Where raceway-type service masts are used, all raceway fittings must be identified for use with service masts.
NEC Section 230.28

ROOF CLEARANCES
NEC Section 230.24(A) plus Exceptions

ELECTRIC METER
NEC Section 230.82(2)

MAX. AND MIN. DISTANCES PER LOCAL UTILITY STANDARDS

TO MAIN SERVICE PANEL

Figure 16 NEC® sections governing service mast installations.

3.2.0 Panelboard Locations

The main service disconnect or panelboard is normally located in a portion of an unfinished basement or utility room on an outside wall so that the service cable coming from the electric meter can terminate immediately into the switch or panelboard when the cable enters the building. In the example home, however, there is no basement and the utility room is located in the center of the house with no outside walls. Consequently, a somewhat different arrangement will have to be used. A load center is a type of panelboard that is normally located at the service entrance of a residential installation. The load center usually contains a main circuit breaker, which is the main disconnect. Circuit breakers are provided for equipment such as electric water heaters, ranges, dryers, air conditioning and heating units, and breakers that feed sub-panels such as lighting panels.

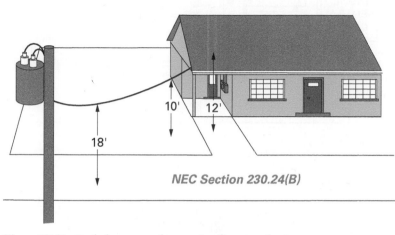

Figure 17 Vertical clearances for service drop conductors.

NEC Section 230.70(A)(1) requires that the service disconnecting means be installed in a readily accessible location—either outside or inside the building. If located inside the building, it must be located nearest the point of entrance of the service conductors. In the sample home, there are at least two methods of installing the panelboard in the utility room that will comply with this *NEC®* regulation, as well as the requirements in *NEC Sections 110.26 and 240.24*.

The first method utilizes a weatherproof 100A disconnect (safety switch or circuit breaker enclosure) mounted next to the meter base on the outside of the building. With this method, service conductors are provided with overcurrent protection; the neutral conductor is also grounded at this point, as this becomes the main disconnect switch. Three-wire cable with an additional grounding wire is then routed from this main disconnect to the panelboard in the utility room. All three current-carrying conductors (two ungrounded and one neutral) must be insulated with this arrangement; the equipment ground, however, may be bare. The panelboard containing overcurrent protection devices for the branch circuits, which is located in the utility room, now becomes a subpanel (*Figure 18*).

Note that *NEC Section 230.85* requires an outdoor emergency disconnect on all one- and two-family dwelling unit services.

An alternate method utilizes conduit from the meter base that is routed under the concrete slab and then up to a main panelboard located in the utility room. *NEC Section 230.6* considers conductors to be outside of a building when they are installed under not less than 2" (50 mm) of concrete beneath a building or installed in conduit under not less than 18" (450 mm) of earth beneath a building. The sample residence has a 4" (100 mm) reinforced concrete slab—well within the *NEC®* regulations. Therefore, the service conductors from the meter base that are installed under the concrete slab in conduit are considered to be outside the house, and no disconnect is required at the meter base. When this conduit emerges in the utility room, it will run straight up into the bottom of the panelboard, again meeting the *NEC®* requirement that the panel be located nearest the point of entrance of the service conductors. Always check with your local authority having jurisdiction for specific requirements about where services are to be located. Details of this service arrangement are shown in *Figure 19*.

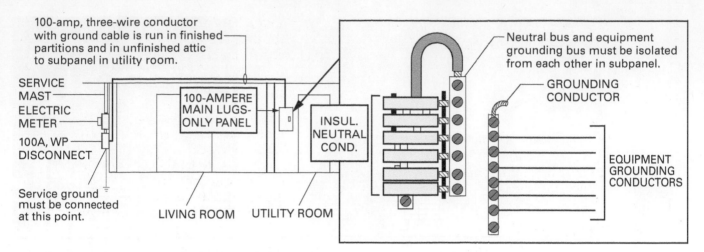

Figure 18 One method of wiring a panelboard for the sample residence.

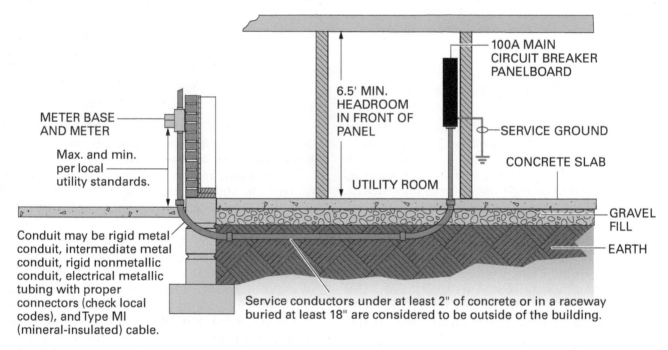

Figure 19 Alternate method of service installation for the sample residence.

3.0.0 Section Review

1. Per *NEC Section 230.24(B)*, if 120V/240V service conductors pass over residential property and driveways or commercial property that is not subject to truck traffic, the conductors must be at least _____.

   a. 12' (3.7 m) above the ground
   b. 15' (4.5 m) above the ground
   c. 18' (5.5 m) above the ground
   d. 20' (6.0 m) above the ground

2. A residence has conductors installed under a 1" (25 mm) slab. According to *NEC Section 230.6*, these conductors are considered _____.

   a. inside the house, and no disconnect is required at the meter base
   b. inside the house, and a disconnect is required at the meter base
   c. outside the house, and no disconnect is required at the meter base
   d. outside the house, and a disconnect is required at the meter base

4.0.0 WIRING METHODS

Objective

Identify wiring methods for various types of residences.

   a. Select and install cable systems.
   b. Select and install raceways.

Trade Terms

Metal-clad (Type MC) cable: A factory assembly of one or more insulated circuit conductors with or without optical fiber members enclosed in an armor of interlocking metal tape, or a smooth or corrugated metallic sheath.

Nonmetallic-sheathed (Type NM and NMC) cable: A factory assembly of two or more insulated conductors enclosed within an overall nonmetallic jacket. Type NM contains insulated conductors enclosed within an overall nonmetallic jacket; Type NMC contains insulated conductors enclosed within an overall, corrosion-resistant, nonmetallic jacket.

Romex®: General Cable's trade name for Type NM cable; however, it is often used generically to refer to any nonmetallic-sheathed cable.

Branch circuits and feeders are used in residential construction to provide power wiring to operate components and equipment and to control wiring to regulate the equipment. Wiring may be further subdivided into either open or concealed wiring.

In open wiring systems, the cable and/or raceways are installed on the surface of the walls, ceilings, columns, and other areas where they are in view and are readily accessible. Open wiring is often used in areas where appearance is not important, such as in unfinished basements, attics, and garages.

Concealed wiring systems are installed inside walls, partitions, ceilings, columns, and behind baseboards or moldings where they are out of view and are not readily accessible. This type of wiring is generally used in all new construction with finished interior walls, ceilings, and floors, and it is the preferred type of wiring where appearance is important.

In general, there are two basic wiring methods used in the majority of modern residential electrical systems. They are:

- Sheathed cables of two or more conductors
- Raceway (conduit) systems

The method used on a given job is determined by the requirements of the *NEC®*, any amendments made by local authorities, the type of building construction, and the location of the wiring in the building. In most applications, either of the two methods may be used, and both methods are frequently used in combination.

4.1.0 Cable Systems

Several types of cable are used in wiring systems to feed or supply power to equipment. These include nonmetallic-sheathed cable, metal-clad (MC) cable, underground feeder cable, and service-entrance cable.

4.1.1 Nonmetallic-Sheathed Cable

Nonmetallic-sheathed (Type NM and NMC) cable (*NEC Article 334*) is manufactured in two- or three-wire configurations with varying sizes of conductors. In both two- and three-wire cables, conductors are color-coded: one conductor is black while the other is white in two-wire cable; in three-wire cable, the additional conductor is red. Both types also have a grounding conductor, which is usually bare, but it is sometimes covered with green plastic insulation, depending upon the manufacturer. Type NMC is basically the same as Type NM, but the outer nonmetallic sheath is corrosion-resistant. The jacket or covering consists of rubber, plastic, or fiber. Jacket markings typically provide the manufacturer's name or trademark, wire size, and number of conductors (*Figure 20*). For example, NM 12-2 W/GRD indicates that the jacket contains two No. 12 AWG conductors along with a grounding wire; NM 12-3 W/GRD indicates three conductors plus a grounding wire. Type NM cable is often referred to as **Romex®**.

NEC Section 334.10 permits Type NM and NMC cable to be used in the following applications:

- One- and two-family dwelling units and attached or detached garages or storage buildings
- Multi-family dwellings when they are of Types III, IV, and V construction (see *NEC Informative Annex E*)
- Other structures if concealed behind a 15-minute thermal finish barrier and are of Types III, IV, and V construction

Per *NEC Section 334.10(A)*, Type NM cable is permitted as follows:

- For both exposed and concealed work in normally dry locations
- To be fished in voids in masonry block or tile walls

Per *NEC Section 334.10(B)*, Type NMC cable is permitted as follows:

- For both exposed and concealed work in dry, moist, damp, and corrosive locations
- In outside and inside walls of masonry block or tile
- In a shallow chase in masonry, concrete, or adobe protected against nails or screws by a steel plate at least $\frac{1}{16}$" (1.59 mm) thick and covered with plaster, adobe, or similar finish

NEC Section 334.12 prohibits the use of Type NM and NMC cable in the following applications:

- In any dwelling or structure not specifically permitted in *NEC Section 334.10*
- As open runs in dropped or suspended ceilings in other than one- and two-family and multi-family dwellings
- As service-entrance cable
- In commercial garages with hazardous (classified) areas
- In theaters and similar locations, except as permitted by *NEC Section 518.4(B)*
- In motion picture studios
- In storage battery rooms
- In hoistways or on elevators or escalators
- Embedded in poured cement, concrete, or aggregate
- In hazardous (classified) areas

In addition, Type NM cable is prohibited in the following areas:

- Where exposed to corrosive fumes or vapors
- Embedded in masonry, adobe, fill, or plaster
- In a shallow chase in masonry, concrete, or adobe and covered with plaster, adobe, or similar finish
- In wet or damp locations

Type NM cable is the most common type of cable for residential use. *Figure 21* shows additional *NEC*® regulations pertaining to the installation of Type NM cable.

4.1.2 Metal-Clad Cable

Metal-clad (Type MC) cable is manufactured in two-, three-, and four-wire assemblies with varying sizes of conductors and is used in locations similar to those for Type NM and NMC cable. Unlike Type NM and NMC, it can also be used as service-entrance cable and in other locations permitted by *NEC Section 330.10*.

The metallic spiral covering on Type MC cable offers a greater degree of mechanical protection than Type NM cable. Type MC cable may be embedded in plaster finish, brick, or other masonry, except in damp or wet locations. It may also be run in the air voids of masonry block or tile walls, except where such walls are exposed or subject to excessive moisture or dampness. It may be used in wet locations if the conditions of *NEC Section 330.10(A)(11)* are met. (Refer to *Figure 22* and *Figure 23*.) It may not be used where subject to physical damage.

> **NOTE**
>
> In the past, armored cable (Type AC), also called *BX*® cable, was commonly used in residential applications. Today, Type MC cable is used because it has a plastic wrapping to protect the conductors and does not require an insulating bushing at cable terminations.

4.1.3 Underground Feeder Cable

Underground feeder (Type UF) cable (*NEC Article 340*) may be used underground, including direct burial in the earth, as a feeder or branch circuit cable when provided with overcurrent protection at the rated ampacity as required by the *NEC*®. When Type UF cable is used above grade where it will come in direct contact with the rays of the sun, its outer covering must be sun-resistant.

Furthermore, where Type UF cable emerges from the ground, some means of mechanical protection must be provided. This protection may be in the form of conduit or guard strips. *NEC Section 300.5(D)(1)* requires that the protection extend from the minimum burial depth below grade to a point at least 8' (2.5 m) above grade. *NEC Section 300.5(D)(4)* states that if conduit is used as protection, the permitted types are EMT, RMC, IMC, RTRC-XW, and Schedule 80 PVC or equivalent. Type UF cable resembles Type NM cable; however, the jacket is constructed of weather-resistant material to provide the required protection for direct-burial wiring installations.

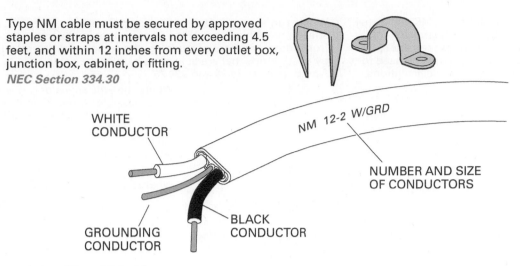

Type NM cable must be secured by approved staples or straps at intervals not exceeding 4.5 feet, and within 12 inches from every outlet box, junction box, cabinet, or fitting.
NEC Section 334.30

WHITE CONDUCTOR

NM 12-2 W/GRD

NUMBER AND SIZE OF CONDUCTORS

BLACK CONDUCTOR

GROUNDING CONDUCTOR

Figure 20 Characteristics of Type NM cable.

4.1.4 Service-Entrance Cable

Service-entrance (Type SE) and underground service-entrance (Type USE) cable, when used for electrical services, must be installed as specified in *NEC Articles 230 and 338*. Service-entrance cable is available with the grounded conductor bare for outside service conductors, and also with an insulated grounded conductor for interior wiring systems.

Type SE cable is permitted for use on branch circuits or feeders provided that all current-carrying conductors are insulated; this includes the grounded or neutral conductor. Where a conductor in the cable is not insulated, it is permitted to be used only as an equipment grounding conductor for branch circuits or feeders. Where used as an interior wiring method, the installation requirements of *NEC Article 334* must be followed, except for determining the ampacity of the cable. Where installed as exterior wiring, the requirements of *NEC Article 225* must be met, with the supports for the cable in accordance with *NEC Section 334.30*. SE Style R (SER) cable is used in residential applications for subfeeds for ranges. Installation rules for Type SE cable for both exterior and interior wiring are summarized in *Figure 24*.

4.2.0 Raceways

A raceway is any channel that is designed and used solely for the purpose of holding wires, cables, or busbars. Types of raceways include rigid metal conduit, intermediate metal conduit, rigid nonmetallic conduit, flexible metallic conduit, electrical metallic tubing, and auxiliary gutters. Raceways are constructed of either metal or insulating material, such as polyvinyl chloride or PVC (plastic). Metal raceways are joined using threaded, compression, or setscrew couplings; nonmetallic raceways are joined using cement-coated couplings. Where a raceway terminates in an outlet box, junction box, or other enclosure, an approved connector must be used.

Raceways provide mechanical protection for the conductors that run in them and also prevent accidental damage to insulation and the conducting material. They also protect conductors from corrosive atmospheres and prevent fire hazards to life and property by confining arcs and flames that may occur due to faults in the wiring system. Conduits or raceways are used in residential applications for service masts, underground wiring embedded in concrete, and sometimes in unfinished basements, shops, or garage areas.

Another function of metal raceways is to provide a continuous equipment grounding system throughout the electrical system. To maintain this feature, it is extremely important that all raceway systems be securely bonded together into a continuous conductive path and properly connected to the system ground.

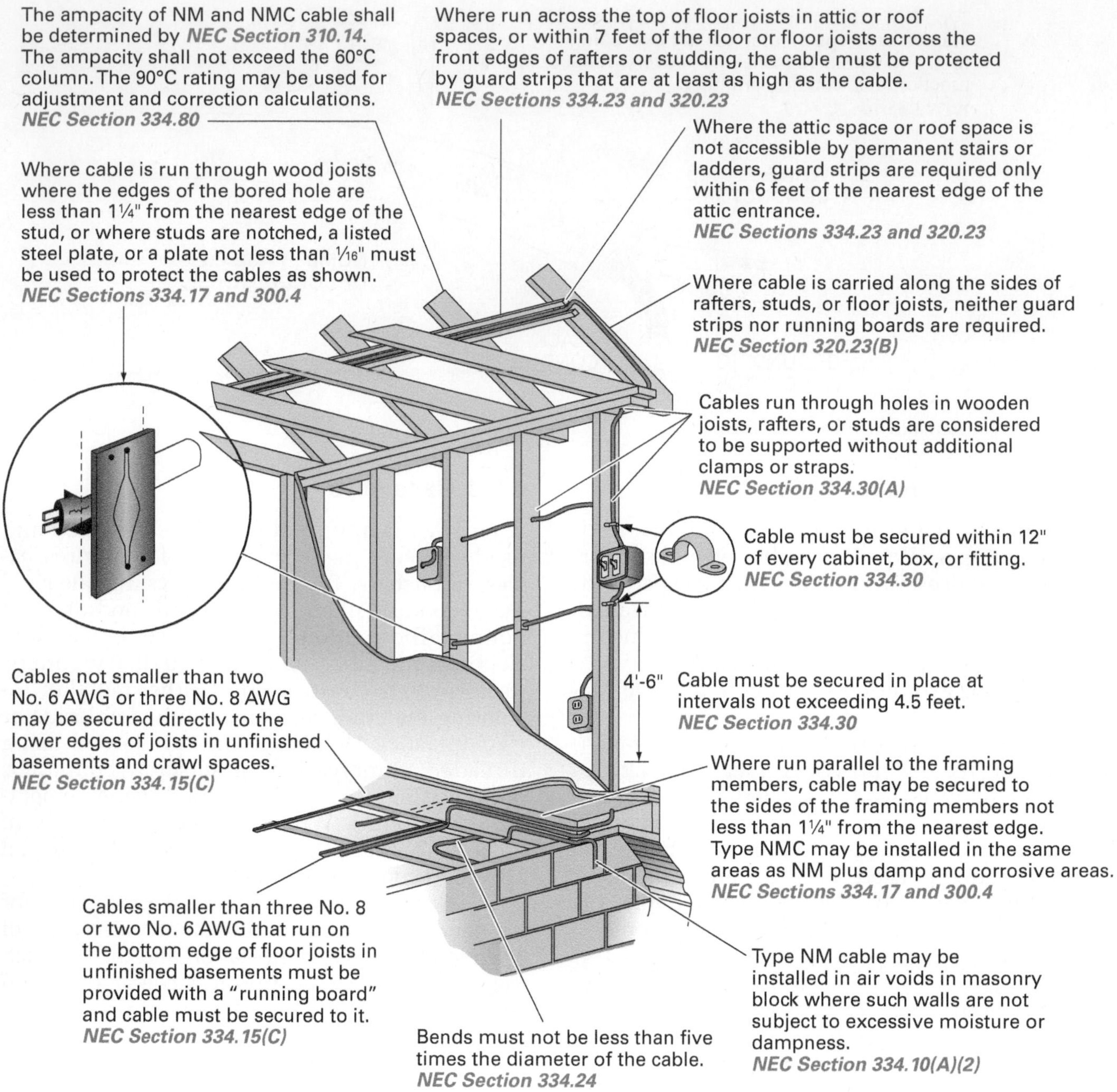

The ampacity of NM and NMC cable shall be determined by *NEC Section 310.14*. The ampacity shall not exceed the 60°C column. The 90°C rating may be used for adjustment and correction calculations. *NEC Section 334.80*

Where cable is run through wood joists where the edges of the bored hole are less than 1¼" from the nearest edge of the stud, or where studs are notched, a listed steel plate, or a plate not less than ¹⁄₁₆" must be used to protect the cables as shown. *NEC Sections 334.17 and 300.4*

Where run across the top of floor joists in attic or roof spaces, or within 7 feet of the floor or floor joists across the front edges of rafters or studding, the cable must be protected by guard strips that are at least as high as the cable. *NEC Sections 334.23 and 320.23*

Where the attic space or roof space is not accessible by permanent stairs or ladders, guard strips are required only within 6 feet of the nearest edge of the attic entrance. *NEC Sections 334.23 and 320.23*

Where cable is carried along the sides of rafters, studs, or floor joists, neither guard strips nor running boards are required. *NEC Section 320.23(B)*

Cables run through holes in wooden joists, rafters, or studs are considered to be supported without additional clamps or straps. *NEC Section 334.30(A)*

Cable must be secured within 12" of every cabinet, box, or fitting. *NEC Section 334.30*

Cables not smaller than two No. 6 AWG or three No. 8 AWG may be secured directly to the lower edges of joists in unfinished basements and crawl spaces. *NEC Section 334.15(C)*

4'-6" Cable must be secured in place at intervals not exceeding 4.5 feet. *NEC Section 334.30*

Where run parallel to the framing members, cable may be secured to the sides of the framing members not less than 1¼" from the nearest edge. Type NMC may be installed in the same areas as NM plus damp and corrosive areas. *NEC Sections 334.17 and 300.4*

Cables smaller than three No. 8 or two No. 6 AWG that run on the bottom edge of floor joists in unfinished basements must be provided with a "running board" and cable must be secured to it. *NEC Section 334.15(C)*

Bends must not be less than five times the diameter of the cable. *NEC Section 334.24*

Type NM cable may be installed in air voids in masonry block where such walls are not subject to excessive moisture or dampness. *NEC Section 334.10(A)(2)*

Figure 21 NEC® sections governing the installation of Type NM cable.

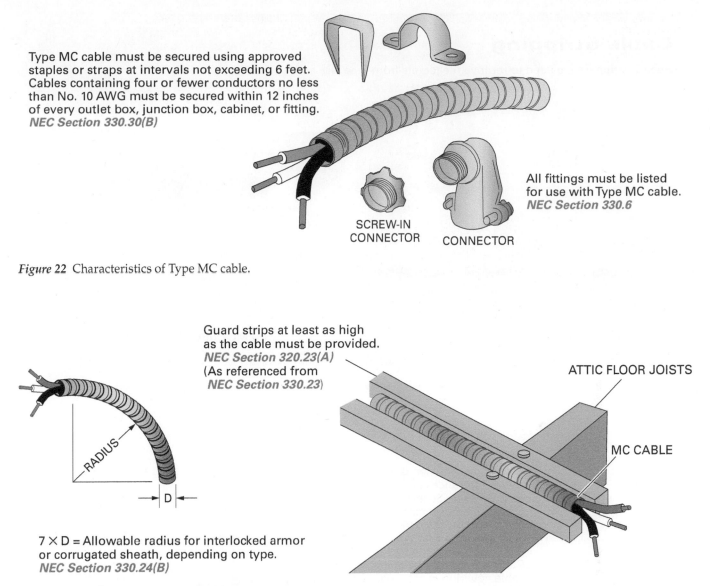

Type MC cable must be secured using approved staples or straps at intervals not exceeding 6 feet. Cables containing four or fewer conductors no less than No. 10 AWG must be secured within 12 inches of every outlet box, junction box, cabinet, or fitting. *NEC Section 330.30(B)*

All fittings must be listed for use with Type MC cable. *NEC Section 330.6*

SCREW-IN CONNECTOR

CONNECTOR

Figure 22 Characteristics of Type MC cable.

Guard strips at least as high as the cable must be provided. *NEC Section 320.23(A)* (As referenced from *NEC Section 330.23*)

ATTIC FLOOR JOISTS

MC CABLE

RADIUS

D

7 × D = Allowable radius for interlocked armor or corrugated sheath, depending on type. *NEC Section 330.24(B)*

Figure 23 NEC® sections governing the installation of Type MC cable.

Cable Stripping

Special strippers are used to remove the jackets from Type NM and Type MC cable.

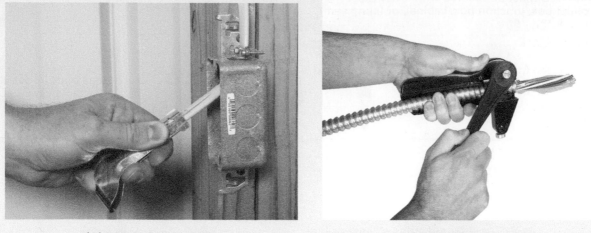

(A) NM CABLE RIPPER

(B) MC CABLE CUTTER

Figure Credit: Greenlee / A Textron Company

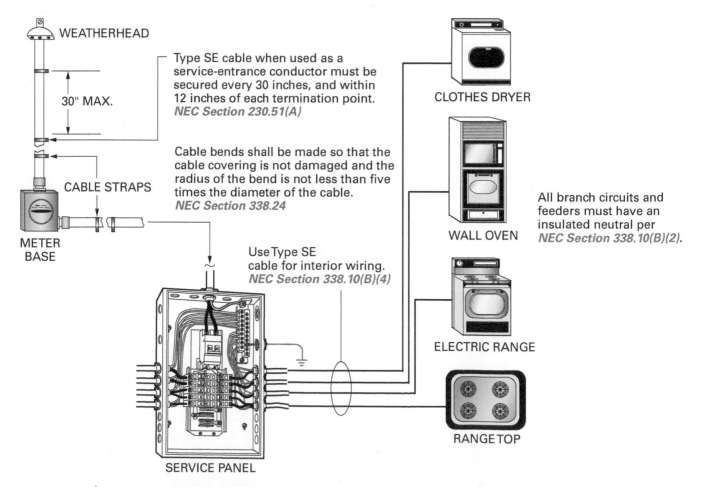

WEATHERHEAD

30" MAX.

CABLE STRAPS

METER BASE

Type SE cable when used as a service-entrance conductor must be secured every 30 inches, and within 12 inches of each termination point. *NEC Section 230.51(A)*

Cable bends shall be made so that the cable covering is not damaged and the radius of the bend is not less than five times the diameter of the cable. *NEC Section 338.24*

Use Type SE cable for interior wiring. *NEC Section 338.10(B)(4)*

SERVICE PANEL

CLOTHES DRYER

WALL OVEN

All branch circuits and feeders must have an insulated neutral per *NEC Section 338.10(B)(2).*

ELECTRIC RANGE

RANGE TOP

Figure 24 NEC® sections governing Type SE cable.

4.0.0 Section Review

1. In three-wire NM cable, the color code of the conductors is _____.

    a. black, white, and red, with a bare or green grounding conductor
    b. black, black, and white, with a bare or green grounding conductor
    c. black, blue, and white with a bare or green grounding conductor
    d. black, black, and black, with a bare or green grounding conductor

2. Which of the following areas of a residence would most likely use conductors run in conduit?

    a. Bedrooms
    b. Bathrooms
    c. Kitchens
    d. Unfinished basements

5.0.0 BRANCH CIRCUITS AND SIZING OUTLET BOXES

Objective

Lay out branch circuits and size outlet boxes.
a. Complete the branch circuit layout for power.
b. Complete the branch circuit layout for lighting.
c. Install outlet boxes.

Performance Task

3. Select the proper type and size outlet box needed for a given set of wiring conditions.

Trade Term

Switch leg: A circuit routed to a switch box for controlling electric lights.

The point at which electrical equipment, appliances, and devices are connected to the wiring system is commonly called an *outlet*. There are many classifications of outlets: lighting, receptacle, motor, appliance, and so forth. This section discusses the power and lighting outlets typically found in residential electrical wiring systems.

5.1.0 Branch Circuit Layout for Power

When viewing an electrical drawing, outlets are indicated by symbols (usually a small circle with appropriate markings to indicate the type of outlet). The most common symbols for receptacles are shown in *Figure 25*.

5.1.1 Branch Circuits and Feeders

The conductors that extend from the panelboard to the outlets are called *branch circuits* and are defined by the *NEC®* as the point of a wiring system that extends beyond the final overcurrent device protecting the circuit (*Figure 26*).

A feeder consists of all conductors between the service equipment and the final overcurrent device. *Figure 27* shows a feeder being used to supply a subpanel from the main service panel.

In general, the size of the branch circuit conductors varies depending upon the load requirements of the electrically operated equipment connected to the outlet. For residential use, most branch circuits consist of either No. 14 AWG, No. 12 AWG, No. 10 AWG, or No. 8 AWG conductors.

The basic branch circuit requires two wires or conductors to provide a continuous path for the flow of electric current, plus a third wire for equipment grounding. The usual receptacle branch circuit operates at 120V.

Fractional horsepower motors and small electric heaters usually operate at 120V and are connected to 120V branch circuits by means of a receptacle, junction box, or direct connection.

With the exception of very large residences and tract-development houses, the size of the average residential electrical system of the past has not been large enough to justify the expense of preparing complete electrical working drawings and specifications. Such electrical systems were usually laid out by the architect in the form of a sketchy outlet arrangement or else laid out by the electrician on the job, often only as the work progressed. However, many technical developments in residential electrical use—such as electric heat with sophisticated control wiring, increased use of electrical appliances, various electronic alarm systems, new lighting techniques, and the need for energy conservation techniques—have greatly expanded the demand and extended the complexity of today's residential electrical systems.

Each year, the number of homes with electrical systems designed by consulting engineering firms increases. Such homes are provided with complete electrical working drawings and specifications, similar to those frequently provided for commercial and industrial projects. Still, these are more the exception than the rule. Most residential projects will not have a complete set of drawings.

Circuit layout is provided on the drawings to follow for several reasons:

- They provide a visual layout of house wiring circuitry.
- They provide a sample of electrical residential drawings that are prepared by consulting engineering firms, although the number may still be limited.
- They introduce the method of showing electrical systems on working drawings to provide a foundation for tackling advanced electrical systems.

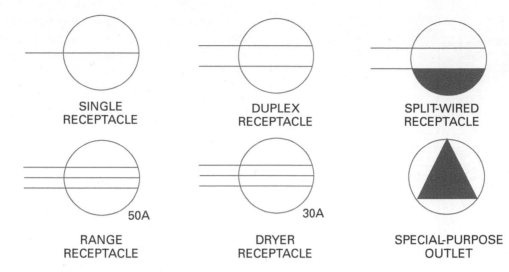

SINGLE
RECEPTACLE

DUPLEX
RECEPTACLE

SPLIT-WIRED
RECEPTACLE

50A

RANGE
RECEPTACLE

30A

DRYER
RECEPTACLE

SPECIAL-PURPOSE
OUTLET

Figure 25 Typical outlet symbols appearing in electrical drawings.

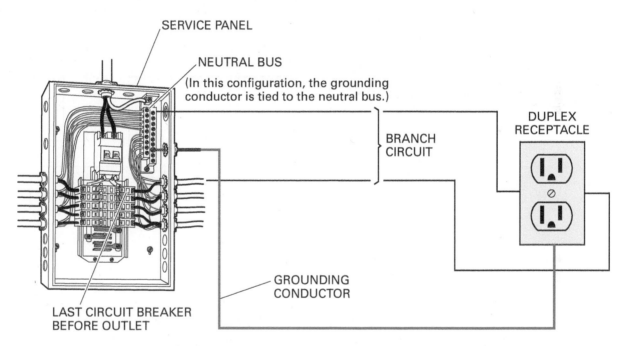

Figure 26 Components of a duplex receptacle branch circuit.

Branch circuits are shown on electrical drawings by means of a single line drawn from the panelboard (or by homerun arrowheads indicating that the circuit goes to the panelboard) to the outlet or from outlet to outlet where there is more than one outlet on the circuit.

The lines indicating branch circuits can be solid to show that the conductors are to be run concealed in the ceiling or wall; dashed to show that the conductors are to be run in the floor or ceiling below; or dotted to show that the wiring is to be run exposed. *Figure 28* shows examples of these three types of branch circuit lines.

In *Figure 28*, No. 12 indicates the wire size. The slash marks shown through the circuits indicate the number of current-carrying conductors in the circuit. Although two slash marks are shown, in actual practice, a branch circuit containing only two conductors usually contains no slash marks; that is, any circuit with no slash marks is assumed to have two conductors. However, three or more conductors are always indicated on electrical working drawings—either by slash marks for each conductor, or else by a note.

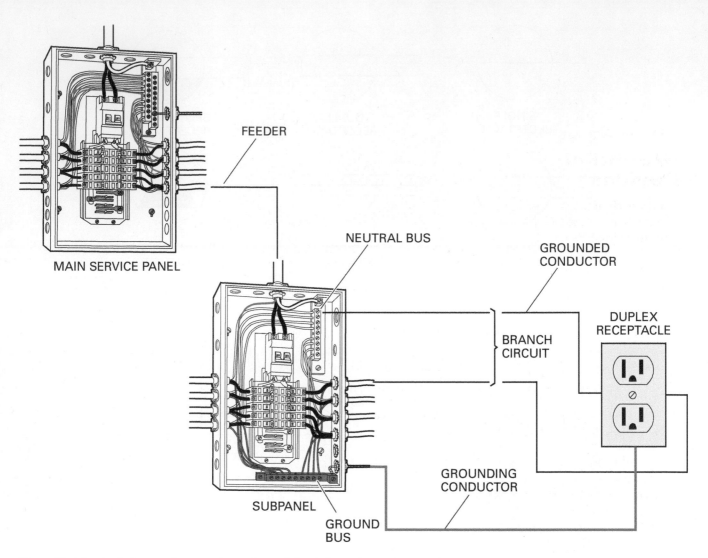

FEEDER

NEUTRAL BUS

GROUNDED CONDUCTOR

MAIN SERVICE PANEL

BRANCH CIRCUIT

DUPLEX RECEPTACLE

SUBPANEL

GROUNDING CONDUCTOR

GROUND BUS

Figure 27 A feeder being used to supply a subpanel from the main service panel.

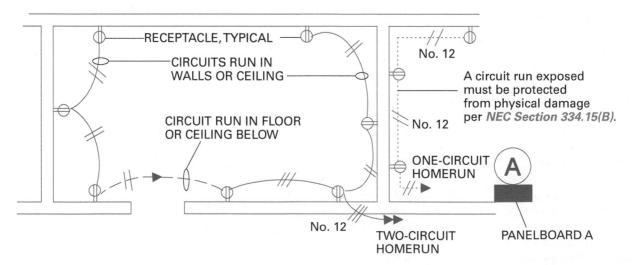

RECEPTACLE, TYPICAL

CIRCUITS RUN IN WALLS OR CEILING

CIRCUIT RUN IN FLOOR OR CEILING BELOW

No. 12

A circuit run exposed must be protected from physical damage per *NEC Section 334.15(B)*.

No. 12

ONE-CIRCUIT HOMERUN

A

No. 12

TWO-CIRCUIT HOMERUN

PANELBOARD A

Figure 28 Types of branch circuit lines shown on electrical working drawings.

Never assume that you know the meaning of any electrical symbol. Although great efforts have been made in recent years to standardize drawing symbols, architects, consulting engineers, and electrical drafters still modify existing symbols or devise new ones to meet their own needs. Always consult the symbol list or legend on electrical working drawings for an exact interpretation of the symbols used.

5.1.2 Locating Receptacles

NEC Section 210.52 states the minimum requirements for the location of receptacles in dwelling units. It specifies that in each kitchen, family room, bedroom, living room, dining room, or similar area, receptacle outlets shall be installed so that no point along the floor line in any wall space is more than 6' (1.8 m), measured horizontally, from an outlet in that space, including any wall space 2' (600 mm) or more in width and the wall space occupied by fixed panels in exterior walls, but excluding sliding panels. Receptacle outlets shall, insofar as practicable, be spaced equal distances apart. Receptacle outlets in floors shall not be counted as part of the required number of receptacle outlets unless located within 18" (450 mm) of the wall.

The *NEC*® defines wall space as a wall that is unbroken along the floor line by doorways, fireplaces, or similar openings. Each wall space that is 2' (600 mm) or more in width must be treated individually and separately from other wall spaces within the room.

The purpose of *NEC Section 210.52* is to minimize the use of cords across doorways, fireplaces, and similar openings.

With this *NEC*® requirement in mind, outlets for our sample residence will be laid out (*Figure 29*). In laying out these receptacle outlets, the floor line of the wall is measured (also around corners), but not across doorways, fireplaces, passageways, or other spaces where a flexible cord extended across the space would be unsuitable.

In general, duplex receptacle outlets must be no more than 12' (3.7 m) apart. When spaced in this manner, a 6' (1.8 m) extension cord will reach a receptacle from any point along the wall line.

Note that at no point along the wall line are any receptacles more than 12' (3.7 m) apart or more than 6' (1.8 m) from any door or room opening. Where practical, no more than eight receptacles are connected to one circuit. However, this is just a design consideration because general-purpose receptacles in dwelling units are sized on the basis of 3VA per square foot (33VA per square meter) of dwelling space. A 15A branch circuit is rated at 1,800VA (15A x 120V = 1,800VA) and the *NEC*® requires that for every 600 square feet (1,800VA ÷ 3VA/sq. ft. = 600 sq. ft.), a circuit to supply lighting and receptacles must be installed. Always check with the local authorities about the requirements for the number of branch circuits in a dwelling.

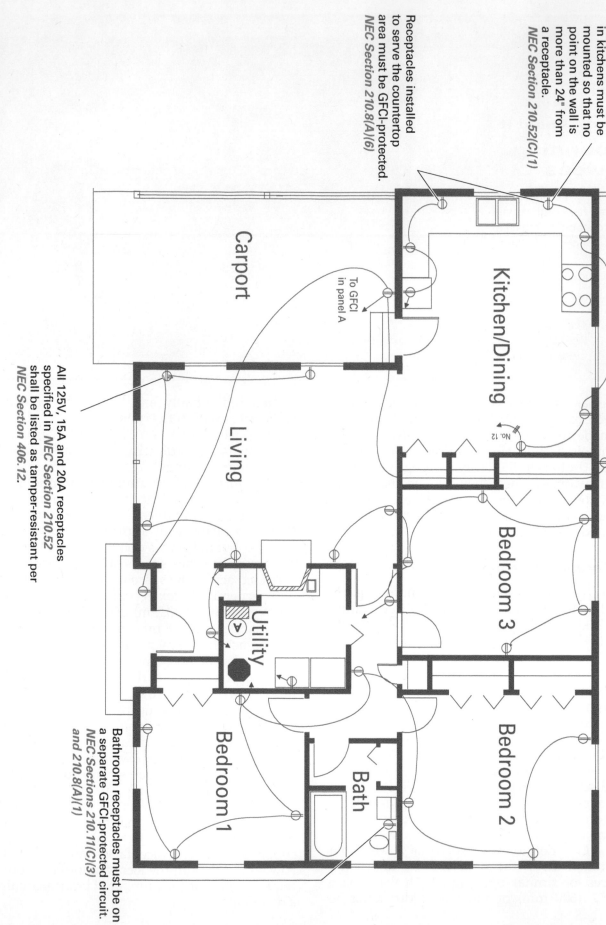

Receptacles located above countertops in kitchens must be mounted so that no point on the wall is more than 24" from a receptacle. *NEC Section 210.52(C)(1)*

Receptacles installed to serve the countertop area must be GFCI-protected. *NEC Section 210.8(A)(6)*

To GFCI in panel A

Carport

Kitchen/Dining

No. 12

Living

Bedroom 3

Bedroom 2

Utility

Bath

Bedroom 1

All 125V, 15A and 20A receptacles specified in *NEC Section 210.52* shall be listed as tamper-resistant per *NEC Section 406.12*.

Bathroom receptacles must be on a separate GFCI-protected circuit. *NEC Sections 210.11(C)(3) and 210.8(A)(1)*

Figure 29 Floor plan of the sample residence.

NCCER – *Electrical*

The utility room has at least one receptacle for the laundry on a separate circuit in order to comply with *NEC Sections 210.11(C)(2) and 210.52(F)*.

One duplex receptacle is located in the vestibule for cleaning purposes, such as feeding a portable vacuum cleaner or similar appliance. It is connected to the living room circuit. An additional duplex receptacle is required per *NEC Section 210.52(H)* in hallways of 10' (3.0 m) or more.

Although this is not shown in the figure, the living room outlets could be split-wired (the lower half of each duplex receptacle is energized all the time, while the upper half can be switched on or off). The reason for this is that a great deal of the illumination for this area will be provided by portable table lamps, and the split-wired receptacles provide a means to control these lamps from several locations, such as at each entry to the living room, if desired. Split receptacles are discussed in more detail in the next section.

To comply with *NEC Sections 210.11(C)(1) and 210.52(B)*, the kitchen receptacles are laid out as follows. In addition to the number of branch circuits determined previously, two or more 20A small appliance branch circuits must be provided to serve all receptacle outlets (including refrigeration equipment) in the kitchen, pantry, breakfast room, dining room, or similar area of the house. Such circuits, whether two or more are used, must have no other outlets connected to them. All receptacles serving a kitchen countertop require GFCI protection. No small appliance branch circuit shall serve more than one kitchen.

To comply with *NEC Sections 210.11(C)(3) and 210.52(D)*, bathroom receptacle(s) must be on a separate branch circuit supplying only bathroom receptacles or on a circuit supplying a single bathroom with no loads other than that bathroom. All receptacles located within a bathroom require GFCI protection. GFCI protection is also required on garage and exterior receptacles. All other branch circuits that supply the lighting and general-purpose receptacles in dwelling units must have arc fault circuit interrupter protection to comply with *NEC Section 210.12*.

5.1.3 Split-Wired Duplex Receptacles

In modern residential construction, it is common to have duplex wall receptacles that have one of the outlets wired as a standard duplex outlet (hot all the time) and the other half controlled by a wall switch. This allows table or floor lamps to be controlled by a wall switch and leaves the other outlet available for items that are not to be switched. This wiring method is commonly referred to as a split receptacle. Note that switched receptacles are installed to provide lighting. Dimmer switches are not permitted to be used per *NEC Section 404.14(E)*.

Most duplex 15A and 20A receptacles are provided with a breakoff tab that permits each of the two receptacle outlets to be supplied from a different source or polarity. For example, one outlet would be supplied from the hot leg of a series of outlets and the other outlet supplied from the **switch leg** of a light switch. A diagram of this arrangement is shown in *Figure 30*.

Another application of split receptacles is shown in *Figure 31*. In this example, one outlet connected from a double-pole circuit breaker supplies 240V for an appliance, such as a window air conditioning unit, while the other outlet is connected from one pole of the double-pole circuit breaker and the other side is connected to the neutral or grounded conductor to supply 120V for an appliance, such as a lamp. *NEC Section 210.4(B)* requires the use of a two-pole breaker when two circuits are connected to one duplex receptacle so that all ungrounded conductors of the circuit are disconnected simultaneously. This circuit and the split receptacle mentioned above are both considered multiwire branch circuits.

5.1.4 Multiwire Branch Circuits

NEC Article 100 defines a multiwire branch circuit as "a branch circuit that consists of two or more ungrounded conductors that have a voltage between them, and a grounded conductor that has equal voltage between it and each ungrounded conductor of the circuit and that is connected to the neutral or grounded conductor of the system."

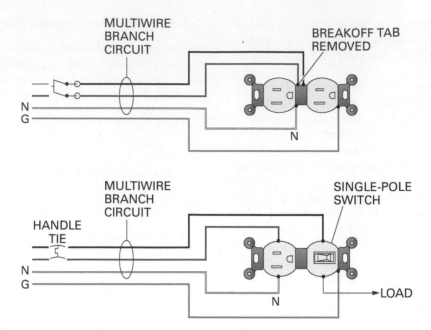

Figure 30 Two 120V receptacle outlets supplied from different sources.

5.1.5 240-Volt Circuits

The electric range, clothes dryer, and water heater in a residence all operate at 240VAC. Each will be fed by a separate circuit and connected to a two-pole circuit breaker of the appropriate rating in the panelboard. To determine the conductor size and overcurrent protection for the range, proceed as follows:

Step 1 Find the nameplate rating of the electric range. This has previously been determined to be 12kVA.

Step 2 Refer to *NEC Table 220.55*. Because Column C of this table applies to ranges rated at 12kVA (12kW) and under, this will be the column to use in this example.

Step 3 Under the Number of Appliances column, locate the appropriate number of appliances (one in this case), and find the maximum demand given for it in Column C. Column C states that the circuit should be sized for 8kVA (not the nameplate rating of 12kVA).

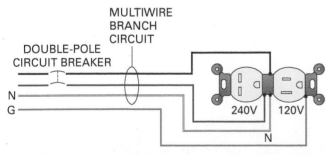

Figure 31 Combination receptacle.

Step 4 Calculate the required conductor ampacity as follows:

$$\frac{8{,}000\text{VA}}{240\text{V}} = 33.33\text{A (round to 33A)}$$

> **NOTE**
>
> The *NEC®* requires rounding of the final value for amps to the nearest whole ampere. Values under 0.5A are dropped. See *NEC Section 220.5(B)*.

The minimum branch circuit must be rated at 40A because common residential circuit breakers are rated in steps of 15A, 20A, 30A, 40A, and so forth. A 30A circuit breaker is too small, so a 40A circuit breaker is selected. The conductors must have a current-carrying capacity that is equal to or greater than the overcurrent protection. Therefore, No. 8 AWG conductors will be used.

If a cooktop and wall oven were used instead of the electric range, the circuit would be sized similarly. The *NEC®* specifies that a branch circuit for a counter-mounted cooking unit and not more than two wall-mounted ovens, all supplied from a single branch circuit and located in the same room, is computed by adding the nameplate ratings of the individual appliances and treating this total as equivalent to one range. Therefore, two appliances of 6kVA each may be treated as a single range with a 12kVA nameplate rating.

Figure 32 shows how the electric range circuit may appear on an electrical drawing. The connection may be made directly to the range junction box, but more often a 50A range receptacle is mounted at the range location and a range cord-and-plug set is used to make the connection. This facilitates moving the appliance later for maintenance or cleaning.

Figure 33 shows several types of receptacle configurations used in residential wiring applications. You will eventually recognize these configurations at a glance.

The branch circuit for the water heater in the sample residence must be sized for its full capacity because there is no diversity or demand factor for this appliance. Because the nameplate rating on the water heater indicates two heating elements of 4,500W each, the first inclination would be to size the circuit for a total load of 9,000W (volt-amperes). However, only one of the two elements operates at a time (*Figure 34*). Note that each element is controlled by a separate thermostat. The lower element becomes energized when the thermostat calls for heat, and at the same time, the thermostat opens a set of contacts to prevent the upper element from operating. When the lower element's thermostat is satisfied, the lower contacts open, and at the same time, the thermostat closes the contacts for the upper element to become energized to maintain the water temperature.

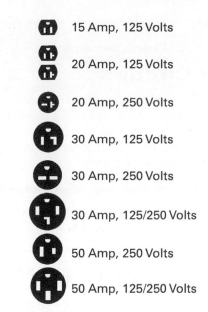

Figure 33 Residential receptacle configurations.

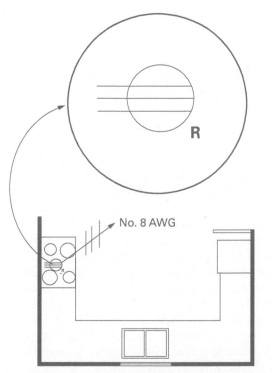

Figure 32 Range circuit shown on an electrical drawing.

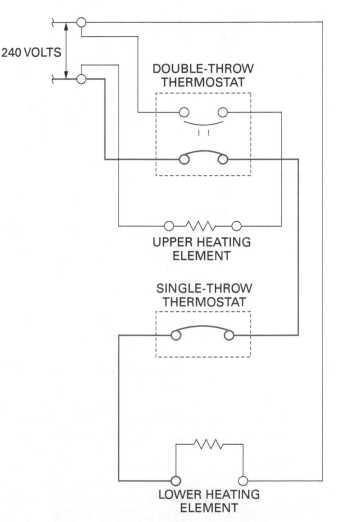

Figure 34 Wiring diagram of water heater controls.

Using this information, the circuit for the water heater may be sized as follows:

$$\frac{4,500VA}{240V} = 18.75A \times 1.25 = 23.44A$$

NEC Section 422.13 requires that the branch circuits that supply storage-type water heaters having a capacity of 120 gallons (450 L) or less be rated as continuous loads and therefore sized at not less than 125% of the nameplate rating. Our calculation shows this to not be less than 23A. Normally, this would require a maximum rating for the branch circuit to be not more than 25A. (See standard ratings of overcurrent devices in *NEC Section 240.6*.) However, *NEC Section 422.11(E)(3)* permits a single nonmotor-operated appliance to be protected by overcurrent devices rated up to 150% of the nameplate rating of the appliance. In this case, 4,500VA ÷ 240V = 18.75A x 150% = 28.125A. Because the next standard rating is 30A, the water heater will be wired with No. 10 AWG conductors protected by a 30A overcurrent device.

The *NEC®* specifies that electric clothes dryers must be rated at 5kVA or the nameplate rating, whichever is greater. In this case, the dryer is rated at 5.5kVA, and the conductor current-carrying capacity is calculated as follows:

$$\frac{5,500VA}{240V} = 22.92A$$

A three-wire, 30A circuit will be provided (No. 10 AWG wire). It is protected by a 30A circuit breaker. The dryer may be connected directly, but a 30A dryer receptacle is normally provided for the same reasons as mentioned for the electric range.

Large appliance outlets rated at 240V are frequently shown on electrical drawings using lines and symbols to indicate the outlets and circuits. In some cases, no drawings are provided.

5.1.6 Electric Heating Circuits

Electric baseboard heating offers an advantage in that its installation costs are lower than those for other types of heating systems. In general, electric baseboard heaters should be located on the outside wall near the areas of greatest heat loss, such as under windows. The controls for wall-mounted thermostats should be located on an interior wall, about 50" (1.3 m) above the floor to sense the average room temperature. *Figure 35* shows an electric heating arrangement for the sample residence.

5.2.0 Branch Circuit Layout for Lighting

A simple lighting branch circuit requires two conductors to provide a continuous path for current flow. The usual lighting branch circuit operates at 120V; the white (grounded) circuit conductor is therefore connected to the neutral bus in the panelboard, while the black (ungrounded) circuit conductor is connected to an overcurrent protection device.

Lighting branch circuits and outlets are shown on electrical drawings by means of lines and symbols; that is, a single line is drawn from outlet to outlet and then terminated with an arrowhead to indicate a homerun to the panelboard. Several methods are used to indicate the number and size of conductors, but the most common is to indicate the number of conductors in the circuit by using slash marks through the circuit lines and then indicate the wire size by a notation adjacent to these slash marks. For example, two slash marks indicate two conductors; three slash marks indicate three conductors. Some electrical designers omit slash marks for two-conductor circuits. In this case, the conductor size is usually indicated in the symbol list or legend.

The circuits used to feed residential lighting must conform to standards established by the *NEC®* as well as by local and state ordinances. Most of the lighting circuits should be calculated to include the total load, although at times this is not possible because the electrician cannot be certain of the exact wattage that might be used by the homeowner. For example, an electrician may install four porcelain lampholders for the unfinished basement area, each to contain one 100W incandescent lamp. However, the homeowners may eventually replace the original lamps with others rated at 150W or even 200W. Thus, if the electrician initially loads the lighting circuit to full capacity, the circuit will probably become overloaded in the future.

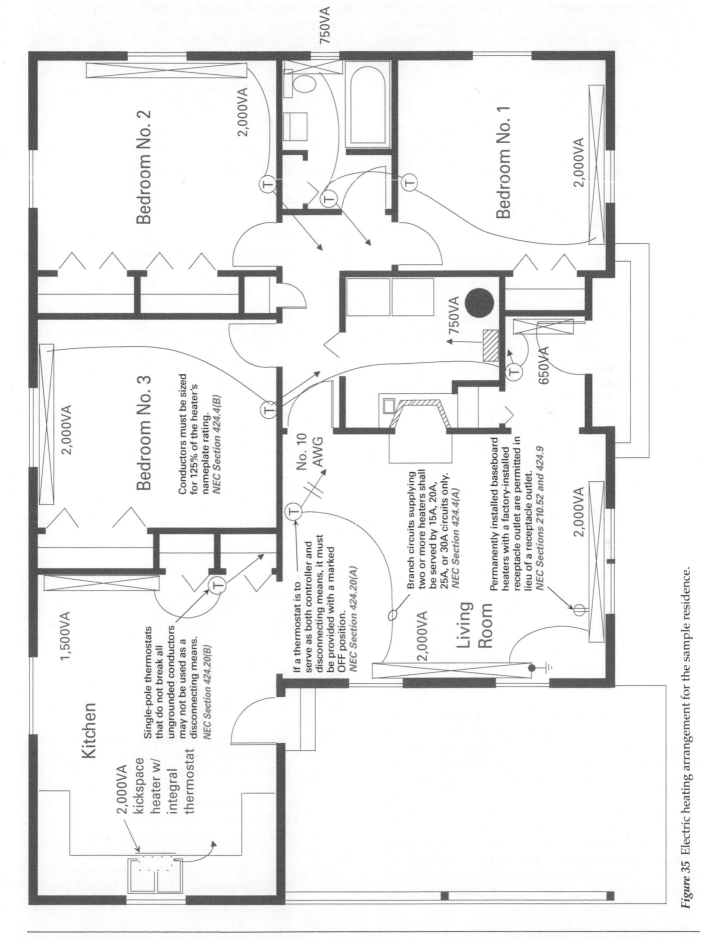

750VA

2,000VA

Bedroom No. 2

Bedroom No. 1

2,000VA

Bedroom No. 3

750VA

Conductors must be sized for 125% of the heater's nameplate rating.
NEC Section 424.4(B)

2,000VA

650VA

2,000VA

No. 10 AWG

Single-pole thermostats that do not break all ungrounded conductors may not be used as a disconnecting means.
NEC Section 424.20(B)

If a thermostat is to serve as both controller and disconnecting means, it must be provided with a marked OFF position.
NEC Section 424.20(A)

Branch circuits supplying two or more heaters shall be served by 15A, 20A, 25A, or 30A circuits only.
NEC Section 424.4(A)

2,000VA

Living Room

Permanently installed baseboard heaters with a factory-installed receptacle outlet are permitted in lieu of a receptacle outlet.
NEC Sections 210.52 and 424.9

1,500VA

Kitchen

2,000VA kickspace heater w/ integral thermostat

Figure 35 Electric heating arrangement for the sample residence.

It is recommended that no residential branch circuit be loaded to more than 80% of its rated capacity. Because most circuits used for lighting are rated at 15A, the total ampacity (in volt-amperes) for the circuit is as follows:

$$15A \times 120V = 1{,}800VA$$

Therefore, if the circuit is to be loaded to only 80% of its rated capacity, the maximum initial connected load should be no more than 1,440VA, as shown by the calculation below:

$$80\% \text{ of } 1{,}800VA =$$
$$0.8 \times 1{,}800VA =$$
$$1{,}440VA$$

Figure 36 shows one possible lighting arrangement for the sample residence. All lighting fixtures are shown in their approximate physical location as they should be installed. In actual practice, the location of lighting fixtures (luminaires) and their related switches will probably be the extent of the information shown on working drawings. The circuits shown are meant to illustrate how lighting circuits are routed, not to imply that such drawings are typical for residential construction. If fixtures are used in a closet, they must meet the requirements of *NEC Section 410.16*.

Electrical symbols are used to show the fixture types. Switches and lighting branch circuits are also shown by appropriate lines and symbols. The meanings of the symbols used on this drawing are explained in the symbol list in *Figure 37*.

5.3.0 Outlet Boxes

Electricians installing residential electrical systems must be familiar with outlet box capacities, means of supporting outlet boxes, and other requirements of the *NEC*®. Boxes were discussed in detail in an earlier module, but a general review of the rules and necessary calculations is provided here.

The maximum numbers of conductors of the same size permitted in standard outlet boxes are listed in *NEC Table 314.16(A)*. These figures apply where no fittings or devices such as fixture studs, cable clamps, switches, or receptacles are contained in the box and where no grounding conductors are part of the wiring within the box. Obviously, in all modern residential wiring systems there will be one or more of these items contained in every outlet box installed. Therefore, where one or more of the above-mentioned items are present, the total number of conductors will

be less than that shown in the table. Also, if the box contains a looped, unbroken conductor 12" (300 mm) or more in length, it must be counted twice.

For example, a deduction of two conductors must be made for each strap containing a wiring device entering the box (based on the largest size conductor connected to the device) such as a switch or duplex receptacle; a further deduction of one conductor must be made for up to four equipment grounding conductors entering the box (based on the largest size grounding conductor). For instance, a 3" x 2" x 2¾" (75 mm x 50 mm x 70 mm) box is listed in the table as containing a maximum of six No. 12 wires. If the box contains cable clamps and a duplex receptacle, three wires will have to be deducted from the total of six, providing for only three No. 12 wires. If a ground wire is used, which is always the case in residential wiring, only two No. 12 wires may be used.

For example, to size a metallic outlet box for two No. 12 AWG conductors with a ground wire, cable clamp, and receptacle, proceed as follows:

Step 1 Calculate the total number of conductors and equivalents [*NEC Section 314.16(B)*]. One ground wire plus one cable clamp plus one receptacle (two wires) plus two No. 12 conductors equals a total of six No. 12 conductors.

Step 2 Determine the amount of space required for each conductor. *NEC Table 314.16(B)* gives the box volume required for each conductor. No. 12 AWG equals 2.25 cubic inches (36.9 cubic cm).

Step 3 Calculate the outlet box space required by multiplying the number of cubic inches required for each conductor by the total number of conductors:

US Measure:
$$6 \times 2.25in^3 = 13.5in^3$$
Metric:
$$6 \times 36.9cm^3 = 221cm^3$$

Step 4 Once you have determined the required box capacity, again refer to *NEC Table 314.16(A)* and note that a 3" x 2" x 2¾" (75 mm x 50 mm x 70 mm) box comes closest to our requirements. This box is rated for 14 cubic inches (230 cubic cm).

Now, size the box for two additional conductors. Where four No. 12 conductors enter the box with two ground wires, only the two additional No. 12 conductors must be added to our previous count for a total of 8 conductors (6 + 2 = 8).

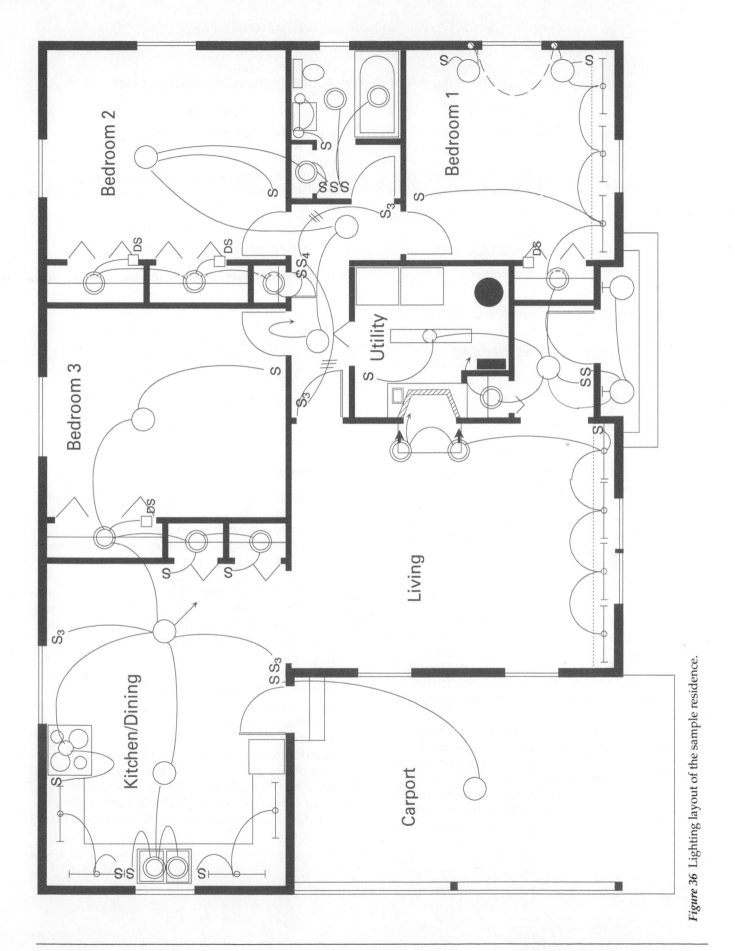

Figure 36 Lighting layout of the sample residence.

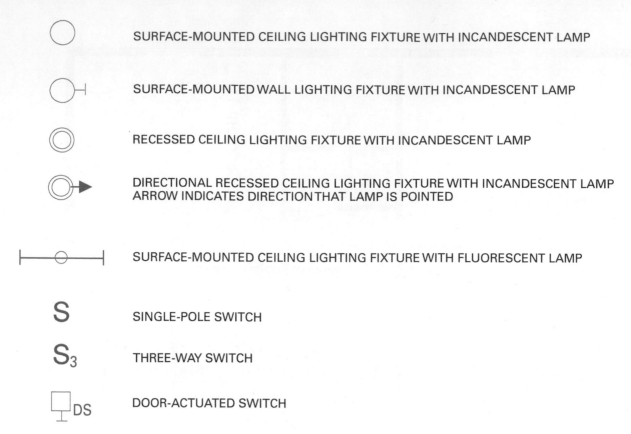

SURFACE-MOUNTED CEILING LIGHTING FIXTURE WITH INCANDESCENT LAMP

SURFACE-MOUNTED WALL LIGHTING FIXTURE WITH INCANDESCENT LAMP

RECESSED CEILING LIGHTING FIXTURE WITH INCANDESCENT LAMP

DIRECTIONAL RECESSED CEILING LIGHTING FIXTURE WITH INCANDESCENT LAMP
ARROW INDICATES DIRECTION THAT LAMP IS POINTED

SURFACE-MOUNTED CEILING LIGHTING FIXTURE WITH FLUORESCENT LAMP

S   SINGLE-POLE SWITCH

S_3 THREE-WAY SWITCH

DS   DOOR-ACTUATED SWITCH

Figure 37 Symbols.

Remember, up to four ground wires in a box counts as only one conductor; any number of cable clamps also counts as only one conductor. Therefore, the box size required for use with two additional No. 12 conductors may be calculated as follows:

US Measure:
$$8 \times 2.25\,in^3 = 18\,in^3$$
Metric:
$$8 \times 36.9\,cm^3 = 295\,cm^3$$

Again, refer to *NEC Table 314.16(A)* and note that a 3" x 2" x $3\frac{1}{2}$" (75 mm x 50 mm x 90 mm) device box with a rated capacity of 18.0 cubic inches (295 cubic cm) is the closest device box that meets *NEC®* requirements. An alternative is to use a square 4" x $1\frac{1}{4}$" (100 mm x 32 mm) square box with a single-gang plaster ring, as shown in *Figure 38*. This box also has a capacity of 18.0 cubic inches (295 cubic cm).

Other box sizes are calculated in a similar fashion. When sizing boxes for different size conductors, remember that the conductor volume varies as shown in *NEC Table 314.16(B)*.

5.3.1 Mounting Outlet Boxes

Outlet box configurations are almost endless, and if you research the various methods of mounting these boxes, you will be astonished. In this section, some common outlet boxes and their mounting considerations will be reviewed.

The conventional metallic device box, which is used for residential duplex receptacles and switches for lighting control, may be mounted to wall studs using 16d (16-penny) nails placed through the round mounting holes passing through the interior of the box. The nails are then driven into the wall stud. When nails are used for mounting outlet boxes in this manner, the nails must be located within $\frac{1}{4}$" (6 mm) of the back or ends of the enclosure.

Nonmetallic boxes normally have mounting nails fitted to the box for mounting. Other boxes have mounting brackets. When mounting outlet boxes with brackets, use either wide-head roofing nails or box nails about $1\frac{1}{4}$" (32 mm) in length. *Figure 39* shows various methods of mounting outlet boxes.

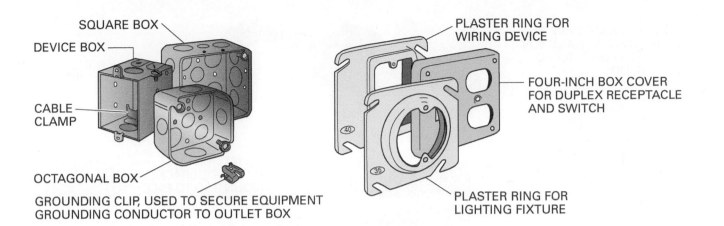

SQUARE BOX

DEVICE BOX

CABLE CLAMP

OCTAGONAL BOX

GROUNDING CLIP, USED TO SECURE EQUIPMENT
GROUNDING CONDUCTOR TO OUTLET BOX

PLASTER RING FOR
WIRING DEVICE

FOUR-INCH BOX COVER
FOR DUPLEX RECEPTACLE
AND SWITCH

PLASTER RING FOR
LIGHTING FIXTURE

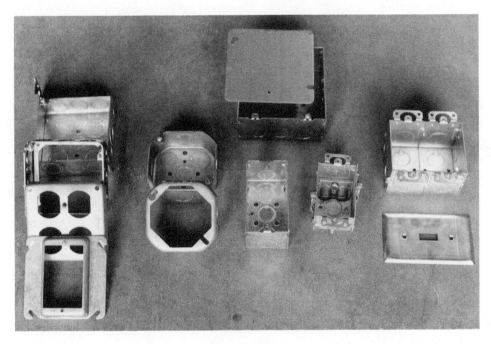

Figure 38 Typical metallic outlet boxes with extension (plaster) rings.

Before mounting any boxes during the rough wiring process, first find out which type and thickness of finish will be used on the walls. This will dictate the depth to which the boxes must be mounted to comply with *NEC®* regulations. For example, the finish on plastered walls or ceilings is normally $\frac{1}{2}$" (13 mm) thick; gypsum board or drywall is either $\frac{1}{2}$" (13 mm) or $\frac{5}{8}$" (16 mm) thick; and wood paneling is normally only $\frac{1}{4}$" (6 mm) thick. (Some tongue-and-groove wood paneling is thicker.) Always check the material thickness of final finishes before installing boxes.

The *NEC®* specifies the amount of space permitted from the edge of the outlet box to the finished wall. When a noncombustible wall finish (such as plaster, masonry, or tile) is used, the box may be recessed $\frac{1}{4}$" (6 mm). However, when combustible finishes are used (such as wood paneling), the box must be flush (even) with the finished wall or ceiling. Refer to *Figure 40* and *NEC Section 314.20*.

When Type NM cable is used in either metallic or nonmetallic outlet boxes, the cable assembly, including the sheath, must extend into the box by not less than $\frac{1}{4}$" (6 mm) per *NEC Section 314.17(B)(2)*. In all instances, all permitted wiring methods must be secured to the boxes by means of either cable clamps or approved connectors. The one exception to this rule is where Type NM cable is used with 2 $\frac{1}{4}$" x 4" (57 mm x 100 mm) or smaller nonmetallic boxes where the cable is fastened within 8" (200 mm) of the box. In this case, the cable does not have to be secured to the box. See *NEC Section 314.17(B)(2), Exception*.

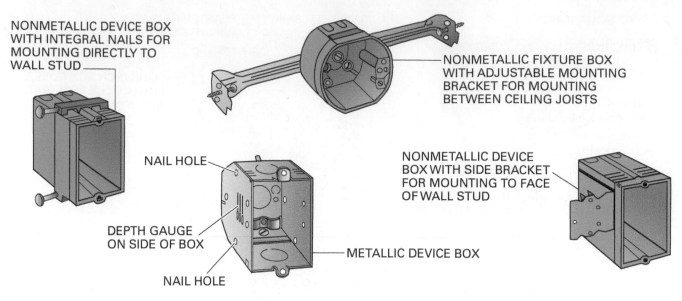

NONMETALLIC DEVICE BOX WITH INTEGRAL NAILS FOR MOUNTING DIRECTLY TO WALL STUD

NONMETALLIC FIXTURE BOX WITH ADJUSTABLE MOUNTING BRACKET FOR MOUNTING BETWEEN CEILING JOISTS

NAIL HOLE

NONMETALLIC DEVICE BOX WITH SIDE BRACKET FOR MOUNTING TO FACE OF WALL STUD

DEPTH GAUGE ON SIDE OF BOX

METALLIC DEVICE BOX

NAIL HOLE

Figure 39 Several methods of mounting outlet boxes.

5.3.2 Ceiling Fan Boxes

A box used at fan outlets is not permitted to be used as the sole support for ceiling (paddle) fans, unless it is listed for the application as the sole means of support. Where a ceiling fan does not exceed 70 pounds (32 kg) in weight, it is permitted to be supported by outlet boxes listed and identified for such use. Boxes designed to support more than 35 pounds (16 kg) must be marked with the maximum weight to be supported. See *NEC Section 314.27(C)*. These boxes must be rigidly supported from a structural member of the building. *NEC Section 314.27(C)* also requires that a ceiling-fan rated box be used if the location of the box is acceptable for a fan installation.

Mounting Boxes

To quickly mount each box at the same height from the floor, make a simple height template (story pole) and mark it with the receptacle and switch heights. The story pole consists of an L-shaped jig made out of 2 x 2s or 2 x 4s. After installing the boxes, make sure to push the wires well back into the box so that the sheetrock installers will not damage the wires when they rout out a hole for the receptacle.

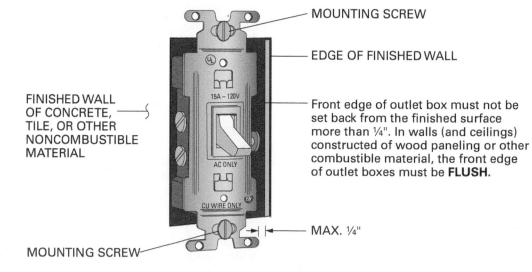

MOUNTING SCREW

EDGE OF FINISHED WALL

FINISHED WALL OF CONCRETE, TILE, OR OTHER NONCOMBUSTIBLE MATERIAL

Front edge of outlet box must not be set back from the finished surface more than ¼". In walls (and ceilings) constructed of wood paneling or other combustible material, the front edge of outlet boxes must be **FLUSH**.

MAX. ¼"

MOUNTING SCREW

Figure 40 Outlet box installation.

Calculating Conductors

In a square 4" x 1½" (100 mm x 38 mm) metal box, one 14/3 cable with ground feeds three 14/2 cables with ground wires. The red wire of the 14/3 cable feeds a receptacle, and the black wire feeds the 14/2 black wires. All of the white wires are spliced together, with one brought out to the receptacle terminal. The ground wires are all spliced, with one brought out to the grounding terminal on the receptacle and one to the ground clip on the box. All four cables are connected with box connectors rather than internal clamps. Using *NEC Section 314.16*, determine whether this wiring violates the code.

5.0.0 Section Review

1. A branch circuit that supplies a storage-type water heater with a capacity of 120 gallons (450 L) or less and a nameplate rating of 20A must be rated at a minimum of _____.

   a. 18A
   b. 20A
   c. 22A
   d. 25A

2. A branch circuit for lighting has a capacity of 1,800VA. It should be loaded to no more than _____.

   a. 1,440VA
   b. 1,800VA
   c. 2,250VA
   d. 2,700VA

3. A metallic outlet box for two No. 12 AWG conductors with a ground wire, cable clamp, a switch, and a looped conductor over 12" (300 mm) would count as _____.

   a. two conductors
   b. four conductors
   c. five conductors
   d. eight conductors

6.0.0 WIRING DEVICES

Objective

Select and install various wiring devices.

   a. Select and install receptacles.
   b. Select and install switches.
   c. Install devices near residential swimming pools, spas, and hot tubs.

Wiring devices include various types of receptacles and switches, the latter being used for lighting control. Switches are covered in *NEC Article 404*, while regulations for receptacles may be found in *NEC Article 406*.

6.1.0 Receptacles

Receptacles are rated by voltage and amperage capacity. *NEC Section 406.3* requires that receptacles connected to a 15A or 20A circuit have the correct specifications for the application and be of the grounding type. *NEC Section 406.12* requires that all 15A and 20A, 125V receptacles installed in dwelling units be listed as tamper-resistant.

NEC Section 406.4(D) has several requirements for replacement receptacles:

- *NEC Section 406.4(D)(4)* – The replacement needs to be AFCI-protected if specified elsewhere.
- *NEC Section 406.4(D)(5)* – The replacement shall be tamper-resistant if specified elsewhere.
- *NEC Section 406.4(D)(6)* – The replacement must be weather-resistant if specified elsewhere.

Where there is only one outlet on a circuit, the receptacle's rating must be equal to or greater than the capacity of the conductors feeding it per *NEC Section 210.21(B)(1)*. For example, if one receptacle is connected to a 20A residential laundry circuit, the receptacle must be rated at 20A or more. When more than one outlet is on a circuit, the total connected load must be equal to or less than the capacity of the branch circuit conductors feeding the receptacles.

Refer to *Figure 41* for some of the characteristics of a standard 125V, 15A duplex receptacle. Note that the terminals are color coded as follows:

- *Green* – Connection for the equipment grounding conductor

- *Silver* – Connection for the neutral or grounded conductor
- *Brass* – Connection for the ungrounded conductor

A standard 125V, 15A receptacle is also typically imprinted with the following symbols:

- *UL* – Underwriters Laboratories, Inc., listing
- *CSA* – Canadian Standards Association
- *CO/ALR* – Designed for use with both copper and aluminum wire
- *15A* – Receptacle rated for a maximum of 15A
- *125V* – Receptacle rated for a maximum of 125V

The UL label means that the receptacle has undergone testing by Underwriters Laboratories, Inc., and meets minimum safety requirements. Underwriters Laboratories, Inc., was created by the National Board of Fire Underwriters to test electrical devices and materials. The UL label is a safety rating only and does not mean that the device or equipment meets any type of quality standard. The CSA label means that the receptacle is approved by the Canadian Standards Association, the Canadian equivalent to Underwriters Laboratories, Inc. The CSA label means that the receptacle is acceptable for use in Canada.

The CO/ALR symbol means that the device is suitable for use with copper, aluminum, or copper-clad aluminum wire. The CO in the symbol stands for copper while ALR stands for aluminum revised. The CO/ALR symbol replaces the earlier CU/AL mark, which appeared on wiring devices that were later found to be inadequate for use with aluminum wire in the 15A to 20A range. Therefore, any receptacle or wall switch marked with the CU/AL configuration or anything other than CO/ALR should be used only for copper wire.

These same configurations also apply to wall switches used for lighting control. These will be discussed next.

6.2.0 Lighting Control

A single-pole snap-action switch consists of a device containing two stationary current-carrying elements, a moving current-carrying element, a toggle handle, a spring, and a housing. When the contacts are open, as shown in *Figure 42*, the circuit is broken and no current flows. When the moving element is closed by manually flipping the toggle handle, the contacts complete the circuit and the lamp is energized (*Figure 43*).

PLASTER EARS WITH
BREAKOFF TABS

SLOT FOR MOUNTING
SCREWS

LONGER SLOT INDICATES
NEUTRAL OR GROUNDED
CONDUCTOR

NEUTRAL OR GROUNDED
TERMINALS ARE INDICATED
BY SILVER-COLORED SCREWS

AMPERAGE/VOLTAGE RATING

INCORPORATES TAMPER-RESISTANT
SHUTTER MECHANISM (HIDDEN)

GROUNDING CONNECTION
HAS GREEN SCREW HEAD

UNDERWRITERS
LABORATORIES, INC., LISTING

CO/ALR DESIGNATION INDICATES THAT
THE SWITCH IS DESIGNED FOR USE WITH
BOTH COPPER AND ALUMINUM WIRE

SMALL SLOT INDICATES
UNGROUNDED CONDUCTOR
Ungrounded conductors are connected to
the brass screws on opposite side from
grounded conductor screws.

CANADIAN STANDARDS ASSOCIATION

GROUNDING
SLOT

TAMPER-RESISTANT
MECHANISM
(INSIDE RECEPTACLE)

Figure 41 Standard 125V, 15A duplex receptacle.

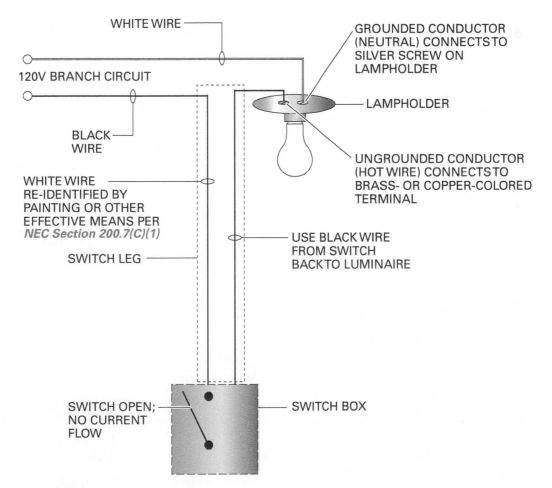

WHITE WIRE

120V BRANCH CIRCUIT

BLACK
WIRE

WHITE WIRE
RE-IDENTIFIED BY
PAINTING OR OTHER
EFFECTIVE MEANS PER
NEC Section 200.7(C)(1)

SWITCH LEG

SWITCH OPEN;
NO CURRENT
FLOW

GROUNDED CONDUCTOR
(NEUTRAL) CONNECTS TO
SILVER SCREW ON
LAMPHOLDER

LAMPHOLDER

UNGROUNDED CONDUCTOR
(HOT WIRE) CONNECTS TO
BRASS- OR COPPER-COLORED
TERMINAL

USE BLACK WIRE
FROM SWITCH
BACK TO LUMINAIRE

SWITCH BOX

Figure 42 Switch operation, contacts open.

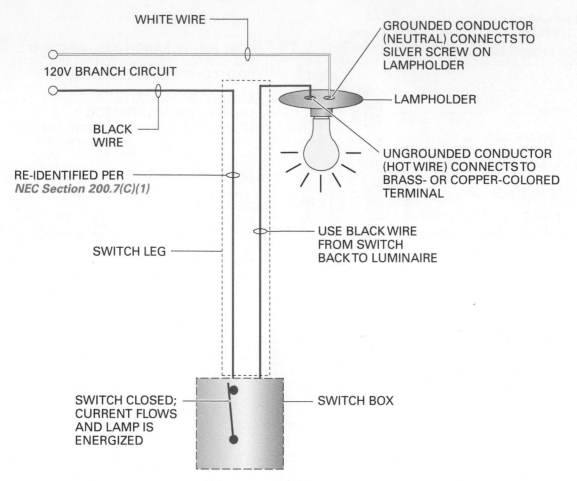

WHITE WIRE

GROUNDED CONDUCTOR (NEUTRAL) CONNECTS TO SILVER SCREW ON LAMPHOLDER

120V BRANCH CIRCUIT

LAMPHOLDER

BLACK WIRE

UNGROUNDED CONDUCTOR (HOT WIRE) CONNECTS TO BRASS- OR COPPER-COLORED TERMINAL

RE-IDENTIFIED PER *NEC Section 200.7(C)(1)*

USE BLACK WIRE FROM SWITCH BACK TO LUMINAIRE

SWITCH LEG

SWITCH CLOSED; CURRENT FLOWS AND LAMP IS ENERGIZED

SWITCH BOX

Figure 43 Switch operation, contacts closed.

The quiet switch (*Figure 44*) is the most common switch for use in lighting applications. Its operation is much quieter than the snap-action switch.

The quiet switch consists of a stationary contact and a moving contact that are close together when the switch is open. Only a short, gentle movement is required to open and close the switch, producing very little noise. This type of switch may be used only on alternating current.

Quiet switches are common for loads from 10A to 20A, and are available in single-pole, three-way, and four-way configurations.

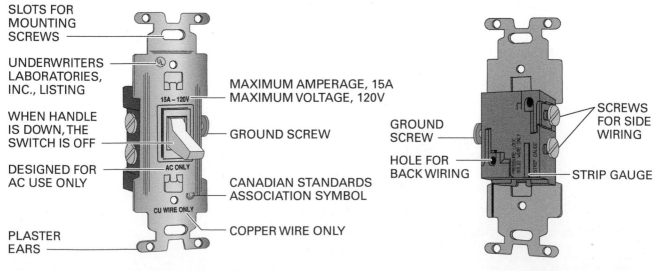

SLOTS FOR MOUNTING SCREWS

UNDERWRITERS LABORATORIES, INC., LISTING

MAXIMUM AMPERAGE, 15A
MAXIMUM VOLTAGE, 120V

15A – 120V

WHEN HANDLE IS DOWN, THE SWITCH IS OFF

GROUND SCREW

DESIGNED FOR AC USE ONLY

AC ONLY

CANADIAN STANDARDS ASSOCIATION SYMBOL

CU WIRE ONLY

COPPER WIRE ONLY

PLASTER EARS

GROUND SCREW

SCREWS FOR SIDE WIRING

HOLE FOR BACK WIRING

STRIP GAUGE

Figure 44 Characteristics of a single-pole quiet switch.

6.2.1 Three-Way Switches

Three-way switches are used to control one or more lamps from two different locations, such as at the top and bottom of stairways, in a room that has two entrances, etc. A typical three-way switch is shown in *Figure 45*.

A three-way switch has three terminals. The single terminal at one end of the switch is called the *common* or *hinge point*. This terminal is easily identified because it is darker than the other two terminals. The feeder (hot wire) or switch leg is always connected to the common dark or black terminal. The two remaining terminals are called *traveler terminals*. These terminals are used to connect three-way switches together.

The connection of two three-way switches is shown in *Figure 46*. By means of the two switches, it is possible to control the lamp from two locations. By tracing the circuit, it may be seen how these three-way switches operate.

A 120V circuit emerges from the left side of the drawing. The white or neutral wire connects directly to the neutral terminal of the lamp. The hot wire carries current, in the direction of the arrows, to the common terminal of the three-way switch on the left. Because the handle is in the Up position, the current continues to the top traveler terminal and is carried by this traveler to the other three-way switch. Note that the handle is also in the Up position on this switch; this picks up the current flow and carries it to the common

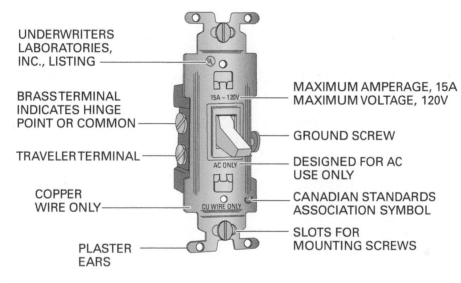

Figure 45 Typical three-way switch.

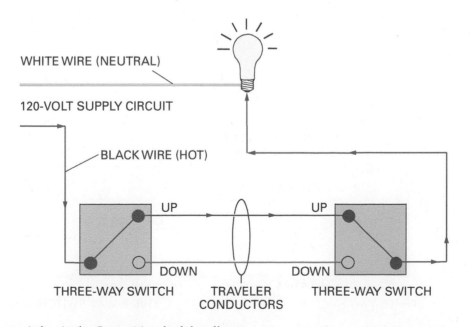

Figure 46 Three-way switches in the On position; both handles are up.

---

point, which continues to the ungrounded terminal of the lamp to make a complete circuit. The lamp is energized.

Moving the handle to a different position on either three-way switch will break the circuit, which in turn de-energizes the lamp. For example, let's say a person leaves the room at the point of the three-way switch on the left, and the switch handle is flipped down, as shown in *Figure 47*. Note that the current flow is now directed to the bottom traveler terminal, but because the handle of the three-way switch on the right is still in the Up position, no current will flow to the lamp.

If another person enters the room at the location of the three-way switch on the right, and the handle is flipped downward, as shown in *Figure 48*, this change provides a complete circuit to the lamp, which causes it to be energized. In this example, current flow is on the bottom traveler. Again, changing the position of the switch handle (pivot point) on either three-way switch will de-energize the lamp.

In actual practice, the exact wiring of the two three-way switches to control the operation of a lamp will be slightly different from the routing shown in these three diagrams. There are several ways that two three-way switches may be connected. One solution is shown in *Figure 49*. In this case, two-wire, Type NM cable is fed to the three-way switch on the left.

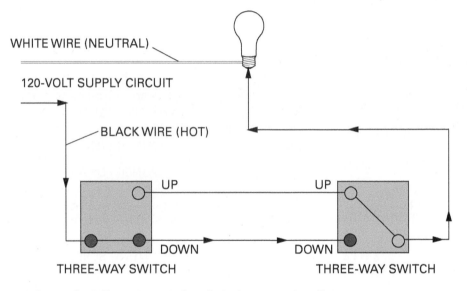

Figure 47 Three-way switches in the Off position; one handle is down, one handle is up.

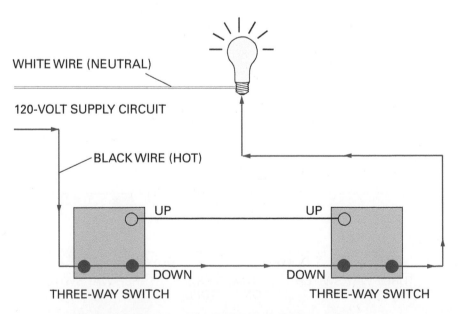

Figure 48 Three-way switches with both handles down; the light is energized.

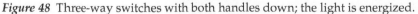

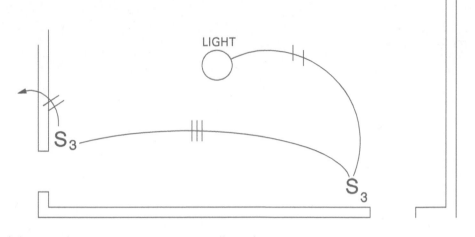

Figure 49 Method of showing the wiring arrangement on a floor plan.

The black or hot conductor is connected to the common terminal on the switch, while the white or neutral conductor is spliced to the white conductor of the three-wire, Type NM cable leaving the switch. This three-wire cable is necessary to carry the two travelers plus the neutral to the three-way switch on the right. At this point, the black and red wires connect to the two traveler terminals, respectively. The white or neutral wire is again spliced—this time to the white wire of another two-wire, Type NM cable. The neutral wire is never connected to the switch itself. The black wire of the two-wire, Type NM cable connects to the common terminal on the three-way switch. This cable, carrying the hot and neutral conductors, is routed to the luminaire outlet for connection to the fixture.

Another solution is to feed the luminaire outlet with two-wire cable. Run another two-wire cable carrying the hot and neutral conductors to one of the three-way switches. A three-wire cable is pulled between the two three-way switches, and then another two-wire cable is routed from the other three-way switch to the luminaire outlet.

Some electricians use a shortcut method that eliminates one of the two-wire cables in the preceding method. In this case, a two-wire cable is run from the lighting fixture outlet to one three-way switch. Three-wire cable is pulled between the two three-way switches—two of the wires for travelers and the third for the common point return. This method is shown in *Figure 50*.

6.2.2 Four-Way Switches

Two three-way switches may be used in conjunction with any number of four-way switches to control a lamp, or a series of lamps, from any number of positions. When connected correctly, the actuation of any one of these switches will change the operating condition of the lamp (i.e., turn the lamp either on or off).

Figure 51 shows how a four-way switch may be used in combination with two three-way switches to control a device from three locations. In this example, note that the hot wire is connected to the common terminal on the three-way switch on the left. Current then travels to the top traveler terminal and continues on the top traveler conductor to the four-way switch. Because the handle is Up on the four-way switch, current flows through the top terminals of the switch and into the traveler conductor going to the other three-way switch.

Again, the switch is in the Up position. Therefore, current is carried from the top traveler terminal to the common terminal and then to the lighting fixture to energize it.

If the position of any one of the three switch handles is changed, the circuit will be broken and no current will flow to the lamp. For example, assume that the four-way switch handle is flipped downward. The circuit will now appear as shown in *Figure 52*, and the light will be out.

Remember, any number of four-way switches may be used in combination with two three-way switches, but two three-way switches are always necessary for the correct operation of one or more four-way switches.

6.2.3 Photoelectric Switches

The chief application of the photoelectric switch is to control outdoor lighting, especially the dusk-to-dawn lights found in suburban areas. This switch has an endless number of possible uses and is a great tool for electricians dealing with outdoor lighting situations.

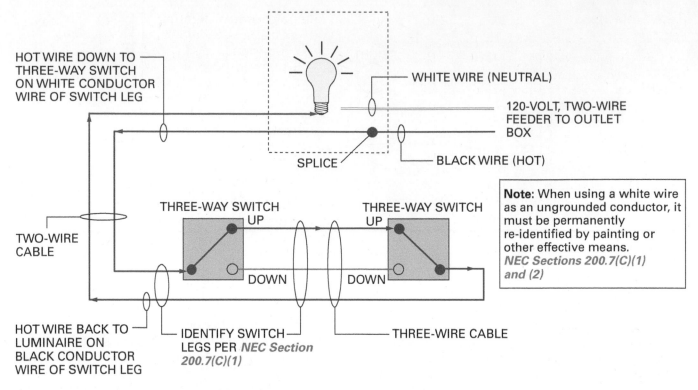

HOT WIRE DOWN TO
THREE-WAY SWITCH
ON WHITE CONDUCTOR
WIRE OF SWITCH LEG

WHITE WIRE (NEUTRAL)

120-VOLT, TWO-WIRE
FEEDER TO OUTLET
BOX

SPLICE

BLACK WIRE (HOT)

THREE-WAY SWITCH
UP

THREE-WAY SWITCH
UP

DOWN

DOWN

TWO-WIRE
CABLE

Note: When using a white wire as an ungrounded conductor, it must be permanently re-identified by painting or other effective means. *NEC Sections 200.7(C)(1) and (2)*

HOT WIRE BACK TO
LUMINAIRE ON
BLACK CONDUCTOR
WIRE OF SWITCH LEG

IDENTIFY SWITCH
LEGS PER *NEC Section
200.7(C)(1)*

THREE-WIRE CABLE

Figure 50 One way to connect a pair of three-way switches to control one luminaire.

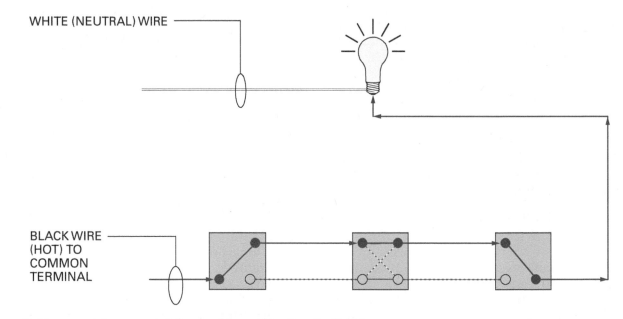

WHITE (NEUTRAL) WIRE

BLACK WIRE
(HOT) TO
COMMON
TERMINAL

Figure 51 Three- and four-way switches used in combination; the light is on.

Think About It

Wiring Three-Way Switches

Using a schematic drawing, explain the actual wiring of two different three-way switches, one in which the load and supply come in from different boxes, and the other in which the load and supply come in from the same box. Be specific about which wires connect to which terminals.

6.2.4 Relays

Next to switches, relays play the most important part in the control of light. However, the design and application of relays is a study in itself, and they are far beyond the scope of this module. Still, a brief mention of relays is necessary to round out your knowledge of lighting controls.

An electric relay is a device that uses an electric current to cause the opening or closing of one or more pairs of contacts. These contacts are usually capable of controlling much more power than is necessary to operate the relay itself. This is one of the main advantages of relays.

One popular use of the relay in residential lighting systems is that of remote control lighting. In this type of system, all relays are designed to operate on a 24V circuit and are used to control 120V lighting circuits. They are rated at 20A, which is sufficient to control the full load of a normal lighting branch circuit, if desired.

Remote control switching makes it possible to install a switch wherever it is convenient and practical to do so or wherever there is an obvious need for a switch, no matter how remote it is from the lamp or lamps it is to control. This method enables lighting designs to achieve new advances in lighting control convenience at a reasonable cost. Remote control switching is also ideal for rewiring existing homes with finished walls and ceilings.

One relay is required for each luminaire or each group of luminaires that are controlled together. Switch locations for remote control follow the same rules as for conventional direct switching. However, because it is easy to add switches to control a given relay, no opportunities should be overlooked for adding a switch to improve the convenience of control.

Remote control lighting also has the advantage of using selector switches at central locations. For example, selector switches located in the master bedroom or in the kitchen of a home enable the owner to control every luminaire on the property from this location. For example, the selector switch may be used to control outside or basement lights that might otherwise be left on inadvertently.

6.2.5 Dimmers

Dimming a lighting system provides control of the quantity of illumination. It may be done to create certain moods or to blend the lighting from different sources for various lighting effects.

For example, in homes with formal dining rooms, a chandelier mounted directly above the dining table and controlled by a dimmer switch becomes the centerpiece of the room while providing general illumination. The dimmer adds versatility because it can set the mood for the activity—low brilliance (candlelight effect) for formal dining or bright for an evening of playing cards. When chandeliers with exposed lamps are used, the dimmer is essential to avoid a garish and uncomfortable atmosphere. The chandelier should be sized in proportion to the dining area.

> **NOTE**
>
> It is very important that dimmers be matched to the wattage of the application. Check the manufacturer's data.

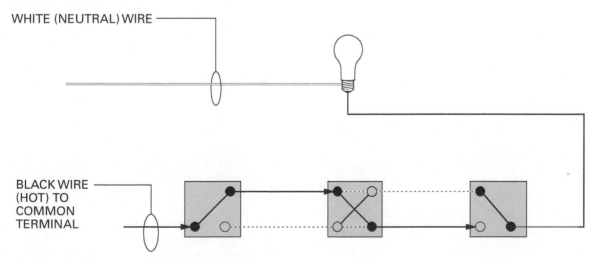

Figure 52 Three- and four-way switches used in combination; the light is off.

6.2.6 Switch Locations

Although the location of wall switches is usually provided for convenience, the *NEC*® also stipulates certain mandatory locations for lighting fixtures and wall switches (also referred to as *listed wall-mounted control devices*). See *NEC Section 210.70(A)* for specific locations in dwelling units. These locations are deemed necessary for added safety in the home for both the occupants and service personnel.

For example, the *NEC*® requires adequate light in areas where heating, ventilating, and air conditioning (HVAC) equipment is placed. Furthermore, these lights must be conveniently controlled so that homeowners and service personnel do not have to enter a dark area where they might come in contact with dangerous equipment. Three-way switches are required under certain conditions. The *NEC*® also specifies regulations governing lighting fixtures in clothes closets, along with those governing lighting fixtures that may be mounted directly to the outlet box without further support. *Figure 53* summarizes some of the *NEC*® requirements for light and switch placement in the home. For further details, refer to *NEC Article 210*.

6.2.7 Low-Voltage Electrical Systems

Conventional lighting systems operate and are controlled by the same system voltage, generally 120V in residential lighting circuits. The *NEC*® permits the use of low-voltage systems to control lighting circuits. There are some advantages to low-voltage systems. One advantage is that the control of lighting from several different locations is more easily accomplished, such as with the remote control system discussed earlier. For example, outside flood lighting can be controlled from several different rooms in a house. The cost of the control wiring is less in that it is rated for a lower voltage and only carries a minimum amount of current compared to a standard lighting system. When extensive or complex lighting control is required, low-voltage systems are preferred. Also, because these circuits are low-energy circuits, circuit protection is not required.

NEC Article 725 governs the installation of low-voltage system wiring. These provisions apply to remote control circuits, low-voltage relay switching, low-energy power circuits, and low-voltage circuits. The *NEC*® divides these circuits into three categories:

• Remote control
• Signaling
• Power-limited circuits

As mentioned earlier, circuit protection of the low-voltage circuit is not required; however, the high-voltage side of the transformer that supplies the low-voltage system must be protected. *NEC Chapter 9, Tables 11(A) and 11(B)* cover circuits that are inherently limited in power output and therefore require no overcurrent protection or are limited by a combination of power source and overcurrent protection.

There are a number of requirements of the power systems described in *NEC Chapter 9, Tables 11(A) and 11(B)* and the notes preceding the tables. You should read and study all applicable portions of the *NEC*® before installing low-voltage power systems. (Low-voltage systems are described in more detail in later modules of this curriculum.)

6.3.0 Residential Swimming Pools, Spas, and Hot Tubs

The *NEC*® recognizes the potential danger of electric shock to persons in swimming pools, wading pools, and therapeutic pools, or near decorative pools or fountains. This shock could occur from electric potential in the water itself or as a result of a person in the water or a wet area touching an enclosure that is not at ground potential. Accordingly, the *NEC*® provides rules for the safe installation of electrical equipment and wiring in or adjacent to swimming pools and similar locations. *NEC Article 680* covers the specific rules governing the installation and maintenance of swimming pools, spas, and hot tubs.

The electrical installation procedures for hot tubs and swimming pools are too vast to be covered in detail in this module. However, the general requirements for the installation of outlets, overhead fans and lighting fixtures, and other items are summarized in *Figure 54*.

Besides *NEC Article 680* (*Figure 55*), another good source for learning more about electrical installations in and around swimming pools is from manufacturers of swimming pool equipment, including those who manufacture and distribute underwater luminaires. Many of these manufacturers offer pamphlets detailing the installation of their equipment with helpful illustrations, code explanations, and similar details. This literature is usually available at little or no cost to qualified personnel. You visit manufacturer websites to request information or browse available literature or contact your local electrical supplier or contractor who specializes in installing residential swimming pools.

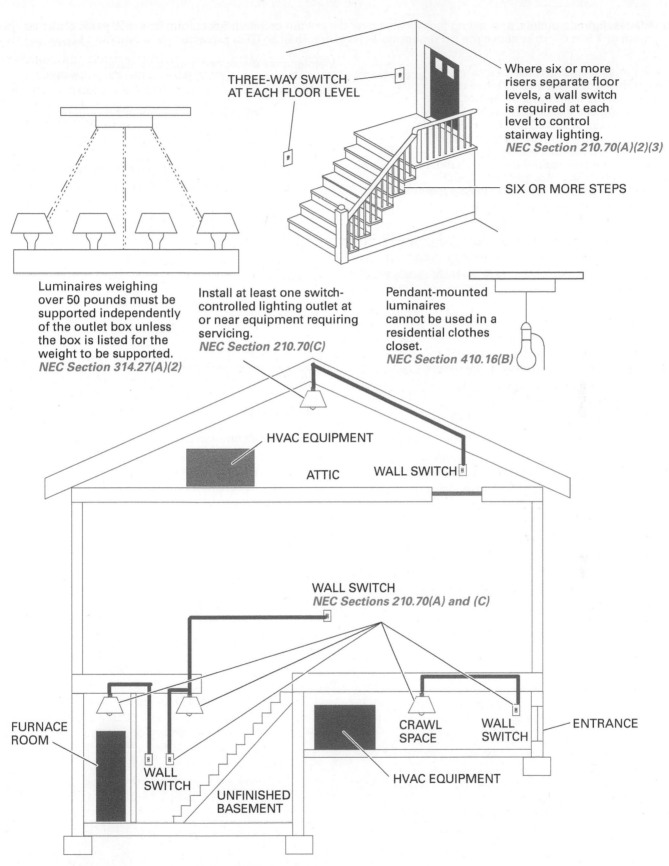

THREE-WAY SWITCH AT EACH FLOOR LEVEL

Where six or more risers separate floor levels, a wall switch is required at each level to control stairway lighting. NEC Section 210.70(A)(2)(3)

SIX OR MORE STEPS

Luminaires weighing over 50 pounds must be supported independently of the outlet box unless the box is listed for the weight to be supported. *NEC Section 314.27(A)(2)*

Install at least one switch-controlled lighting outlet at or near equipment requiring servicing. *NEC Section 210.70(C)*

Pendant-mounted luminaires cannot be used in a residential clothes closet. *NEC Section 410.16(B)*

HVAC EQUIPMENT

ATTIC

WALL SWITCH

WALL SWITCH *NEC Sections 210.70(A) and (C)*

FURNACE ROOM

WALL SWITCH

UNFINISHED BASEMENT

CRAWL SPACE

WALL SWITCH

ENTRANCE

HVAC EQUIPMENT

Figure 53 NEC® requirements for light and switch placement.

Luminaires, lighting outlets, and ceiling fans located over the hot tub or within 5 feet from its inside walls shall be a minimum of 7 feet 6 inches above the maximum water level and shall be GFCI-protected [*NEC Section 680.43(B)*].

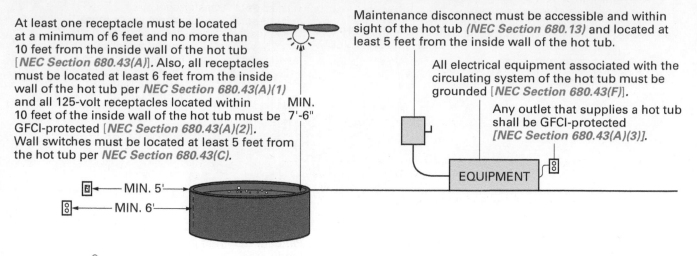

At least one receptacle must be located at a minimum of 6 feet and no more than 10 feet from the inside wall of the hot tub [*NEC Section 680.43(A)*]. Also, all receptacles must be located at least 6 feet from the inside wall of the hot tub per *NEC Section 680.43(A)(1)* and all 125-volt receptacles located within 10 feet of the inside wall of the hot tub must be GFCI-protected [*NEC Section 680.43(A)(2)*]. Wall switches must be located at least 5 feet from the hot tub per *NEC Section 680.43(C)*.

Maintenance disconnect must be accessible and within sight of the hot tub *(NEC Section 680.13)* and located at least 5 feet from the inside wall of the hot tub.

All electrical equipment associated with the circulating system of the hot tub must be grounded [*NEC Section 680.43(F)*].

Any outlet that supplies a hot tub shall be GFCI-protected [*NEC Section 680.43(A)(3)*].

MIN. 7'-6"

MIN. 5'

MIN. 6'

EQUIPMENT

Figure 54 NEC® requirements for packaged indoor hot tubs.

Think About It
Residential Wiring

Make a mental wiring tour of your home. Picture several rooms, including the kitchen and utility/laundry room. How is each device connected to the power source, and what is the probable amperage and overcurrent protection? Which other devices might or might not be included in the circuit? How many branch circuits serve each room? Later, examine the panelboard and exposed wiring to see how accurately you identified the branch circuits.

Pools

Pools are a common site for homeowners to add lights, receptacles, or heaters, and code violations are common. If you are called in for a service problem, ask the homeowner about any do-it-yourself wiring.

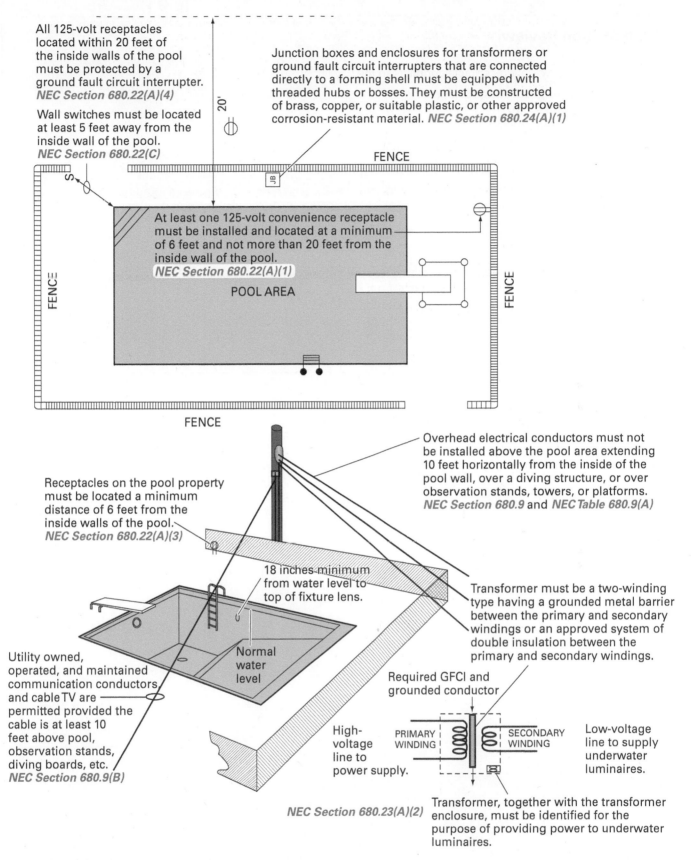

All 125-volt receptacles located within 20 feet of the inside walls of the pool must be protected by a ground fault circuit interrupter. *NEC Section 680.22(A)(4)*

Wall switches must be located at least 5 feet away from the inside wall of the pool. *NEC Section 680.22(C)*

Junction boxes and enclosures for transformers or ground fault circuit interrupters that are connected directly to a forming shell must be equipped with threaded hubs or bosses. They must be constructed of brass, copper, or suitable plastic, or other approved corrosion-resistant material. *NEC Section 680.24(A)(1)*

FENCE

JB

20'

At least one 125-volt convenience receptacle must be installed and located at a minimum of 6 feet and not more than 20 feet from the inside wall of the pool. *NEC Section 680.22(A)(1)*

POOL AREA

FENCE

FENCE

FENCE

Receptacles on the pool property must be located a minimum distance of 6 feet from the inside walls of the pool. *NEC Section 680.22(A)(3)*

Overhead electrical conductors must not be installed above the pool area extending 10 feet horizontally from the inside of the pool wall, over a diving structure, or over observation stands, towers, or platforms. *NEC Section 680.9* and *NEC Table 680.9(A)*

18 inches minimum from water level to top of fixture lens.

Normal water level

Transformer must be a two-winding type having a grounded metal barrier between the primary and secondary windings or an approved system of double insulation between the primary and secondary windings.

Required GFCI and grounded conductor

Utility owned, operated, and maintained communication conductors and cable TV are permitted provided the cable is at least 10 feet above pool, observation stands, diving boards, etc. *NEC Section 680.9(B)*

High-voltage line to power supply.

PRIMARY WINDING

SECONDARY WINDING

Low-voltage line to supply underwater luminaires.

NEC Section 680.23(A)(2)

Transformer, together with the transformer enclosure, must be identified for the purpose of providing power to underwater luminaires.

Figure 55 NEC® requirements for typical swimming pool installations.

6.0.0 Section Review

1. The brass screw on a receptacle connects to the _____.

   a. equipment grounding conductor
   b. neutral conductor
   c. grounded conductor
   d. ungrounded conductor

2. The correct operation of a four-way switch requires at least _____.

   a. one three-way switch
   b. two three-way switches
   c. three three-way switches
   d. four three-way switches

3. When installing a hot tub, GFCI protection is required for all receptacles located within _____.

   a. 10' (3.0 m) of the inside wall of the hot tub
   b. 12' (3.7 m) of the inside wall of the hot tub
   c. 15' (4.5 m) of the inside wall of the hot tub
   d. 20' (6.0 m) of the inside wall of the hot tub

1. The minimum number of small appliance circuits required by the *NEC*® in a kitchen area is _____.

   a. 1
   b. 2
   c. 3
   d. 4

2. When sizing residential electrical services, the laundry load is calculated as _____.

   a. 1,200VA
   b. 1,500VA
   c. 1,800VA
   d. 2,400VA

3. When sizing the general lighting load for electric services, the first 3,000VA is rated at _____.

   a. 50%
   b. 65%
   c. 85%
   d. 100%

4. According to *NEC Section 220.53*, a demand factor of 75% may be applied to four or more _____.

   a. electric stoves
   b. fixed appliance loads
   c. heating loads
   d. air conditioning loads

5. The minimum service disconnect rating for a one-family dwelling is _____.

   a. 60A
   b. 100A
   c. 200A
   d. 250A

6. The general lighting load in a residence is 3,600VA at 120V. What is the amperage?

   a. 20A
   b. 30A
   c. 45A
   d. 60A

7. A single-pole GFCI breaker is rated at _____.

   a. 120VAC
   b. 240VAC
   c. 120/240VAC
   d. 240/480VAC

8. Exposed noncurrent-carrying metal parts of fixed equipment must be grounded if the equipment operates with any terminal over _____.

   a. 60V to ground
   b. 120V to ground
   c. 150V to ground
   d. 240V to ground

9. Which section of the *NEC*® requires that raceway fittings be identified for use with service masts?

   a. *NEC Section 230.28*
   b. *NEC Section 230.40*
   c. *NEC Section 250.46*
   d. *NEC Section 250.84*

10. Service drop conductors must be located at a minimum height of _____.

   a. 6' above the ground or other accessible surface
   b. 8' above the ground or other accessible surface
   c. 10' above the ground or other accessible surface
   d. 12' above the ground or other accessible surface

11. The clearance between a service conductor without an overall jacket and a porch or balcony must not be less than _____.

   a. 2' (600 mm)
   b. 3' (900 mm)
   c. 8' (2.5 m)
   d. 10' (3.0 m)

12. *NEC Section 230.6* considers conductors to be outside the building when installed under at least _____.

   a. 1" (25 mm) of concrete
   b. 2" (50 mm) of concrete
   c. 4" (100 mm) of concrete
   d. 5" (125 mm) of concrete

13. Type NM cable is prohibited _____.
    a. in shallow chases of masonry, concrete, or adobe
    b. in the framework of a building
    c. where used with protective strips
    d. in attic spaces

14. Type MC cable is prohibited _____.
    a. in concrete or plaster where dry
    b. in dry masonry
    c. in attic spaces
    d. where subject to physical damage

15. When used for interior wiring systems, Type SE cable is available with _____.
    a. a non-insulated ground or neutral conductor
    b. an insulated grounded conductor
    c. no ground conductor
    d. guard strips

16. Type SER cable may be used _____.
    a. in overhead applications
    b. underground
    c. as a subfeed under certain conditions
    d. in hazardous locations

17. The symbol shown in Figure RQ01 indicates a _____.

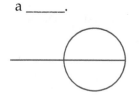

Figure RQ01

    a. special purpose outlet
    b. single receptacle
    c. range receptacle
    d. duplex receptacle

18. The symbol shown in Figure RQ02 indicates a _____.

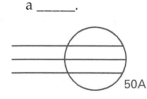

Figure RQ02

    a. duplex receptacle
    b. range receptacle
    c. dryer receptacle
    d. split-wired receptacle

19. The symbol shown in Figure RQ03 indicates a _____.

Figure RQ03

    a. duplex receptacle
    b. single receptacle
    c. dryer receptacle
    d. special-purpose outlet

20. The symbol shown in Figure RQ04 indicates a _____.

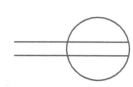

Figure RQ04

    a. range receptacle
    b. single receptacle
    c. duplex receptacle
    d. split-wired receptacle

21. The symbol shown in Figure RQ05 indicates a _____.

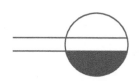

Figure RQ05

    a. special-purpose outlet
    b. split-wired receptacle
    c. duplex receptacle
    d. range receptacle

22. The symbol shown in Figure RQ06 indicates a _____.

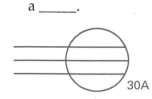

Figure RQ06

    a. dryer receptacle
    b. split-wired receptacle
    c. duplex receptacle
    d. special-purpose outlet

23. Conductors used for electric water heaters must be sized at _____.

    a. 70% of the heater's nameplate rating
    b. 95% of the heater's nameplate rating
    c. 100% of the heater's nameplate rating
    d. 125% of the heater's nameplate rating

24. A wall switch is required to control stairway lighting at each level when the levels are separated by _____.

    a. three or more risers
    b. four or more risers
    c. five or more risers
    d. six or more risers

25. In a residential pool area, receptacles _____.

    a. are not allowed within the fenced area
    b. must be approved for providing power to underwater lighting fixtures
    c. must be at least 6 feet from the inside walls of the pool
    d. are only allowed if required by the pool manufacturer

Trade Terms Quiz

Fill in the blank with the correct term that you learned from your study of this module.

1. A cable that contains insulated circuit conductors enclosed in armor made of metal is _____.

2. A factory-assembled cable with two or more insulated conductors and a nonmetallic jacket is called _____.

3. _____ are pieces of equipment that have been designed for a particular purpose.

4. A _____ is used for turning an electrical circuit On and Off.

5. The circuit that is routed to a switch box for controlling electric lights is known as a _____.

6. A _____ is equipped with a conductor terminal to accept a bonding jumper.

7. A _____ is a bare or green insulated conductor used to ensure conductivity between metal parts that are required to be electrically connected.

8. The _____ is comprised of the conductors that extend from the last power company pole to the point of connection at the service facilities.

9. The _____ is the point where power is supplied to a building.

10. The _____ lie between the point of termination of the overhead service drop or underground service lateral and the main disconnecting device in the building.

11. The _____ mainly provides overcurrent protection to the feeder and service conductors.

12. A _____ is comprised of the underground conductors through which service is supplied between the power company's distribution facilities and their first point of connection to the building.

13. The portion of a wiring system that extends beyond the final overcurrent device is the _____.

14. A _____ is a circuit conductor between the service equipment, the source of a separately derived system, or other power supply source and the final branch circuit overcurrent device.

15. Normally located at the service entrance of a residential installation, a _____ sometimes contains the main disconnect.

16. Raceway, cable, wires, boxes, and other equipment are installed during _____.

17. The term _____ is General Cable's trade name for Type NM cable, but it is often used to refer to any nonmetallic-sheathed cable.

Trade Terms

Appliances
Bonding bushing
Bonding jumper
Branch circuit
Feeder
Load center

Metal-clad (Type MC) cable
Nonmetallic-sheathed (Type NM and NMC) cable
Romex®

Roughing in
Service drop
Service entrance
Service-entrance conductors

Service-entrance equipment
Service lateral
Switch
Switch leg

Supplemental Exercise

1. Because all residential electrical outlets are never used at the same time, the *NEC®* allows a diversity or _____ to be used when sizing the general lighting load for electric services.

2. True or False? In a residential electric service, the service can be grounded to the underground gas piping system.

3. True or False? GFCI protection is required for receptacles in residential bathrooms.

4. In a noncombustible wall, the outlet box may be recessed _____.

5. A duplex receptacle outlet must be installed in all hallways longer than _____.

6. Electric ranges, clothes dryers, and water heaters that operate at 240V require a(n) _____ circuit breaker.

7. In general, there are two basic wiring methods used in the majority of modern residential electrical systems. What are they?

   _____

8. In a standard 125V receptacle, which wire is connected to the brass terminal?

   _____

9. Three-way switches are used to control lamps from _____ different locations.

10. Using *NEC Table 314.16(A)*, calculate the cubic inches required for the receptacle outlets shown in the table below. Then, indicate the size of the metallic box that should be used.

Number and Size of Conductors in Box	Free Space within Box for Each Conductor	Total Cubic Inches of Box Space Required	Which Size Metallic Box May Be Used?
A. Six No. 12 conductors and three ground wires	2.25	_____	_____
B. Seven No. 12 conductors and three ground wires with one receptacle	2.25	_____	_____
C. Two No. 14 conductors and one ground wire	2.0	_____	_____
D. Four No. 14 conductors and two ground wires	2.0	_____	_____
E. Six No. 14 conductors and three ground wires with one receptacle	2.0	_____	_____

Dan Lamphear
Associated Builders and Contractors, Inc.

Like many other people, Dan Lamphear just fell into his career as an electrician. But once he discovered the electrical trade, he knew he had found a home. Since then, he has progressed from a helper to an apprentice, a journeyman, an independent contractor, an inventor, and finally, a teacher.

It was as much luck as anything else that led Dan toward a career as a professional electrician more than two decades ago. He wasn't sure what he wanted to do with his life after graduating from high school. However, after watching an electrician perform a commercial wiring job at a friend's business—and providing a helping hand—he was hooked. "It seemed like a challenging career, and I was curious to learn more about how electricity works," he recalls.

Dan was hired by that same electrician, under whom he apprenticed for several years before hearing about the NCCER program. He jumped at the chance to further his skills through the program. "Like they say, knowledge is money," he smiles.

After graduating from the program, Dan struck out on his own as an independent electrician, specializing in plant maintenance, industrial, and commercial work. His ability to diagnose and repair electrical problems in factory machinery soon made him a valuable contractor in Milwaukee's industrial sector.

He also discovered his knack for invention, and he has designed and built specialized machinery for a company that hired him as its full-time electrical maintenance supervisor. "Knowing the electrical side of machinery allowed me to understand how they operate mechanically," he says of his work as an inventor.

Dan later returned to the Associated Builders and Contractors (ABC) chapter, which trains out of a local community college, to repay the favor that helped him embark on his career. He teaches Electrical Level 2 courses for students who represent the next generation of professional electricians.

"Knowing how to use test instruments is perhaps the most important aspect of the job," he notes. "I still have some of the same meters I started out with."

David Lewis
Instructor
Putnam Career & Technical Center

David Lewis started his career working in coal mines. After a few years he opened his own electrical business. Now he is an electrical instructor and works with the State Department of Education on curriculum development. He also serves on the NCCER revision team for the Electrical curriculum.

How did you first get interested in the field?

After graduating from high school in 1972 and attending college for a while, I decided that I wanted to work in the coal mines. Electricity interested me, so I became a maintenance foreman/electrician. Eventually, I started my own business.

What kind of training have you been through?

While working in the coal mines, I attended several electrical training classes and obtained my underground electrical license. While in business for myself I got my Master Electrician license and attended many update classes. Since I have started teaching, I have attended classes on PLCs and other topics. I also went back to college and obtained a Bachelor of Science degree in Career and Technical Education.

What work have you done in your career?

In the coal mines I worked on all types of mining equipment. After starting my own business, I worked mainly in residential and light commercial wiring.

Tell us about your present job and what you like about it.

I enjoy being an instructor in Electrical Technology. I work mostly with high school students and during the two years they are with me, it is great to see them grasp the knowledge of electricity.

Which factors have contributed most to your success?

Hard work and the willingness to learn from experienced electricians.

What advice would you give to those new to the field?

Try to learn all you can. Work with an experienced electrician and learn from them. Attend any training or classes that you can. There is always something new to learn.

Trade Terms Introduced in This Module

Appliances: Equipment designed for a particular purpose (for example, using electricity to produce heat, light, or mechanical motion). Appliances are usually self-contained, are generally available for applications other than industrial use, and are normally produced in standard sizes or types.

Bonding bushing: A special conduit bushing equipped with a conductor terminal to accept a bonding jumper. It also has a screw or other sharp device to bite into the enclosure wall to bond the conduit to the enclosure without a jumper when there are no concentric knockouts left in the wall of the enclosure.

Bonding jumper: A bare or green insulated conductor used to ensure the required electrical conductivity between metal parts required to be electrically connected. Bonding jumpers are frequently used from a bonding bushing to the service-equipment enclosure to provide a path around concentric knockouts in an enclosure wall, and they may also be used to bond one raceway to another.

Branch circuit: The portion of a wiring system extending beyond the final overcurrent device protecting a circuit.

Feeder: Any circuit conductor between the service equipment, the source of a separately derived system, or other power supply source and the final branch circuit overcurrent device.

Load center: A type of panelboard that is normally located at the service entrance of a residential installation. It sometimes contains the main disconnect.

Metal-clad (Type MC) cable: A factory assembly of one or more insulated circuit conductors with or without optical fiber members enclosed in an armor of interlocking metal tape, or a smooth or corrugated metallic sheath.

Nonmetallic-sheathed (Type NM and NMC) cable: A factory- assembly of two or more insulated conductors enclosed within an overall nonmetallic jacket. Type NM contains insulated conductors enclosed within an overall nonmetallic jacket; Type NMC contains insulated conductors enclosed within an overall, corrosion-resistant, nonmetallic jacket.

Romex®: General Cable's trade name for Type NM cable; however, it is often used generically to refer to any nonmetallic-sheathed cable.

Roughing in: The first stage of an electrical installation, when the raceway, cable, wires, boxes, and other equipment are installed. This is the electrical work that must be done before any finishing work can be done.

Service drop: The overhead conductors, through which electrical service is supplied, between the last power company pole and the point of their connection to the service facilities located at the building.

Service entrance: The point where power is supplied to a building (including the equipment used for this purpose). The service entrance includes the service main switch or panelboard, metering devices, overcurrent protective devices, and conductors/raceways for connecting to the power company's conductors.

Service-entrance conductors: The conductors between the point of termination of the overhead service drop or underground service lateral and the main disconnecting device in the building.

Service-entrance equipment: Equipment that provides overcurrent protection to the feeder and service conductors, a means of disconnecting the feeders from energized service conductors, and a means of measuring the energy used.

Service lateral: The underground conductors through which service is supplied between the power company's distribution facilities and the first point of their connection to the building or area service facilities located at the building.

Switch: A mechanical device used for turning an electrical circuit on and off.

Switch leg: A circuit routed to a switch box for controlling electric lights.

Appendix

Other Codes and Standards That Apply to Electrical Installations

Until 2000, there were three model building codes:

- *Standard Building Code (SBC)* – Published by the Southern Building Code Congress International.
- *BOCA National Building Code (NBC)* – Published by the Building Officials and Code Administrators.
- *Uniform Building Code (UBC)* – Published by the International Conference of Building Officials.

The three code writing groups, SBCCI, BOCA, and UBC, combined into one organization called the *International Code Council* with the purpose of writing one nationally accepted family of building and fire codes. It is known as the *International Building Code*.

The *International Residential Code (IRC)* is adopted as part of the electrical code requirements in many areas of the country. The IRC covers one- and two-family dwellings of three stories or less. The IRC includes requirements for such things as ventilating fans for bathrooms, requirements for smoke detectors, and other items not specified by the *NEC*®. The *IRC* covers all trades, including building, plumbing, mechanical, gas, energy, and electrical.

The NFPA also publishes its own building code, NFPA 5000.

To be thoroughly competent in the electrical trade, you should become familiar with the contents of these codes and the terminology used in them.

Additional Resources

This module presents thorough resources for task training. The following resource material is suggested for further study.

National Electrical Code® Handbook, Latest Edition. Quincy, MA: National Fire Protection Association.

Figure Credits

Section Review Answer Key

SECTION 1.0.0

Answer	Section Reference	Objective
1. c	1.1.1	1a
2. a	1.2.0; *NEC Table 220.54*	1b
3. d	1.3.1	1c
4. b	1.4.0	1d

SECTION 2.0.0

Answer	Section Reference	Objective
1. d	2.1.0	2a
2. b	2.2.0	2b
3. b	2.3.0	2c

SECTION 3.0.0

Answer	Section Reference	Objective
1. a	3.1.1	3a
2. b	3.2.0	3b

SECTION 4.0.0

Answer	Section Reference	Objective
1. a	4.1.1	4a
2. d	4.2.0	4b

SECTION 5.0.0

Answer	Section Reference	Objective
1. d	5.1.5	5a
2. a	5.2.0	5b
3. d	5.3.0	5c

SECTION 6.0.0

Answer	Section Reference	Objective
1. d	6.1.0	6a
2. b	6.2.2	6b
3. a	6.3.0; Figure 54	6c

1.0.0 Section Review

Question 1

Since the *NEC®* specifies 3 volt-amperes per square foot of living space, multiply the square feet by 3VA:

$3VA \times 900ft^2 = 2,700VA$

The residence has a general lighting load of **2,700VA**.

Question 4

Divide the total general lighting load by 120V to find the amperage:

$2,700VA \div 120V = 22.5A$

The amperage is **22.5A**.

5.0.0 Section Review

Question 1

Since the *NEC®* states that a branch circuit supplying storage-type water heaters having a capacity of 120 gallons (450 L) or less must be rated a minimum of 125% of the nameplate rating:

125% of 20A =
$20A \times 1.25 = 25A$

The circuit must be rated at **25A**.

Question 2

Since a circuit should only be loaded to 80% of its rated capacity, find 80% of 1,800VA:

80% of 1,800VA =
$0.8 \times 1,800VA =$
1,440VA

The circuit should be loaded to no more than **1,440VA**.

NCCER CURRICULA — USER UPDATE

NCCER makes every effort to keep its textbooks up-to-date and free of technical errors. We appreciate your help in this process. If you find an error, a typographical mistake, or an inaccuracy in NCCER's curricula, please fill out this form (or a photocopy), or complete the online form at **www.nccer.org/olf**. Be sure to include the exact module ID number, page number, a detailed description, and your recommended correction. Your input will be brought to the attention of the Authoring Team. Thank you for your assistance.

Instructors – If you have an idea for improving this textbook, or have found that additional materials were necessary to teach this module effectively, please let us know so that we may present your suggestions to the Authoring Team.

NCCER Product Development and Revision

13614 Progress Blvd., Alachua, FL 32615

Email: curriculum@nccer.org
Online: www.nccer.org/olf

❏ Trainee Guide   ❏ Lesson Plans   ❏ Exam   ❏ PowerPoints   Other _____

Craft / Level: _____ Copyright Date: _____

Module ID Number / Title: _____

Section Number(s): _____

Description: _____

_____

_____

_____

Recommended Correction: _____

_____

_____

_____

Your Name: _____

Address: _____

_____

Email: _____ Phone: _____

This page is intentionally left blank.

Electrical Test Equipment

OVERVIEW

The test equipment selected for a specific task depends on the type of measurement and the level of accuracy required. This module covers the applications of various types of electrical test equipment. It also describes meter safety precautions and category ratings.

Module 26112-20

26112-20 V10.0

ELECTRICAL TEST EQUIPMENT

Objectives

When you have completed this module, you will be able to do the following:

1. Identify various types of electrical test equipment.
   a. Identify the applications of a voltmeter.
   b. Identify the applications of an ohmmeter.
   c. Identify the applications of a clamp-on ammeter.
   d. Identify the applications of a multimeter.
   e. Identify the applications of other meters.
2. Identify meter category ratings and safety requirements.
   a. Select a meter with the correct category rating for an application.
   b. Identify electrical test equipment safety hazards.

Performance Tasks

Under the supervision of the instructor, you should be able to do the following:

1. Measure the voltage in the classroom from line to neutral and neutral to ground.

2. Use an ohmmeter to measure the value of various resistors.

Trade Terms

Backfeed
Coil
Continuity
d'Arsonval meter movement
Frequency

Industry Recognized Credentials

If you are training through an NCCER-accredited sponsor, you may be eligible for credentials from NCCER's Registry. The ID number for this module is 26112-20. Note that this module may have been used in other NCCER curricula and may apply to other level completions. Contact NCCER's Registry at 888.622.3720 or go to **www.nccer.org** for more information.

> **NOTE**
>
> NFPA 70®, *National Electrical Code*® and *NEC*® are registered trademarks of the National Fire Protection Association, Quincy, MA.

Contents

Figures and Tables

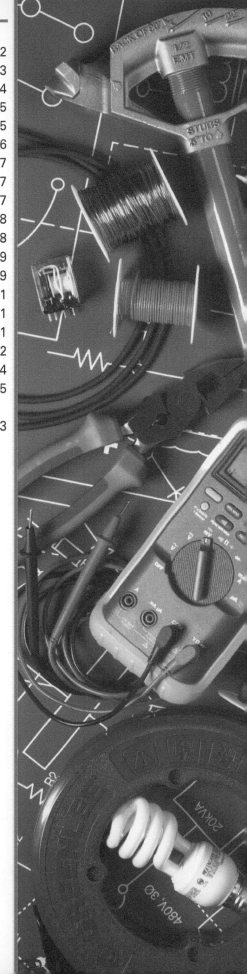

This page is intentionally left blank.

1.0.0 IDENTIFYING METERS

Objective

Identify various types of electrical test equipment.

 a. Identify the applications of a voltmeter.
 b. Identify the applications of an ohmmeter.
 c. Identify the applications of a clamp-on ammeter.
 d. Identify the applications of a multimeter.
 e. Identify the applications of other meters.

Trade Terms

Coil: A number of turns of wire, especially in spiral form, used for electromagnetic effects or for providing electrical resistance.

Continuity: An electrical term used to describe a complete (unbroken) circuit that is capable of conducting current. Such a circuit is also said to be closed.

d'Arsonval meter movement: A meter movement that uses a permanent magnet and moving coil arrangement to move a pointer across a scale.

Frequency: The number of cycles completed each second by a given AC voltage; usually expressed in hertz. One hertz equals one cycle per second.

Electrical test instruments and meters are generally used for the following tasks:

* Troubleshooting electrical/electronic circuits and equipment
* Verifying proper operation of circuits and components

The test equipment selected for a specific task depends on the type of measurement and the level of accuracy required. This module will focus on some of the test equipment used by electricians. Upon completion of this module, you should be able to select the appropriate test equipment for a specific application and identify the applicable safety hazards.

In 1882, a Frenchman named Arsene d'Arsonval invented the galvanometer. This meter used a stationary permanent magnet and a coil that moved to indicate current flow on a calibrated scale. The early galvanometer was very accurate but could measure only very small currents. Over the following years, many improvements were made that extended the range of the meter and increased its ruggedness. The d'Arsonval meter movement (*Figure 1*) is the basis for analog meters.

A moving-coil meter movement operates on the electromagnetic principle. In its simplest form, the moving-coil meter uses a coil of very fine wire wound on a light aluminum frame. A permanent magnet surrounds the coil. The aluminum frame is mounted on pivots to allow it and the coil to rotate freely between the poles of the permanent magnet. When current flows through the coil, it becomes magnetized, and the polarity of the coil is repelled by the field of the permanent magnet. This causes the coil frame to overcome the force of a spring and rotate on a pivot. The distance it rotates is determined by the amount of current that flows through the coil. By attaching a pointer to the coil frame and adding a calibrated scale, the amount of current flowing through the meter can be measured. Multiplier resistors are used to extend the range of the meter movement for voltage measurements, while shunt resistors are used to extend the range of the meter movement for current measurements.

Today, most meters use solid-state digital components because they are easier to read than mechanical (analog) meters and have no meter movement or moving parts. However, analog meters still have advantages in certain applications, such as when testing certain unique systems or those operating at less than 50V, when determining transformer polarity, or in applications where a fast response is needed. In addition, while digital meters provide an instantaneous value, the moving needle of an analog meter provides a more complete picture of the measured value and its fluctuations.

1.1.0 Voltmeter

A voltmeter is used to measure voltage, also known as *potential difference* or *electromotive force (emf)*. It is connected in parallel with the circuit or component being measured. An analog meter uses the basic d'Arsonval meter movement with

Phantom Readings

The sensitivity of a digital meter can sometimes produce a low reading known as a *phantom* or *ghost reading*. This is due to the induction from the electrical field around the energized conductors in close proximity to the meter.

SCALE

POINTER

SHUNT AND
MULTIPLIER
RESISTORS

RANGE
SELECTOR
SWITCH

ROUND PERMANENT MAGNET
(LOCATED BEHIND PLATE)

RETAINING PINS

FUNCTION
SELECTOR SWITCH

MOVING COIL

Figure 1 d'Arsonval meter movement.

internally switched resistors to measure different voltage ranges. A digital meter uses an analog-to-digital converter chip to convert the sensed values into a digital or graphic display.

Many digital voltmeters are autoranging, which means that the meter will automatically search for the correct scale. When using a voltmeter that is not autoranging, always start with the highest voltage range and work down until the indication reads somewhere between half and three-quarter scale. This will provide a more accurate reading and prevent damage to the meter. On many meters, a DC value is indicated by a straight line with three dashes beneath it, while an AC value is indicated by a sine wave.

1.1.1 Voltage Tester

A voltmeter is used when the exact value of the voltage is required. However, electricians are often concerned with identifying only whether voltage

is present, and if so, the general range of the voltage. In other words, is it energized, and if so, is it at 120V, 240V, or 480V? In these cases, a voltage tester is used. The range of voltage and the type of current (AC and/or DC) that a voltage tester is capable of measuring are usually indicated on the scales that display the reading (*Figure 2*).

Advanced voltage testers offer additional features, such as a digital readout, GFCI test capability, and even the ability to switch between use as a contact and noncontact detector (*Figure 3*).

A voltage tester must be checked before each use to make sure that it is in good condition and is operating correctly. The external check of the tester should include a careful inspection of the insulation on the leads for cracks or frayed areas. Faulty leads constitute a safety hazard, so they must be replaced. As a check to make sure that the voltage tester is operating correctly, the probes of the tester are first connected to a known energized source. The voltage indicated on the tester should match the voltage of the source. If there is no indication, the voltage tester is not operating correctly, and it must be repaired or replaced. It must also be repaired or replaced if it indicates a voltage different from the known voltage of the source.

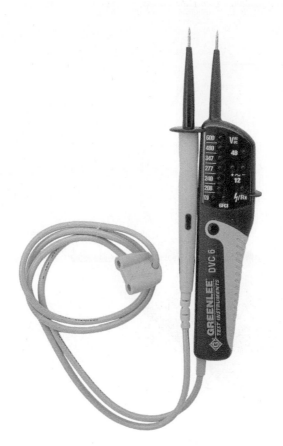

Figure 2 Voltage tester.

Voltage testers are used to make sure that voltage is available when it is needed and to ensure that power has been cut off when it should have been. In a troubleshooting situation, it might be necessary to verify that power is available in order to be sure that lack of power is not the problem. For example, if there were a problem with a power tool, such as a drill, a voltage tester might be used to make sure that power is available to run the drill. A voltage tester might also be used to verify that there is power available to a three-phase motor that will not start.

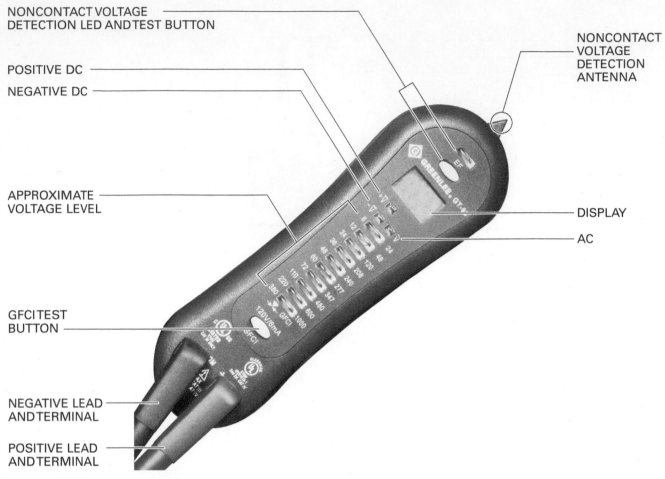

NONCONTACT VOLTAGE
DETECTION LED AND TEST BUTTON

NONCONTACT
VOLTAGE
DETECTION
ANTENNA

POSITIVE DC

NEGATIVE DC

APPROXIMATE
VOLTAGE LEVEL

DISPLAY

AC

GFCI TEST
BUTTON

NEGATIVE LEAD
AND TERMINAL

POSITIVE LEAD
AND TERMINAL

Figure 3 Multifunction voltage tester.

Voltage Detector

A simple noncontact (proximity) voltage detector can also be used to indicate the presence of voltage within its specified range rating but does not discriminate between ranges of values in the same way as a voltage tester. A voltage detector is handy for quickly scanning for the presence of voltage in junction boxes or termination cabinets and can even be used to trace circuits through walls. The voltage detector shown here glows in the presence of voltages between 50V and 1,000V.

Figure Credit: Greenlee / A Textron Company

1.1.2 Proving Unit

A proving unit supplies an electronic voltage source that can be used to validate that a meter is operating properly prior to using it on an actual circuit. To operate a proving unit, the leads of the meter to be tested are connected to the positive and negative terminals on the proving unit, and then the meter is checked to make sure that it displays the expected value. The proving unit shown in *Figure 4* is capable of providing a 240V AC or DC test point.

In addition to testing for meter functionality, proving units are also used in live-dead-live tests as a safe known energized (live) source.

1.2.0 Ohmmeter

An ohmmeter measures the resistance of a circuit or component. It can also be used to locate open circuits or shorted circuits. An ohmmeter consists of a DC current meter movement, a low-voltage DC power source (usually a battery), and current-limiting resistors, all of which are connected in series with the meter (*Figure 5*).

Before measuring the resistance of an unknown resistor or electrical circuit, connect the test leads together. This zeroes out or nulls the resistance of the leads. Some analog ohmmeters have a zero adjustment knob. With the leads connected together, turn the adjustment knob until the meter registers zero ohms. This adjustment must be made each time a different range is selected.

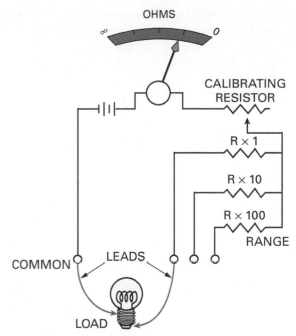

Figure 5 Ohmmeter schematic.

Many digital ohmmeters are autoranging. The correct scale is internally selected and the reading will indicate the range (ohms, K-ohms, or M-ohms). Analog ohmmeters require the user to select the desired range. Most digital meters also have an audible tone when the measured value is very low or at zero ohms. This indicates a closed circuit and is useful when using the meter as a continuity tester. A continuity test is used to determine if a circuit is complete.

> **WARNING!**
>
> Prior to taking a reading with an ohmmeter, verify that both sides of the circuit are de-energized by using a voltmeter. If the circuit were energized, its voltage could cause a current to flow through the meter. This can damage the meter and/or circuit and cause personal injury.

Think About It

Resistance

Why does the resistance vary when holding a resistor by pinching the meter leads against the resistor with your fingers when measuring it, versus measuring it while holding it in clips?

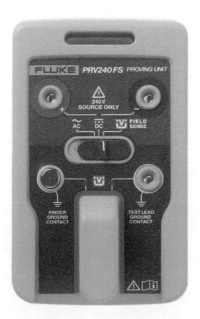

Figure 4 Proving unit.

When making resistance measurements in circuits, each component in the circuit can be tested individually by removing the component from the circuit and connecting the ohmmeter leads across it. However, the component does not have to be totally removed from the circuit. Usually, the part can be effectively isolated by disconnecting one of its leads from the circuit. Note that this method can still be somewhat time consuming.

1.3.0 Ammeter

A clamp-on ammeter, also known as a *clamp meter*, can measure current without having to make contact with uninsulated wires (*Figure 6*). This type of meter operates by sensing the strength of the electromagnetic field around the wire(s).

Clamp-on ammeters measure current by using simple transformer principles. The conductor(s) being measured would be the primary and the jaws (clamp) of the meter would be the secondary. The current in the primary winding induces a current in the secondary winding. If the ratio of the primary winding to the secondary winding is 1,000, then the secondary current is $^1/_{1000}$ of the current flowing in the primary. The smaller secondary current is connected to the meter's input. For example, a 1A current in the conductor will produce 0.001A (1mA) in the meter.

To measure the current, open the jaws of the meter and close them around the conductor(s) to be measured. Make sure that the jaws are clean and close tightly. Then read the magnitude of the current on the meter display.

Many meters have a Hold function. This is useful in tight locations when it is hard to read the meter while it is clamped around the conductor(s). Just press the Hold button when the value is measured, then remove and read the meter.

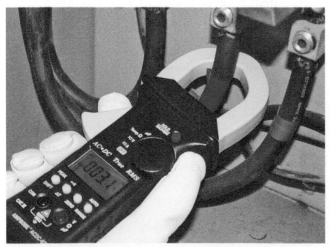

Figure 6 Clamp-on ammeter.

Some meters also have a Min/Max or Peak function. This allows the technician to record the maximum inrush, as with a motor start.

> **CAUTION**
>
> When using a clamp-on ammeter, make sure that the range of the meter is at least as high as the current to be measured. If the meter is digital and the current is too high, the display will read OL (overload). This means that the meter has been overloaded. If the meter is an analog meter, the indicating needle will peg (move) above the maximum limit on the scale, which might damage the meter.

> **WARNING!**
>
> Using a clamp-on ammeter may expose you to energized systems and equipment. Never use this type of meter unless you are qualified and are following NFPA, OSHA, and company/institutional safety procedures.

1.4.0 Multimeter

The multimeter is also known as a *volt-ohm-milliammeter (VOM)*. An analog VOM is shown in *Figure 7*. It is a multipurpose instrument that combines the three previous meters discussed. When using an analog meter, you must select the proper voltage (DC or AC), function (volts, amps, or ohms), and range. When using a digital VOM (*Figure 8*), you must select the proper voltage. Most have an autoranging feature for the magnitude.

> **WARNING!**
>
> Before use, a VOM must be checked on a known power source, such as that provided by a proving unit. Always read and follow the meter manufacturer's instructions. Failure to do so can result in equipment damage, injury, or even death.

Figure 7 Analog VOM.

Use of a VOM is the same as using the individual voltmeter, ohmmeter, and clamp-on ammeter. Current clamps can be used with most multimeters for measuring AC and DC currents above the milliamp level. These can be plugged into either the amp or voltage test lead connections, depending on the current clamp. Always refer to the manufacturer's instructions for the device in use. Current clamps are available in various current ranges, from 50A to several thousand amps. The jaws are available in different shapes and sizes to suit various applications, from round to rectangular, and even flexible.

Some multimeters can also measure the **frequency** of an AC waveform—this function is useful for diagnosing harmonic problems in an electrical distribution system.

Another feature is the Min/Max memory function. It will record the minimum and maximum readings over the time period selected. Other common multimeter functions include capacitance measurement, diode and transistor testers, temperature measurement, and true RMS measurement for accurate voltage and current readings at different frequencies. Refer to the manufacturer's instructions for the meter in use. In addition to the current clamps used with standard multimeters, clamp-on multimeters are also available with test leads (*Figure 9*). They are used to measure AC and DC current, AC and DC voltage, resistance, and other values.

1.5.0 Specialty Meters

In addition to the more common meters, a variety of special meters are available to complete various electrical measurements. Some of these meters are megohmmeters, motor and phase rotation testers, circuit testers, irradiance meters, recording instruments, and remote readout meters.

1.5.1 Megohmmeter

An ordinary ohmmeter cannot be used for measuring resistances of several million ohms, such as those found in conductor insulation or between motor or transformer windings. The instrument used to measure very high resistances is known as a *megohmmeter*, *Megger®*, or *insulation resistance tester*. They can be powered by alternating current, battery (*Figure 10*), or hand cranking (*Figure 11*).

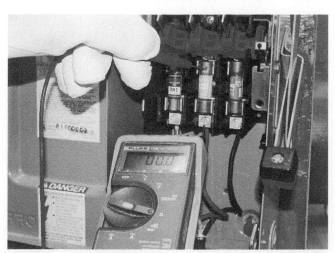

Figure 8 Digital VOM.

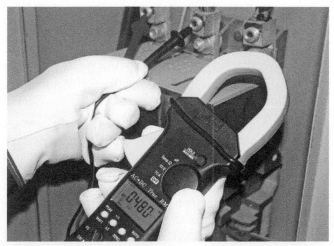

Figure 9 Clamp-on multimeter.

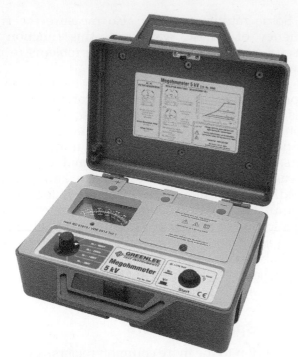

Figure 10 Battery-operated megohmmeter.

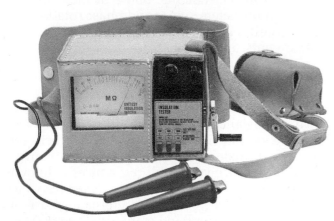

Figure 11 Hand-crank megohmmeter.

Improper use of a megohmmeter can result in personal injury or equipment damage. Always read and follow the meter manufacturer's instructions for safe use. Some of the minimum safety precautions for megohmmeter use include the following:

• High voltages are present when using a megohmmeter. For example, in 600V class systems, applied megohmmeter voltages are typically 500V and 1,000V. Only qualified individuals may use this equipment. Always wear appropriate personal protective equipment when approaching energized parts.
• De-energize and verify the de-energization of the circuit before connecting the meter. Make sure all capacitors are discharged.

• If possible, disconnect the item being checked from the other circuit components before using the meter.
• Do not exceed the manufacturer's recommended voltage test levels for the cable or equipment under test. Many manufacturers have different test levels based on the age of the cable or equipment being tested.
• Never touch the test leads when the meter is energized or powered. Insulation resistance testers generate high voltage, and touching the leads could result in injury or electrical shock.
• After the test, discharge any energy that may be left in the circuit by grounding the conductor or equipment for a period of time equal to the duration of the test.
• When conducting insulation resistance tests on cables or bus ducts where there are exposed parts that are remote from the testing position, safely secure or barricade the exposed end to protect others from inadvertent contact with the test voltage.

CAUTION	If a megohmmeter is used to test switchgear, all of the electronics must be disconnected prior to testing the switchgear. The voltage produced by the megohmmeter may damage electronic equipment.

Meter manufacturers supply detailed manuals for testing various devices and equipment. Always follow these instructions. The insulation resistance should be within the range of values specified in the manufacturer's instructions and specifications.

1.5.2 *Motor and Phase Rotation Testers*

Before connecting a three-phase motor to a circuit, the legs or windings of the motor (T1, T2, and T3) must be matched to the phases of the circuit (L1, L2, and L3). This ensures that the motor rotates in the proper direction. Improper connections may damage both the motor and any connected equipment.

A motor rotation tester can be used to identify the legs of the motor (*Figure 12*), while a phase rotation tester can be used to identify the phases of the circuit (*Figure 13*).

WARNING!	Do not connect a motor rotation tester to energized equipment. This can result in injury and equipment damage.

Meter Care

Like all meters, a megohmmeter is a sensitive instrument. Treat it with care and keep it in its case when not in use.

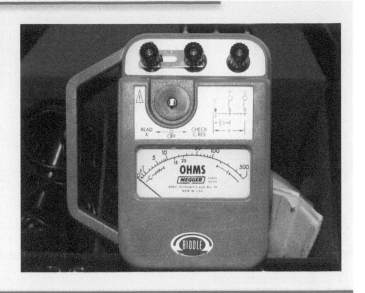

To use a motor rotation tester, connect the three motor wires to the T1, T2, and T3 leads on the tester, then rotate the motor shaft a half-turn while pressing the Test button (the direction of rotation depends on the tester in use; always follow the manufacturer's instructions). Either the clockwise or counterclockwise LED will light up. If the required rotation is clockwise and the clockwise LED lights up, tag the motor wires to correspond to the motor rotation leads. If the required rotation is clockwise and the counterclockwise LED lights up, switch a pair of leads and retest. Non-contact rotation testers are also available.

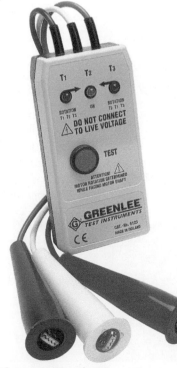

Figure 12 Motor rotation tester.

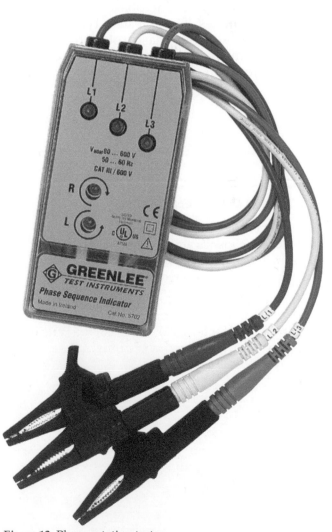

Figure 13 Phase rotation tester.

Phase Rotation Tester

The leads for a phase rotation tester may not correspond to typical circuit color coding. A good idea is to put phase tape on the meter leads to correspond to the circuit color coding. This will help to ensure correct connections.

A phase rotation tester, also called a *phase sequence indicator*, is used on three-phase electrical systems to indicate the phase sequence rotation of the voltages. These testers typically have LEDs to indicate the phase rotation. A phase sequence is measured as clockwise or counter-clockwise rotation.

A phase rotation tester is used when it is necessary to ensure the same phase rotation throughout a facility. If three-phase equipment is connected incorrectly, damage can occur. To test phase rotation, de-energize and lock out power to the circuit, then connect the three leads of the tester to the phase conductors in the circuit. Next, safely energize the circuit and observe the meter. Make note of the color scheme of the connected leads to the system, along with the phase sequences as indicated on the meter. This is necessary to ensure that the added equipment follows the same phase rotation. De-energize and lock out the circuit before disconnecting the leads.

1.5.3 Circuit Tester

A circuit tester can be used to verify the condition of a grounded circuit. To use the tester, it is simply plugged into a grounded outlet. The different lights on the unit indicate various circuit conditions: open ground, open neutral, open hot, reversed polarity, or a good circuit. See *Figure 14.* GFCI circuit testers are also available.

Non-Contact Phase Rotation Testers

Non-contact phase rotation testers are also available. Their leads are clamped around the equipment's insulated conductors, rather than having to make contact with bare conductors.

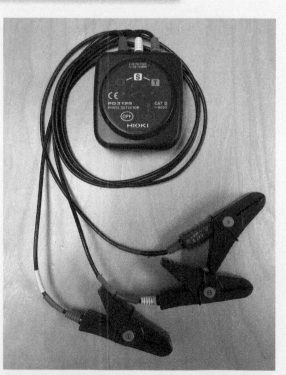

Figure Credit: Josiah Schuh

Figure 14 Circuit tester.

1.5.4 Irradiance Meter

Irradiance meters are used when installing and testing solar photovoltaic systems (*Figure 15*). They measure the solar radiation energy on a specified surface area over a given time. Irradiance is typically expressed in watts per square meter, British thermal units (BTUs) per square foot per hour, or kilowatt-hours per square meter per day. Some irradiance meters include additional features, such as the ability to measure solar transmission through windows. Others include an inclinometer to measure roof pitch, a compass to measure roof orientation, a thermometer to measure ambient air and module temperature, and wireless data logging capability.

1.5.5 Recording Instruments

The term *recording instrument* describes many instruments that make a permanent record of measured quantities over a period of time. Recording instruments use either a paper strip or electronic accessible memory. Those using electronic memory are usually called *data loggers*. These instruments record electrical quantities, including potential difference, current, power, resistance, and frequency. They can also record nonelectrical quantities by electrical means, such as a temperature recorder that uses a potentiometer system to record thermocouple output.

Figure 15 Irradiance meter.

It is often necessary to know the conditions that exist in an electrical circuit over a period of time to determine such things as peak loads, voltage fluctuations, and so on. An automatic recording instrument can be connected to take readings at specified intervals for later review and analysis. Some meters can upload data to a PC for real-time data logging and graphing (*Figure 16*).

1.5.6 Remote Readout Meters

Emerging meter technologies include various types of remote readout meters (*Figure 17*). Some meters have a removable screen that allows for readings at a safe distance from the measured circuit or equipment. Other meters pair with smartphone applications that allow the use of Bluetooth meter readings. This feature is especially useful for sharing direct readings with remote engineering or maintenance personnel.

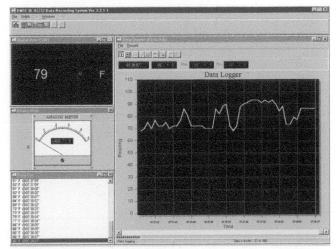

Figure 16 Data recording system.

(A) METER WITH DETACHABLE DISPLAY (B) METER READOUT USING SMARTPHONE APP

Figure 17 Remote readout meters.

1.0.0 Section Review

1. A voltmeter is used to measure the value of a circuit's _____.
   a. electromagnetic field
   b. resistance
   c. electromotive force
   d. current

2. A closed circuit is indicated by an ohmmeter reading of _____.
   a. zero
   b. infinity
   c. ERR
   d. ten

3. A meter that operates by sensing the strength of the electromagnetic field around the wire is a(n) _____.
   a. clamp-on ammeter
   b. megohmmeter
   c. motor rotation tester
   d. ohmmeter

4. Current measurements in milliamps are typically read using a(n) _____.
   a. clamp-on ammeter
   b. VOM
   c. megohmmeter
   d. insulation resistance tester

5. An expected resistance of 1,000,000 ohms would be measured using a _____.
   a. recording instrument
   b. clamp-on ammeter
   c. multimeter
   d. megohmmeter

2.0.0 CATEGORY RATINGS AND SAFETY REQUIREMENTS

Objective

Identify meter category ratings and safety requirements.

a. Select a meter with the correct category rating for an application.
b. Identify electrical test equipment safety hazards.

Performance Tasks

1. Measure the voltage in the classroom from line to neutral and neutral to ground.

2. Use an ohmmeter to measure the value of various resistors.

Trade Term

Backfeed: The reverse flow of electrical power in a distribution system caused by power induced into the system from outside sources, such as when using a generator during outages or when using a solar photovoltaic (PV) system. Backfeeds can also be caused by equipment overloads.

It is essential to select the correct meter for the application and to use it safely and in accordance with the manufacturer's instructions in order to prevent injury or equipment damage. One of the most important considerations in meter selection is the test equipment category rating.

2.1.0 Meter Category Ratings

Transient power spikes represent a serious hazard. Distribution systems and loads are becoming more complex, increasing the risk of transient power spikes. Lightning strikes on outdoor transmission lines and switching surges from normal switching operations can also produce dangerous high-energy transients. Motors, capacitors, variable speed drives, and power conversion equipment can also generate power spikes.

Safety systems are built into test equipment to protect electricians from transient power spikes. The International Electrotechnical Commission (IEC) developed a safety standard, *IEC 1010*, for test equipment that was adapted as *UL Standard UL 3111-1*. These standards define four overvoltage installation categories, often abbreviated as CAT I, CAT II, CAT III, and CAT IV (*Table 1*). These categories identify the hazards posed by transients; the higher the category number, the greater the risk to the electrician. A higher category number refers to an installation with higher power available and higher-energy transients. See *Figure 18*.

2.1.1 Selecting a Meter

When selecting a meter, remember that the potential danger from transients increases as you get closer to the utility power source. In addition, the greater the available short-circuit current, the greater the risk, and the higher the category rating. Keep in mind that an installation may have areas with different category ratings—always choose a meter rated for the highest category you will be working in. Then select the appropriate voltage level, amperage, and frequency. In addition, make sure that the test leads are rated as high as the meter.

Table 1 Overvoltage Installation Categories

Overvoltage Category	Installation Examples
CAT I	Electronic equipment and circuitry
CAT II	Single-phase loads such as small appliance tools, outlets at more than 30 feet from a CAT III source or 60 feet from a CAT IV source
CAT III	Three-phase motors, single-phase commercial or industrial lighting, switchgear, busduct and feeders in industrial plants
CAT IV	Three-phrase power at meter, service-entrance, or utility connection, any outdoor conductors, including signs, a pump, or exterior power generation (solar, wind, etc.).

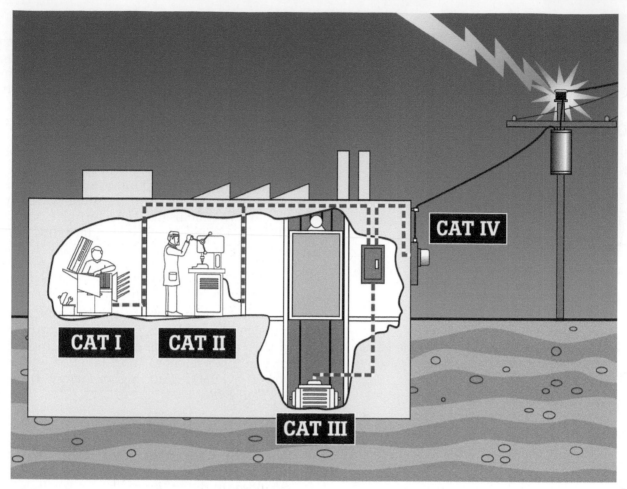

Figure 18 Typical overvoltage installation category locations.

Choose meters that are independently tested and certified by UL, CSA, or another recognized testing organization. Certified meters are marked with the category rating on the meter housing (*Figure 19*).

> **WARNING!**
>
> Make sure to select a meter with the correct category rating for the environment. Always check test leads for wear, cracks in the insulation, or signs of damage before using.

2.1.2 Meter Calibration Requirements

In addition to selecting a meter with the correct category rating, it is also essential that all meters be calibrated in accordance with the manufacturer's instructions. The accuracy of the calibrating standard must be greater than that of the instrument tested (typically ten times higher). Per the American National Standards Institute (ANSI), field and laboratory instruments must be calibrated at a minimum of once per year.

Dated calibration labels must be visible on all test equipment. In addition, written records must be maintained that show the dates and results of all instrument calibration or testing procedures.

2.2.0 Electrical Test Equipment Safety Hazards

Safety must be the primary responsibility of all personnel on a job site. The safe installation, maintenance, and operation of electrical equipment requires strict adherence to local and national codes and safety standards, as well as facility and company safety policies. Carelessness can result in serious injury or death due to electrical shock, burns, falls, flying objects, etc. After an accident has occurred, investigation almost invariably shows that it could have been prevented by the exercise of simple safety precautions and procedures. It is your personal responsibility to identify and eliminate unsafe conditions and unsafe acts that cause accidents.

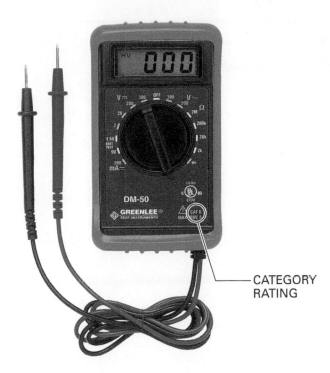

CATEGORY
RATING

Figure 19 Category rating on a typical meter.

It is important to remember that de-energizing main supply circuits by opening supply switches will not necessarily de-energize all circuits in a given piece of equipment. A source of danger that has often been neglected or ignored, sometimes with tragic results, is the input to electrical equipment from other sources, such as a backfeed. The rescue of a victim shocked by the power input from a backfeed is often hampered because of the time required to determine the source of power and isolate it. Always turn off all power inputs before working on equipment and lock out and tag, then check with an operating voltage tester to be sure that the equipment is safe to work on.

Safety can never be stressed enough. There are times when your life literally depends on it. Always observe the following precautions:

- Remember that the common 120V power supply is not a low, relatively harmless voltage but is a voltage that has caused more deaths than any other.
- Thoroughly inspect all test equipment before each use. Check for broken leads or knobs, damaged plugs, or frayed cords. Do not use equipment that is wet or damaged.
- Make sure the rating of any leads or accessories meets or exceeds the rating of the meter.
- Do not work with energized equipment unless you are both qualified and approved by your supervisor.
- Never shortcut safety; strictly adhere to all energized work policies and procedures.
- When testing circuits, test at higher ranges first, then work your way down to lower ranges.
- Always have a standby person present during hot work. This person should know whom to contact in case of emergency and how to disconnect the power.

Think About It

Putting It All Together

Which test instrument would be required for each of the following applications?

- Identify a short circuit in house wiring
- Measure the secondary voltage of an AC transformer
- Identify a blown fuse in a circuit
- Identify the contact configuration of a three-way switch or multi-pole relay

2.0.0 Section Review

1. Which category rating covers installations that present the lowest risk?
   a. CAT I
   b. CAT II
   c. CAT III
   d. CAT IV

2. Which of the following is true with regard to meter safety?
   a. 120V power is relatively harmless.
   b. All test leads carry the same category rating.
   c. When testing circuits, test at the lowest ranges first, then work your way up.
   d. Thoroughly inspect all test equipment before each use.

1. Which of the following is true regarding analog meters?

   a. Analog meters are solid-state instruments.
   b. Analog meters provide a digital readout.
   c. Analog meters rely on a moving-coil meter movement.
   d. Analog meters provide an autoranging setting.

2. A voltmeter is used to test _____.

   a. exact voltages
   b. voltage ranges
   c. power
   d. sine waves

3. In order to ensure safety, before measuring low voltages, you should first test for _____.

   a. resistance
   b. current
   c. vibration
   d. higher voltages

4. An ammeter is used to measure _____.

   a. current
   b. voltage
   c. resistance
   d. insulation value

5. Clamp-on ammeters operate by _____.

   a. using d'Arsonval meter movement
   b. sensing the strength of the electromagnetic field around the wire
   c. measuring the high resistance end of a power transformer
   d. using a resistive shunt

6. A meter with applied voltages of 500V or 1,000V is a(n) _____.

   a. megohmmeter
   b. ammeter
   c. multimeter
   d. continuity tester

7. A motor rotation tester _____.

   a. tests an energized motor
   b. gets connected to the motor supply conductors
   c. is used to identify the motor legs
   d. works only in a clockwise rotation

8. The highest level of protection is provided by instruments rated as _____.

   a. CAT I
   b. CAT II
   c. CAT III
   d. CAT IV

9. The International Electrotechnical Commission (IEC) developed a safety standard for overvoltage installation categories of CAT I, CAT II, CAT III, and CAT IV for _____.

   a. wiring
   b. electrical equipment
   c. test equipment
   d. signaling circuits

10. What is the lowest acceptable overvoltage category rating for a multimeter that will be used to test small appliances?

    a. CAT I
    b. CAT II
    c. CAT III
    d. CAT IV

Trade Terms Quiz

Fill in the blank with the correct term that you learned from your study of this module.

1. Used for electromagnetic effects or for providing electrical resistance, a _____ is a number of turns of wire.

2. A _____ uses a permanent magnet and moving coil arrangement to move a pointer across a scale.

3. Usually expressed in hertz, _____ is the number of cycles completed each second by a given AC voltage.

4. _____ is an uninterrupted electrical path for current flow.

5. Power from a solar PV system can cause a dangerous _____ in equipment assumed to be de-energized.

Trade Terms

Backfeed
Coil
Continuity

d'Arsonval meter movement
Frequency

1. The measurement of the electromotive force of a circuit is accomplished using a(n) _____.
   a. ammeter
   b. wattmeter
   c. voltmeter
   d. ohmmeter

2. When using a voltmeter that is not autoranging, start with the highest setting and work down until the meter reads somewhere between

   _____.

3. A(n) _____ is used to extend the range of a meter movement for current measurements.

4. True or False? Always connect an ohmmeter in parallel with a load.

5. The voltage range of a meter movement can be extended by adding a(n) _____ in series.

6. Short circuits can be detected by using a(n) _____.

7. What type of test equipment would you use to check the resistance between motor windings?

   _____.

8. The phase sequence of a circuit is identified using a(n) _____.

9. True or False? A voltage tester is used for precise voltage measurements.

10. A(n) _____ is used to take electrical readings at specified intervals.

Clarence "Ed" Cockrell

HR/Safety Manager
Vector Electric & Controls, Inc.

How did you get started in the construction industry?

I worked as a summer electrical helper during high school and college and found it very rewarding. I was looking for a job that would hold my interest for more than a year. I studied electrical engineering at Louisiana State University for two years.

Who inspired you to enter the industry? Why?

My brother-in-law and father-in-law inspired me to enter the industry.

What do you enjoy most about your job?

I started off as an electrician's apprentice, which offered many potential job opportunities. I enjoyed seeing the work progress from dirt to a functional building and finding new challenges as new technology changes the way we install the electrical components. My electrical training also opened the door to the possibility of being a field superintendent, project manager, senior office manager, human resource/training manager, and then human resource/safety/training manager.

Do you think training and education are important in construction? If so, why?

Training gives an apprentice the opportunity to become a great electrician and not an electrical laborer, by that I mean not just a conduit or cable tray installer or wire puller, but a well-rounded electrician. It also allows the apprentice to advance beyond the limits of being an electrician. Almost all of our supervisors, estimators, and project managers have been trained by an apprenticeship program and followed it up with more NCCER training.

How important are NCCER credentials to your career?

I went through a state certified apprenticeship and then completed the CSST training. Without these certifications, I believe I would be no more than a second-rate electrician.

How has training/construction impacted your life and your career?

It has allowed me to raise a family, buy a house and raise a child in a comfortable fashion. I have advanced many times at work, which has led to increased wages. Through my job, I have met and had dealings with many interesting people from all walks of life.

Would you suggest construction as a career to others? If so, why?

Yes. It offers a rewarding career opportunity to anyone willing to take pride and ownership in their learning and work.

How do you define craftsmanship?

I believe that craftsmanship is always delivering a great quality job in whatever you do. It takes pride and self-esteem to deliver work deemed to meet this definition of craftsmanship. This pride in their work helps build life qualities that become the building blocks of a truly honorable life. They walk through their community as a positive contributor as well as helping to build a secure and prosperous business. They teach their children through their actions how those who build contribute to the well-being of their community and country.

Trade Terms Introduced in This Module

Backfeed: The reverse flow of electrical power in a distribution system caused by power induced into the system from outside sources, such as when using a generator during outages or when using a solar photovoltaic (PV) system. Backfeeds can also be caused by equipment overloads.

Coil: A number of turns of wire, especially in spiral form, used for electromagnetic effects or for providing electrical resistance.

Continuity: An electrical term used to describe a complete (unbroken) circuit that is capable of conducting current. Such a circuit is also said to be closed.

d'Arsonval meter movement: A meter movement that uses a permanent magnet and moving coil arrangement to move a pointer across a scale.

Frequency: The number of cycles completed each second by a given AC voltage; usually expressed in hertz. One hertz equals one cycle per second.

Additional Resources

This module presents thorough resources for task training. The following reference material is recommended for further study.

ABCs of DMMs, Multimeter Features and Functions Explained. Everett, WA: Fluke Corporation.
ABCs of Multimeter Safety. Everett, WA: Fluke Corporation.
Clamp Meter ABCs. Everett, WA: Fluke Corporation.
Electronics Fundamentals: Circuits, Devices, and Applications, Thomas L. Floyd. New York, NY: Pearson.
Power Quality Analyzer Uses for Electricians. Everett, WA: Fluke Corporation.
Principles of Electric Circuits, Thomas L. Floyd. New York, NY: Pearson.

Figure Credits

Section Review Answer Key

SECTION 1.0.0

Answer	Section Reference	Objective
1. c	1.1.0	1a
2. a	1.2.0	1b
3. a	1.3.0	1c
4. b	1.4.0	1d
5. d	1.5.1	1e

SECTION 2.0.0

Answer	Section Reference	Objective
1. a	2.1.0; Table 1	2a
2. d	2.2.0	2b

This page is intentionally left blank.

NCCER CURRICULA — USER UPDATE

NCCER makes every effort to keep its textbooks up-to-date and free of technical errors. We appreciate your help in this process. If you find an error, a typographical mistake, or an inaccuracy in NCCER's curricula, please fill out this form (or a photocopy), or complete the online form at **www.nccer.org/olf**. Be sure to include the exact module ID number, page number, a detailed description, and your recommended correction. Your input will be brought to the attention of the Authoring Team. Thank you for your assistance.

Instructors – If you have an idea for improving this textbook, or have found that additional materials were necessary to teach this module effectively, please let us know so that we may present your suggestions to the Authoring Team.

NCCER Product Development and Revision

13614 Progress Blvd., Alachua, FL 32615

Email: curriculum@nccer.org
Online: www.nccer.org/olf

❏ Trainee Guide   ❏ Lesson Plans   ❏ Exam   ❏ PowerPoints   Other _____

Craft / Level: _____   Copyright Date: _____

Module ID Number / Title: _____

Section Number(s): _____

Description: _____

_____

_____

_____

_____

Recommended Correction: _____

_____

_____

_____

_____

Your Name: _____

Address: _____

_____

Email: _____   Phone: _____

This page is intentionally left blank.

Glossary

90° bend: A bend that changes the direction of the conduit by 90°.

Accessible: Able to be reached, as for service or repair.

Ammeter: An instrument for measuring electrical current.

Ampacity: The maximum current in amperes a conductor can carry continuously under the conditions of use without exceeding its temperature rating.

Amperes (A): The basic unit of measurement for electrical current, represented by the letter A

Appliances: Equipment designed for a particular purpose (for example, using electricity to produce heat, light, or mechanical motion). Appliances are usually self-contained, are generally available for applications other than industrial use, and are normally produced in standard sizes or types.

Approved: Meeting the requirements of an appropriate regulatory agency.

Arc flash boundary (AFB): An approach limit at a distance from exposed energized electrical conductors or circuit parts within which a person could receive a second-degree burn if an electrical arc flash were to occur.

Arc flash risk assessment: A study investigating a worker's potential exposure to arc flash energy, conducted for the purpose of injury prevention and the determination of safe work practices and appropriate levels of PPE.

Arc rating: The maximum incident energy resistance demonstrated by a material (or a layered system of materials) prior to material breakdown, or at the onset of a second-degree skin burn. Expressed in joules/cm^2 or calories/cm^2.

Architect's scale: A special ruler with various measurement scales that can be used when drafting or making measurements on architectural drawings. Architect's scales measure distances in inches.

Architectural drawings: Working drawings consisting of site plans, floor plans, elevations, sectional views, details, and other information necessary for the construction of a building.

Articles: The articles are the main topics of the *NEC®*, beginning with *NEC Article 90, Introduction*, and ending with *NEC Article 840, Premises-Powered Broadband Communications Systems*.

As-built drawings: A marked-up set of drawings, also called red-lines, is made at the completion of the project, showing all changes made during the construction project.

Atoms: The smallest particles to which an element may be divided and still retain the properties of the element.

Back-to-back bend: Any bend formed by two 90° bends with a straight section of conduit between the bends.

Backfeed: The reverse flow of electrical power in a distribution system caused by power induced into the system from outside sources, such as when using a generator during outages or when using a solar photovoltaic (PV) system. Backfeeds can also be caused by equipment overloads.

Battery: A DC voltage source consisting of two or more cells that convert chemical energy into electrical energy.

Block diagram: A single-line diagram used to show electrical equipment and related connections. See power-riser diagram.

Blueprint: An exact copy or reproduction of an original drawing.

Bonding bushing: A special conduit bushing equipped with a conductor terminal to accept a bonding jumper. It also has a screw or other sharp device to bite into the enclosure wall to bond the conduit to the enclosure without a jumper when there are no concentric knockouts left in the wall of the enclosure.

Bonding jumper: A bare or green insulated conductor used to ensure the required electrical conductivity between metal parts required to be electrically connected. Bonding jumpers are frequently used from a bonding bushing to the service-equipment enclosure to provide a path around concentric knockouts in an enclosure wall, and they may also be used to bond one raceway to another.

Bonding wire: A wire used to make a continuous grounding path between equipment and ground.

Branch circuit: The portion of a wiring system extending beyond the final overcurrent device protecting a circuit.

Cable trays: Rigid structures used to support electrical conductors.

Capstan: The turning drum of the cable puller on which the rope is wrapped and pulled.

Change order: A formal document from the project manager or owner, specifying one or more changes to the drawings, specifications, or project scope. It will also state any changes in project cost.

Chapters: Chapters contain a group of articles related to a broad category. Nine chapters form the broad structure of the *NEC®*.

Charge: A quantity of electricity that is either positive or negative.

Circuit: A complete path for current flow.

Coil: A number of turns of wire, especially in spiral form, used for electromagnetic effects or for providing electrical resistance.

Concentric bends: 90° bends made in two or more parallel runs of conduit with the radius of each bend increasing from the inside of the run toward the outside.

Conductors: Materials through which it is relatively easy to maintain an electric current.

Conduit: A round raceway, similar to pipe, that houses conductors.

Connector: Device used to physically connect conduit or cable to an outlet box, cabinet, or other enclosure.

Continuity: An electrical term used to describe a complete (unbroken) circuit that is capable of conducting current. Such a circuit is also said to be closed.

Contour lines: Curving lines on a site plan, following a given elevation. The space between contour lines tells the slope of the property, such as steep when they are close together or fairly level when they are widely separated.

Coulomb: A unit of electrical charge equal to 6.25×10^{18} electrons (or 6.25 quintillion electrons). A coulomb is the common unit of quantity used for specifying the size of a given charge.

Current: The movement, or flow, of electrons in a circuit. Current (I) is measured in amperes.

d'Arsonval meter movement: A meter movement that uses a permanent magnet and moving coil arrangement to move a pointer across a scale.

Detail drawing: An enlarged, detailed view taken from an area of a drawing and shown in a separate view.

Developed length: The actual length of the conduit that will be bent.

Dimensions: Sizes or measurements printed on a drawing.

Double-insulated/ungrounded tools: Electrical tools that are constructed so that the case is insulated from electrical energy. The case is made of a nonconductive material.

Electrical drawing: A means of conveying a large amount of exact, detailed information in an abbreviated language. Consists of lines, symbols, dimensions, and notations to accurately convey an engineer's designs to electricians who install the electrical system on a job.

Electrical service: The electrical components that are used to connect the serving utility to the premises wiring system.

Electrons: Negatively charged particles that orbit the nucleus of an atom.

Elevation view: An architectural drawing showing height, but not depth; usually the front, rear, and sides of a building or object.

Engineer's scale: A special ruler with various measurement scales that can be used when drafting or making measurements on architectural drawings. Engineer's scales measure distances in decimal units.

Error precursors: Situations that put a worker at risk due to the demands of the task, conditions, worker attitude, and/or environment.

Exceptions: Exceptions follow the applicable sections of the *NEC®* and allow alternative methods to be used under specific conditions.

Explosion-proof: Designed and constructed to withstand an internal explosion without creating an external explosion or fire.

Exposed location: Not permanently closed in by the structure or finish of a building; able to be installed or removed without damage to the structure.

Feeder: Any circuit conductor between the service equipment, the source of a separately derived system, or other power supply source and the final branch circuit overcurrent device.

Fibrillation: Very rapid irregular contractions of the muscle fibers of the heart that result in the muscle being unable to contract and pump blood properly.

Fish tape: A hand device used to pull a wire through a conduit run.

Floor plan: A drawing of a building as if a horizontal cut were made through a building at about window level, and the top portion removed. The floor plan is what would appear if the remaining structure were viewed from above.

Frequency: The number of cycles completed each second by a given AC voltage; usually expressed in hertz. One hertz equals one cycle per second.

Gain: Because a conduit bends in a radius and not at right angles, the length of conduit needed for a bend will not equal the total determined length. Gain is the distance saved by the arc of a 90° bend.

Ground fault circuit interrupter (GFCI): A protective device that functions to de-energize a circuit or portion thereof within an established period of time when a current to ground exceeds some predetermined value. This value is less than that required to operate the overcurrent protective device of the supply circuit.

Grounded tool: An electrical tool with a three-prong plug at the end of its power cord or some other means to ensure that stray current travels to ground without passing through the body of the user. The ground plug is bonded to the conductive frame of the tool.

Handy box: Single-gang outlet box used for surface mounting to enclose receptacles or wall switches on concrete or concrete block construction of industrial and commercial buildings; nongangable; also made for recessed mounting; also known as a utility box.

Hot stick: An insulated tool designed for the manual operation of disconnecting switches, fuse removal and insertion, and the application and removal of temporary grounds.

Incident energy: The amount of thermal energy impressed on a surface at a certain distance from the source of an electrical arc. Incident energy is typically expressed in calories per square centimeter (cal/cm^2).

Informational Note: Explanatory material that follows specific *NEC®* sections. Informational notes are not enforceable.

Institute for Electrical and Electronics Engineers (IEEE): A professional organization that develops international standards impacting electronics, telecommunications, information technology, and power-generation products and services.

Insulator: A material through which it is difficult to conduct an electric current.

International Electrotechnical Commission (IEC): An international organization that develops consensus standards for all electrical and electronic technologies.

Joule (J): A unit of measurement for doing work, represented by the letter J. One joule is equal to one newton-meter (Nm).

Junction box: An enclosure where one or more raceways or cables enter, and in which electrical conductors can be, or are, spliced.

Kick: A bend in a piece of conduit, usually less than 45°, made to change the direction of the conduit.

Kilo: A prefix used to indicate one thousand (for example, one kilowatt is equal to one thousand watts).

Kirchhoff's current law: The statement that the total amount of current flowing through a parallel circuit is equal to the sum of the amounts of current flowing through each current path.

Kirchhoff's voltage law: The statement that the sum of all the voltage drops in a circuit is equal to the source voltage of the circuit.

Limited approach boundary: An approach limit at a distance from an exposed energized electrical conductor or circuit part within which a shock hazard exists.

Load center: A type of panelboard that is normally located at the service entrance of a residential installation. It sometimes contains the main disconnect.

Matter: Any substance that has mass and occupies space.

Mega: A prefix used to indicate one million; for example, one megawatt is equal to one million watts.

Metal-clad (Type MC) cable: A factory assembly of one or more insulated circuit conductors with or without optical fiber members enclosed in an armor of interlocking metal tape, or a smooth or corrugated metallic sheath.

Mouse: A cylinder of foam rubber that fits inside the conduit and is then propelled by compressed air or vacuumed through the conduit run, pulling a line or tape.

National Electrical Manufacturers Association (NEMA): The association that maintains and improves the quality and reliability of electrical products.

National Fire Protection Association (NFPA): The publishers of the *NEC®*. The NFPA develops codes and standards to minimize the possibility and effects of fire.

Nationally Recognized Testing Laboratories (NRTLs): Product safety certification laboratories that are responsible for testing and certifying electrical equipment.

Neutrons: Electrically neutral particles (neither positive nor negative) that have the same mass as a proton and are found in the nucleus of an atom.

Nonmetallic-sheathed (Type NM and NMC) cable: A factory assembly of two or more insulated conductors enclosed within an overall nonmetallic jacket. Type NM contains insulated conductors enclosed within an overall nonmetallic jacket; Type NMC contains insulated conductors enclosed within an overall, corrosion-resistant, nonmetallic jacket.

Nucleus: The center of an atom. It contains the protons and neutrons of the atom.

Occupational Safety and Health Administration (OSHA): The federal government agency established to ensure a safe and healthy environment in the workplace.

Offset: An offset is two bends placed in a piece of conduit to change elevation to go over or under obstructions or for proper entry into boxes, cabinets, etc.

Ohms (Ω): The basic unit of measurement for resistance, represented by the symbol Ω.

Ohm's law: A statement of the relationships among current, voltage, and resistance in an electrical circuit: current (I) equals voltage (E) divided by resistance (R). Generally expressed as a mathematical formula: $I = E/R$.

Ohmmeter: An instrument used for measuring resistance.

On-the-job learning (OJL): Job-related learning an apprentice acquires while working under the supervision of journey-level workers. Also called *on-the-job training (OJT)*.

One-line diagram: A drawing that shows, by means of lines and symbols, the path of an electrical circuit or system of circuits along with the various circuit components. Also called a *single-line diagram*.

Outlet box: A metallic or nonmetallic box installed in an electrical wiring system from which current is taken to supply some apparatus or device.

Parallel circuits: Circuits containing two or more parallel paths through which current can flow.

Parts: Certain articles in the *NEC®* are subdivided into parts that cover a specific topic. Parts have Roman numeral designations (e.g., *NEC Article 250, Part IX, Instruments, Meters, and Relays*).

Plan: A drawing made as though the viewer were looking straight down (from above) on an object.

Polychlorinated biphenyls (PCBs): Toxic chemicals that may be contained in liquids used to cool certain types of large transformers and capacitors.

Power: The rate of doing work, or the rate at which energy is used or dissipated. Electrical power is measured in watts.

Power-riser diagram: A single-line block diagram used to indicate the electric service equipment, service conductors and feeders, and subpanels. Notes are used on power-riser diagrams to identify the equipment; indicate the size of conduit; show the number, size, and type of conductors; and list related materials. A panelboard schedule is usually included with power-riser diagrams to indicate the exact components (panel type and size), along with fuses, circuit breakers, etc., contained in each panelboard.

Proton: The smallest positively charged particle of an atom. Protons are contained in the nucleus of an atom.

Pull box: A sheet metal box-like enclosure used in conduit runs to facilitate the pulling of cables from point to point in long runs, or to provide for the installation of conduit support bushings needed to support the weight of long riser cables, or to provide for turns in multiple conduit runs.

Qualified person: One who has demonstrated the skills and knowledge related to the construction and operation of the electrical equipment and installations and has received safety training to identify and avoid the hazards involved.

Raceway system: An enclosure that houses the conductors in an electrical system (such as fittings, boxes, and conduit).

Raceways: Enclosed channels designed expressly for holding wires, cables, or busbars, with additional functions as permitted in the *NEC®*.

Raintight: Constructed or protected so that exposure to a beating rain will not result in the entrance of water under specified test conditions.

Relays: Electromechanical devices consisting of a coil and one or more sets of contacts. Used as a switching device.

Request for information: A formal document from the contractor, seeking clarification when errors, omissions, and discrepancies are found in or between the construction documents.

Resistance: An electrical property that opposes the flow of current through a circuit. Resistance (R) is measured in ohms.

Resistors: Any devices in a circuit that resist the flow of electrons.

Restricted approach boundary: An approach limit at a distance from an exposed energized electrical conductor or circuit part within which there is an increased likelihood of electric shock.

Rise: The length of the bent section of conduit measured from the bottom, centerline, or top of the straight section to the end of the conduit being bent.

Romex®: General Cable's trade name for Type NM cable; however, it is often used generically to refer to any nonmetallic-sheathed cable.

Rough-in: The beginning stage of wiring that involves the installation of the panelboard, raceway system, wiring, and boxes.

Roughing in: The first stage of an electrical installation, when the raceway, cable, wires, boxes, and other equipment are installed. This is the electrical work that must be done before any finishing work can be done.

Scale: On a drawing, the size relationship between an object's actual size and the size it is drawn. Scale also refers to the measuring tool used to determine this relationship.

Schedule: A systematic method of presenting equipment lists on a drawing in tabular form.

Schematic: A type of drawing in which symbols are used to represent the components in a system.

Schematic diagram: A detailed diagram showing complicated circuits, such as control circuits.

Sectional view: A cutaway drawing that shows the inside of an object or building.

Sections: Parts and articles are subdivided into sections. Sections have numeric designations that follow the article number and are preceded by a period (e.g., *NEC Section 501.4*).

Segment bend: A large bend formed by multiple short bends or shots.

Series circuit: A circuit with only one path for current flow.

Series-parallel circuits: Circuits that contain both series and parallel current paths.

Service drop: The overhead conductors, through which electrical service is supplied, between the last power company pole and the point of their connection to the service facilities located at the building.

Service entrance: The point where power is supplied to a building (including the equipment used for this purpose). The service entrance includes the service main switch or panelboard, metering devices, overcurrent protective devices, and conductors/raceways for connecting to the power company's conductors.

Service-entrance conductors: The conductors between the point of termination of the overhead service drop or underground service lateral and the main disconnecting device in the building.

Service-entrance equipment: Equipment that provides overcurrent protection to the feeder and service conductors, a means of disconnecting the feeders from energized service conductors, and a means of measuring the energy used.

Service lateral: The underground conductors through which service is supplied between the power company's distribution facilities and the first point of their connection to the building or area service facilities located at the building.

Shop drawing: A drawing that is usually developed by manufacturers, fabricators, or contractors to show specific dimensions and other pertinent information concerning a particular piece of equipment and its installation methods.

Site plan: A drawing showing the location of a building or buildings on the building site. Such drawings frequently show topographical lines, electrical and communication lines, water and sewer lines, sidewalks, driveways, and similar information.

Solenoids: Electromagnetic coils used to control a mechanical device such as a valve.

Splice: Connection of two or more conductors.

Stub-up: Another name for the rise in a section of conduit at 90°. Also, a term used for conduit penetrating a slab or the ground.

Substation: An enclosed assembly of high-voltage equipment, including switches, circuit breakers, buses, and transformers, that connects the power generation facility to the grid and through which electrical energy is passed in order to change its characteristics, such as stepping voltage up or down, changing control frequency, or other characteristics.

Switch: A mechanical device used for turning an electrical circuit on and off.

Switch leg: A circuit routed to a switch box for controlling electric lights.

Tap: Intermediate point on a main circuit where another wire is connected to supply electrical current to another circuit.

Transformers: Devices consisting of one or more coils of wire wrapped around a common core. Transformers are commonly used to step voltage up or down.

Trim-out: The final stage of wiring that involves the installation and termination of devices and fixtures.

Trough: A long, narrow box used to house electrical connections that could be exposed to the environment.

Underwriters Laboratories, Inc. (UL): An agency that evaluates and approves electrical components and equipment.

Unqualified person: A person who is not a qualified person.

Valence shell: The outermost ring of electrons that orbit about the nucleus of an atom.

Voltage drop: The change in voltage across a component that is caused by the current flowing through it and the amount of resistance opposing it.

Voltage: The driving force that makes current flow in a circuit. Voltage, often represented by the letter E, is also referred to as *electromotive force (emf)*, *difference of potential*, or *electrical pressure*.

Voltmeter: An instrument for measuring voltage. The resistance of the voltmeter is fixed. When the voltmeter is connected to a circuit, the current passing through the meter will be directly proportional to the voltage at the connection points.

Volts (V): The unit of measurement for voltage, represented by the letter V. One volt is equivalent to the force required to produce a current of one ampere through a resistance of one ohm.

Watertight: Constructed so that moisture will not enter the enclosure under specified test conditions.

Watts (W): The basic unit of measurement for electrical power, represented by the letter W.

Weatherproof: Constructed or protected so that exposure to the weather will not interfere with successful operation.

Wire grip: A device used to link pulling rope to cable during a pull.

Wireways: Steel troughs designed to carry electrical wire and cable.

Written specifications: A written description of what is required by the owner, architect, and engineer in the way of materials and workmanship. Together with working drawings, the specifications form the basis of the contract requirements for construction.

This page is intentionally left blank.

This page is intentionally left blank.

This page is intentionally left blank.

This page is intentionally left blank.